Coxsackieviruses

A GENERAL UPDATE

INFECTIOUS AGENTS AND PATHOGENESIS

Series Editors: Mauro Bendinelli, *University of Pisa*
Herman Friedman, *University of South Florida*

COXSACKIEVIRUSES
A General Update
Edited by Mauro Bendinelli and Herman Friedman

MYCOBACTERIUM TUBERCULOSIS
Interactions with the Immune System
Edited by Mauro Bendinelli and Herman Friedman

Coxsackieviruses

A GENERAL UPDATE

Edited by
MAURO BENDINELLI
University of Pisa
Pisa, Italy
and
HERMAN FRIEDMAN
University of South Florida
Tampa, Florida

Springer Science+Business Media, LLC

Library of Congress Cataloging in Publication Data

Coxsackieviruses: a general update.

(Infectious agents and pathogenesis)
Includes bibliographies and index.
1. Coxsackievirus infections. 2. Coxsackieviruses. I. Bendinelli, Mauro. II. Friedman, Herman, 1931- . III. Series. [DNLM: 1. Coxsackievirus Infections. 2. Coxsackieviruses. QW 168.5.P4 C879]
QR201.C65C69 1988 616'.0194 88-4234

ISBN 978-1-4757-0249-1 ISBN 978-1-4757-0247-7 (eBook)
DOI 10.1007/978-1-4757-0247-7

Originally published by Plenum Press New York in 1988
Softcover reprint of the hardcover 1st edition 1988

Contributors

FAKHRY ASSAAD[†] • Communicable Diseases Division, World Health Organization, Geneva, Switzerland

FULVIO BASOLO • Department of Pathology, University of Pisa, I-56100 Pisa, Italy

KIRK W. BEISEL • Department of Pathology and Laboratory Medicine, Emory University School of Medicine, Atlanta, Georgia 30322

ELEANOR J. BELL • Enterovirus Reference (Scotland) Laboratory, Regional Virus Laboratory, Ruchill Hospital, Glasgow G20 9NB, Scotland

MAURO BENDINELLI • Institute of Epidemiology, Hygiene and Virology, University of Pisa, I-56100 Pisa, Italy

MARIA R. CAPOBIANCHI • Institute of Virology, University of Rome, 00185 Rome, Italy

NANDO K. CHATTERJEE • Wadsworth Center for Laboratories and Research, New York State Department of Health, Albany, New York 12201

PIER GIULIO CONALDI • Institute of Epidemiology, Hygiene and Virology, University of Pisa, I-56100 Pisa, Italy

RICHARD L. CROWELL • Department of Microbiology and Immunology, Hahnemann University School of Medicine, Philadelphia, Pennsylvania 19102

FERDINANDO DIANZANI • Institute of Virology, University of Rome, 00185 Rome, Italy

RÜDIGER DÖRRIES • Institute for Virology and Immunobiology, University of Würzburg, D-8700 Würzburg, Federal Republic of Germany

[†]Deceased.

KARIN ESTEVES • Epidemiology and Management Support Services, World Health Organization, Geneva, Switzerland

GIOVANNI FEDERICO • Department of Pediatrics, University of Pisa, I-56100 Pisa, Italy

CHARLES J. GAUNTT • Department of Microbiology, The University of Texas Health Science Center at San Antonio, San Antonio, Texas 78284

NORMAN R. GRIST • Department of Infectious Diseases, University of Glasgow, Glasgow G12 8QQ, Scotland; Communicable Diseases Unit, Ruchill Hospital, Glasgow G20 9NB, Scotland

G. HAMMOND • Cadham Provincial Laboratory, Winnipeg, Manitoba R3C 3Y1, Canada

KUO-HOM LEE HSU • Department of Microbiology and Immunology, Hahnemann University School of Medicine, Philadelphia, Pennsylvania 19102

SALLY ANN HUBER • Department of Pathology, University of Vermont, Burlington, Vermont 05405

REINHARD KANDOLF • Department of Virology, Max Planck Institute for Biochemistry, D-8033 Martinsried; Department of Internal Medicine I, Klinikum Grosshadern, University of Munich, D-8000 Munich 70, Federal Republic of Germany

MARK H. KAPLAN • Division of Infectious Disease and Immunology, North Shore University Hospital, Manhasset, New York 11030; Cornell University Medical College, New York, New York 10021

BURTON J. LANDAU • Department of Microbiology and Immunology, Hahnemann University School of Medicine, Philadelphia, Pennsylvania 19102

A. MARTIN LERNER • Wayne State University, Division of Infectious Diseases, Hutzel Hospital, Detroit, Michigan 48201

ROGER M. LORIA • Departments of Microbiology, Immunology and Pathology, Virginia Commonwealth University; School of Basic Health Sciences, Medical College of Virginia, Richmond, Virginia 23298

BRIAN W. J. MAHY • Animal Virus Research Institute, Pirbright, Woking, Surrey GU24 ONF, England

DONATELLA MATTEUCCI • Institute of Epidemiology, Hygiene and Virology, University of Pisa, I-56100 Pisa, Italy

TAKASHI ONODERA • Laboratory of Viral Immunology, National Institute of Animal Health, Tsukuba, Ibaraki 305, Japan

MARK A. PALLANSCH • Division of Viral Diseases, Center for Infectious Diseases, Centers for Disease Control, U.S. Public Health Service, U.S. Department of Health and Human Services, Atlanta, Georgia 30333

BELLUR S. PRABHAKAR • Laboratory of Oral Medicine, National Institute of Dental Research, National Institutes of Health, Bethesda, Maryland 20892

DANIEL REID • Communicable Diseases Unit, Ruchill Hospital, Glasgow G20 9NB, Scotland

MILAGROS P. REYES • Wayne State University, Division of Infectious Diseases, Hutzel Hospital, Detroit, Michigan 48201

GIOVANNI ROCCHI • Infectious Diseases Clinic, Department of Public Health, Second University of Rome, 00191 Rome, Italy

NOEL R. ROSE • Department of Immunology and Infectious Diseases, The Johns Hopkins University, School of Hygiene and Public Health, Baltimore, Maryland 21205

NATHALIE J. SCHMIDT • Viral and Rickettsial Disease Laboratory, Division of Laboratories, California State Department of Health Services, Berkeley, California 94704

DAVID P. SCHNURR • Center for Advanced Medical Technology, San Francisco State University, San Francisco, California 94132; Viral and Rickettsial Disease Laboratory, Division of Laboratories, California State Department of Health Services, Berkeley, California 94704

MAGGIE SCHULTZ • Department of Microbiology and Immunology, Hahnemann University School of Medicine, Philadelphia, Pennsylvania 19102

ANTONIO TONIOLO • Institute of Microbiology and Virology, University of Sassari Medical School, 07100 Sassari, Italy

STEVEN TRACY • Department of Pathology and Microbiology, University of Nebraska Medical Center, Omaha, Nebraska 68105

ANTONIO VOLPI • Infectious Diseases Clinic, Department of Public Health, Second University of Rome, 00191 Rome, Italy

Preface to the Series

The mechanisms of disease production by infectious agents are presently the focus of an unprecedented flowering of studies. The field has undoubtedly received impetus from the considerable advances recently made in the understanding of the structure, biochemistry, and biology of viruses, bacteria, fungi, and other parasites. Another contributing factor is our improved knowledge of immune responses and other adaptive or constitutive mechanisms by which hosts react to infection. Furthermore, recombinant DNA technology, monoclonal antibodies, and other newer methodologies have provided the technical tools for examining questions previously considered too complex to be successfully tackled. The most important incentive of all is probably the regenerated idea that infection might be the initiating event in many clinical entities presently classified as idiopathic or of uncertain origin.

Infectious pathogenesis research holds great promise. As more information is uncovered, it is becoming increasingly apparent that our present knowledge of the pathogenic potential of infectious agents is often limited to the most noticeable effects, which sometimes represent only the tip of the iceberg. For example, it is now well appreciated that pathologic processes caused by infectious agents may emerge clinically after an incubation of decades and may result from genetic, immunologic, and other indirect routes more than from the infecting agent in itself. Thus, there is a general expectation that continued investigation will lead to the isolation of new agents of infection, the identification of hitherto unsuspected etiologic correlations, and, eventually, more effective approaches to prevention and therapy.

Studies on the mechanisms of disease caused by infectious agents demand a breadth of understanding across many specialized areas, as well as much co-operation between clinicians and experimentalists. The series *Infectious Agents and Pathogenesis* is intended not only to document the state of the art in this fascinating and challenging field but also to help lay bridges among diverse areas and people.

M. Bendinelli
H. Friedman

Preface

It is now just 40 years since coxsackieviruses were first isolated by Dalldorf and Sickles in the "eponymous" town of Coxsackie, New York. Yet the overall contribution of coxsackieviruses to clinically evident disease of humans is still largely an open problem. Following their discovery, coxsackieviruses were under intense clinical and laboratory scrutiny for a long time. Because of their relationship to polioviruses, the understanding of their structure, biochemistry, biology, and epidemiology advanced rapidly as a result of the formidable efforts that eventually led to the defeat of poliomyelitis. The ability of these viruses to infect mice permitted dissection of their pathogenicity in an experimental host and elucidation of conditions that influence its expression. Coxsackieviruses have been progressively associated with an increasing array of widely diverse human diseases. However, only some of the suggested causal correlations have been substantiated with satisfactory certainty. For others, conclusive evidence has so far resisted investigation. Most important, among the latter are chronic maladies, such as dilated cardiomyopathy and juvenile diabetes, that demand consideration.

In recent times, there has been a partial eclipse of the subject of coxsackieviruses in the medical literature. In addition to the difficulties encountered in pinpointing their pathogenic potential, possible reasons include the general decline of interest in enteroviruses, which ensued after the conquest of poliomyelitis, and the continuous appearance in the limelight of new, more esoteric, and therefore more "appealing" viruses. An additional factor was probably the realization that all distinctions within the *Enterovirus* genus of *Picornaviridae* are arbitrary and the consequent naming of newly identified serotypes as enterovirus (followed by a number). Although taxonomically correct, this decision has dumped the coxsackieviruses, already numerous, into a company so crowded that this may have discouraged further research in the area.

Nevertheless, in the last few years there has been a number of significant forward steps in our understanding of coxsackievirus infection. To mention a few: the genomes of members of the group have been cloned, new sensitive tools for the laboratory diagnosis of infection have been developed, and, somewhat unexpectedly, coxsackieviruses have been seen to persist for long periods *in vivo* and *in vitro* and to induce immunopathology and autoimmunity. Furthermore, an increased frequency of severe coxsackieviral diseases has been noted in newborns and in immunocompromised patients, and new diseases that might have a coxsackieviral etiology have been recognized. All of these advances are important and are also excellent indicators that a resurgence of clinical and experimental research on these viruses may be extremely rewarding.

This leads us to the two main purposes of this book. One is to present recent advances in our understanding of coxsackievirus biology and pathogenesis. Most leading investigators active in this area are represented as authors of one or more chapters. The second is to stimulate renewed interest concerning these elusive pathogens. As coxsackieviruses are so widespread, it is essential to gain a full appreciation of their clinical importance. This is, for example, necessary for evaluation of the public health usefulness of developing appropriate vaccination procedures. The exciting developments that are taking place in the area of viral vaccines make it feasible to prepare a vaccine capable of simultaneously protecting against multiple coxsackieviruses as well as other viruses.

A possible criticism of this book is that coxsackieviruses are dealt with separately from other enteroviruses. Apart from properties typical of these viruses, we see at least another sound reason for doing so. Parting enemies and attacking them in small groups has always been strategically wise.

M. Bendinelli
H. Friedman

Contents

1. Classification and General Properties

BRIAN W. J. MAHY

1. History and Classification 1
2. Isolation and Propagation 5
3. Occurrence in Nonhuman Species 7
4. Physicochemical Properties 8
5. Hemagglutination 9
6. Virion Structure and Morphology 10
References 13

2. The Genome of Group B Coxsackieviruses

STEVEN TRACY

1. Introduction 19
2. General Remarks on the CVB Genome 20
3. The 5′ Nontranslated Region 22
4. The P1 Region 24
5. The P2 Region 26
6. The P3 Region 26
7. The 3′ Nontranslated Region 27
8. Summary 28
References 30

3. Replication

NANDO K. CHATTERJEE

1. Introduction 35
2. Virus Growth, Purification, and Assay 35
3. Viral Replication 36
4. Concluding Remarks 47
References 48

4. Perspectives on Cellular Receptors as Determinants of Viral Tropism

RICHARD L. CROWELL, BURTON J. LANDAU, KUO-HOM LEE HSU, AND MAGGIE SCHULTZ

1. Introduction 51
2. Pathogenesis of Coxsackievirus Infections 52
3. Structural Studies Identifying the Virion Attachment Site 55
4. Specific Cellular Receptors for Coxsackieviruses 55
5. Future Directions 58
References 59

5. The Role of Interferon in Picornavirus Infections

FERDINANDO DIANZANI, MARIA R. CAPOBIANCHI, DONATELLA MATTEUCCI, AND MAURO BENDINELLI

1. The Interferon System 65
2. The Interferon in Picornavirus Infections 68
3. Conclusions 76
References 77

6. Interactions with the Immune System

MAURO BENDINELLI, PIER GIULIO CONALDI, AND DONATELLA MATTEUCCI

1. Introduction 81
2. Nonspecific Mechanisms of Resistance to Coxsackieviruses 82
3. Antibody Response to Coxsackieviruses 83

4. Cell-Mediated Immune Response to Coxsackieviruses 88
5. Coxsackievirus Replication in Immunocompetent Cells 91
6. Immunodepression by Coxsackieviruses 93
7. Histopathology of Lymphoid Organs in Coxsackievirus-Infected Hosts 95
8. Mechanisms and Significance of Damage to the Immune System 96
9. Concluding Remarks 99
References 99

7. The Role of Immune Mechanisms in Pathogenesis

SALLY ANN HUBER

1. Introduction 103
2. Role of Virus in Disease Pathogenesis 103
3. Evidence for Immune Pathogenesis in Picornavirus Infections 105
4. Mechanisms of Pathogenesis 107
5. Hypothetical Model for Picornavirus-Induced Autoimmunity in Myocarditis 112
References 114

8. Application of Monoclonal Antibodies to the Study of Coxsackieviruses

BELLUR S. PRABHAKAR

1. Introduction 117
2. Generation and Characterization of Monoclonal Antibodies .. 118
3. Antigenic Variants among Naturally Occurring Clinical Isolates 118
4. Selection of Antigenic Variants 121
5. Characterization of Antigenic Variants 121
6. Demonstration of Conserved and Nonconserved Epitopes on CVB4 124
7. Neutralization Epitope Diversity among Laboratory Isolates .. 125
8. Selection of Antigenic Variants with Changes in Their Biologic Properties 127
9. Characterization of Viral Polypeptides 128
10. Anti-idiotypic Antibodies to Anti-CVB4 Monoclonal Antibodies 129

11. Molecular Mimicry and Autoimmunity 131
12. Perspective 131
References 133

9. Host Conditions Affecting the Course of Coxsackievirus Infections

ROGER M. LORIA

1. Introduction 135
2. Infection: Route, Dose, Site of Virus Localization 136
3. Temperature 138
4. Effect of Age and Aging on Coxsackievirus Infection 139
5. Fetal and Neonatal Infections 142
6. Genetics 145
7. Nutrition 148
8. Effect of Hormones on Infection 149
9. Other Agents 150
10. Summary 151
References 151

10. The Possible Role of Viral Variants in Pathogenesis

CHARLES J. GAUNTT

1. Introduction 159
2. Origin of Variants 159
3. Types of Coxsackievirus Group B Variants 161
4. Coxsackievirus Group B Variants and Animal Models of Diseases 165
5. Coxsackievirus Group A Variants 170
6. Unaccounted-for Coxsackievirus Variants 170
7. Future Goals 171
References 173

11. Persistent Infections

DAVID P. SCHNURR AND NATHALIE J. SCHMIDT

1. Introduction 181
2. *In Vitro* Persistent Infections 181

3. *In Vivo* Persistent Infections 193
4. Consequences of Persistent Infection 194
5. Discussion 198
6. Conclusion 199
References 200

12. New Approaches to Laboratory Diagnosis
RÜDIGER DÖRRIES

1. Introduction 203
2. Improvements of Established Techniques 205
3. New Technical Approaches 206
4. Concluding Remarks 216
References 216

13. General Pathogenicity and Epidemiology
NORMAN R. GRIST AND DANIEL REID

1. General Pathogenicity of Coxsackieviruses 221
2. Epidemiologic Observations 223
3. Role of Coxsackieviruses in Various Disease Syndromes 234
4. Conclusions 236
References 237

14. Coxsackievirus Infection in Children under Three Months of Age
MARK H. KAPLAN

1. Introduction 241
2. Mild Coxsackie B Infection Acquired from a Sick Household Member or Nosocomially 242
3. Intrapartum (Congenital) Fulminant Coxsackie B Infection of the Newborn 244
4. Differential Diagnosis of Fulminant Coxsackie B Viral Infection 248
5. Viral Diagnosis 249
6. Treatment of Fulminant Infection 249
7. Prevention 249
References 250

15. Myocarditis: Clinical and Experimental Correlates

MILAGROS P. REYES AND A. MARTIN LERNER

1. Introduction 253
2. Coxsackievirus-Induced Cardiopathies 254
3. Specific Models of Heart Muscle Disease 257
4. Conclusions 268
References 269

16. Relationship of Coxsackievirus to Cardiac Autoimmunity

KIRK W. BEISEL AND NOEL R. ROSE

1. Viruses and Autoimmunity 271
2. Clinical Evidence for Coxsackievirus Infection and Autoimmunity in Heart Disease 272
3. Experimental Studies of Virus-Associated Autoimmune Myocarditis 273
4. Discussion and Conclusions 289
References 290

17. The Impact of Recombinant DNA Technology on the Study of Enteroviral Heart Disease

REINHARD KANDOLF

1. Introduction 293
2. Diagnostic Problems in Suspected Enterovirus Heart Disease 294
3. Molecular Cloning and Characterization of Full-Length Reverse-Transcribed Coxsackievirus B3 Genomic RNA 295
4. Cloned Coxsackievirus B3 cDNA as a Diagnostic Reagent for the Detection of Enteroviruses 298
5. Detection of Enteroviral RNA in Infected Cells by *In Situ* Nucleic Acid Hybridization 299
6. Detection of Enterovirus Sequences in the Myocardium from Patients with Myocarditis and Dilated Cardiomyopathy by *In Situ* Hybridization 303
7. Expression of Coxsackievirus B3 Proteins in *E. Coli* and Generation of Virus-Specific Antibodies 305
8. Simulation of Coxsackievirus Cardiac Disease in Cultured Human Heart Cells 308

9. Persistent Coxsackievirus B3 Carrier-State Infection of Cultured Human Myocardial Fibroblasts 309
10. Antiviral Activity of Human Fibroblast Interferon in Coxsackievirus B3-Infected Cultured Human Heart Cells 310
11. Downregulation of Interferon Receptors 312
12. Summary and Outlook 312
References 315

18. Neurologic Disorders

ELEANOR J. BELL, FAKHRY ASSAAD, AND KARIN ESTEVES

1. Introduction 319
2. Epidemiologic Analysis 320
3. Neurologic Syndromes 327
4. New Virus Diagnostic Techniques 331
5. Discussion 334
References 336

19. Mucocutaneous Syndromes

GIOVANNI ROCCHI AND ANTONIO VOLPI

1. Introduction 339
2. Hand-Foot-and-Mouth Disease 339
3. Herpangina 343
4. Acute Lymphonodular Pharyngitis 345
5. Erythematous Rashes 345
6. Epidemic Conjunctivitis 347
7. Concluding Remarks 348
References 348

20. Diabetes Mellitus

ANTONIO TONIOLO, GIOVANNI FEDERICO, FULVIO BASOLO, AND TAKASHI ONODERA

1. Introduction 351
2. Epidemiology of Insulin-Dependent Diabetes Mellitus 352
3. Seroepidemiologic Studies 353
4. Follow-Up Studies 360
5. Clinical Observations 361

6. Postmortem Studies 363
7. Experimental Models 365
8. Summary and Conclusions 375
References 376

21. Epidemiology

G. HAMMOND

1. Introduction 383
2. Routes of Virus Transmission 383
3. Factors That Influence the Distribution of Coxsackievirus Infections 384
4. A Model for Epidemiologic Patterns of Coxsackievirus Infections 385
5. Coxsackievirus Infections within Families 386
6. Coxsackievirus Infections within Institutions 387
7. Seroepidemiologic Surveys of Coxsackievirus Infections 389
8. Geographic Patterns of Coxsackievirus Infections by Virus Identification 391
9. Summary 396
References 396

22. Epidemiology of Group B Coxsackieviruses

MARK A. PALLANSCH

1. Introduction 399
2. Surveillance 400
3. Association between Infection and Clinical Disease 403
4. Future Techniques and Studies 408
References 412

Index 419

1

Classification and General Properties

BRIAN W. J. MAHY

The coxsackieviruses (CV) are small, icosahedral positive single-stranded RNA viruses that cause common enteric infections in human populations throughout the world. Depending on virus type and factors such as age at infection, they may cause a wide spectrum of clinical diseases, including aseptic meningitis, common colds, epidemic myalgia (Bornholm disease), myocarditis, herpangina, pharyngitis, conjunctivitis, hand-foot-and-mouth disease, and possibly some cases of juvenile diabetes mellitus. CV were originally identified because of their pathogenicity for newborn mice and hamsters and were subsequently divided into two groups, A (CVA) and B (CVB), on the basis of differences in the tissue damage induced in these animals.

1. HISTORY AND CLASSIFICATION

In the summer of 1947, Gilbert Dalldorf and Grace Sickles were investigating five small outbreaks of poliomyelitis in upstate New York.[1] Laboratory mice were inoculated intracerebrally with 20% fecal suspensions prepared from two paralyzed boys. Suckling mice, 3–7 days of age, became paralyzed, while no paralysis could be induced in mice over 12 days of age, and it proved impossible to adapt the virus to weanling mice. The virus was repeatedly covered from the two patients' fecal spec-

BRIAN W. J. MAHY • Animal Virus Research Institute, Pirbright, Woking, Surrey GU24 ONF, England.

imens; the serum of both patients developed neutralizing activity against both virus isolates. Dalldorf and Sickles pointed out that although these observations showed that the virus is capable of infecting humans, it had not been shown to induce the paralysis, and so might be a coincidental infection.

During the following 2 years, further virus isolates that induced severe muscular paralysis in suckling mice and hamsters were described,[2,3] although the pathogenicity of these viruses for man was unknown. Dalldorf[4] suggested that a name for the group would be desirable. Since the original isolate was made from the Hudson River village of Coxsackie, New York State, Dalldorf suggested that the group be named for its place of origin.

As more isolates were described, it became clear not only that CV existed as multiple antigenic types,[3,5] but also that there were differences in the type of tissue damage induced in suckling mice. Some isolates caused flaccid paralysis and generalized myopathy, whereas others induced spastic paralysis with a more focal myopathy together with lesions in the central nervous system (CNS) and in fat tissue. On the basis of these differences in pathogenesis, the CV were divided into groups A and B, respectively.[6,7] Clinical and epidemiologic observations did not, however, suggest that either group of viruses was the cause of paralytic poliomyelitis in the patients from which the isolates were made.[8]

The validity of the classification of CV as A or B was questioned by Melnick,[9] who proposed a system based on antigenicity alone. However, the original distinction between groups A and B was supported by subsequent studies on tissue culture tropism[10,11] and disease syndromes with which the viruses were associated in humans. CVB readily multiply in monkey kidney cell cultures, in which CVA are difficult to propagate, even though some can be adapted to human amnion cells.[11] Although both CVA and CVB cause febrile illness or aseptic meningitis in man, herpangina is invariably associated with CVA infections and with epidemic myalgia (Bamble or Bornholm disease) with CVB infections.[12,13]

Tables I and II list the prototype strains of CV with reference to their original isolation. The differentiation of types depends on at least a 20-fold difference in neutralization; two viruses are said to belong to the same serotype if they crossreact by more than 5% compared with the reaction with homologous antiserum. The preparation and standardization of reference antisera for all CVA and CVB was described by Wenner and his group in two papers published in 1965.[24,25]

The CV were identified because of their pathogenicity for suckling mice and can be defined as those enteroviruses that produce flaccid or spastic paralysis and recognizable pathologic lesions in suckling mice but not adult mice. The year after their first isolation, Enders *et al.*[26] suc-

TABLE I
Prototype Strains of Coxsackievirus Group A

Type	Prototype strain	Year	Place of origin	Comments[a]	Reference
A1	Tomkins	1947	Coxsackie, New York		Dalldorf *et al.*[1]
A2	Fleetwood	1947	Wilmington, Delaware		Sickles *et al.*[5]
A3	Olson	1948	New York		Sickles *et al.*[5]
A4	High Point	1948	North Carolina	Isolated from sewage	Melnick[14]
A5	Swartz	1950	New York		Dalldorf *et al.*[15]
A6	Gdula	1949	Silver Creek, New York		Dalldorf[16]
A7	Parker	1949	New York		Dalldorf[16]
A8	Donovan	1949	Levittown, New York		Dalldorf[16]
A9	Bozek	1950	Glen Cove, New York		Dalldorf[16]
A10	Kowalik	1950	New York		Dalldorf[16]
A11	Belgium 1	1951	Brussels		Godenne *et al.*[17]
A12	Texas 12	1948	Texas	Isolated from flies	Contreras *et al.*[18]
A13	Flores		Mexico		Sickles *et al.*[10]
A14	G14	1950	South Africa		Gear[19]
A15	G9	1950	South Africa		Gear[19]
A16	G10	1950	South Africa		Gear[19]
A17	G12	1950	South Africa		Gear[19]
A18	G13	1950	South Africa		Gear[19]
A19	Dohi	1952	Japan		Sickles *et al.*[10]
A20	IH Pool 35	1955	Albany, New York		Sickles *et al.*[11]
A20a	Tulane 1623	1951	Bayou Bouif, Louisiana		Sickles *et al.*[11]
A20b	Cecil P2647	1951	Germiston, South Africa		Sickles *et al.*[11]
A20c	Thai C-18	1959	Thailand		Abraham and Cheever[94]
A21	Kuykendall	1952	San Leandro, California		Sickles *et al.*[11]
A22	Chulman	1955	Albany, New York		Sickles *et al.*[11]
A23	Vispo	1955	Ancona, Italy	Now called echovirus 9	Sickles *et al.*[11]
A24	Joseph	1952	South Africa		Sickles *et al.*[11]

[a]All isolates from human stools, except A4 and A12.

TABLE II
Prototype Strains of Coxsackievirus Group B

Type	Prototype strain	Year	Place of origin	Reference
B1	Conn-5	1948	Connecticut	Melnick *et al.*[3]
B2	Ohio-1	1947	Ohio	Melnick *et al.*[20]
B3	Nancy	1949	Connecticut	Melnick and Ledinko[21]
B4	JVB (Benschoten)	1951	New York	Dalldorf[6]
B5	Faulkner	1952	Kentucky	Steigman[22]
B6	Schmitt	1953	Luzon, Philippine Islands	Hammon *et al.*[23]

ceeded in cultivating poliovirus in tissue culture, revolutionizing the isolation and study of viruses. In particular, enteric viruses were isolated that produced cytopathic changes, when inoculated into cell cultures, but that were nonpathogenic for newborn mice or subhuman primates. These were termed enteric cytopathogenic human orphan, or ECHO, viruses.[27] Although these viruses do not normally induce disease in suckling mice, it became clear that they may rarely cause aseptic meningitis, encephalitis, respiratory disease, exanthema, gastrointestinal (GI) symptoms, pericarditis, and myocarditis. In addition, variants that induce paralytic symptoms in newborn mice have been reported, particularly with echovirus type 9.[28] Originally, 34 echovirus types were described, but echovirus 10 was reclassified on morphologic grounds as reovirus 1, echovirus 28 has been reclassified as rhinovirus 1A, and echovirus 34 as a variant of CVA24.[29]

Of the original 23 strains of CVA, one type, A9, is highly cytopathogenic for monkey kidney cell cultures, but is difficult to grow in suckling mice. It has been suggested that this virus would have been classified as an echovirus had it originally been isolated in cell culture,[30] but it also shares an antigen with the CVB group. For these reasons, it is clear that the echoviruses and CV can be difficult to distinguish, and their classification has depended closely on the historic development of techniques for their study.

In 1957, the CV, echovirus, and poliovirus were brought together under the name enteroviruses, having in common the human alimentary tract as their natural habitat, and similar physicochemical properties.[31] By 1962, the CV comprised 30 members, 24 in group A and six in group B. The Committee on Enteroviruses then proposed that any new enterovirus isolates should not be assigned as CV, echovirus, or poliovirus, but given a number.[32] A modification of this original proposal was accepted in 1970,[29] when 67 human enteroviruses had been identified and divided into four groups: three polio, 24 CVA, six CVB, and 34 echoviruses. It should be noted, however, that CVA23 proved to be identical serologically to echovirus 9 and is no longer classified as a CV, so that the CV group has only 29 members (A1–22 and 24 and B1–6). Over the subsequent 16 years, only five further isolates not allotted to a group have been described; these are named enterovirus 68–72.

Enterovirus 68 was isolated from patients with pneumonia and bronchiolitis in California.[33] Enterovirus 69 was isolated from a rectal swab of a healthy child in Toluca, Mexico, in 1959, and has not been associated with any disease in humans.[34] Enterovirus 70 was isolated in 1971 from epidemics of acute hemorrhagic conjunctivitis in Japan, Singapore, and Morocco. These outbreaks were part of a pandemic that involved Africa, South East Asia, Japan, India, and England during 1969–1971.[35]

Enterovirus 71 is the only one of the five viruses known to share an antigen with a CV (A16), but there is no cross-neutralization between the two viruses.[36] It was originally isolated from the brain of a fatal case of encephalitis in California in 1970 and was associated with an epidemic of meningitis and paralysis in children in Bulgaria in 1975.[37] It has also been associated with a hand-foot-and-mouth disease epidemic in Japan,[38] indicating that it may have CV-like biologic properties.[39]

Finally, enterovirus 72 is now assigned to the causative agent of human hepatitis A.[40] It shares no particular features with the CV and probably should no longer be included in the enterovirus genus.

The enteroviruses form a recognized genus within the family Picornaviridae. Included in the genus are enteroviruses of other species, such as simian, porcine, and bovine enteroviruses, but the type species of the enteroviruses is human poliovirus type 1. The other established genera within the Picornaviridae are the rhinoviruses, which bear close relationships to the enteroviruses, and the cardiovirus and aphthovirus genera. In all, more than 230 serotypes are recognised within the Picornaviridae. The currently known groups are listed in Table III.

2. ISOLATION AND PROPAGATION

Coxsackieviruses may be isolated from stools, rectal swabs, throat swabs, cerebrospinal fluid (CSF), nasal secretions, and other body fluids although generally not from urine. They are also found in sewage and in sewage-contaminated water and have been isolated from flies. A variety of primate-derived cell cultures may be used for isolation of CVB types and CVA9, but other CVA types may require isolation by injection into suckling mice. Standard laboratory procedures for isolation involve inoculation of tube cell cultures of Hep 2, WI-38, RD, and rhesus monkey

TABLE III
Picornaviruses

Genus	Members
Enterovirus	Human polioviruses 1–3
	Human coxsackieviruses A1–A22, A24
	Human coxsackieviruses B1–B6
	Human echoviruses 1–9, 11–27, 29–33
	Human enteroviruses 68–71
	Human enterovirus 72 (hepatitis A virus)
	Simian enteroviruses 1–18
	Porcine enteroviruses 1–11[96]
	Bovid enteroviruses 1–2[97,98]
	Avian enteroviruses[99] (avian encephalomyelitis virus, duck hepatitis virus type 1 and 3, avian nephritis virus)
Cardiovirus	Encephalomyocarditis virus group (e.g., mengovirus, Columbia SK, EMC, Maus-Elberfeld viruses)
	Theiler's encephalomyelitis viruses[100]
Rhinovirus	Human rhinovirus 1–113
	Bovine rhinovirus 1–3[103]
Aphthovirus	Foot-and-mouth disease virus (types O, A, C, SAT1, SAT2, SAT3, and Asia 1)
Unassigned to genera	Equine rhinovirus 1–3[102]
	Acid-stable equine picornaviruses[103–105]
	Cricket paralysis virus[106]
	Drosophila C virus
	Gonometa virus plus at least 15 candidate insect picornaviruses[106]

kidney, as well as scoring for cytopathic effects (CPE). If no CPE is seen, the specimens are inoculated into groups of suckling mice by the intracerebral, intraperitoneal, and subcutaneous routes and passed into other suckling mice if pathologic changes are seen in the recipients.

A number of CVA types, which failed to cause a CPE in cell cultures on first inoculation, have since been adapted to infect *in vitro* cell cultures, such as human amnion or WI-38 cells. However, types 1, 4, 5, 6, 19, and 22 have not been adapted to *in vitro* cultivation. CVA21, which produces common coldlike symptoms, can be isolated directly in cell culture and only becomes pathogenic for suckling mice after repeated tissue culture passage. CVB can also be propagated in hamster kidney and pig kidney cells.

The characteristic CPE in susceptible cell cultures include cell rounding, with pyknotic nuclei, followed by degeneration and death, leaving holes in the cell sheet that can easily be visualized under low-power microscopy. Identification of virus isolates normally depends on neutralization of the cytopathic effects by specific antisera.[24,25,41] Where the isolate fails to cause a recognizable change in cell culture, neutralization of pathogenicity for suckling mice may be used instead. Since there are a large number of possible enteroviruses to consider in typing any isolate, a simplified system involving pools of antisera was developed[41,42] and has come into general use.[44,45] The neutralization may be made more clear cut by prior treatment of the crude viral suspension with sodium deoxycholate or chloroform.[46] The virus neutralization test is nevertheless laborious and time consuming; new identification methods based on techniques such as enzyme-linked immunosorbent assay (ELISA) are being developed and are reviewed in Chapter 12, this volume.

The classic neutralization test relies on polyclonal antisera to differentiate virus types. It has been known for many years that variants can appear during tissue culture or mouse passage that can be detected by incomplete or intratypic neutralization and classically termed prime antigenic variants.[47] Now that panels of monoclonal antibodies are becoming available, it is possible to quantify these antigenic differences; this approach exhibits considerable diversity in neutralization epitopes on different isolates of the same virus type.[48] The possibility that these represent functionally important differences affecting properties such as pathogenicity is being explored.[49,50] As more information becomes available, it may be necessary to redefine the classification of CV types to include subtypes identified on the basis of specific epitopes, which also define important biologic properties.

3. OCCURRENCE IN NONHUMAN SPECIES

Coxsackieviruses are known to be widely distributed in human beings, where they may be associated with a wide range of disease syndromes. From an epidemiologic point of view, it is necessary to assess the possible significance of animal infections, particularly in domestic species, which come into frequent contact with man. Although few investigations have been reported, these suggest that human CV infections of animals are not infrequent.

Among nondomesticated animals, the isolation of CVA4 from the blood of a cottontail rabbit[51] and of CVA5 from the brain of a fox[52] has been reported. CVB3 has been isolated from the intestines of a bat and CVB3 and CVB6 from rats in the Philippines.[53] The occurrence of CV in flies is also well documented,[18,54–58] and it is likely that all these infections reflect direct contact with human sewage.

Coxsackieviruses have also been found in domesticated dogs and pigs. Lundgren *et al.*[57] recovered 136 strains of CVB type 1 (118), type 3 (4), and type 5 (14), from a colony of 69 healthy beagle dogs in Albuquerque for a 1-year period. Low titers of serum antibody were present in some of the dog sera; it was concluded that the dogs became subclinically infected by contact with infected humans. However, a later survey of the same dog colony[58] failed to yield CV, although rectal swabs of six dogs in other environments in Albuquerque, New Mexico, and Salt Lake City, Utah, yielded strains of CVB1. In a detailed study of nine households in San Isidro de Heredia, Costa Rica, 56 humans and 46 animals were sampled.[59] Twenty-five virus isolates were made from humans, including 12 CVA20 and one CVA17. Of the 33 virus isolates from animals, 3 were CVA20 (2 dogs and 1 pig) and one CVA9 (dog). In a survey of human viruses present in animals in West Bengal, CVB3, CVB5, and CVB6 could be isolated from dogs and CVB4 from chickens.[60]

The occurrence of CVA in pigs has also been documented in the Netherlands, where type A5 was recovered from sick pigs.[61] Eighteen cases of mild febrile illness were observed in the children of six families who had had contact with the sick pigs, but no virus was recovered from the affected children. The virus was shown to reproduce the disease when passed to healthy pigs, which developed a serologic response postinfection.

The virus causing swine vesicular disease, first recognized in 1966 in Lombardy, Italy, is closely related to CVB5, from which it was probably derived.[62] Subsequent outbreaks occurred in Hong Kong in 1971 and in Europe and Japan in 1972–1975. The symptoms of the disease are similar to foot-and-mouth disease in pigs—with a fever as well as vesicular lesions on the feet and snout. However, even though the infectivity of swine vesicular disease can be neutralized by CVB5 antiserum,[63] CVB5 causes no disease when experimentally injected into pigs. The two viruses are closely related by RNA hybridization and polypeptide composition[64–66]; RNA sequence studies will be necessary to provide further information on their possible evolutionary relationship.

4. PHYSICOCHEMICAL PROPERTIES

Coxsackieviruses share a number of physicochemical properties with other members of the Picornaviridae. The family was named in 1963 from an acronym that included reference to each virus group and the common property of lack of an essential lipid component (*p*oliovirus, *i*nsensitivity to ether, *c*oxsackievirus, *o*rphan virus, *r*hinovirus, *r*ibo*n*ucleic *a*cid), the two rr's being contracted to one: the fact that pico also means very small supported the etymology.[67] Picornaviruses have an external diameter of about 30 nm and a total molecular weight of $\sim 8.5 \times 10^6$, of which 30% by weight is ribonucleic acid, the rest protein.[68]

The four principal genera can be distinguished by various physical properties.[69] The buoyant density of enteroviruses and cardioviruses in CsCl is 1.34, whereas rhinoviruses equilibrate at densities of 1.39–1.42, and aphthoviruses at 1.43–1.45. Enteroviruses, together with cardioviruses, are relatively stable at low pH and maintain infectivity for many hours over the range pH 3–9 at physiologic ionic strength. In one early study, CV infectivity was maintained over pH range 2.3–9.4 for 1 day, and pH 4–8 for 7 days.[70] Cardioviruses rapidly lose infectivity at pH 4.5–6.5 in the presence of chloride or bromide ions,[68] however, and at low ionic strength CVA13 was reported to be unstable, even at neutral pH.[71]

Coxsackieviruses are also relatively heat-stable and even at 50°C will survive for about 30 min; when suspended in milk, cream, or ice cream, they may survive at considerably higher temperatures for quite long periods.[72] Like other enteroviruses, CV are more heat-stable in the presence of Mg^{2+} and other divalent cations and can survive up to 3 hr at 50°C in 1 M $MgCl_2$.

Inactivation of CV infectivity occurs rapidly in the presence of 0.1 N HCl or 0.3% formaldehyde, but infectivity survives treatment with ether or chloroform, 70% ethanol, or 5% lysol.[74] The growth of CV in cell culture is sensitive to guanidine–hydrochloride at concentrations that do not affect cell metabolism or growth rate.[75,76] Mutation to guanidine resistance is rapid, with resistant mutants appearing after only a few passages; guanidine-dependent mutants may develop as well.[76] Most enteroviruses are also inhibited by 2(α-hydroxybenzyl)benzimadazole (HBB), but CVA types 7, 11, 13, 16, and 18 were unaffected by the drug, even though CVA types 9, 21, and CVB types 3 and 5 were inhibited. The molecular basis of these differences in HBB sensitivity is unknown.

5. HEMAGGLUTINATION

Only some CV types will agglutinate erythrocytes; even within types, the property can be strain specific and lost on tissue-culture passage. Types A20, A21, A24 and B1, B3, and B5 have been reported to agglutinate human type O erythrocytes.[78,79] Agglutination is inhibited by antibody, and absorption of virus suspensions with erythrocytes removes both the hemagglutinating ability and infectivity, proving that the hemagglutinin is associated with the virus particle. CVA7 causes agglutination of those fowl erythrocytes that are sensitive to vaccinia hemagglutinin.[80] In this case, the hemagglutinin appears to be soluble and not virion associated. Although it was originally hoped that hemagglutination–inhibition might provide an easy and sensitive serologic test for antibody, too few strains hemagglutinate consistently for the test to be generally useful.

6. VIRION STRUCTURE AND MORPHOLOGY

The genetic material of CV is contained in a single-stranded RNA molecule, some 7400 nucleotides long, which is infectious[81,82] and therefore of messenger-sense polarity (positive stranded). As with other picornaviruses, there is a genome-linked protein (VPg) at the 5′ end and the 3′ terminal is polyadenylated. The first 740 nucleotides (about 10% of the genome) at the 5′ end are noncoding, and from the AUG initiator codon a continuous open reading frame encodes a polyprotein, which in CVB3 is 2185 amino acids long.[83] A number of laboratories have begun to study the nucleotide sequences of CVB3 isolates and to compare them with other picornavirus genome sequences[83–86]; these studies are reviewed in Chapter 2, this volume, and are not discussed here.

Apart from the VPg at the 5′ end of the genome, all picornaviruses contain four capsid protein subunits, VP1, VP2, VP3, and VP4, termed 1D, 1B, 1C, and 1A, respectively, in the proposed systematic nomenclature of picornavirus polypeptides.[87] The sizes of these capsid proteins in CVB3 are VP1, 310, VP2, 328, and VP3 and VP4, 69 amino acids.[81,83] The capsid proteins and all virus-induced nonstructural proteins are generated from the polyprotein by a series of proteolytic cleavages catalyzed by virus-encoded proteases. These replicative events are broadly similar in all enteroviruses and are described in Chapter 3, this volume.

Early electron micrographs of CV showed small round icosahedral particles with a diameter close to 280 Å, similar to that of poliovirus[81,88] (Fig. 1). Crystals of CV were obtained some 30 years ago,[89] but only during the past 2 years have the three-dimensional structures of three picornaviruses been solved, those of human rhinovirus 14,[90] poliovirus type 1,[91] and mengovirus, a member of the cardiovirus genus.[92]

Overall, the structures of all three viruses are strikingly similar, although poliovirus and rhinovirus are more similar to each other than

FIGURE 1. Coxsackie B5 (Faulkner) virus particles cultured in Vero cells, concentrated and purified by differential centrifugation, sucrose gradient fractionation and pelleting. Resuspended particles were negatively stained with 2% methylamine tungstate at pH 7.2 for examination in the electron microscope. (**A**) Approximate magnification ×210,000. (**B**) Approximate magnification ×53,000.

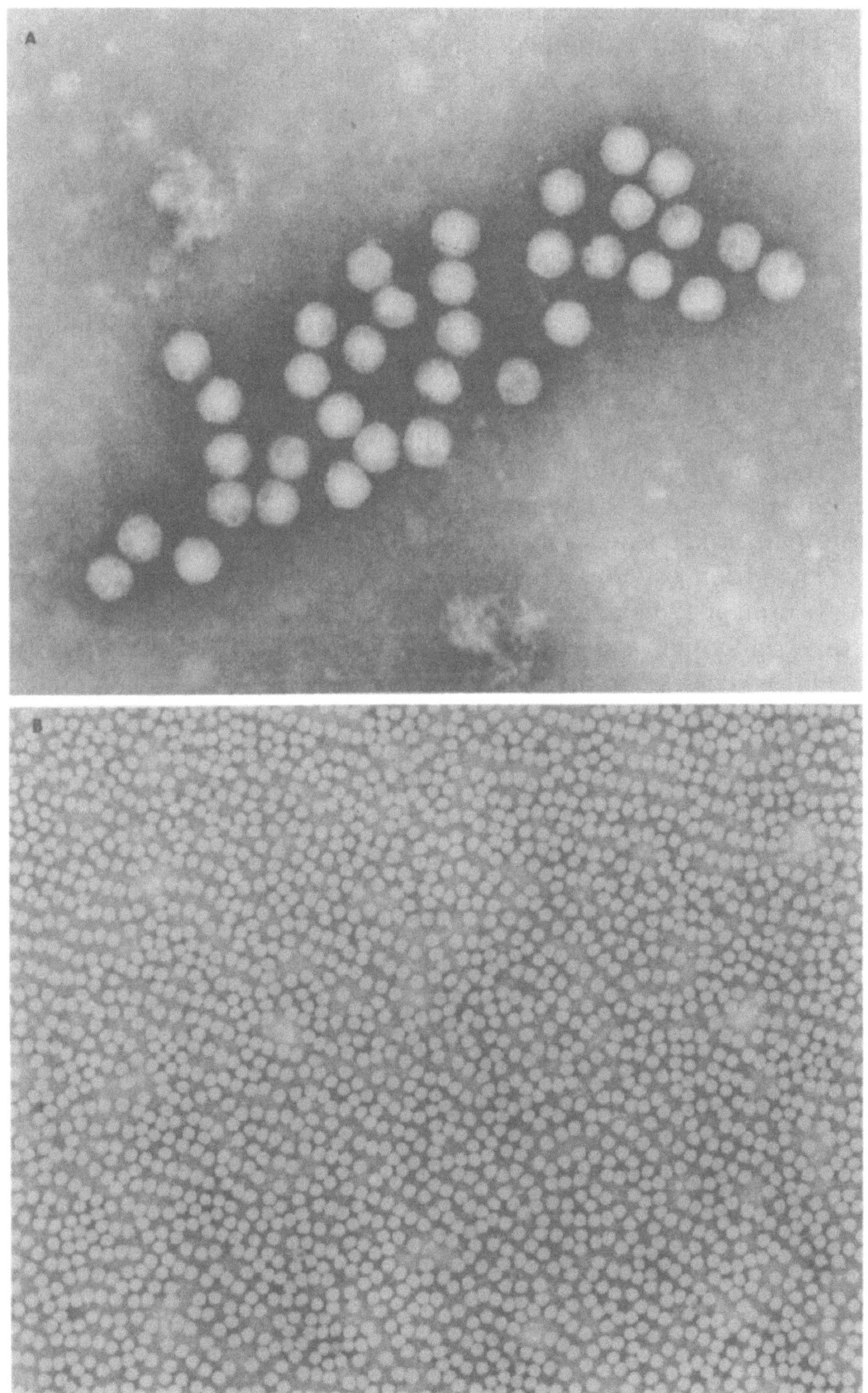
A
B

they are to mengovirus. Since poliovirus is the closest of these to CV, the poliovirus structure is summarized here.

The icosahedral virion consists of 60 molecules each of VP1, VP2, VP3, and VP4, which form a protein shell surrounding the RNA genome. A variety of studies have shown that the most antigenic capsid protein is VP1, although antigenic sites are also present in VP2 and VP3, whereas VP4 is internal and probably interacts directly with the RNA.[68] The X-ray diffraction data permitted identification of most of the residues in all four capsid proteins, except for residues at the amino-terminal of VP1, VP2, and VP4.

The positions of the genome RNA and VPg were not identified either. Each of the three largest capsid proteins consists of a core composed of an eight-stranded antiparallel β-barrel with two flanking β-helical regions. The structures are similar, differing mainly in the extent and conformations at their terminals and in the loops connecting β-strands. By contrast, VP4 has an extended conformation apart from a short antiparallel beta sheet near its amino-terminal.

The virion contains 60 copies of each protein, arranged on the icosahedron at 12 pentamers. Each pentamer consists of five protomers, in turn composed of one molecule each of VP1, VP2, VP3, and VP4. During virus assembly, the promoters are formed from the polyprotein as one molecule each of VP1, VP3, and a precursor protein VP0, which is later cleaved by an autocatalytic mechanism to form VP2 and VP4.[93] Five such protomers come together to form a disc-shaped pentamer with the five VP1 molecules at the centre and VP0 and VP3 at the periphery. The complete capsid is formed from 12 such pentamers. The central VP1 molecules are prominent on the virion surface and are thus accessible to antibody binding. Around this VP1 peak is a canyon separating the five VP1 subunits clustered about the pentamer axis from the surrounding VP2 and VP3 subunit. The canyon is 25 Å deep and 12–30 Å wide in rhinovirus 14 and has been postulated to be a cell receptor-binding site on the surface of the virus.

The availability of neutralizing monoclonal antibodies has permitted selection of virus mutants that resist neutralization (escape mutants) and can be used to locate antigenically important epitopes. In all cases, such mutations map with major external surface features of the virion.[91] As more information becomes available on such antigenic sites in CV, the three-dimensional poliovirus structure will provide an important framework for their interpretation. It is also likely, now that the principles of atomic structure determination have been established with three picornaviruses, that the three-dimensional structure of a CV will be determined. This should provide greater insight into the structural features of importance in virus neutralization and, it is hoped, into the nature of tissue-specific interactions, which underlie differences in virus pathogenicity, the *raison d'être* for the study of CV.

ACKNOWLEDGMENTS. I thank Nick Knowles for much helpful discussion and criticism, Chris Smale for Fig. 1, and Beryl Hutchinson for her skillful typing of the manuscript.

REFERENCES

1. Dalldorf, G., and Sickles, G. M., 1948, An unidentified, filtrable agent isolated from the feces of children with paralysis, *Science* **108:**61–62.
2. Dalldorf, G., Sickles, G. M., Plager, H., and Gifford, R., 1949, A virus recovered from the feces of 'poliomyelitis' patients pathogenic for suckling mice, *J. Exp. Med.* **89:**567–582.
3. Melnick, J. L., Shaw, E. W., and Curnen, E. C., 1949, A virus from patients diagnosed as non-paralytic poliomyelitis or aseptic meningitis, *Proc. Soc. Exp. Biol. Med.* **71:**344–349.
4. Dalldorf, G., 1949, The coxsackie group of viruses, *Science* **110:**594.
5. Sickles, G. M., and Dalldorf, G., 1949, Serologic differences among strains of the coxsackie group of viruses, *Proc. Soc. Exp. Biol. Med.* **72:**30–31.
6. Dalldorf, G., 1950, The Coxsackie viruses, *Bull. NY Acad Med.* **26:**329–335.
7. Gifford, R., and Dalldorf, G., 1951, The morbid anatomy of experimental Coxsackie virus infection, *Am. J. Pathol.* **27:**1047–1064.
8. Dalldorf, G., and Gifford, R., 1951, Clinical and epidemiologic observations of coxsackie virus infection, *N. Engl. J. Med.* **244:**868–873.
9. Melnick, J. L., 1953, The coxsackie group of viruses, *Ann. NY Acad. Sci.* **56:**587–595.
10. Sickles, G. M., Mutterer, M., Feorino, P., and Plager, H., 1955, Recently classified types of coxsackie virus, group A. Behaviour in tissue culture, *Proc. Soc. Exp. Biol. Med.* **90:**529–531.
11. Sickles, G. M., Mutterer, M., and Plager, H., 1959, New types of coxsackie virus group A. Cytopathogenicity in tissue culture, *Proc. Soc. Exp. Biol. Med.* **102:**742–743.
12. Grist, N. R., Bell, E. J., and Assaad, F., 1978, Enteroviruses in human disease, *Prog. Med. Virol.* **24:**114–157.
13. Gamble, D. R., 1984, Enteroviruses: Polio-, ECHO-, and coxsackie viruses, in: *Topley and Wilson's Principles of Bacteriology, Virology and Immunity,* Vol. 4, 7th ed. (F. Brown, and G. Wilson, eds.), pp. 394–419, Arnold, London.
14. Melnick, J. L., 1950, Studies on the coxsackie viruses: Properties, immunological aspects and distribution in nature, *Bull. NY Acad. Med.* **26:**342–356.
15. Dalldorf, G., and Sickles, E. M., 1956, The cocksackieviruses, in: *Diagnostic Procedures for Virus and Rickettsial Diseases,* 2nd ed., pp. 153–168, American Public Health Association, New York.
16. Dalldorf, G., 1953, The coxsackie virus group, *Ann. NY Acad. Sci.* **56:**583–586.
17. Godenne, M. O., 1954, Premiers isolements de virus coxsackie chez deux enfants pendant l'épidémie de myalgie epidemique à Bruxelles en 1951, *Acta Paediatr. Belg.* **8:**29–42.
18. Contreras, G., Barnett, V. H., and Melnick, J. L., 1952, Identification of coxsackie viruses by immunological methods and their classification into 16 antigenically distinct types, *J. Immunol.* **69:**395–414.
19. Gear, J. H. S., 1952, Poliomyelitis in Southern Africa, in: *Poliomyelitis,* Papers and

Discussions presented at *the Second International Poliomyelitis Conference,* pp. 437–441, Lippincott, Philadelphia.
20. Melnick, J. L., Ledinko, N., Kaplan, A. S., and Kraft, L. M., 1950, Ohio strains of a virus pathogenic for infant mice (Coxsackie group). Simultaneous occurrence with poliomyelitis virus in patients with "summer grippe," *J. Exp. Med.* **91:**185–195.
21. Melnick, J. L., and Ledinko, N., 1950, Immunological reactions of the coxsackie viruses I. The neutralisation test: Technic and application, *J. Exp. Med.* **92:**463–482.
22. Steigman, A. J., 1957, Viruses in search of disease, *Ann. NY Acad. Sci.* **67:**249–250.
23. Hammon, W. McD., Yohn, D. S., and Pavia, R. A., 1960, Isolation and characterisation of prototype viruses ECHO-26, ECHO-27 and Coxsackie B6, *Proc. Soc. Exp. Biol. Med.* **103:**164–168.
24. Kamitsuka, P. S., Lon, T. Y., Fabiyi, A., and Wenner, H. A., 1965, Preparation and standardisation of coxsackievirus reference antisera. I. For twenty-four group A viruses, *Am. J. Epidemiol.* **81:**283–305.
25. Wenner, H. A., Behbehani, A. M., and Kamitsuka, P. S., 1965, Preparation and standardization of coxsackievirus reference antisera. II. For six group B viruses, *Am. J. Epidemiol.* **82:**27–39.
26. Enders, J. F., Weller, T. H., and Robbins, F. C., 1949, Cultivation of the Lansing strain of poliomyelitis virus in cultures of various human embryonic tissues, *Science* **109:**85–87.
27. Committee on the ECHO Viruses, 1955, Enteric cytopathogenic human orphan (ECHO) viruses, *Science* **122:**1187–1188.
28. Eggers, H. J., and Sabin, A. B., 1959, Factors determining pathogenicity of variants of ECHO 9 virus for newborn mice, *J. Exp. Med.* **110:**951–967.
29. Rosen, L., Melnick, J. L., Schmidt, N. J., and Wenner, H. A., 1970, Subclassification of enteroviruses and ECHO virus type 34, *Arch. Ges. Virusforsch.* **30:**89–92.
30. Melnick, J. L., 1965, Echoviruses, in: *Viral and Rickettsial Infections of Man,* 4th ed. (F. L. Horsfall and I. Tamm, eds.), pp. 513–545, JB Lippincott, Philadelphia.
31. Melnick, J. L., Dalldorf, G., Enders, J. F., Hammon, W. McD., Sabin, A. B., Syverton, J. T., and Wenner, H. A., 1957, The enteroviruses, *Am. J. Public Health* **47:**1556–1566.
32. Melnick, J. L., Dalldorf, G., Enders, J. F., Gelfaud, H. M., Hammon, W. McD., Huebner, R. J., Rosen, L., Sabin, A. B., Syverton, J. T., and Wenner, H. A., 1962, Classification of human enteroviruses, *Virology* **16:**501–504.
33. Schieble, J. H., Fox, F. L., and Lennette, E. H., 1967, A probable new human picornavirus associated with respiratory disease, *Am. J. Epidemiol.* **85:**297–310.
34. Melnick, J. L., Tagaya, I., and von Magnus, H., 1974, Enteroviruses 69, 70 and 71, *Intervirology* **4:**369–370.
35. Yoshi, T., 1977, Replication of enterovirus 70 in non-human primate cell cultures, *J. Gen. Virol.* **36:**377–384.
36. Hagiwara, A., Tagaya, I., and Yoneyama, T., 1978, Common antigen between coxsackievirus A16 and enterovirus 71, *Microbiolog. Immunol.* **22:**81–88.
37. Melnick, J. L., Schmidt, N. J., Mirkovic, R. R., Chumakov, M. P., Lavrova, I. K., and Voroshilova, M. K., 1979, Identification of Bulgarian strain 258 of enterovirus 71, *Intervirology* **12:**297–302.
38. Hagihara, A., Tagaya, I., and Yoneyama, T., 1978, Epidemic of hand, foot and mouth disease associated with enterovirus 71 infection, *Intervirology* **9:**60–63.
39. Melnick, J. L., 1984, Enterovirus type 71 infections: A varied clinical pattern sometimes mimicking paralytic poliomyelitis, *Rev. Infect. Dis.* **6:**5387–5390.
40. Melnick, J. L., 1983, Classification of human hepatitis A virus as enterovirus 72 and of hepatitis B virus as hepadnavirus type 1, *Intervirology* **18:**105–106.
41. Melnick, J. L., Rennick, V., Hampil, B., Schmidt, N. J., and Ho, H. H., 1973,

Lyophilised combination pools of enterovirus equine antisera: Preparation and test procedures for the identification of field strains of 42 enteroviruses, *Bull. WHO* **48:**263–268.

42. Lim, K. A., and Benyesh-Melnick, M., 1960, Typing of viruses by combinations of antiserum pools. Application to typing of enteroviruses (Coxsackie and ECHO), *J. Immunol.* **84:**309–317.
43. Schmidt, N. J., Melnick, J. L., Wenner, H. A., Ho, H. H., and Burkhardt, M. A., 1971, Evaluation of enterovirus immune horse serum pools for identification of virus field strains, *Bull. WHO* **45:**317–330.
44. Nelson, D., Hiemstra, H., Minor, T., and D'Alessio, D., 1979, Non-polio enterovirus activity in Wisconsin based on a 20-year experience in a diagnostic virology laboratory, *Am. J. Epidemiol.* **109:**352–361.
45. Melnick, J. L., and Wimberly, I., 1985, Lyophilised combination pools of enterovirus equine antisera: New LBM pools prepared from reserves of antisera stored frozen for two decades, *Bull. WHO* **63:**543–550.
46. Kapsenberg, J. G., Ras, A., and Korte, J., 1979, Improvement of enterovirus neutralization by treatment with sodium deoxycholate or chloroform, *Intervirology* **12:**329–334.
47. Wigand, R., and Sabin, A. B., 1962, Intratypic antigenic heterogeneity of coxsackie B viruses, *Arch. Ges. Virusforsch.* **12:**29–41.
48. Prabhakar, B. S., Haspel, M. V., McClintock, P. R., and Notkins, A. L., 1982, High frequency of antigenic variants among naturally occurring human coxsackievirus B4 virus isolates identified by monoclonal antibodies, *Nature (Lond.)* **300:**374–376.
49. Cao, Y., Schnurr, D. P., and Schmidt, N. J., 1984, Monoclonal antibodies for study of antigenic variation in coxsackievirus type B4: Association of antigenic determinants with myocarditic properties of the virus, *J. Gen. Virol.* **65:**925–932.
50. Webb, S. R., Kearse, K. P., Foulke, C. L., Hartig, P. C., and Prabhakar, B. S., 1986, Neutralisation epitope diversity of coxsackievirus B4 isolates detected by monoclonal antibodies, *J. Med. Virol.* **20:**9–15.
51. O'Connor, J. R., and Morris, J. A., 1955, Recovery of Texas-1 type Coxsackie virus from the blood of a wild rabbit and from sewage contaminating rabbits' feeding ground, *Am. J. Hyg.* **61:**314–320.
52. Makower, H., and Skurska, Z., 1957, Badania rad wirusami Coxsackie Dioniesiebie III, Izolacja virus a Coxsackie 2 mozgu lisa, *Arch. Immunol. Ther.* **5:**219–224.
53. Gregorio, S. B., Nakao, J. C., and Beran, G. W., 1972, Human enteroviruses in animals and arthropods in the central Philippines, *Southeast Asian J. Trop. Med. Public Health* **3:**45–51.
54. Melnick, J. C., and Dow, R. P., 1953, Poliomyelitis in Hidalgo County, Texas, 1948. Poliomyelitis and Coxsackie viruses from flies, *Am. J. Hyg.* **58:**288–309.
55. Ward, R., 1952. Poliomyelitis and Coxsackie virus from Egyptian flies, *Fed. Proc.* **111:**486.
56. Gelfand, H. M., 1961, The occurrence in nature of the Coxsackie and ECHO viruses, *Prog. Med. Virol.* **3:**193–244.
57. Lundgren, D. L., Clapper, W. E., and Sanchez, A., 1968, Isolation of human enteroviruses from Beagle dogs, *Proc. Soc. Exp. Biol. Med.* **128:**463–467.
58. Lundgren, D. L., Sanchez, A., Magnuson, M. G., and Clapper, W. E., 1970, A survey for human enteroviruses in dogs and man, *Arch. Ges. Virusforsch.* **32:**229–235.
59. Grew, N., Gohd, R. S., Arguedas, J., and Kato, J. I., 1970, Enteroviruses in rural families and their domestic animals, *Am. J. Epidemiol.* **91:**518–526.
60. Graves, I. L., and Oppenheimer, F. R., 1975, Human viruses in animals in West Bengal: An ecological analysis, *Hum. Ecol.* **3:**105–130.

61. Verlinde, J. D., and Versteeg, J., 1958, Coxsackie viruspneumoniae bij biggen als smetstofbron voor de meus, *Tijdschr. Diergeneesk.* **83:**459–468.
62. McKercher, P. B., and Graves, J. H., 1981, Swine vesicular disease, in: *CRC Handbook Series in Zoonoses,* Section B: *Viral Zoonoses,* Vol. II (J-H. Steel and G. W. Beran, eds), pp. 161–167, CRC Press, Boca Raton, Florida.
63. Graves, J. H., 1973, Serological relationship of swine vesicular disease virus and Coxsackie B5 virus, *Nature (London)* **245:**314–315.
64. Brown, F., Wild, T. F., Rowe, L. W., Underwood, B. O., and Harris, T. J. R., 1976, Comparison of swine vesicular disease virus and coxsackie B5 virus by serological and RNA hybridization methods, *J. Gen. Virol.* **31:**231–237.
65. Harris, T. J. R., and Brown, F., 1975, Correlation of polypeptide composition with antigenic variation in the swine vesicular disease and coxsackie B5 viruses, *Nature (London)* **258:**758–760.
66. Harris, T. J. R., Doel, T. R., and Brown, F., 1977, Molecular aspects of the antigenic variation of swine vesicular disease virus and coxsackie B5 viruses, *J. Gen. Virol.* **35:**299–315.
67. Melnick, J. L., 1983, Portraits of viruses: The Picornaviruses, *Intervirology* **20:**61–100.
68. Rueckert, R. R., 1985, Picornaviruses and their replication, in: *Virology* (B. N. Fields, ed.), pp. 705–738, Raven Press, New York.
69. Newman, J. F. E., Rowlands, D. J., and Brown, F., 1973, A physicochemical subgrouping of the mammalian picornaviruses, *J. Gen. Virol.* **18:**171–180.
70. Robinson, L. K., 1950, Effect of heat and pH on strains of coxsackievirus, *Proc. Soc. Exp. Biol. Med.* **75:**580–582.
71. Cords, C. E., James, C. G., and McLaren, L. C., 1975, Alteration of capsid proteins of coxsackievirus A13 by low ionic concentrations, *J. Virol.* **15:**244–252.
72. Kaplan, A. S., and Melnick, J. L., 1954, Effect of milk and other dairy products on the thermal inactivation of Coxsackie viruses, *Am. J. Public Health* **44:**1174–1184.
73. Wallis, C., and Melnick, J. L., 1961, Cationic stabilization—A new property of enteroviruses, *Virology* **16:**683–700.
74. Kaplan, A. S., and Melnick, J. L., 1952, Effect of milk and cream on the thermal inactivation of human poliomyelitis virus, *Am. J. Public Health* **42:**525–534.
75. Rightzel, W. A., Dice, J. R., McAlpine, R. J., Timm, F. A., McLean, I. W., Dixon, G. J., and Schabel, F. M., 1961, Antiviral effect of guanidine, *Science* **134:**558–559.
76. Sergiescu, D., Horodniceanu, F., and Aubert-Combiescu, A., 1972, The use of inhibitors in the study of picornavirus genetics, *Prog. Med. Virol.* **14:**123–199.
77. Tamm, I., and Eggers, H. J., 1962, Differences in the selective virus inhibitory action of 2-(α-hydroxybenzyl)-benzimidaxole and guanidine–HCl, *Virology* **18:**439–447.
78. Goldfield, M., Srihongse, S., and Fox, F. P., 1957. Hemagglutinin associated with certain human enteric viruses, *Proc. Soc. Exp. Biol. Med.* **96:**788–791.
79. Rosen, L., and Kern, J., 1961, Hemagglutination and hemagglutination-inhibition with coxsackie B viruses, *Proc. Soc. Exp. Biol. Med.* **107:**626–628.
80. Grist, N. R., 1960, Isolation of coxsackie A7 virus in Scotland, *Lancet* **1:**1054–1060.
81. Mattern, C. F. T., 1962, Some physical and chemical properties of coxsackieviruses A9 and A10, *Virology* **17:**520–532.
82. Sprunt, K., Redman, W. M., and Alexander, H. E., 1959, Infectious ribonucleic acid derived from enteroviruses, *Proc. Soc. Exp. Biol. Med.* **101:**604–608.
83. Lindberg, A. M., Stalhandske, P. O. K., and Pettersson, U., 1987, Genome of coxsackievirus B3, *Virology* **156:**50–63.
84. Stalhandske, P. O. K., Lindberg, M., and Pettersson, U., 1984, Replicase gene of coxsackievirus B3, *J. Virol.* **51:**742–746.
85. Tracy, S., Liu, H. L., and Chapman, N. M., 1985, Coxsackievirus B3: Primary struc-

ture of the 5′ non-coding and capsid protein-coding regions of the genome, *Virus Res.* **3:**263–270.

86. Kandolf, R., and Hofschneider, P. H., 1985, Molecular cloning of the genome of a cardiotropic coxsackie B3 virus: Full-length reverse-transcribed recombinant cDNA generates infectious virus in mammalian cells, *Proc. Natl. Acad. Sci. USA* **82:**4818–4822.
87. Rueckert, R. R., and Wimmer, E., 1984, Systematic nomenclature of picornavirus proteins, *J. Virol.* **50:**957–959.
88. Finch, J. T., and Klug, A., 1959, Structure of poliomyelitis virus, *Nature (Lond.)* **183:**1709–1714.
89. Mattern, C. F. T., and du Buy, H. G., 1956, Purification and crystallization of Coxsackie virus, *Science* **123:**1037–1038.
90. Rossmann, M. G., Arnold, E., Erickson, J. W., Frankenberger, E. A., Griffiths, J. P., Hecht, H-J., Johnson, J. E., Kamer, G., Luo, M., Mosser, A. G., Rueckert, R. R., Sherry, B., and Vriend, G., 1985, Structure of a human common cold virus and functional relationship to other picornaviruses, *Nature (Lond.)* **317:**145–153.
91. Hogle, J. M., Chow, M., and Filman, D. J., 1985, Three-dimensional structure of poliovirus at 2.9 A resolution, *Science* **229:**1538–1365.
92. Luo, M., Vriend, G., Kamer, G., Minor, I., Arnold, E. Rossmann, M. G., Boege, U., Scraba, D. G., Duke, G. M., and Palmenberg, A. C., 1987, The atomic structure of mengo virus at 3.0 resolution, *Science* **235:**182–191.
93. Arnold, E., Luo, M., Vriend, G., Rossmann, M. G., Palmenberg, A. C., Parks, G. D., Nicklin, M. J. H., and Wimmer, E., 1987, Implication of the picornavirus capsid structure for polyprotein processing, *Proc. Natl. Acad. Sci. USA* **84:**21–25.
94. Abraham, A. S., and Cheever, F. S., 1963, Virus isolation studies of stool specimens obtained from patients with cholera, *Proc. Soc. Exp. Biol. Med.* **112:**981–987.
95. Matthews, R. E. F., 1982, Classification and nomenclature of viruses, *Intervirology* **17:**1–199.
96. Knowles, N. J., Buckley, L. S., and Pereira, H. G., 1979, Classification of porcine enteroviruses by antigenic analysis and cytopathic effects in tissue culture: Description of 3 new serotypes, *Arch. Virol.* **62:**201–208.
97. Knowles, N. J., and Barnett, I. T. R., 1985, A serological classification of bovine enteroviruses, *Arch. Virol.* **83:**141–155.
98. Hamblin, C., Knowles, N. J., and Hedger, R. S., 1985, Isolation and identification of bovid enteroviruses from free-living wild animals in Botswana, *Vet. Rec.* **116:**237–238.
99. McFerran, J. B., and McNulty, M. S., 1986, Recent advances in Enterovirus infections of birds, *A Seminar in the CEC Agricultural Research Programme, Brussels, June 1985* (J. B. McFerran and M. S. McNulty, eds.), Veterinary Research Laboratories, Belfast, Northern Ireland.
100. Ozden, S., Tangy, F., Chamorro, M., and Brahic, M., 1986, Theiler's virus genome is closely related to that of encephalomyocarditis virus, the prototype cardiovirus, *J. Virol.* **60:**1163–1165.
101. Yamashita, H., Akashi, H., and Inaba, Y., 1985, Isolation of a new serotype of bovine rhinovirus from cattle, *Arch. Virol.* **83:**113–116.
102. Steck, F., Hofer, B., Schaeren, B., Nicolet, J., and Gerber, H., 1978, Equine rhinoviruses: New serotypes, in: *Proceedings of the Fourth International Conference on Equine Infectious Diseases, Lyon, September 1976* (J. T. Bryans and H. Berber, eds.), Veterinary Publications, Inc., Princeton, New Jersey.
103. Bohm, H. O., 1965, Uber die isolierung und charakterisierung eines picornavirus vom pferd, *Zentralbl. Vet. Med.* **11B:**240–250.

104. Mumford, J. A., and Thomson, G. R., 1978, Studies on picornaviruses isolated from the respiratory tract of horses, in: *Proceedings of the Fourth International Conference on Equine Infect. Diseases, Lyon, Sept. 1976* (J. T. Bryans and H. Gerber, eds.), Veterinary Publications, Inc., Princeton, New Jersey.

105. Fukunaga, Y., Kumanomido, T., Imagawa, H., Ando, Y., Kamada, M., Wada, R., and Akiyama, Y., 1981, Isolation of picornavirus from horses associated with Getah virus infection, *Jpn. J. Vet. Sci.* **43**:569–572.

106. King, L. A., Pullin, J. S. K., Stanway, G., Almond, J. W., and Moore, N. F., 1987, Cloning of the genome of cricket paralysis virus: Sequence of the 3′ end, *Virus Res.* **6**:331–344.

2

The Genome of the Group B Coxsackieviruses

STEVEN TRACY

1. INTRODUCTION

The six serotypes of the coxsackievirus B group (CVB1–6) and the 23 serotypes of the coxsackievirus A group (CVA1–23) represent a large proportion of the recognized enteroviruses. Recent interest in the molecular biology of the CVB group reflects the growing realization that the CVBs probably are involved in important human diseases such as myocarditis[1] and have been implicated as putative causative agents in a variety of other diseases, such as diabetes mellitus,[2,3] birth defects of the heart,[4] and hydroencephalopathy.[5] This chapter presents aspects of the CVB genome and the relationship of this genome to other, well-characterized enteroviral and picornaviral genomes. This discussion bases its remarks on the genomes of CVB1, CVB3, and CVB4 as the best-characterized CVB genomes to date. For general information on the organization and molecular biology of the picornaviral and enteroviral genomes, the reader is directed to several recent reviews.[6–8]

The CVB3 (strain Nancy) was the first CVB genome to be cloned[9–11] and extensively characterized through hybridization analyses and partial sequencing of the genome.[12–14] Recently, the genomes of CVB1[15] and CVB4[16] have been cloned and fully sequenced. In addition, the remainder of CVB3 has been sequenced[17] and portions of the CVB6 P3 coding region (U. Fortmueller, N. Chapman, and S. Tracy,

---

STEVEN TRACY • Department of Pathology and Microbiology, University of Nebraska Medical Center, Omaha, Nebraska 68105.

unpublished data) have been cloned and characterized by both sequence and nucleic acid hybridization studies. The CVB2 genome has been reported as cloned,[18] but no supportive data have been published to date. Comparison of CVB genomes to other enterovirus genomes must employ the extremely well characterized prototype enteroviral genome, that of poliovirus types 1–3 (PV1–3).[19]

2. GENERAL REMARKS ON THE CVB GENOME

The complete characterization of the PV1–3[19–21] and human rhinovirus 2 and 14 (HRV2 and 14)[22–24] genomes by nucleotide sequence analysis have demonstrated that the CVB genome belongs to this group (for convenience termed PCR, for polio–coxsackie–rhino) of picornaviral genomes on the basis of identical genome organization (location of noncoding and protein-coding regions) and similar nucleotide sequences resulting in coded proteins of similar amino acid sequence, content, and function. At this juncture, one can expect few, if any, unique observations to be gained from extensively characterizing the remaining CVB genomes. This has been further demonstrated by the primer extension sequence analyses of the 5′ nontranslated regions (NTRs) of several strains in each of the six CVB genomes (M. Pallansch, unpublished data). It would still be useful to clone genomes of specific members of the CVA and ECHO groups as they remain the last enteroviral groups for which there exists no extensive molecular information. However, on the basis of homology data gained from past[25] hybridization analyses, it is likely that these two enteroviral groups also will share, perhaps with the odd exception (e.g., echovirus 22[26]), the PCR genome design.

The CVB3 genome structure (Fig. 1) is identical in format to that of the PV group. The viral genome consists of a single-stranded, message-sense RNA molecule which is approximately 7400 ribonucleotides long, polyadenylated at the 3′ end and has, at the 5′ terminal, a covalently bound, virally coded, very small protein, VPg. The naked viral RNA is infectious as is the full-length complementary DNA (cDNA) clone.[11] The genome consists of about 740 bases at the 5′ end (the nontranslated region or NTR) that code for no known protein followed by a single open reading frame of approximately 2182 codons. The open reading frame is terminated by a stop codon and followed by a polyriboadenylate (poly A) tract of 50–80 bases. The coding capacity and coding regions are similar to that of the PV group. Recently, the genomes of CVB1,

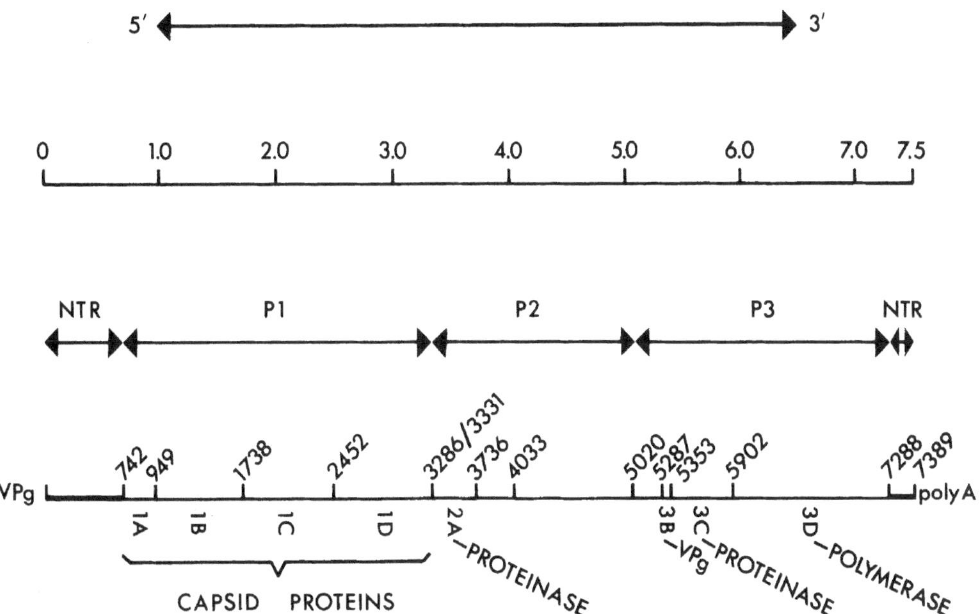

FIGURE 1. Schematic diagram of the CVB genome. Nontranslated regions (NTR) at the 5′ and the 3′ ends of the genome are depicted with heavier lines. Major subdivisions and products of the genome are shown, as are postulated cleavage sites of the viral polyprotein.[14–17]

CVB3, and CVB4 have been fully sequenced (14–17) and shown to be about 7390 nucleotides long (discounting the poly A tract at the 3′ end). As well, these data have demonstrated an almost identical length to each of the major genomic regions of poliovirus: the 5′ nontranslated region (NTR), the P1-, P2-, and P3-coding regions. A primary difference at the sequence level is the presence of 32 bases unique to the CVB genome immediately following the translation termination codon in the 3′ NTR.[9]

The overall PCR genome design tolerates some variation, however. Both HRV2 and 14 lack about 100 bases immediately prior to the start of translation in the 5′ NTR relative to the PV and CVB genomes. Kuge and Nomoto[27] constructed cDNA clones of PV1, which also lack these approximate 100 nucleotides upstream from the functional AUG; these workers showed such clones produce infectious virus in cell culture. The significance of these findings *in vivo* is not yet known, however. Such variations within a group should not be considered trivial, even though they are located in a NTR. Given the mutability of a viral RNA genome,[28] it must be assumed that nontranslated regions likely are conserved, either at the sequence level or simply as a sequence block, for reasons significant to replication of the virus *in vivo*.

3. THE 5′ NONTRANSLATED REGION

In all enteroviral genomes, there exists a 5′ sequence of varying length that begins with the 5′-terminal nucleotide, ends at the translational start AUG codon, is approximately 8–10% of the total viral genome in length, to which no known protein coding function has been ascribed (Fig. 1). The length of the 5′ NTRs in the CVB group is approximately 740 bases.[10,15,16] Unlike the heavily G + C weighted 5′ NTR of the HRV group, the nucleotide composition of the CVB 5′ NTR is unremarkable. The virally coded small protein, VPg, is covalently bound to the 5′-terminal U residue.[29] It has been shown with *in vitro* synthesized full-length PV RNA that the RNA does not require bound VPg in order to be infectious.[30]

It seems clear that the 5′ NTR is a highly structural RNA sequence and maintenance of this RNA structure is vital to the virus' replicative strategy. A potentially stable stem-and-loop structure, almost identical to that observed in the PV genomes in the same location, can be described in the 5′-terminal 40 bases of all CVB genomes sequenced in this region (CVB1, CVB3, and CVB4). In addition, computer folding of the RNA sequences suggests the CVB 5′ region to be similar to the PV genomes in the generalized cruciform secondary structure existing at the 5′ ends of PCR RNAs (A. Palmenberg, personal communication).

The purpose of the well-conserved and intricate secondary structures in the 5′ NTR is unclear. The high level of 5′ NTR nucleotide conservation within the CVB group (about 80–85%, similar to the PV and HRV genomes) argues for a role(s) critical to the virus' survival. The experiments of Racaniello and Baltimore[31] showed the virus required the major portion of the 5′-end sequence of the NTR as the lack of the 5′-terminal 115 bases from an otherwise complete cDNA copy of PV1 resulted in an uninfectious clone. Further work with an infectious PV1 cDNA demonstrated that various insertions, deletions, or rearrangements within the 5′ NTR resulted in loss of infectivity (V. Racaniello, personal communication). However, the 5′ terminal two U residues were deleted during cloning without loss of infectivity of an otherwise full-length CVB3 cDNA,[11] termination of the 5′ ends with a short homopolymeric tail derived during cloning resulted in infectious RNA when transcribed from an RNA transcription vector,[30] and the approximately 100 or so 5′ bases of the translational start at position N743 in PV1 have been deleted without loss of infectivity in cell culture of a full-length PV cDNA.[27] Regions in the 5′ NTR have been shown which are conserved between groups (e.g.. between PV and CVB genomes; termed PCR similar tracts) interspersed with regions of little or no nucleotide homology (PCR dissimilar tracts).[14] These interspersed tracts are not observed within a group of genomes because the 5′ NTR within any group (e.g., the CVB) are highly conserved relative to each member in the group. The exception to this are the 100 or so bases prior to the translational start site; this region shows the most variation within any group of genomes in the 5′ NTR.[19] However, the rhinoviruses have done away with this region altogether, which, with the observation that PV may also lose this region without loss of infectivity in tissue culture,[27] suggests that this region is not necessarily vital to basic replication of the PCR genome. Thus, it would appear that the well-conserved 5′ NTR interior sequence (about N>2 – N640) and its organization is of utmost importance and that certain deletions at either of the ends may be tolerated at least in cell culture systems.

The importance of the structural integrity of the 5′ NTRs is supported by experimental data from genomic cDNA constructions. Racaniello and Baltimore[31] demonstrated that manipulation of these sequences abolished infectivity of cDNA clones. Semler *et al.*[32] demonstrated that CVB3 sequences substituted in place of the homologous PV1 sequences in an infectious PV1 cDNA clone (while maintaining the PV1 5′ NTR structure and length) resulted in an infectious poliovirus. It is important to note that these chimeric viruses, which are completely PV1 in the coding regions, replicated with nucleotide sequences in the 5′ NTR derived from CVB3, some of them being tracts of PCR dissimilarity mentioned above. The observation that dissimilar tracts in the 5′ NTR (well conserved within a group but divergent when compared between groups of genomes) of one genome can function in another dissimilar genome may imply that such tracts play a (passive?) spacer role as well as having some function. This function may be at the level of protein translation, RNA transcription, or perhaps in how the viral RNA and capsid proteins interact to promote packaging in the infected cell. What the function is, how the similar and dissimilar tracts relate to one another in terms of secondary and tertiary structure and function, and why different viral genotypes maintain partially unique sequences in this region are intriguing questions for further research.

In view of the rapidly changeable nature of viral RNA populations,[28] it will be interesting to observe the fate of the CVB3 sequences that are not homologous to the replaced PV1 sequences as a function of repeated passage of the chimeric viruses[32] in cell culture. The conservation of these (PCR dissimilar) sequences within a group of genomes may imply specific groups have a functional preference for specific sequences. When introduced into a similar, yet foreign, genome and allowed to be pressured by repeated passage in cell culture, it might be expected that polymerase-induced nucleotide change in this region will occur until the CVB3 sequences are again homologous with the original (replaced) PV1 sequences at which point they would be stably maintained. Such a result would point to a functional, but not critical, role in basic viral replication for such sequences. Similar experiments, using constructions based on the CVB3 rather than on the PV genome, will permit observations about the effect of such alterations on the ability of the chimeric virus to replicate, as well as its pathogenicity in the mouse host.

4. THE P1 REGION

The P1-coding region contains the RNA sequences which code for the four viral capsid proteins called proteins 1A–1B–1C–1D[33] or VP4, VP2, VP3, and VP1. Like the PV group, the CVB group codes for its capsid proteins in the 5′–3′ order 1A–1B–1C–1D (Fig. 1). The P1 region and translation of the viral polyprotein begin at the AUG codon located approximately at N740. Although other AUG codons exist in the 5′ NTR, only the N740 position allows an open reading frame of about 6546 nucleotides coding for the polyprotein of approximately 2182 amino acids.[15–17] The size of this reading frame is similar to that described for other members of the PCR group.

Protein 1A (VP4) is the smallest of the capsid proteins, consisting of 69 amino acids and, as for the same protein coded for by PV and HRV genomes, begins with the conserved amino acid sequence MGAQVS. Given that protein 1A lies internally in the viral capsid and, by analogy to PV and HRV capsids,[34–37] presents none of its sequence on the virion outer surface, it is not surprising that this protein is the most highly conserved (relative to other PCR VP4 proteins) of all the capsid proteins. No cysteine residues have been shown to exist in the 1A protein of PV1 or CVB3,[38] and the predicted amino acid sequences in CVB1, CVB3, and CVB4 contain none.

Protein 1B (VP2) is cleaved from protein 1A at an N/S dipeptide junction. This cleavage is postulated to occur inside the capsid more or less simultaneously in all 60 capsid subunits following the formation of the virion by an autocatalytic event.[35,39] At 263 amino acids, protein 1B is slightly smaller than that of PV1 and shares only 50–53% amino acid homology with its PV1 homologue. Comparison of the CVB1, CVB3, and CVB4 protein 1B sequences, however, shows 80–85% overall homology. Protein 1B is the second largest of the capsid proteins and, together with proteins 1C and 1D, helps shape the antigenic character of the virus. Within this coding region begins the nucleotide sequence shown to be useful as a CVB3-specific hybridization probe.[10,12,25] In the same genomic region of CVB4, there exists a CVB4-specific sequence (i.e., no cross-hybridization with other non-CVB4 enteroviral RNAs) (S. Tracy, unpublished data).

Capsid protein 1C (VP3) contains 238 amino acids, identical in size to, and sharing 131 amino acids in common with, the VP3 of PV1. The cleavage site between the 1B and 1C proteins is a Q/G dipeptide. Capsid proteins 1C and 1D share the least homology with the homologous proteins of the PV group as well as with other members of the CVB group.[16] Proteins 1C and 1D are predominant on the exterior of the virion and present between them the primary viral antigenic epitopes.

The major capsid protein, 1D (VP1), is cleaved from the 1C protein at a Q/G dipeptide. The length of the CVB3 protein 1D is uncertain because of the ambiguity of the cleavage site predicted between 1D and protein 2A, although CVB4 apparently conserves the Y/G site also present in PV and HRV14.[16] The 1D/2A cleavage site for CVB3 has been tentatively placed between an I/R pair or a Y/R pair based on analogy to known sequences from the PCR group.[14] Iizuka and colleagues placed this site at a T/G pair for the CVB1 genome.[15] Comparison of the CVB1-, CVB3-, and CVB4-predicted amino acid sequence about this region shows that, following the Y/R (Y/K in CVB4) site, the two proteins are held closely in common, and prior to this site, the amino acid sequences of CVB1 and CVB4 vary significantly from CVB3 (although the CVB1 and CVB4 sequences are quite similar). Analysis of the sequences for PV1–3[19] also show excellent homology in the carboxyl terminal of protein 1D. However, because of the uncertainty of the site due to the lack of an actual amino acid sequence, the site is represented as tentative in Fig. 1.

The exciting results of Mapoles *et al.*,[40] describing the isolation of an HeLa cell receptor for the CVB group, and the elegant crystallographic work describing the three dimensional atomic structures of two similar PCR viruses, PV1 and HRV14[34–37] as well as meningovirus (a cardiovirus),[41] will be extremely useful in studying how these viruses attach to and penetrate the cell and should point the way to novel antiviral drugs in the future. The probable receptor on the viral outer shell has been characterized as a kidney-shaped canyon with sides composed of capsid proteins 1C and 1D.[41] Presumably, a projection on the cellular receptor protein interacts with this canyon when the virus binds to the cell surface. Due to the similarities within the PCR genomes and their coding, it may safely be assumed that the poliovirus paradigm holds for CVB as well. It has been reported[35] that two separate amino acid sequences in the capsid proteins 1C and 1D that map in the region of the canyon depicted by crystallographic analysis are highly conserved in HRV14, PV, and foot-and-mouth-disease viruses. Likewise, similar conserved sequences can be found in the 1C and 1D proteins coded by the CVB genomes. Design of low-molecular-weight compounds capable of binding with high affinity to these structurally similar regions inside the cleft of the viral receptor might be an effective way to inhibit interaction of the cellular receptor protein moiety with the canyon, thereby preventing viral entry. Rossman *et al.*[36] presented evidence that specific compounds can penetrate into a small porelike structure at the base of this canyon and bind well, inhibiting uncoating of the virus. Such approaches are receiving intense scrutiny.

5. THE P2 REGION

The P2 region is located in the middle of the CVB genome (Fig. 1). Toyoda *et al.*[42] showed the poliovirus 2A protein to have a proteinase activity, cleaving the Y/G dipeptide separating the P1 and P2 regions of the polyprotein (between capsid protein 1D and the protein 2A). The predicted amino acid sequence of 2A and CVB1, CVB3, and CVB4 are nearly identical and share four blocks (5–14 amino acids in length) of high amino acid homology with the same predicted sequence for the PV group as well, strongly suggesting that a similar proteinase activity is associated with the CVB protein 2A. Protein 2A may be a thiol proteinase, similar to the enteroviral protein 3C, in that the active site amino acid sequence for thiol proteinases is conserved to some degree (GDCGG) within a highly conserved amino acid sequence block.[42] Comparison of the predicted amino acid homologies at the 1D/2A junction for CVB1, CVB3, CVB4, and PV1–3 may suggest Y/G (PVs), Y/R, and Y/K (CVBs) as the junction dipeptides, possibly implying a significance for the presence of a tyrosine residue at the cleavage site for this enzyme. As the 2A protein may well require a specific tertiary structure surrounding its cleavage site, however, prediction of its cleavage site in the CVB group based on presumed junction dipeptides and homologies about the terminal of 2A may have only limited value.

The function of the remainder of the P2 region is not well understood. A strong homology between the CVB1 and PV1 internal sequences for protein 2C (X protein) has been noted.[15] No mutations resulting in a viable virus have been mapped in the 2C protein region of PV.[6] This protein plays an important but as yet unknown role in enteroviral replication and may tolerate little significant change. It is not a polymerase.[8]

6. THE P3 REGION

The P3-coding region of the CVB genome codes for three primary products: the 5′ genome-end-linked VPg, the viral specific proteinase, and the RNA polymerase. Following translation of the polymerase-coding region, translation is terminated with a single stop codon.

The smallest of the viral proteins is VPg (protein 3B) that is cleaved from the precursor protein 3AB at a Q/G site. The protein 3B is 22 amino acids long and is substantially different from that of the PV group. The VPg is attached to the 5′ end of the viral RNA with a covalent linkage but is not necessary for the infectivity of the RNA in transfection assays using synthetic RNAs.[30]

The viral proteinase (protein 3C) is the primary proteinase cleaving the polyprotein synthesized from the viral RNA. Protein 3C cleaves at Q/G sites in the polyprotein as well as cleaving itself out of the nascent polyprotein.[43,44] The amino acid sequences predicted for the CVB proteinases are very similar. Recent work with site-directed mutagenesis in the protein 3C coding region of PV1[45] and the availability of full-length clones of CVB3[11] (N. Chapman and S. Tracy, unpublished results) point the way to a clearer understanding of how this proteinase and the protein 2A activity interact during productive infections. The use of CVB : PV cDNA constructions, as well as expression vector studies similar to those performed with PV1 protein 2A,[42] should provide a wealth of knowledge regarding the specificity of these enzymatic activities for their specific cleavage sites.

The RNA polymerase (protein 3D) is 462 amino acids long and replicates the viral RNA. The mechanism of RNA replication is not clear but may involve an intermediate with twice the length of the viral RNA—one positive strand and one negative strand joined at the 5′ end of the positive strand[46]—although this interpretation is disputed.[47] The predicted amino acid sequences for the CVB1, CVB3, and CVB4 polymerases are similar to each other, as they are to the homologous protein in the PV group. Given the similarities between the CVB and the PV genomes, it is not unreasonable to assume that an *in vitro* construction substituting, for example, the CVB3 polymerase for the PV1 polymerase in the PV genome, similar in principle to other PV1 : CVB3 constructions,[32] will produce an infectious virus in cell culture. Such constructions may prove useful in assessing the role(s) of the 5′ and 3′ NTRs of the genome in viral RNA replication.

7. THE 3′ NONTRANSLATED REGION

Approximately 1% of the enteroviral genome is contained in the 3′ nontranslated region (NTR). The 3′ NTR is bounded on the 5′ side by the termination codon in the open reading frame and on the 3′ end by the poly A tract of approximately 60–90 residues. Sequence analysis of the CVB3 3′ NTR showed it to be about 98 nucleotides long, 32 bases longer than that of the PV genomes.[9] These 32 bases occur immediately following the translational stop codon and have not been detected in other PCR sequences with the exception of a partial similar sequence in swine vesicular disease virus,[48] thought to be a member of the CVB group in swine. Interestingly, there are 13 nucleotide differences between the CVB1 and CVB3 sequences versus only one to three changes in the sequences of PV1–3.[19] Of the 13 changes, eight occur within the 32 base sequence specific to CVB genomes and 75% of these are transversions. It has been suggested elsewhere that maintenance of the secondary structure in the 5′ NTR might be facilitated by allowing transitional, rather than transversional, change.[14] This region has the predominant number of base changes when CVB1 and CVB3 are compared, an observation which has been noted as well for the 5′ NTR sequence prior to the initiation codon. By analogy to the 5′ NTR in enteroviruses, wherein the sequence of about 100 nucleotides prior to the start of translation differs significantly between viruses,[19] is absent in the HRV genomes,[22–24] and may not be necessary for PV replication in cell culture,[27] it is interesting to speculate that the 32 base sequence in the CVB 3′ NTR may be dispensable as well for the replication of CVB.

Immediately following the 32-base sequence peculiar to the CVB 3′ NTR, a stem-and-loop structure can be described in the CVB1, 3, and 4 sequences. Interestingly, a transition occurs in the CVB1 sequence that is not countered by a similar transition on the opposite side, making the structure less energetically stable than the CVB3 stem.[15] Comparison of these structures to that in PV1 by these same workers demonstrated a similar (two base) mismatch in the PV1 stem. Both changes occur in the upper half of the stems, suggesting that it may be the lower half of the stems that is structurally important to maintain through compensatory base change.[49]

8. SUMMARY

Analysis of CVB1, CVB3, and CVB4 nucleotide sequence data by comparison to the sequences of the PV and HRV genomes has shown that the CVB genome varies little in basic structure and probable function from the well-characterized poliovirus genome. The major difference in overall sequence is that the CVB genome has an additional 32 bases immediately following the translational termination codon in the 3′ NTR. This short sequence has not been found in other (non-CVB) PCR genomes to date. Whether these 32 bases are important for the life cycle of the CVB genome is unclear; experiments are under way in this laboratory using CVB3 : PV1 cDNA constructions for the study of

CVB3-induced myocarditis in mice that should answer this question.

Why do certain CVB3 variants induce myocarditis in specific strains of mice while other variants do not? Despite major efforts to elucidate the molecular grounds for either neurovirulence or vaccinelike character in PV strains,[50–52] current data suggest only that differences between virulent and nonvirulent genomes may be at the single nucleotide level, often located in the 5′ half of the genome. Although induction of a specific disease is clearly coded at the genetic level in the specific virus, the host's own genetics are also involved, playing a significant, if not the major, role in the induction of the disease. Because single base changes in the viral RNA are predicted to cause significant changes in the secondary (and by inference, tertiary) structure of the viral RNA molecule,[50] alteration of the viral nucleic acid through the use of *in vitro* designed, cDNA-based genomes will be an effective tool to ask specific questions of viral replication and of the host–virus interaction at the molecular level.

The similarity of the CVB group to the PV and HRV groups suggests that current work investigating potential antiviral drugs for the latter viruses may be applicable to the CVB group. The structural similarities between the probable virus receptor canyon[35,36] and that which can be inferred from sequence data in the CVB group suggest a single compound may be effective against many different enteroviruses. Such compounds would bind in regions of the capsid proteins that tend to be conserved, probably for structural reasons. Thus, it might be expected that such compounds would remain effective even given the rapid genomic changes that RNA virus-coding sequences can undergo. Such drugs would be of immense value should the CVB group unequivocably be shown to be involved with human heart disease and/or other serious human diseases.

Study of the CVB genome has supplied useful information for the design of hybridization probe sequences which are able to detect many enteroviruses in general or specifically a genotype, e.g., CVB3.[12,25] In addition, knowledge and comparison of various enteroviral nucleotide sequences have made it possible to predict specific oligonucleotides which might be useful in a similar manner.[53] For example, it will be interesting to determine whether the well-conserved, apparently CVB-specific, sequence in the CVB 3′ NTR can be used to predict oligonucleotides for a CVB group-specific hybridization probe. The availability of well-characterized hybridization probes, when coupled with efficient accelerated-rate hybridization systems (D. Kohne, personal communication), should soon make possible specific identification of members of the CVB group in samples within a working day at sensitivities approaching the 10–100 infectious particle range.

The CVB group has been implicated in acute myocarditis, as well as perhaps dilated cardiomyopathy, in humans. A recent report describing the detection of coxsackie B virus RNA in human heart[18] is suggestive of such a connection. The data in this report do not, however, support the authors' statements that coxsackievirus B-specific sequences were found in these hearts as the hybridization probe used would have detected a wide gamut of enteroviruses, not just the six coxsackie B genotypes, under the hybridization criteria reported.[12,25] Direct implication of these viruses with the above diseases by nucleic acid hybridization must await experiments with viral probes which lack contaminating plasmid sequences which may result in false positive hybridization signals. Such an approach, using *in situ* hybridization techniques, is currently in progress in this laboratory.

ACKNOWLEDGMENTS. I thank David Kohne, Owen Jenkins, Akio Nomoto, Mark Pallansch, Ann Palmenberg, and Haruka Toyoda for graciously and generously communicating data prior to publication.

REFERENCES

1. Woodruff, J., 1980, Viral myocarditis. A Review, *Am. J. Pathol.* **101**:425–483.
2. Jenson, A., and Rosenberg, H., 1984, Multiple viruses in diabetes mellitus, *Prog. Med. Virol.* **29**:197–217.
3. Rayfield, E., and Seto, Y., 1978, Viruses and the pathogenesis of diabetes mellitus, *Diabetes* **27**:1126–1140.
4. Brown, G., and Evans, T., 1967, Serologic evidence of coxsackievirus etiology of congenital heart disease, *JAMA* **199**:183–187.
5. Gauntt, C., Jones, D., Huntingdon, H., Arizpe, H., Gudvangen, R., and DeShambo, R., 1984, Murine forebrain anomalies induced by coxsackievirus B3 variants, *J. Med. Virol.* **14**:341–355.
6. Koch, F., and Koch, G., 1985, *The Molecular Biology of Poliovirus*, Springer-Verlag, New York.
7. Rueckert, R., 1985, Picornaviruses and their replication, in: *Fundamental Virology* (B. Fields and D. Knipe, eds.), pp. 357–390, Raven, New York.
8. Palmenberg, A., 1987, Comparative organization and genome structure in picornaviruses, in: *Positive Strand RNA Viruses* (M. Brinton and R. Rueckert, eds.), pp. 25–34, Alan Liss, New York.
9. Stalhandkse, P., Lindberg, M., and Petterson, U., 1984, Replicase gene of coxsackievirus B3, *J. Virol.* **51**:742–746.
10. Tracy, S., Chapman, N., and Liu, H., 1985, Molecular cloning and partial characterization of the coxsackievirus B3 genome, *Arch. Virol.* **85**:157–163.
11. Kandolf, R., and Hofschneider, P., 1986, Molecular cloning of the genome of a cardiotropic coxsackie B3 virus: Full-length reverse transcribed recombinant cDNA generates infectious virus in mammalian cells, *Proc. Natl. Acad. Sci. USA* **82**:4818–4822.
12. Tracy, S., 1984, A comparison of genomic homologies among the coxsackievirus B group: Use of fragments of the cloned coxsackievirus B3 genome as probes, *J. Gen. Virol.* **65**:2167–2172.
13. Tracy, S., 1985, Comparison of genomic homologies in the coxsackievirus B group by use of cDNA : RNA dot-blot hybridization, *J. Clin. Microbiol.* **21**:371–374.

14. Tracy, S., Liu, H., and Chapman, N., 1985, Coxsackievirus B3: Primary structure of the 5′ non-coding and capsid protein coding regions of the genome, *Virus Res.* **3:**2663–270.
15. Iizuka, N., Kuge, S., and Nomoto, A., 1987, Complete nucleotide sequence of the genome of coxsackievirus B1, *J. Virol.* **156:**64–73.
16. Jenkins, O., Booth, J., Minor, P., and Almond, J., 1987, The complete nucleotide sequence of coxsackievirus B4 and its comparison to other members of the picornaviridae, *J. Gen. Virol.* **68:**1835–1848.
17. Lindberg, A., Stalhandske, P., and Petterson, U., 1987, Genome of coxsackievirus B3, *Virology* **156:**50–63.
18. Bowles, N., Richarson, P., Olson, E., and Archard, L., 1986, Detection of coxsackie-B-virus specific RNA sequences in myocardial biopsy samples from patients with myocarditis and dilated cardiomyopathy, *Lancet* **1:**1120–1123.
19. Toyoda, H., Kohara, M., Kataoka, Y., Suganuma, T., Omata, T., Imura, N., and Nomota, A., 1984, Complete nucleotide sequences of all three poliovirus serotype genomes, *J. Mol. Biol.* **174:**561–585.
20. Kitamura, N., Semler, B., Rothberg, P., Larsen, G., Adler, C., Dorner, A., Amini, E., Hanecak, R., Lee, J., van der Werf, S., Anderson, C., and Wimmer, E., 1981, Primary structure, gene organization, and polypeptide expression of poliovirus RNA, *Nature (Lond.)* **291:**547–553.
21. Stanway, G., Hughes, P., Mountford, R., Reeve, P., Minor, P., Schild, G., and Almond, J., 1984, Comparison of the complete nucleotide sequences of the genomes of the neurovirulent poliovirus P3/Leon/37 and its attenuated Sabin vaccine derivative P3/Leon/12alb, *Proc. Natl. Acad. Sci. USA* **81:**1539–1543.
22. Callahan, P., Mizutani, S., and Colonno, C., 1985, Molecular cloning and sequence determination of human rhinovirus 14 genome RNA, *Proc. Natl. Acad. Sci. USA* **82:**732–736.
23. Stanway, G., Hughes, P., Mountford, R., Minor, P., and Almond, J., 1984, The complete nucleotide sequence of a common cold virus: Human rhinovirus 14, *Nucleic Acids Res.* **12:**7859–7875.
24. Skern, T., Sommergruber, W., Blass, D., Gruendler, P., Fraundorfer, F., Pider, C., Fogy, I., and Knochler, E., 1985, Human rhinovirus 2: Complete nucleotide sequence and proteolytic processing signals in the capsid protein region, *Nucleic Acids Res.* **13:**2111–2126.
25. Rotbart, H., Levin, M., Villareal, L., Tracy, S., Semler, B., and Wimmer, E., 1985, Factors affecting the detection of enteroviruses in cerebrospinal fluid with coxsackievirus B3 and poliovirus 1 cDNA probes, *J. Clin. Microbiol.* **22:**220–224.
26. Seal, L., and Jamison, R., 1984, Evidence for secondary structure within the virion RNA of echovirus 11, *J. Virol.* **50:**641–644.
27. Kuge, S., and Nomoto, A., 1987, Construction of virable deletion and insertion mutants of the Sabin strain of type 1 poliovirus: Function of the 5′ noncoding sequence in viral replication, *J. Virol.* **61:**1478–1487.
28. Holland, J., Spindler, K., Horodyski, F., Grabau, B., Nichol, S., and Vandepol, S., 1982, Rapid evolution of RNA genomes, *Science* **215:**1577–1585.
29. Hewlett, M., and Florkiewicz, R., 1980, Sequence of picornavirus RNAs containing a radioiodinated 5′ linked peptide reveals a conserved 5′ sequence, *Proc. Natl. Acad. Sci. USA* **77:**303–307.
30. van der Werf, S., Bradley, J., Wimmer, E., Studier, F., and Dunn, J., 1986, Synthesis of infectious poliovirus RNA by purified T7 RNA polymerase, *Proc. Natl. Acad. Sci. USA* **83:**2330–2334.
31. Racaniello, V., and Baltimore, D., 1981, Cloned poliovirus complementary DNA is infectious in mammalian cells, *Science* **214:**916–919.

32. Semler, B., Johnson, V., and Tracy, S., 1986, A chimeric plasmid from cDNA clones of poliovirus and coxsackievirus produces a recombinant virus that is temperature-sensitive, *Proc. Natl. Acad. Sci. USA* **83:**1777–1781.
33. Rueckert, R., and Wimmer, E., 1984, Systematic nomenclature of picornavirus proteins, *J. Virol.* **50:**957–959.
34. Hogle, J., Chow, M., and Filman, D., 1985, Three-dimensional structure of poliovirus at 2.9 Angstrom resolution, *Science* **229:**1358–1365.
35. Rossman, M., Arnold, E., Erickson, J., Frankenberger, E., Griffith, J., Hecht, H., Johnson, J., Kamer, G., Luo, M., Mosser, A., Ruckert, R., Sherry, B., and Vriend, G., 1985, Structure of a human cold virus and functional relationship to other picornaviruses, *Nature (Lond.)* **317:**145–153.
36. Rossman, M., Arnold, E., Kamer, G., Kremer, M., Luo, M., Smith, T., Vriend, G., Rueckert, R., Mosser, A., Sherry, B., Boege, U., Scraba, D., McKinlay, M., and Diana, G., 1987, Structure and function of human rhinovirus 14 and mengo virus: Neutralizing antigenic sites, putative receptor binding site, neutralization by drug binding, in: *Positive Strand RNA Viruses* (M. Brinton and R. Rueckert, eds.), pp. 59–77, Alan R. Liss, New York.
37. Hogle, J., Chow, M., and Filman, D., 1987, The three-dimensional structure of poliovirus: Implications for assembly and immune recognition, in: *Positive Strand RNA Viruses* (M. Brinton and R. Rueckert, eds.), pp. 79–92, Alan R. Liss, New York.
38. Philipson, L., Beatrice, S., and Crowell, R., 1973, A structural model for picornaviruses as suggested from an analysis of urea-degraded virions and procapsids of coxsackie B3, *Virology* **54:**69–79.
39. Parks, G., and Palmenberg, A., 1985, Hydrazine-induced in vitro cleavage of EMC VPO precursor, presented at the *Fourth Meeting of the European Group of Molecular Biology of Picornaviruses, Seillac, France,* (abst. B5).
40. Mapoles, J., Krah, D., and Crowell, R., 1985, Purification of a HeLa cell receptor protein for group B coxsackieviruses, *J. Virol.* **55:**560–566.
41. Luo, M., Vriend, G., Karrer, G., Minor, I., Arnold, E., Rossman, M., Boege, U., Scraba, D., Duke, G., and Palmenberg, A., 1987, The atomic structure of Mengo virus at the 3.0 Angstrom Resolution. *Science* **235:**182–191.
42. Toyoda, H., Nicklin, M., Murray, M., Anderson, C., Dunn, J., Studier, F., and Wimmer, E., 1986, A second virus encoded proteinase involved in proteolytic processing of poliovirus polyprotein, *Cell* **45:**761–770.
43. Hanecak, R., Semler, B., Anderson, C., and Wimmer, E., 1982, Proteolytic processing of poliovirus polypeptides: Antibodies to polypeptide P3-7c inhibit cleavage at glutamine–glycine pairs, *Proc. Natl. Acad. Sci. USA* **79:**3973–3977.
44. Hanecak, R., Semler, B., Ariga, H., Anderson, C., and Wimmer, E., 1984, Expression of a cloned gene segment of poliovirus in *E. Coli:* Evidence of a cloned gene segment of poliovirus in *E. coli:* Evidence for autocatalytic production of the viral proteinase, *Cell* **37:**1063–1073.
45. Dewalt, P., and Semler, B., 1987, Site-directed mutagenesis of proteinase 3C results in a poliovirus deficient in synthesis of viral RNA polymerase, *J. Virol.* **61:**2162–2170.
46. Young, D., Dunn, B., Tobin, G., and Flanegan, J., 1986, Anti-VPg antibody precipitation of product RNA synthesized in vitro by the poliovirus polymerase and host factor is mediated by VPg on the poliovirion RNA template, *J. Virol.* **58:**715–723.
47. Richards, O., Hey, T., and Ehrenfeld, E., 1987, Poliovirus snapback double-stranded RNA isolated from infected HeLa cells is deficient in poly(A), *J. Virol.* **61:**2307–2310.
48. Porter, G., and Fellner, P., 1978, 3′-Terminal nucleotide sequences in the genome RNA of picornaviruses, *Nature (Lond.)* **276:**298–301.
49. Gutell, R., Noller, H., and Woese, G., 1986, Higher order structure in robosomal RNA, *EMBO J.* **5:**1111–1113.

50. Evans, D., Dunn, G., Minor, P., Schild, G., Cann, A., Stanway, G., Almond, J., Currey, K., and Maizel, J., Jr., 1985, Increased neurovirulence associated with a single nucleotide change in a non-coding region of the Sabin type 3 poliovaccine genome, *Nature (Lond.)* **314:**548–550.
51. Kohara, M., Omata, T., Kameda, A., Semler, B., Itoh, H., Wimmer, E., and Nomoto, A., 1985, In vitro phenotypic markers of a poliovirus recombinant constructed from infectious cDNA clones of the neurovirulent Mahony strain and the attenuated Sabin 1 strain, *J. Virol.* **53:**782–786.
52. Omata, T., Kohara, M., Kuge, S., Komatsu, T., Abe, S., Semler, B., Kameda, A., Itoh, H., Arita, M., Wimmer, E., and Nomoto, A., 1986, Genetic analysis of the attenuation phenotype of poliovirus type 1, *J. Virol.* **58:**348–358.
53. Chapman, N., Gauntt, C., and Tracy, S., 1987, Identification of coxsackie B viruses and enteroviruses in general using nucleic acid hybridization, *Eur. Heart J.* (Suppl. Inflammatory Heart Disease) (in press).

3

Replication

NANDO K. CHATTERJEE

1. INTRODUCTION

The purpose of this chapter is to summarize the development of our current understanding of the replication and molecular biology of coxsackieviruses (CV). To balance the discussion, the properties of the virion, its structural proteins, its genome, and some of its translational products are presented. As will become evident, most of the studies reviewed were conducted in group B coxsackieviruses (CVB) probably because CVB are easier to grow than group A coxsackieviruses (CVA) in tissue-culture cells. I hope that this contribution, along with the others presented in this volume, will help explain the diverse pathogenicities of these viruses.

2. VIRUS GROWTH, PURIFICATION, AND ASSAY

Coxsackieviruses have been grown in egg yolk sacs; in monkey, hamster, or pig kidney cell cultures; in human amnion or diploid cells; and in HeLa cells, Vero cells, and Buffalo green monkey kidney cells.

Coxsackievirus A is difficult to grow. Although several serotypes grow rapidly in primate, mouse, and other cell cultures, most do not. A guinea pig embryo cell-culture system has been reported suitable for isolation and propagation of several CVA serotypes from clinical specimens.[1] The RD cell line, derived from human rhabdomyosarcoma, sup-

NANDO K. CHATTERJEE • Wadsworth Center for Laboratories and Research, New York State Department of Health, Albany, New York 12201.

ports replication of several CVA, while CVB replicate poorly there or not at all.[2]

In monkey kidney cell cultures CBV grow rapidly. Temperature-sensitive mutants (e.g., deficient in genomic RNA synthesis or virion assembly) have been isolated by growing CVB3 in HeLa cells at various temperatures.[3] CVB also readily establish persistent infections in human lymphoid cell lines.[4] Persistence is maintained by a carrier culture mechanism, which involves spread of the virus through the medium and replication among a minority of cells at any given time. Replication of CVB3 occurs for days in cultures of murine neonatal skin fibroblasts in the absence of cytopathology and results in alteration of the plasma membrane.[5] Similar plasma membrane changes are seen also in virus-inoculated cultures of adolescent mouse heart fibroblasts.

Several excellent procedures have been described to purify CV from infected cells and to assay them.[6–9] Our procedure was devised to isolate virus from the early and late stages of infection of HeLa cells with different CVB. This procedure uses Nonidet P40 (NP40) to extract the infected cells, which increases the yield of purified virus. More importantly, NP40 extraction of freshly harvested cells releases two distinct populations of particles, virions, and membrane-bound virions (MBV).[10,11]

For plaque assay of CVB1–6,[9] we have used Vero cells with relative ease and great success. The addition of 2–5 mM $MgCl_2$ in the overlay and staining media increases plaque size considerably (from pinhead to 2–4 mm) for CVB1 and CVB2 (N. Chatterjee and C. Tuchowski, unpublished observations). Fungizone (1%) (Grand Island Biological) was occasionally added to these media for plaque assay of nonsterile gradient fractions with no noticeable detrimental effect on the assay.

3. VIRAL REPLICATION

For replication in infected cells, the virions first bind to specific receptors on the cell membrane and enter the cytoplasm. Uncoating then releases the virion RNA from its protein coat. Subsequent processes lead to synthesis of viral proteins, replication of viral RNA, and finally production of progeny virus, identical to that of the parent virions.

3.1. Virus Attachment and Uncoating

The initial event during replication is attachment of the virus to receptors on the host cell surface, a process mediated by a protein called virion attachment protein (VAP). CVA and CVB use different receptors for binding.[12] Each receptor has the capacity to bind one virion.[13] Binding occurs at 4°C and more rapidly at 37°C. Using electron microscopy, Roesing *et al.*[14] located receptors for CVB3 on the microvilli of HeLa cells and on the body of the cell. There are approximately 10^5 receptors per cell for CVB3.[15] CVB receptor may be a glycoprotein.[16]

Virion attachment protein occurs as multiple copies on the viral capsid surface. In CVB3, capsid protein VP2, which induces neutralizing antibodies, has been suggested to be the VAP.[17]

The attached virion next penetrates the cell surface and migrates through the cytoplasm. When it reaches a vesicle, processing or disassembly of the virion results in eclipse and uncoating of the viral genome for replication. No detailed studies of CV penetration and uncoating have as yet been performed. However, protein analysis of radiolabeled CVB3 before and after binding to HeLa cells showed that release of the smallest capsid protein—VP4, located on the surface of the native virion—signals the beginning of processing or disassembly.[6] Penetration, eclipse, and uncoating occur at 37°C. Eclipse is irreversible when no infectious virus can be recovered from the infected cells.

3.2. Protein and RNA Synthesis

The effect of CVB infection on host cell protein and RNA synthesis has been examined by radioactive labeling of mock-infected and CVB4- and CVB5-infected HeLa cells.[9] Host protein synthesis began to decline at 2 hr postinfection and was less than 20% of the control by 6 hr. Actinomycin D-resistant viral RNA synthesis started at about 2 hr postinfection, peaked by 5 hr, and then declined rapidly. Virus-specific protein synthesis began when host protein synthesis was declining, increased during the ensuing period, and declined in late infection. Measurement of acid-precipitable radioactivity from Triton N101-washed nuclei (the detergent was used to prepare intact nuclei free of cytoplasmic tags) indicated a progressive decline in the migration of cytoplasmic proteins into the nucleus in the infected cells only. CVB-induced inhibition of host protein synthesis is therefore similar to that observed in poliovirus-infected cells.[18]

In both CVB4- and CVB5-infected cells, the synthesized proteins at 4 hr postinfection were predominantly viral.[9] The molecular weight of these proteins in the cytoplasm varied from 23,500 to >92,500; some comigrated with virus capsid proteins. Several cytoplasmic proteins of 26,000–69,000 M_r were detected in the nucleus of the infected cells. Synthesis of numerous viral proteins of ~12,000–90,000 M_r has also been reported in CVB3-infected cells.[3]

Holland and Kiehn[19] and Kiehn and Holland[20] showed that all or nearly all viral proteins of several picornaviruses, including CVB1 and CVB5, result from cleavage of protein precursors with large molecular weight, ranging from 100,000 to 190,000. By pulse-chase analysis of [^{35}S]methionine-labeled proteins of CVB5-infected cells, we detected a polypeptide of ~100,000 M_r after 7 min of pulse.[9] This initial polypeptide later cleaved during the chase into intermediate polypeptides of 60,000 to >69,000 M_r. These in turn were cleaved into smaller, prominent, stable proteins of 23,500–38,000 M_r. Thus, virus-specific proteins in coxsackieviruses are generated by post-translational cleavages of precursor polypeptides. This processing of precursor polypeptides into virus-specific proteins is also found in other picornaviruses.[21,22]

3.3. Virion and Membrane-Bound Virions

During our investigations of CVB, we isolated from infected cells, in addition to mature virions, a previously unrecognized virus-specific ribonucleoprotein particle from infected cells.[10] The particle—which we designated MBV, since it contained host-cell cytoplasmic proteins—is present in CVB2-, CVB4-, CVB5-, and CVB6-infected HeLa cells. To liberate MBV, it was necessary to extract the infected cells several times with phosphate-buffered saline containing 0.5–0.8% NP40 in a Dounce homogenizer. Simple freezing and thawing of the infected cell several times, the usual procedure for virus extraction, liberated essentially no MBV. The extraction of freshly harvested cells liberated significantly more MBV than did extraction of frozen cells.

Typical profiles of virions and MBV in CsCl gradients, after radioactive labeling of viral protein and RNA in CVB5- or CVB6-infected HeLa cell cultures, are shown in Fig. 1. Double-labeled CVB5 separated as two peaks in the first gradient (Fig. 1A). After centrifugation in the second gradient, peak I banded at about $\rho = 1.34$ (Fig. 1B) and peak II at about $\rho = 1.30$ (Fig. 1C). Peak I represents virions; peak II represents MBV. Virions and MBV of CVB6 (Fig. 1E,F) and CVB4[11] separated in the same way. For mock-infected cells very little of the radioactivity of methionine-labeled proteins was detected in these regions of the gradients (Fig. 1G–I). After sedimentation in sucrose graidents, CsCl-gradient-purified CVB5 virions sedimented at ~150 S and MBV at ~107 S.

Membrane-bound virions are morphologically distinguishable from virions by electron microscopy after negative staining of the purified particles with uranyl acetate.[10] Nearly all (90%) of the virion particles excluded the stain and appeared spherical to polygonal (Fig. 2A). By contrast, all MBV particles were penetrated by the negative stain and had a pronounced polygonal profile (Fig. 2B). The background of MBV preparations, even after three bandings in CsCl gradients, contained

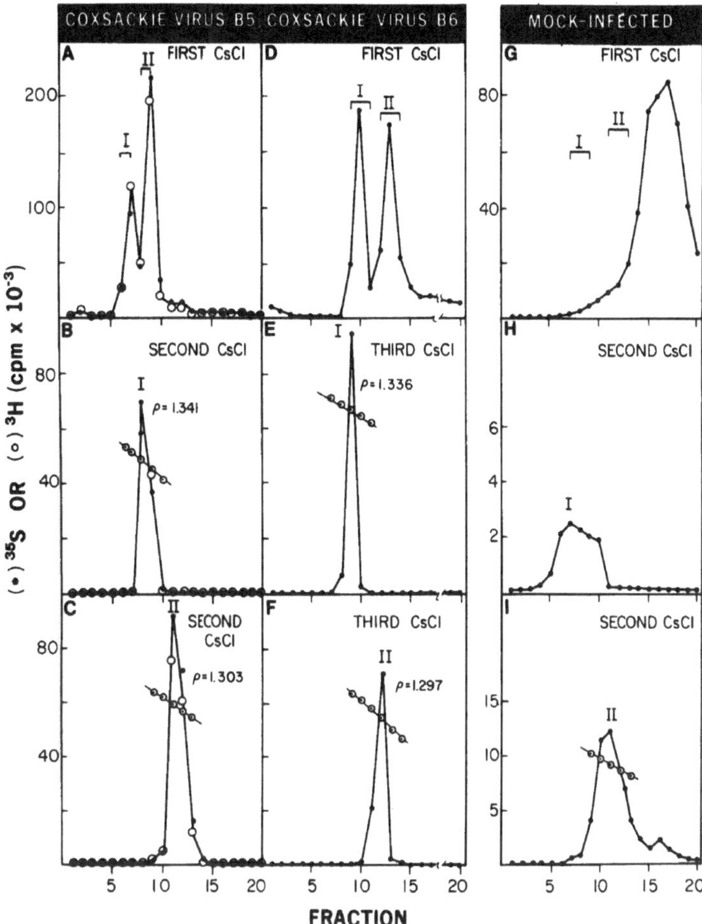

FIGURE 1. CsCl density-gradient profiles of [^{35}S]methionine- and [^{3}H]uridine-labeled CV. Double-labeled CVB5 formed two peaks on centrifugation in the first gradient (A). Each peak, I (virions) and II (MBV), was pooled, dialyzed, and recentrifuged in the second gradient (B,C). Methionine-labeled CVB6 from the two peaks (D) was centrifuged likewise in the second gradient (not shown) and then in the third (E,F). Methionine-labeled material in the two peaks from mock-infected cells (G) was also centrifuged in the second gradient (H,I). (From Chatterjee *et al.*[10])

negatively stained extraneous material that was not detected in virion preparations. This material has not been identified, but it may be fragments of MBV that were broken when the sample was prepared for electron microscopy. Most of the virion particles measured 32–33 nm; MBV had a broader distribution with an average diameter of ~31 nm (Fig. 3).

Membrane-bound virions were considerably less infective than virions in the CVB studied. At about 24 hr postinfection in CVB5-infected cultures, the infectivity of the MBV ranged from 7.6×10^7 to 2.2×10^9 total plaque-forming units (PFU), compared with 13.5×10^8 to $2.3 \times$

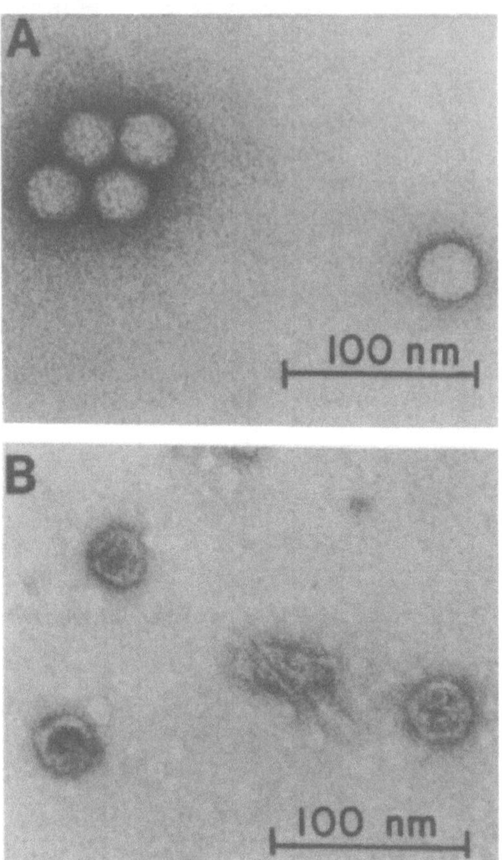

FIGURE 2. Electronmicrographs of CsCl-gradient-purified CVB5 virions (**A**) and MBV (**B**) after staining with 0.5% uranyl acetate. (From Chatterjee *et al.*[10])

10^{11} total PFU for the virions. Thus, virions were up to 100 times more infective than MBV. The $A_{260:280}$ was ~1.22 for MBV, and 1.64 for virions, suggesting that MBV have relatively more protein than RNA. MBV were quite stable. Their infectivity did not decrease significantly when they were stored for 1–2 months in liquid nitrogen.

Membrane-bound virions and virions appeared simultaneously in CVB5-infected cells as early as 3.5 hr postinfection and thereafter up to 24 hr postinfection. The yield of each population increased with the time of infection. However, the yield of MBV at any given time was significantly lower than the yield of virions. Cultures infected with highly purified preparations of virions, MBV, or a combination of the two produced both virions and MBV. Their yield was maximum from cells infected with virions alone, minimum from cells infected only with MBV, and intermediate from cells infected with both. MBV thus appear to inhibit infection caused by the virion, yet MBV can produce infection by themselves. Morphogenesis of virions and MBV probably occurs in rough membranes of the infected cells.[11]

Alkali-dissociated prototype CVB4 releases a protein kinase that actively phosphorylates several proteins of virions and MBV, as well as exogenous phosphate-acceptor proteins such as protamine sulfate.[23] Nearly 20-fold more enzyme activity was detected in MBV than in virions. The activity is cyclic, nucleotide independent, and divalent cation dependent and has a pH optimum of 8.0. The enzyme appears to be located internally in the virus and may be host-cell coded. No such activity was detected in a strain of the virus that produces diabetes in mice.

These observations on the morphology and biology of virions and MBV, and on the genomic RNA and capsid proteins discussed in section 3.6, indicate that MBV are a separate and distinct population. MBV resemble the defective-interfering (DI) particles of other picornaviruses,[24,25] in certain features, including a lower buoyant density than that of virions and an ability to decrease the yield of virions by inter-

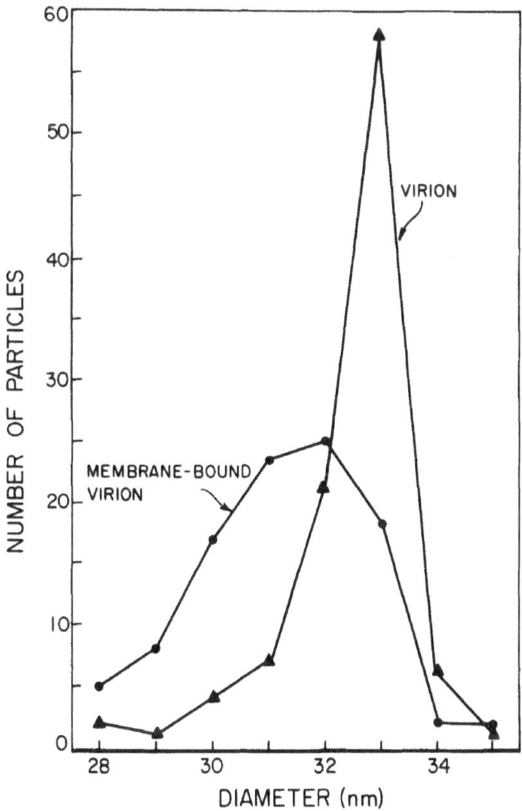

FIGURE 3. Size distribution of CsCl gradient-purified CVB5 virions and MBVs. (From Chatterjee *et al.*[10])

ference. However, MBV differ from DI particles in other respects; e.g., MBV can propagate without helper virus and are associated with several host proteins. Thus, it is unclear whether MBV are true DI particles.

3.4. Other Particles of Dissimilar Densities

In addition to the major infective component of $\rho = 1.34$ in CsCl, a component with $\rho = 1.44$ was found in harvests of several picornaviruses, including CVB5 and swine vesicular disease virus (SVDV).[26] The former was designated the light component and the latter the heavy component. With SVDV, which is related to CVB5 antigenically, structurally and genetically,[8,27] about 98% of the infectivity was in the light component and the rest in the heavy component. The components were morphologically similar, but the heavy component had a smaller diameter (28 versus 30 nm). No interconversion of the two forms was observed, and the components had the same proportions of RNA and protein, with the same polypeptide composition. Each generated a similar proportion of the light and heavy forms on replication, but the light component was fourfold higher in specific infectivity and much more efficient in eliciting neutralizing antibodies in guinea pigs. The two forms may be alternative stable configurations of the virus.

3.5. Virion Structure and Stability

On the basis of analysis of the substructures of urea-degraded CVB3, Philipson *et al.*[28] and Beatrice *et al.*[17] proposed that the structure of CVB3 is similar to that of other picornaviruses: a triangulation number of $T = 3$ with 12 pentamers and 20 hexamers. The pentamer is composed of five units of VPO cleaved to capsid polypeptides VP2 plus VP4, while in the hexamer three units of VP1 combine with three units of VP3. This structure allows for release of a 14 S structure, $(VP1 + VP2 + VP3)_5$, and a 12 S structure $(VP1 + VP2 + VP3)_3$. This structure also permits formation of a pentamer of VP2 (5 S) and a pentadecamer of VP1 plus VP3 (20 S).

CVA10 was crystallized into three distinct crystal habits: rhombic dodecahedral, orthorhombic, and octahedral.[29] The crystal lattice structure and nucleic acid–base ratio of CVA differ significantly from those of polioviruses.

In general, CV are stable at pH 4.0–8.0 for several days, resistant to 5% lysol and 70% ethanol, but inactivated by 0.1 NHCl or 0.3% formaldehyde.[30] CVB, but not CVA, are generally inhibited by 2(α-hydroxylbenzyl)benzimidazole.

Several CVA, but not CVB, were rapidly inactivated in low-ionic-strength solutions at neutral pH.[7] The extent of inactivation was dependent on temperature and molarity. The loss of infectivity was due, not to loss or inactivation of viral RNA, but to particle failure to attach to susceptible cells. Electrophoretic analysis indicated that inactivation of CVA13 in low-ionic-strength solutions reflected specifically the loss of VP4 from the virus particle. A similar inactivation phenomenon, reflecting the loss of VP4 as a result of spontaneous elution of virus particles from susceptible cells, was reported in CVB3.[6]

3.6. The Genome

The genome of CVB is a single-stranded RNA molecule of 7440 nucleotides, which is polyadenylated at the 3′ end.[31] The estimated molecular weight of CVB5 RNA is 2.53×10^6. The RNA contains information for 2480 amino acids and can code for proteins with a total molecular weight of ~273,000.

The genomic RNA directed protein synthesis very efficiently in a cell-free system from rabbit reticulocytes.[31] At a saturating RNA concentration, a five- to sixfold stimulation of protein synthesis over background was observed. Inhibitors of initiation and peptide-chain elongation inhibited protein synthesis almost completely. By contrast, 7-methylguanosine-5′-monophosphate, a cap analogue capable of inhibiting capped messenger RNA (mRNA)-directed protein synthesis, did not inhibit CVB RNA-directed protein synthesis. This observation suggests that the viral RNA is uncapped like other picornavirus RNA. At least 10 polypeptides with an apparent molecular weight of 31,000–98,000 were made after cell-free protein synthesis. Several of these polypeptides resembled virus-specific proteins synthesized in the infected host cells and capsid proteins of the virus; the similarities were established by gel electrophoresis, immunoprecipitation, and peptide mapping.[31]

The genomic RNA of virions and MBV of CVB4[11] (N. Chatterjee and C. Nejman, unpublished observations) and CVB5[10] were indistinguishable by velocity sedimentation in sucrose gradients and by migration in denaturing gels. Each RNA sedimented and migrated as a 35 S molecule. Hybridization experiments showed MBV RNA to be almost completely homologous to virion RNA. However, T_1 oligonucleotide fingerprinting demonstrated minor differences in certain oligonucleotides. For example, one MBV oligonucleotide of CVB4 (Fig. 4C,D, open arrow) could not be detected in the virion RNA, and several oligonucleotide spots of virion RNA (Fig. 4A, arrows) were weak and appeared to occur in submolar quantities in MBV RNA, suggesting some heterogeneity among the MBV. These minor differences between virion and MBV RNA could have originated as small deletions.

The genomic RNA of both virions and MBV contained a small

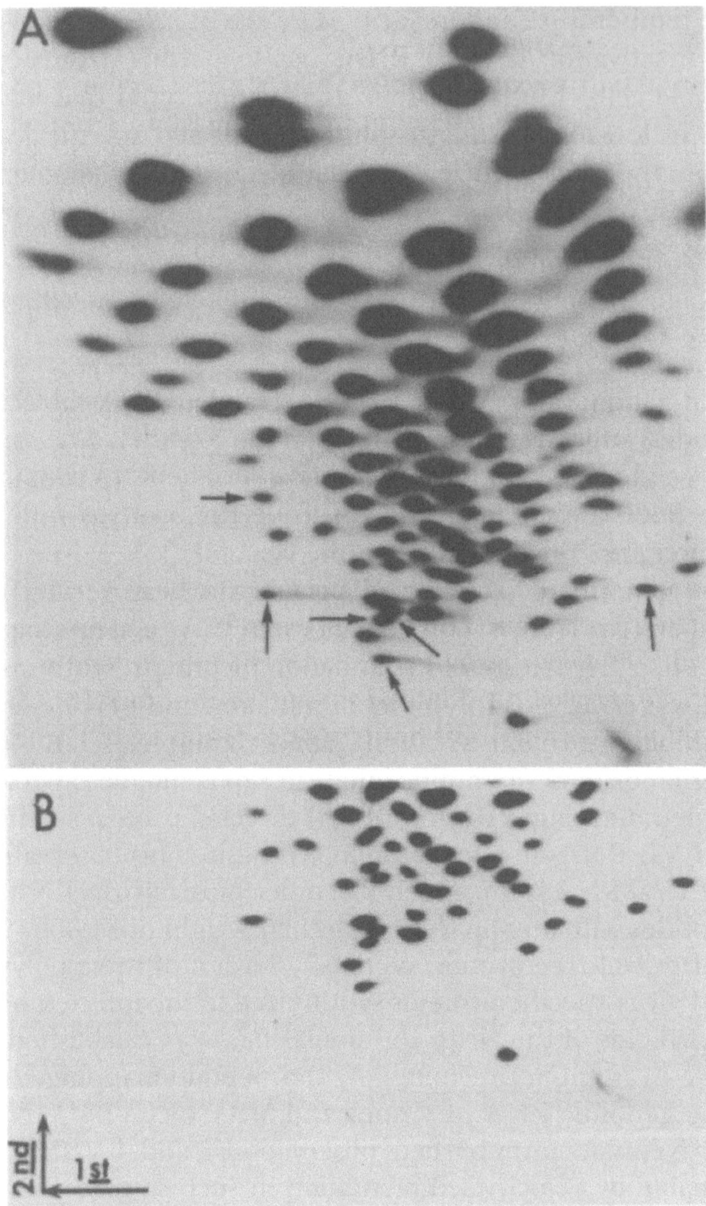

FIGURE 4. Comparison of virion RNA (**A**) and MBV RNA (**C**) of CVB4 by T_1 oligonucleotide fingerprinting. Samples of ^{32}P-labeled virion and MBV RNA were digested exhaustively with RNase T_1, and the products were separated by two-dimensional polyacrylamide gel electrophoresis. Lower portions of the fingerprints are from another preparation of virion (**B**) and MBV RNA (**D**). (From Chatterjee *et al.*[10])

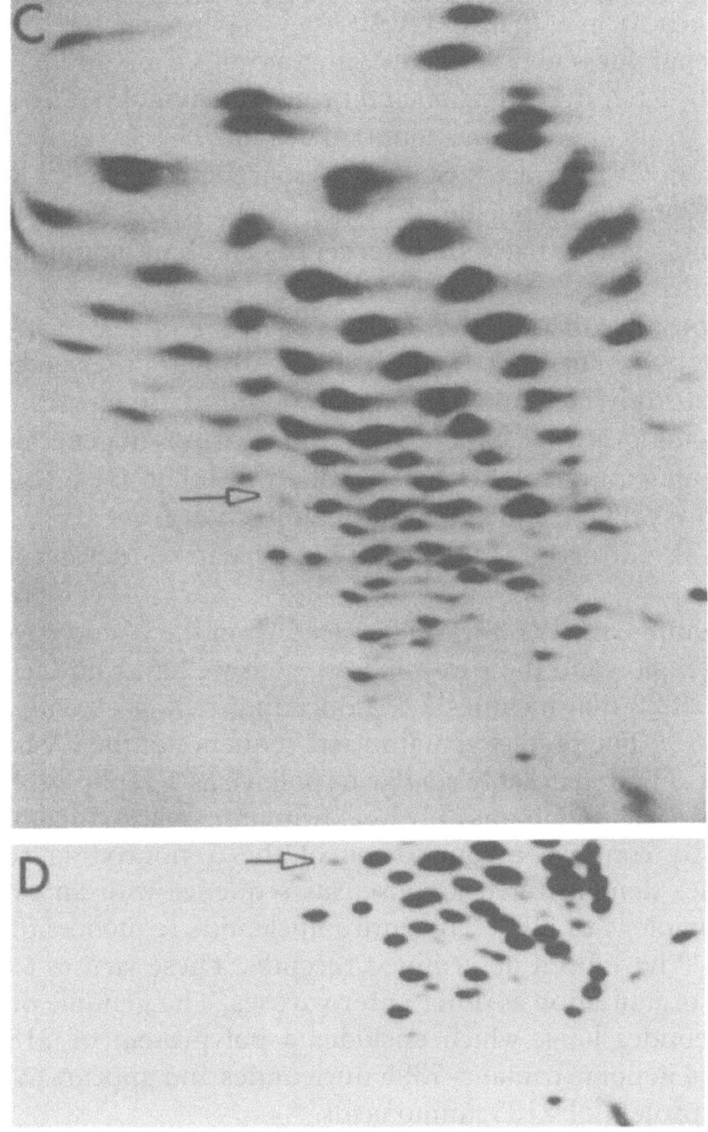

FIGURE 4. *(Cont.)*

protein, designated VPg, that was covalently linked presumably to the 5′ end of the RNA.[11] VPg contained two proteins of different charge. MBV VPg appeared to be considerably smaller than the 5300-M_r virion VPg.

Hybridization of RNA from polioviruses, CVA, CVB, and echoviruses showed that ~30–50% of the nucleotide sequences are shared by various serotypes within each major group, whereas less than 20% homology exists between groups.[32] CVB4 appeared to be more closely related to echoviruses than to CVA.

The relatedness and antigenic variation of CVB5 and SVDV were examined by RNA hybridization and protein analysis.[8,33] The prototype (Faulkner) strain of CVB5 was found to be more closely related to SVDV than to recent CVB5 isolates. Sequences shared by the CBV5 RNA were largely the same as those shared by the SVDV RNA. These and other results demonstrated that wide divergence can occur among CVB5 and SVDV strains.

Hybridization of cloned CVB3 genome[34,35] or cDNA to six CVB serotype RNA[36] with RNA from various CVB and other enteroviruses was used to study genomic homologies among these viruses. Some sequence homology is conserved among enteroviruses in general.[34] CVB5 shared about equivalent homology with the other CVB, but CVB1, CVB2, and CVB3 have diverged in one direction from CVB5.[36] CVB4 and CVB6 also diverged from CVB5, but perhaps earlier in time or at a greater rate.

The genome of CVB3 was sequenced from the 3′ end covering the noncoding region and the viral replicase,[37] as well as from the 5′ end to nucleotide 3822 that includes a 5′-noncoding region extending to nucleotide 738.[38] The predicted amino acid sequence of the CVB3 enzyme was shown to be remarkably similar to poliovirus 1 replicase. However, the 3′-noncoding region was only weakly homologous to the poliovirus 1 sequence. By contrast, a comparison of the 5′-noncoding regions of these viruses demonstrated a consensus sequence with an overall nucleotide homology of 69%. The entire nucleotide sequences of CVB3[39] and CVB4[40] have been determined recently. These viruses exhibit the same gene organization as other enteroviruses. The genome of CVB3 is 7396 nucleotides long, which enclodes a polyprotein of 2185 amino acids. CVB4 genome contains 7395 nucleotides and appears to encode a single polyprotein of 2183 amino acids.

3.7. Capsid Proteins

Like other picornaviruses, CVA and CVB contain four capsid polypeptides, VP1-4. However, estimates of their molecular weight vary. The values for CVB3 capsid polypeptides reported by Crowell and Philipson[6] are lower than those reported by us[9] for all six serotypes of CVB and by Gauntt *et al.*[3] for CVB3.

Our estimates of the molecular weight of the four polypeptides, after comparative analysis of the migration of these proteins from all six serotypes in polyacrylamide gels, are 34,500–37,000 (VP1), 31,000–36,000 (VP2), 26,500–32,500 (VP3), and 5000–5500 (VP4). A trace amount of another polypeptide, VP0, which migrated slightly slower than VP1, could be seen sometimes in CVB1, CVB3, and CVB6; and VP2 appeared as doublet in some gels in CVB4 (N. K. Chatterjee and C. Tuchowski, unpublished observations).

The estimated values proposed by Gauntt *et al.* for CVB3 capsid polypeptides are slightly lower (except for VP1) than ours, while those presented by Crowell and Philipson are considerably lower (except for VP4). Differences in electrophoretic techniques and in methods for virus replication and purification could account for these differences.

MBV of CVB5, CVB6,[10] and CVB4 (N. K. Chatterjee and C. Nejman, unpublished observations) contained more proteins than virions. MBV appeared to contain only three, rather than four, capsid polypeptides plus at least seven additional proteins with an apparent molecular weight of 21,000–92,000. Some of these additional proteins may be of host origin; the rest may be precursors of capsid polypeptides. The missing capsid polypeptide of MBV has not been clearly explained.

Cords *et al.*[7] reported that CVA13 contain five capsid polypeptides: VP1 (36,000), VP2a (27,000), VP2b (25,500), VP3 (21,500), and VP4 (7000). VP2a and VP2b migrated slightly faster than poliovirus VP2. Several possibilities, such as mutant stock virus or cleavage of VP2 precursor at two sites, were adduced to explain the presence of both VP2a and VP2b.

4. CONCLUDING REMARKS

Coxsackieviruses have emerged as an important group of pathogens with the potential to induce a variety of human and animal diseases. One of the distinctive features of these viruses, tissue tropism, may reflect viral variants that occur during replication. Infection with one variant may result in a disease, whereas infection with another may not. Thus, diabetogenic and nondiabetogenic variants of CVB4 have been described.[41–43] The diabetogenic variants are human isolates indistinguishable from the nondiabetogenic prototype virus by neutralization with reference hyperimmune serum. Myocarditic and nonmyocarditic variants of CVB3 have also been reported.[44,45] The role of CV variants in pathogenesis is discussed in detail in Chapter 10, this volume.

Viral variants could originate by mutations. Notkins and co-workers[46] identified a number of naturally occurring antigenic variants of CVB4 using a panel of monoclonal antibodies to the virus. Every variant examined differed from the prototype virus by 2–12 epitopes. The calculated frequency of mutation per antigenic determinant, as high as 10^{-4}, suggests that these variants could have originated by small mutations. It will be interesting to analyze the genomic makeup of some of these variants and to test whether they can actually produce different diseases.

Another possible source of tissue tropism may be differences in the binding affinities of CVB variants to virus-specific receptors in the various tissues. Differences in the capsid proteins—especially VP2, which has been suggested to function as VAP[17]—could lead to conformational changes in the particle. This in turn could alter the binding affinity of the virus to host cell receptors and uncoating, thus influencing pathogenesis.

A promising area of investigation is the function(s) of MBV and the heavy particles of CVB in host–virus interaction and tissue tropism. Both particles resemble typical CV virions in some features, yet they differ in others. One important property of the MBV is their apparent ability to interfere with the growth of virions in co-infected cultures. By this and the other properties (e.g., genetic makeup, within-population heterogeneity, and capsid protein composition), it may be possible for MBV to influence host–virus interaction and pathogenesis.

ACKNOWLEDGMENTS. I thank Nancy Miller for her expert assistance during the preparation of this manuscript. This work was supported by U.S. Public Health Service grant AM 33054 from the National Institutes of Health and by grant D1-002 from the New York State Health Research Council.

REFERENCES

1. Landry, M. L., Madore, H. P., Fong, C. K., and Hsiung, G. D., 1981, Use of guinea pig embryo cell cultures for isolation and propagation of group A coxsackieviruses, *J. Clin. Microbiol.* **13:**588–593.
2. Schmidt, N. J., Ho, H. H., and Lennette, E. H., 1975, Propagation and isolation of group A coxsackieviruses in RD cells, *J. Clin. Microbiol.* **2:**183–185.
3. Gauntt, C. J., Trousdale, M. D., Lee, J. C., and Paque, R. E., 1983, Preliminary characterization of coxsackievirus B3 temperature-sensitive mutants, *J. Virol.* **45:**1037–1047.
4. Matteucci, D., Paglianti, M., Giangregorio, A. M., Capobianchi, M. R., Dianzani, F., and Bendinelli, M., 1985, Group B coxsackieviruses readily establish persistent infection in human lymphoid cell lines, *J. Virol.* **56:**651–656.
5. Lutton, C. W., and Gauntt, C. J., 1986, Coxsackievirus B3 infection alters plasma membrane of neonatal skin fibroblasts, *J. Virol.* **60:**294–296.
6. Crowell, R. L., and Philipson, L., 1971, Specific alterations of coxsackievirus B3 eluted from HeLa cells, *J. Virol.* **8:**509–515.
7. Cords, C. E., James, C. G., and McLaren, L. C., 1975, Alteration of capsid proteins of coxsackievirus A13 by low ionic concentrations, *J. Virol.* **15:**244–252.
8. Harris, T. J. R., Doel, T. R., and Brown, F., 1977, Molecular aspects of the antigenic variation of swine vesicular disease and coxsackie B5 viruses, *J. Gen. Virol.* **35:**299–315.

9. Chatterjee, N. K., and Tuchowski, C., 1981, Comparison of capsid polypeptides of group B coxsackieviruses and polypeptide synthesis in infected cells, *Arch. Virol.* **70**:255–269.
10. Chatterjee, N. K., Samsonoff, W. A., and Tuchowski, C., 1983, Isolation and characterization of a membrane-bound population of group B coxsackieviruses, *J. Virol.* **45**:832–841.
11. Chatterjee, N. K., and Nejman, C., 1985, Membrane-bound virions of coxsackievirus B4: Cellular localization, analysis of the genomic RNA, genome-linked protein, and effect on host macromolecular synthesis, *Arch. Virol.* **84**:105–118.
12. Crowell, R. L., and Landau, B. J., 1983, Receptors in the initiation of picornavirus infections, in: *Comprehensive Virology,* Vol. 18 (H. Fraenkel-Conrat and R. R. Wagner, eds.), pp. 1–42, Plenum Press, New York.
13. Longberg-Holm, K., 1980, Attachment of animal virus to cells, an introduction, in: *Receptors and Recognition,* Series B, Vol. 8: *Virus Receptors,* Part 2 (K. Lonberg-Holm and L. Philipson, eds.), pp. 1–20, Chapman and Hall, London.
14. Roesing, T. G., Toselli, P. A., and Crowell, R. L., 1975, Elution and uncoating of coxsackievirus B3 by isolated HeLa cell plasma membranes, *J. Virol.* **15**:654–667.
15. Crowell, R. L., 1966, Specific cell-surface alteration by enteroviruses as reflected by viral-attachment interference, *J. Bacteriol.* **91**:198–204.
16. Zajac, I., and Crowell, R. L., 1969, Differential inhibition of attachment and eclipse activities of HeLa cells for enteroviruses, *J. Virol.* **3**:422–428.
17. Beatrice, S. T., Katze, M. G., Zajac, B. A., and Crowell, R. L., 1980, Induction of neutralizing antibodies by the coxsackievirus B3 virion polypeptide, VP2, *Virology* **104**:426–438.
18. Ehrenfeld, E., 1984, Picornavirus inhibition of host cell protein synthesis, in: *Comprehensive Virology,* Vol. 19 (H. Fraenkel-Conrat and R. R. Wagner, eds.), pp. 177–221, Plenum Press, New York.
19. Holland, J. J., and Kiehn, E. D., 1968, Specific cleavage of viral proteins as steps in the synthesis and maturation of enteroviruses, *Proc. Natl. Acad. Sci. USA* **60**:1015–1022.
20. Kiehn, E. D., and Holland, J. J., 1970, Synthesis and cleavage of enterovirus polypeptides in mammalian cells, *J. Virol.* **5**:358–367.
21. Rueckert, R. R., 1976, On the structure and morphogenesis of picornaviruses, in: *Comprehensive Virology,* Vol. 6 (H. Fraenkel-Conrat and R. R. Wagner, eds.), pp. 131–213, Plenum Press, New York.
22. Putnak, J. R., and Phillips, B. A., 1981, Picornaviral structure and assembly, *Microbiol. Rev.* **45**:287–315.
23. Chatterjee, N. K., and Nejman, C., 1986, Protein kinase in nondiabetogenic coxsackievirus B4, *J. Med. Virol.* **19**:353–365.
24. Cole, C. N., Smoler, E., Wimmer, E., and Baltimore, D., 1971, Defective interfering particles of poliovirus, I. Isolation and physical properties, *J. Virol.* **7**:478–485.
25. McClure, M. A., Holland, J. J., and Perrault, J., 1980, Generation of defective interfering particles in picornaviruses, *Virology* **100**:408–418.
26. Rowlands, D. J., Shirley, M. W., Sangar, D. V., and Brown, F., 1975, A high density component in several vertebrate enteroviruses, *J. Gen. Virol.* **29**:223–234.
27. Brown, F., Talbot, P., and Burrows, R., 1973, Antigenic differences between isolates of swine vesicular disease and their relationship to coxsackie B5 virus, *Nature (Lond.)* **245**:315–316.
28. Philipson, L., Beatrice, S. T., and Crowell, R. L., 1973, A structural model of picornaviruses as suggested from an analysis of urea-degraded virions and procapsids of coxsackievirus B3, *Virology* **54**:69–79.
29. Mattern, C. F. T., 1962, Some physical and chemical properties of coxsackie virus A9 and A10, *Virology* **17**:520–532.

30. Andrews, C., Pereira, H. G., and Wildy, P., 1978, Picornaviridae, in: *Viruses of Vertebrates,* pp. 1–37, Baillière Tindall, London.
31. Chatterjee, N. K., and Tuchowski, C., 1981, Translation of coxsackievirus RNAs in a rabbit reticulocyte lysate: Characterization of the genome RNA, reaction conditions for translation, and analysis of the products, *Arch. Virol.* **70:**271–283.
32. Young, N. A., 1973, Polioviruses, coxsackieviruses, and echoviruses: Comparison of the genomes by RNA hybridization, *J. Virol.* **11:**832–839.
33. Brown, F., Wild, T. F., Rowe, L. W., Underwood, B. O., and Harris, T. J. R., 1976, Comparison of swine vesicular disease virus and coxsackie B5 virus by serological and RNA hybridization methods, *J. Gen. Virol.* **31:**231–237.
34. Hyypiä, T., Stälhandske, P., Vainionpää, R., and Pettersson, U., 1984, Detection of enteroviruses by spot hybridization, *J. Clin. Microbiol.* **19:**436–438.
35. Tracy, S., Chapman, N. M., and Liu, H. L., 1985, Molecular cloning and partial characterization of coxsackievirus B3 genome, *Arch. Virol.* **85:**157–163.
36. Tracy, S., 1985, Comparison of genomic homologies in the coxsackievirus B group by use of cDNA : RNA dot-blot hybridization, *J. Clin. Microbiol.* **21:**371–374.
37. Stålhandske, P. O. K., Lindberg, M., and Pettersson, U., 1984, Replicase gene of coxsackievirus B3, *J. Virol.* **51:**742–746.
38. Tracy, S., Liu, H. L., and Chapman, N. M., 1985, Coxsackievirus B3: Primary structure of the 5′ non-coding and capsid protein-coding regions of the genome, *Virus Res.* **3:**263–270.
39. Lindberg, A. M., Stålhandske, P. O. K., and Pettersson, U., 1987, Genome of coxsackievirus B3, *Virology* **156:**50–63.
40. Jenkins, O., Booth, J. D., Minor, P. D., and Almond, J. W., 1987, The complete nucleotide sequence of coxsackievirus B4 and its comparison to other members of the picornaviridae, *J. Gen. Virol.* **68:**1835–1848.
41. Yoon, J. W., Austin, M., Onodera, T., and Notkins, A. L., 1979, Virus-induced diabetes mellitus: Isolation of a virus from the pancreas of a child with diabetic ketoacidosis, *N. Engl. J. Med.* **300:**1173–1179.
42. Webb, S. R., Loria, R. M., Madge, G. E., and Kibrick, S., 1979, Coxsackievirus B infection in the mouse: Effects associated with the diabetic gene, db, *Curr. Microbiol.* **3:**15–19.
43. Chatterjee, N. K., Haley, T. M., and Nejman, C., 1985, Functional alterations in pancreatic B cells as a factor in virus-induced hyperglycemia in mice, *J. Biol. Chem.* **260:**12786–12791.
44. Gauntt, C. J., Trousdale, M. D., LaBadie, D. R. L., Paque, R. E., and Nealon, T., 1979, Properties of coxsackievirus B3 variants which are amyocarditic or myocarditic for mice, *J. Med. Virol.* **3:**207–220.
45. Huber, S. A., and Job, L. P., 1983, Differences in cytolytic T cell response of BALB/c mice infected with myocarditic and non-myocarditic strains of coxsackievirus group B, type 3, *Infect. Immun.* **39:**1419–1427.
46. Prabhakar, B. S., Haspel, M. V., McClintock, P. R., and Notkins, A. L., 1982, High frequency of antigenic variants among naturally occurring human coxsackie B4 virus isolates identified by monoclonal antibodies, *Nature (Lond.)* **330:**374–376.

4

Perspectives on Cellular Receptors as Determinants of Viral Tropism

RICHARD L. CROWELL, BURTON J. LANDAU, KUO-HOM LEE HSU, and MAGGIE SCHULTZ

1. INTRODUCTION

Coxsackievirus (CV) infections of humans were recently reviewed[1] and are also discussed elsewhere in this volume. These agents cause a wide variety of syndromes, including meningitis, pleurodynia, myositis, herpangina, myocarditis, pericarditis, pancreatitis, hepatitis, nephritis, orchitis, gastroenteritis, exanthems, acute respiratory disease, and congenitally acquired disease. In addition, inapparent or mild infections may be common and provide a source of virus to help explain the ubiquitous spread of infections. No conclusive evidence is available to explain why some individuals develop acute self-limiting infections, while others develop a severe or chronic disease. Nevertheless, it is predicted that cellular receptors play an important role in determining the tropism of CV in the pathogenesis of infection.

A number of reviews have been written over the past 10 years that attempt to relate the importance of cellular receptors as determinants of viral tropism.[2–8] Although a reasonably convincing case can be made

---

RICHARD L. CROWELL, BURTON J. LANDAU, KUO-HOM LEE HSU, and MAGGIE SCHULTZ • Department of Microbiology and Immunology, Hahnemann University School of Medicine, Philadelphia, Pennsylvania 19102.

for this hypothesis, many experiments remain to be done to confirm it and to place it in perspective with other determinants.

There has been renewed interest in determining the molecular events in the early interactions of viruses with the cell surface that lead to receptor-mediated endocytosis and uncoating of the viral genome. Many current thoughts, findings, and references on a variety of virus–receptor systems, presented in a recent symposium, have been recorded.[9]

The purpose of the present chapter is to provide current information on receptors as determinants of CV tropisms and to suggest a perspective for design of future studies in this area. For example, the application of molecular biologic techniques for the cloning of receptor genes in prokaryotic and eukaryotic cells will provide sufficient amounts of purified receptor proteins for characterization. Of equivalent value will be a genetic analysis of host-range viral variants with differing receptor specificities and disease potential. Ultimately, the interactions of diverse viruses with different inbred mouse strains will serve as models to permit a detailed understanding of the molecular events in the pathogenesis of CV-induced diseases of humans.

2. PATHOGENESIS OF COXSACKIEVIRUS INFECTIONS

An understanding of viral pathogenesis should include a complete description of the gene products of the virus that permit infection at the cellular level. In addition, one must identify the mode of spread of virus between cells, the cell types that become infected in different tissues and organs of the mammalian host, and the cellular responses that lead to disease. Host responses to viral infection, as shown in inbred mice, are varied and greatly influenced by the immune system. The immune response may not only limit viral spread in the host but may cause cellular injury secondary to viral replication as well. It is not inconsequential that viral variants emerge during the replication of millions of virions[10,11] (see Chapter 10, this volume). Those variants best adapted to the environment are selected for further amplification. Thus, the strain of virus used to initiate infection may undergo genotypic and phenotypic changes during viral replication. Monitoring of these changes is necessary to gain insight into the disease process.[12] When one considers the complexity of the animal host, it is no wonder that most studies of viral infections focus on *in vitro* models. Nevertheless, we cannot continue to avoid the issues, and further studies of viral pathogenesis in animal models must be pursued.

The classic studies conducted by Bodian[13] presented a useful model of pathogenesis of poliovirus in monkeys, which has provided the basis of our current understanding of enterovirus infections. In brief, ingestion or inhalation of virus is followed by virus replication in lymphoid tissue and mucosa of the oropharynx and the small intestine. Depending on the stage of infection, virus can be isolated from the nasopharyngeal secretions and from the feces. Prior to onset of illness a transient viremia occurs, which presumably serves as the mechanism by which virus invades the central nervous system (CNS) and other target organs. Polioviruses are neurotropic viruses, with a tropism or predilection for the infection of neurons. Although inflammatory responses may be observed, lytic poliovirus infection of neurons in the anterior horn of the spinal cord leads directly to paralysis. By contrast, the CV of group A (CVA) are selectively myotropic and cause direct injury to the skeletal muscle cells of newborn mice with resultant paralysis and death. Thus, paralysis is caused by the destruction of two different cell types by these different enteroviruses. Their differing cellular tropisms are explained by the recognition that polioviruses and CVA use two different receptors on cultured cells.[2] In fact, it has been known for many years that polioviruses are restricted to primates, as all other animal species are devoid of specific receptors.[14] In general, the presence or absence of these specific cellular receptors restricts the host range of picornaviruses. Exceptions to this generality may be found in relatively few instances[15,16] and in cases in which viral variants have emerged to take advantage of a second receptor.[17,18]

Some of the most exciting studies of picornavirus pathogenesis are being conducted by the use of recombinant DNA technology.[19–22] The strategy is to determine the differences in disease potential that exist between the neurovirulent wild-type polioviruses and the corresponding avirulent Sabin vaccine strains. These studies were made possible as a result of the elucidation of the entire nucleotide sequence of each of the virus strains[23–25] and the preparation of infectious complementary DNA (cDNA),[26] respectively. The preparation of infectious recombinant cDNA clones of the corresponding RNA genomes of poliovirus T1 has permitted testing for neurovirulence in monkeys.[22] The results of these studies suggested that one of the most important factors determining neurovirulence was the extent of viral multiplication in the CNS and that several different biologic characteristics determined avirulence. These findings would have been strengthened by the demonstration that progeny polioviruses in the recombinant studies could be reisolated following monkey passage and shown to be comparable to the inoculum. These results are reminiscent of the earlier findings of Eggers and Sabin,[27] who found that lethal disease in mice inoculated with echovirus

9 would occur only if viral replication exceeded a critical threshold level. Thus, neurovirulence in poliovirus is likely to be a consequence of several gene products, one of which might control virus–receptor interactions,[19] although no direct evidence for this latter possibility has been provided to date. Nevertheless, these are important approaches to the study of viral pathogenesis, and it is anticipated that solutions to this multifaceted problem will be forthcoming in the next several years through the detailed analysis of recombinant viruses.

The CV, among the human enteroviruses, are uniquely suited for animal studies. Whereas the polioviruses are restricted to primates (usually outbred), and echoviruses are mainly without animal hosts, CV can be studied in different strains of inbred mice. The capacity of different mouse strains to respond with different forms of disease following CV infections should permit the future identification of the essential host factors leading to specific diseases. Such studies have already begun in an attempt to sort out the many immunologic factors thought to influence the incidence and severity of myocarditis.[28–30] It is helpful that many parallels exist between the diverse human diseases caused by CV and those produced in mice.[31]

In summary, numerous qualitative and quantitative factors determine the pathogenesis of picornavirus infections (Table I). These factors illustrate the complexities involved in assessing the role of each variable in the development of disease. Thus, studies of polioviruses have provided models for defining cellular tropisms during CV infections. However, significant differences between these virus groups warrant detailed studies of CV as causative agents of severe human diseases.

TABLE I
Quantitative Aspects of Picornavirus Pathogenesis[a,b]

Location of primary sites of replication relative to portal of entry
Amount of inoculum relative to number of available host cells
Yields of virus per host cell per unit time (permissiveness)
Number of host cells per gram tissue
Attainment of excess virions to permit their translocation and to traverse natural barriers to reach secondary targets:
Hematogenous, lymphogenous, neurogenic
Sufficient numbers of virions to exceed natural and acquired inhibitors
Availability of secondary target cells and their concentration
Relative importance of destroyed host cells to normal tissue function
Immunopathology and protective immune response

[a]Modified from Crowell and Landau.[3]
[b]The severity of the disease is influenced by the virulence of the virus strain and the genetic predisposition of the host.

3. STRUCTURAL STUDIES IDENTIFYING THE VIRION ATTACHMENT SITE

The elegant X-ray crystallographic studies of human rhinovirus 14 (HRV-14)[32] and of poliovirus type 1 (PV-1)[33] have exhibited a canyon or cleft surrounding each of the pentameric vertices of the icosahedron. Indirect evidence suggests that these canyons are the virion attachment sites (VAS) that bind to specific cellular receptors on the plasma membrane.[32] The narrow dimensions of the canyon preclude the recognition of epitopes by antibodies. Thus, the use of anti-idiotypic antibody strategies[34–36] are not available to help elucidate the antigenic nature of either the VAS or the receptor. Lack of antibody recognition of the VAS is consistent with the experimental observation that native virions of the group B CV (CVB) do not share a common antigen,[37] even though they share a common receptor on HeLa cells.[2]

We are collaborating with Michael Rossmann and his group in studies whose objective is to determine the crystallographic structure of CVB3 and the location of a purified receptor protein[38] bound to the virion. These studies are aided by the recent determination of the entire nucleotide sequence of the CVB3 genome.[39] Whether the receptor protein would compete with a drug in blocking the ion channel located in the floor of the canyon remains to be determined.[40]

In contrast to many other viruses, no isolated picornavirus virion polypeptide binds with specificity to cellular receptors.[41] Evidently disassembly of the capsid results in a marked conformational change in the VAS to preclude binding to the receptor. Consequently, no virion substructure isolated to date can serve as a specific probe for the cellular receptor.

4. SPECIFIC CELLULAR RECEPTORS FOR COXSACKIEVIRUSES

Multiple copies of specific proteins on the plasma membrane of cells serve as receptors for viruses. The receptor specificity for the different species of picornaviruses sorts these viruses according to their original classification on the basis of disease and histopathology produced in mammalian hosts[42–47] (Table II). This finding is the basis of the hypothesis that cellular receptors are important determinants of viral tropism in the pathogenesis of picornavirus infections.

Studies from our laboratory have determined a large number of characteristics of receptors from HeLa cells for prototype CVB (Table III). The glycoproteins of the receptor may be arranged in a pentameric

TABLE II
Receptor Families for Prototype Picornaviruses Based on Virus Competition for Cell Receptors[a]

Receptor family	Reference
Poliovirus types 1–3	42
Coxsackievirus, group B, types 1–6	42
Coxsackievirus, group A, types 2 and 5; 13,15,18	2, 43
Human rhinovirus	
Types 1A,2,44,49	44
Types 3,5,9,12,14,15,22,32,36,39,41,51,58,59,60,66,67,89, Cox A21	45
Foot-and-mouth disease virus A_{12}, O_{1B}, C_{3Res}, SAT_1, SAT_2, SAT_3	46
Echovirus type 6	42
Cardioviruses	47

[a]From Crowell *et al.*[6]

form (cell membrane pore) to coincide with the postulated circular canyon (VAS) on the virion. The pentameric receptor may be comprised of multimers of Rp-a or of multiple proteins,[3] as has been shown for the acetylcholine (ACh) receptor.[48] Additional studies are needed to establish the nature of the cellular receptor site in the membrane for virus attachment.

Some of the most exciting studies in the field of receptors for picornaviruses are being done by Colonno and his group for HRV-14[49,50] and by Racaniello and Wimmer and colleagues for PV-1.[51] These studies were made possible by the development of monoclonal antibodies against the specific receptors, respectively.[52,53] Much credit is due Campbell and Cords,[54] who were first to obtain monoclonal antibodies against the receptor for several CVB. The identification and purification of a specific receptor protein for the large group of HRV by immunoaffinity chromatography has led to the cloning of the receptor gene in *Escherichia coli*.[50] Nucleotide sequence analysis of the gene is in progress.

The receptor on HeLa cells for attachment of HRV-14 is shared by 74 other serotypes of HRV.[52] This observation has suggested a new strategy for control of rhinovirus infections, namely, to block the receptor by use of a monoclonal antibody. Such an approach may be far more efficient than to prepare a polyvalent vaccine against so many viruses. The results of preliminary human studies, in which the monoclonal antibody was instilled into the nares prior to challenge with infectious HRV, are somewhat encouraging.[50] However, until the function(s) of the different cellular receptors for the different picornavirus species is defined, attempts to block receptors to protect humans from systemic

TABLE III
Properties of Receptors from HeLa Cells for Prototype Group B Coxsackieviruses[a]

Chymotrypsin sensitive; trypsin resistant
Inactivated by periodate, α-, β-glucosidases and α-mannosidase
Regeneration requires mRNA and protein synthesis
Under genetic control of cell
Integral plasma membrane protein, 49,500 M_r, perhaps arranged as pentameric structure
Approximately 10^5 sites/cell
Virus species specificity
Stable at pH 1, 2°C, 10 min; 60°C, 30 min
Antibodies to receptor block virus attachment
Dual function: attach and eclipse virus
Variant receptors select virus variants (hypothesis)

[a]Modified from Crowell *et al.*[6]

infections would appear too risky. Nevertheless, further studies in animal models would appear warranted to extend the concept of receptor blockade as a useful strategy for antiviral prophylaxis.[56]

Monoclonal antibodies against the receptor for polioviruses, unfortunately, have not led to the isolation and characterization of the receptor protein(s).[53,55] These antibodies were insufficient to immunoprecipitate specific receptor proteins from solubilized membranes. They have provided a useful detection system, however, for identifying the receptor protein expressed on mouse L cells transfected with cDNA prepared from a HeLa cell cDNA library.[51] It is only a matter of time until the cloning of the receptor gene for the poliovirus receptor will be identified along with its protein product. The isolation and identification of a receptor protein, Rp-a, from detergent-solubilized HeLa cell plasma membranes for the CVB has been accomplished by use of purified CVB3 as the affinity surface.[38] The virus–receptor complex (VRC) was then purified with methods used for virus purification. The purified VRC was iodinated, repurified and analyzed by SDS-PAGE on slab gels. A protein of 49,500 M_r was identified as the receptor protein (Rp-a). Although an IgG2a class monoclonal antibody that blocks HeLa cell receptors for CVB1, CVB3, and CVB5 has been obtained,[56] this antibody does not recognize Rp-a,[38] and it has been difficult to identify a specific membrane protein by immunoprecipitation and by Western blotting. Furthermore, this monoclonal antibody fails to recognize receptors for CVB on monkey kidney cells. Studies are currently under way to prepare a monospecific rabbit antiserum against the purified Rp-a to permit detection of receptor expression in gene-cloning experiments similar to those conducted by Colonno *et al.*[50] Further studies are needed to identify the relationship between Rp-a and its configuration in the receptor complex on the plasma membrane. It is remarkable that no information is available that relates the virus specificity of receptors on HeLa cells to receptors on murine cells or to any other nonhuman species.[6,57] The rat L_8 myogenic cell line has been shown to have specific receptors for CVA2 and CVA5, which differed from those for binding CVB3.[43] It was unexpected, however, to find that only CVB3, of all of the CVB, replicated in these cells.[6] This was especially surprising, since all of the CVB share the same receptor specificity on HeLa cells.[2] Apparently, there are specific receptors on selected cell types for different virus strains. Perhaps differentiating cell types also express unique receptors that are not shared by all of the CVB.

Serial passage of CVB in the human rhabdomyosarcoma (RD) cell line resulted in the selection of host-range variants of CVB1, CVB3, CVB5, and CVB6.[17] Each variant virus strain (CVB-RD) acquired the capacity for hemagglutination and produced small plaques on HeLa cells, whereas their serologic reactivity and specificty to neutralizing antiserum remained unchanged. CVB3-RD attached well to both RD cells and HeLa cells, whereas parental CVB3 did not attach to RD cells. In virus-competition assays, HeLa cells saturated with CVB3-RD blocked attachment of CVB3, whereas, cells saturated with CVB3 failed to block attachment of CVB3-RD.[6] This one-way receptor blockade suggested that a second VAS was acquired by the CVB3-RD variant for attachment to a second set of receptors on HeLa cells. It is possible that more than one set of receptors may exist on different cell types to account for the selection of host-range viral variants. Reagents to detect the different receptors are needed to sort out their location in nature.

We have found that CVB-RD variants were not only amyocarditic in SJL/J and C_3H/HeJ mice, but also lacked the capacity to produce skeletal muscle disease in young adult SJL/J mice. In addition, it is interesting to note that CVB3 (Woodruff strain) caused skeletal muscle necrosis in 8-week-old SJL/J mice, whereas C_3H/HeJ mice showed no evidence of a comparable disease. Studies are planned to determine whether the altered viral tropisms are receptor mediated.

5. FUTURE DIRECTIONS

Future studies will need to correlate the presence of specific receptors on those cell types that are susceptible to CV infection in a mouse model and in human tissues. There is a real need to develop monoclonal antibodies to the cellular receptors to identify them *in situ*, since no viral protein is suitable and virions are too large to serve this purpose. Cells that are susceptible to infection can be identified by use of labeled antibodies[58] to detect newly synthesized viral proteins. The application of *in vitro* and *in situ* hybridization techniques with labeled cDNA to identify viral RNA genomes in cells is now possible.[59–63]

The expression of receptors on cells does not guarantee that the cell is in a state of susceptibility to viral infection.[43] Furthermore, the identification of receptors on cells may become more complicated if more than one type of receptor aids viral infection.[17] Reagents are needed to identify viral variants and receptor variants, which have been postulated to exist on different strains of mice or even on different tissues in a given animal.[3] One might consider following the spread of virus at intervals postinfection and in normal and immunosuppressed animals of different ages, in an attempt to determine whether specific receptors exist only on infected cells. This latter correlation, however, is likely to be difficult to show convincingly, since many cells might have receptors, but may not be infected for a variety of reasons. Nevertheless, these kinds of studies must be conducted to confirm the hypothesis that specific cellular receptors are important determinants of viral tropism in infection. The specificity of virus–receptor interactions is too high to be dismissed as an artifact or an accident of nature.

ACKNOWLEDGMENTS. Studies from our laboratory were supported by U.S. Public Health Service research grant AI-03771 from the National Institute of Allergy and Infectious Diseases.

REFERENCES

1. Melnick, J. L., 1985, Enteroviruses: Polioviruses, coxsackieviruses, echoviruses, and newer enteroviruses, in: *Virology* (B. Fields, ed.), pp. 739–794, Raven, New York.
2. Crowell, R. L., 1976, Comparative generic characteristics of picornavirus–receptor interactions, in: *Cell Membrane Receptors for Viruses, Antigens, and Antibodies, Polypeptide Hormones and Small Molecules* (R. F. Beers, Jr. and E. G. Bassett, eds.), pp. 179–202, Raven, New York.
3. Crowell, R. L., and Landau, B. J., 1979, Receptors as determinants of cellular tropism in picornavirus infections, in: *Receptors in Human Diseases* (A. G. Bearn and P. W. Choppin, eds.), pp. 1–33, Josiah Macy Foundation, New York.
4. Crowell, R. L., Landau, B. J., and Siak, J-S., 1981, Picornavirus receptors in pathogenesis, in: *Virus Receptors*, Part 2: *Receptors and Recognition*, Series B, Volume 8 (K. Lonberg-Holm and L. Philipson, eds.), pp. 171–184, Chapman and Hall, London.
5. Crowell, R. L., and Landau, B. J., 1983, Receptors in the initiation of picornavirus infections, in: *Comprehensive Virology*, Vol. 18 (H. Frankel-Conrat and R. R. Wagner, eds.), pp. 1–42, Plenum, New York.
6. Crowell, R. L., Reagan, K. J., Schultz, M., Mapoles, J. E., Grun, J. B., and Landau, B.

J., 1985, Cellular receptors as determinants of viral tropism, in: *Genetically Altered Viruses and the Environment*, Banbury Report 22, pp. 147–161, Cold Spring Harbor Laboratory, Cold Spring Harbor, New York.

7. Crowell, R. L., Hsu, K-H. L., Schultz, M., and Landau, B. J., 1986, Cellular receptors in coxsackievirus infections, in: *Positive Strand RNA Viruses*, UCLA Symposium Series (M. A. Brinton and R. R. Rueckert, eds.), pp. 453–466. Alan R. Liss, New York.
8. Sharpe, A. H., and Fields, B. N., 1985, Pathogenesis of viral infections. Basic concepts derived from the reovirus model, *N. Engl. J. Med.* **312:**486–497.
9. Crowell, R. L., and Lonberg-Holm, K. (eds.), 1986, *Virus Attachment and Entry Into Cells*, pp. 1–216, American Society for Microbiology, Washington, D.C.
10. Holland, J. J., 1984, Continuum of change in RNA virus genomes, in: *Concepts of Viral Pathogenesis* (A. L. Notkins and M. B. A. Oldstone, eds.), pp. 137–143, Springer-Verlag, New York.
11. Prabhakar, B. S., Menegus, M. A., and Notkins, A. L., 1985, Detection of conserved and nonconserved epitopes on coxsackievirus B4: Frequency of antigenic change, *Virology* **146:**302–306.
12. Rozhon, E. J., Wilson, A. K., and Jubelt, B., 1984, Characterization of genetic changes occurring in attenuated poliovirus 2 during persistent infection in mouse central nervous systems, *J. Virol.* **50:**137–144.
13. Bodian, D., 1955, Emerging concept of poliomyelitis infection, *Science* **122:**105–108.
14. McLaren, L. C., Holland, J. J., and Syverton, J. T., 1959, The mammalian cell-virus relationship. I. Attachment of poliovirus to cultivated cells of primate and non-primate origin, *J. Exp. Med.* **109:**475–486.
15. Yin, F. H., and Lomax, N. B.. 1983, Host range mutants of human rhinovirus in which nonstructural proteins are altered, *J. Virol.* **48:**410–418.
16. Taylor, M. W., and Chinchar, V. G., 1979, Host restriction of picornavirus infection, in: *The Molecular Biology of Picornaviruses* (R. Perez-Bercoff, ed.), pp. 337–348, Plenum, New York.
17. Reagan, K. J., Goldberg, B., and Crowell, R. L., 1984, Altered receptor specificity of coxsackievirus B3 after growth in rhabdomyosarcoma cells, *J. Virol.* **49:**635–640.
18. Racaniello, V. R., 1984, Poliovirus type II produced from cloned cDNA is infectious in mice, *Virus Res.* **1:**669–675.
19. La Monica, N., Meriam, C., and Racaniello, V. R., 1986, Mapping of sequences required for mouse neurovirulence of poliovirus type 2 Lansing, *J. Virol.* **57:**515–525.
20. Agol, V. I., Drozdov, S. G., Grachev, V. P., Kolesnikova, M. S., Kozlov, V. G., Ralph, N. M., Romanova, L. I., Tolskaya, E. A., Tyufanov, A. V., and Viktorova, E. G., 1985, Recombination between attenuated and virulent strains of poliovirus type 1: Derivation and characterization of recombinants with centrally located crossover points, *Virology* **143:**467–477.
21. Stanway, G., Hughes, P. J., Westrop, G. D., Evans, D. M. A., Dunn, G., Minor, P. D., Schild, G. C., and Almond, J. W., 1986, Construction of poliovirus intertypic recombinants by use of cDNA, *J. Virol.* **57:**1187–1190.
22. Omata, T., Kohara, M., Kuge, S., Komatsu, T., Abe, S., Semler, B. L., Kameda, A., Itoh, H., Arita, M., Wimmer, E., and Nomoto, A., 1986, Genetic analysis of the attenuation phenotype of poliovirus type 1, *J. Virol.* **58:**348–358.
23. Kitamura, N., Semler, B. L., Rothberg, P. G., Larsen, G. R., Adler, C. J., Dorner, A. J., Emini, E. A., Hanecak, R., Lee, J. J., van ler Werf, S., Anderson, C. W., and Wimmer, E., 1981, Primary structure, gene organization, and polypeptide expression of poliovirus RNA, *Nature (Lond.)* **291:**547–553.
24. Toyoda, H., Kohara, M., Kataoka, Y., Suganuma, T., Omata, T., Imura, N., and Nomoto, A., 1984, Complete nucleotide sequences of all three poliovirus serotype

genomes: Implication for genetic relationship, gene function and antigenic determinants. *J. Mol. Biol.* **174**:561–585.

25. Evans, D. M. A., Dunn, G., Minor, P. D., Schild, G. C., Cann, A. J., Stanway, G., Almond, J. W., Currey, K., Maizel, J. V., Jr., 1985, Increased neurovirulence associated with a single nucleotide change in a noncoding region of the Sabin type 3 poliovaccine genome, *Nature (Lond.)* **314**:548–550.
26. Racaniello, V. R., and Baltimore, D., 1981, Cloned poliovirus complementary DNA is infectious in mammalian cells, *Science* **214**:916–919.
27. Eggers, H. J., and Sabin, A. B., 1959, Factors determining pathogenicity of variants of echo 9 virus for newborn mice, *J. Exp. Med.* **110**:951–967.
28. Gauntt, C. J., Gomez, P. T., Duffey, P. S., Grant, J. A., Trent, D. W., Witherspoon, S. M., and Paque, R. E., 1984, Characterization and myocarditic capabilities of coxsackievirus B3 variants in selected mouse strains, *J. Virol.* **52**:598–605.
29. Wolfgram, L. J., Beisel, K. W., Herskowitz, A., and Rose, N. R., 1986, Variations in the susceptibility to coxsackievirus B_3-induced myocarditis among different strains of mice, *J. Immunol.* **136**:1846–1852.
30. Huber, S. A., and Lodge, P. A., 1986, Coxsackievirus B-3 myocarditis. Identification of different pathogenic mechanisms in DBA/2 and BALB/c mice, *Am. J. Pathol.* **122**:284–291.
31. Crowell, R. L., and Landau, B. J., 1978, Picornaviridae: Enterovirus-coxsackieviruses, in: *CRC Handbook Series in Clinical Laboratory Science* (G. D. Hsiung and R. Green, eds.), Vol I, Part 1, pp. 131–155, Boca Raton, Florida.
32. Rossmann, M. G., Arnold, E., Erickson, J. W., Frankenberger, E. A., Griffith, J. P., Hecht, H-J., Johnson, J. E., Kamer, G., Luo, M., Mosser, A. G., Rueckert, R. R., Sherry, B., and Vriend, G., 1985, Structure of a human common cold virus (rhinovirus 14) and functional relationship to other picornaviruses, *Nature (Lond.)* **317**:145–153.
33. Hogle, J. M., Chow, M., and Filman, D. J., 1985, Three-dimensional structure of poliovirus at 2.9 A resolution, *Science* **229**:1358–1365.
34. Co, M. S., Gaulton, G. N., Fields, B. N., and Green, M. I., 1985, Isolation and biochemical characterization of the mammalian reovirus type 3 cell-surface receptor, *Proc. Natl. Acad. Sci. USA* **82**:1494–1498.
35. McClintock, P. R., Prabhakar, B. S., and Notkins, A. L., 1986, Anti-idiotypic antibodies to monoclonal antibodies that neutralize coxsackievirus B_4 do not recognize viral receptors, *Virology* **150**:352–360.
36. Baxt, B., and Morgan, D. O., 1986, Nature of the interaction between foot-and-mouth disease virus and cultured cells, in: *Virus Attachment and Entry into Cells* (R. L. Crowell and K. Lonberg-Holm, eds.), pp. 126–137, American Society for Microbiology, Washington, D.C.
37. Katze, M. G., and Crowell, R. L., 1980, Immunological studies of the group B coxsackieviruses by the sandwich enzyme-linked immunosorbent assay (ELISA) and immunoprecipitation, *J. Gen. Virol.* **50**:357–367.
38. Mapoles, J. E., Krah, D. L., and Crowell, R. L., 1985, Purification of a Hela cell receptor protein for group B coxsackieviruses, *J. Virol.* **55**:560–566.
39. Lindberg, A. M., Stalhandske, P. O. K., and Pettersson. U., 1986, Genome of coxsackievirus B3, *Virology* **156**:50–63.
40. Smith, T. J., Kremer, M. J., Luo, M., Vriend, G., Arnold, E., Kamer, G., Rossmann, M. G., McKinlay, M. A., Diana, G. D., and Otto, M. J., 1986, The site of attachment in human rhinovirus 14 for antiviral agents that inhibit uncoating, *Science* **233**:1286–1293.
41. Beatrice, S. T., Katze, M. G., Zajac, B. A., and Crowell, R. L., 1980, Induction of

neutralizing antibodies by the coxsackievirus B3 virion polypeptide, VP2, *Virology* **104**:426–438.
42. Crowell, R. L., 1966, Specific cell-surface alteration by enteroviruses as reflected by viral-attachment interference, *J. Bacteriol.* **912**:198–204.
43. Schultz, M., and Crowell, R. L., 1983, Eclipse of coxsackievirus infectivity: The restrictive event for a non-fusing myogenic cell line, *J. Gen. Virol.* **64**:1725–1734.
44. Lonberg-Holm, K., Crowell, R. L., and Philipson, L., 1976, Unrelated animal viruses share receptors, *Nature (Lond.)* **259**:679–681.
45. Abraham, G., and Colonno, R. J., 1984, Many rhinovirus serotypes share the same receptors, *J. Virol.* **51**:340–345.
46. Sekiguchi, K., Franke, A. J., and Baxt, B., 1982, Competition for cellular receptor sites among selected aphthoviruses, *Arch. Virol.* **74**:53–64.
47. Burness, A. T. H., and Pardoe, I. U., 1983, A sialoglycopeptide from human erythrocytes with receptor-like properties for encephalomyocarditis and influenza viruses, *J. Gen. Virol.* **64**:1137–1148.
48. Brisson, A., and Unwin, P. N. T., 1985, Quaternary structure of the acetylcholine receptor, *Nature (Lond.)* **315**:474–477.
49. Tomassini, J. E., and Colonno, R. J., 1986, Isolation of a receptor protein involved in attachment of human rhinoviruses, *J. Virol.* **58**:290–295.
50. Colonno, R. J., and Tomassini, J. E., 1987, Viral receptors: A novel approach for the prevention of human rhinovirus infection, presented at the *Proceedings of the Sixth International Symposium on Medical Virology VI,* (L. M. de la Maza and E. M. Peterson, eds.), pp. 331–347, Elsevier, Netherlands.
51. Mendelsohn, C., Johnson, B., Lionetti, K. A., Nobis, P., Wimmer, E., and Racaniello, V. R., 1986, Transformation of a human poliovirus receptor gene into mouse cells, *Proc. Natl. Acad. Sci. USA* **83**: 7845–7849.
52. Colonno, R. J., Callahan, P. L., and Long, W. J., 1986. Isolation of a monoclonal antibody that blocks attachment of the major group of human rhinoviruses, *J. Virol.* **57**:7–12.
53. Nobis, P., Zibirre, R., Meyer, G., Kuhne, J., Warnecke, G., and Koch, G., 1985, Production of a monoclonal antibody against an epitope on HeLa cells that is the functional poliovirus binding site, *J. Gen. Virol.* **66**:2563–2569.
54. Campbell, B. A., and Cords, C. E., 1983, Monoclonal antibodies that inhibit attachment of group B coxsackieviruses, *J. Virol.* **48**:561–564.
55. Minor, P. D., Pipkin, P. A., Hockley, D., Schild, G. C., and Almond, J. W., 1984, Monoclonal antibodies which block cellular receptors of poliovirus, *Virus Res.* **1**:203–212.
56. Crowell, R. L., Field, A. K., Schleif, W. A., Long, W. L., Colonno, R. J., Mapoles, J. E., and Emini, E. A., 1986, Monoclonal antibody that inhibits infection of HeLa and rhabdomyosarcoma cells by selected enteroviruses through receptor blockade, *J. Virol.* **57**:438–445.
57. Much, D. H., and Zajac, I., 1974, The effect of an antireceptor serum on mammalian cell lines, *J. Gen. Virol.* **23**:205–208.
58. Cash, E., Chamorro, M., and Brahic, M., 1986, Quantitation, with a new assay of Theiler's virus capsid protein in the central nervous system of mice, *J. Virol.* **60**:558–563.
59. Kandolf, R., and Hofschneider, P. H., 1985, Molecular cloning of the genome of a cardiotropic coxsackie B3 virus: Full-length reverse-transcribed recombinant cDNA generates infectious virus in mammalian cells, *Proc. Natl. Acad. Sci. USA* **82**:4818–4822.

60. Stalhandske, P. O. K., Lindberg, M., and Pettersson, U., 1984, Replicase gene of coxsackievirus B3, *J. Virol.* **51**:742–746.
61. Tracy, S., Liu, H.-L., and Chapman, N. M., 1985, Coxsackievirus B3: Primary structure of the 5′ non-coding and capsid protein-coding regions of the genome, *Virus Res.* **3**:263–270.
62. Brahic, M., Smith, R. A., Gibbs, C. J., Jr., Garruto, R. M., Tourtellotte, W. W., and Cash, E., 1985, Detection of picornavirus sequences in nervous tissue of amyotrophic lateral sclerosis and control patients, *Ann. Neurol.* **18**:337–343.
63. Bowles, N. E., Olsen, E. G. J., Richardson, P. J., and Archard, L. C., 1986, Detection of coxsackie-B-virus specific RNA sequences in myocardial biopsy samples from patients with myocarditis and dilated cardiomyopathy, *Lancet* **1**:1120–1123.

5

The Role of Interferon in Picornavirus Infections

FERDINANDO DIANZANI, MARIA R. CAPOBIANCHI, DONATELLA MATTEUCCI, and MAURO BENDINELLI

1. THE INTERFERON SYSTEM

Interferons (IFN) are at least three types of functionally related proteins that are newly produced and secreted by cells stimulated with an inducer. Besides viruses, a number of substances are capable of inducing IFN *in vivo* and *in vitro,* including bacterial endotoxin, natural and synthetic RNAs, polysaccharides, polyanions, and mitogens.

Depending on their origin and antigenic specificity, IFNs were originally designed as fibroblast, leukocyte, and immune IFN. They are now called IFN_β, IFN_α, and IFN_γ, respectively. In human cells, there are at least 12 IFN_α genes (in chromosome 9) while only one IFN_β gene has so far been identified. None of the α- or β-genes contains introns. There appears to be a single IFN_γ gene containing three introns on chromosome 12.

β-Interferon is produced by fibroblasts and epithelial cells during viral infections. The induction requires virus internalization and exposure of viral nucleic acid. Viral RNA, either single or double stranded, appears to be the critical factor in triggering the induction process.

FERDINANDO DIANZANI and MARIA R. CAPOBIANCHI • Institute of Virology, University of Rome, 00185 Rome, Italy. DONATELLA MATEUCCI and MAURO BENDINELLI • Institute of Epidemiology, Hygiene and Virology, University of Pisa, I-56100 Pisa, Italy.

Probably because of its poor tendency to diffuse from the site of production, this IFN, even if released in minute amounts, may rapidly reach concentrations sufficiently high in the extracellular fluids to establish antiviral protection quickly in the neighboring cells. The main function of $IFN_β$ seems to be the control and containment of viral replication at the primary site of infection.

Leukocyte or $IFN_α$ is produced by lymphoid cells and is induced not only by viruses, but also by foreign cells, as well as tumor- and virus-infected cells, bacteria and bacterial products, and B mitogens. It is likely that activation of $IFN_α$ production by lymphocytes may occur after some kind of surface interaction, without a requirement for internalization of foreign genetic material.

α-Interferon is readily diffusible and is promptly found at high concentrations in the blood, even during localized viral infections. Its major function may be the protection of leukocytes and endothelial cells during viremia and the reduction of viral spread to the sites of secondary replication.

γ-Interferon is produced by T lymphocytes following stimulation with specific antigens or mitogens in the presence of accessory cells. Oxidation of galactose residues on the membrane of macrophages seems to activate a calcium flux through the producer T lymphocyte membrane, probably after a message molecule is released by the macrophages.

The antiviral activity of $IFN_γ$ is low, in terms of specific activity, as compared with other IFN types. In addition, activation of antiviral mechanisms by this IFN is considerably slower and more complex. However, $IFN_γ$ has been shown to enhance the activity of $IFN_α$ and $IFN_β$. Although its mechanisms are not well defined, such potentiation may reflect a synergistic action between the different mechanisms of cell activation.

Interferons are active in extremely small amounts: one to a few molecules per cell are required to trigger the antiviral state. It is now certain that IFN enters the cell, but the binding to receptors on the plasma membrane appears to be sufficient to trigger the antiviral response. It has not been established, however, whether IFN internalization is required to trigger other IFN-mediated effects on immunomodulation, cell differentiation, and so forth. Human $IFN_α$ and $IFN_β$ share a common receptor coded for by the chromosome 21. $IFN_γ$ has a separate unidentified receptor.

It is generally believed that a second message links the interaction of IFN with its receptor to the induction of the molecules that mediate the antiviral state, but the nature of the message(s) is unknown. In the case of $IFN_α$ and $IFN_β$ the derepressional process is one step and requires a few minutes to occur, while IFN_γ acts through a multi-step derepressional process that takes several hours.

Interferons inhibit the growth of a wide spectrum of DNA and RNA viruses. Virtually all phases of the virus growth cycle have been shown to be affected, at least to some extent; these include virus entry and uncoating, viral nucleic acid synthesis and methylation, protein synthesis, post-translational processing, and virion assembly and release. The mechanism(s) that predominate may depend not only on the specific cell–virus system, but on variables such as IFN concentration and multiplicity of infection (MOI) as well.

The molecular mechanisms that determine the inhibition of viral growth in IFN-treated cells are complex. IFNs induce at least two double-stranded (ds) RNA-dependent enzymatic activities: a 73,000-M_r protein kinase that phosphorylates the small subunit of initiation factor eIF-2 and that is probably responsible for the inhibition of protein synthesis at the initiation level, and an oligonucleotide polymerase that uses adenosine monophosphate (ATP) to synthetise a series of oligonucleotides containing unusual 2′,5′A-phosphodiester bonds, and commonly designated as 2′5′-oligo-A-synthetase. In turn, these products activate an endonuclease responsible for the degradation of RNA. By these pathways, viral multiplication would be inhibited in IFN-treated cells, provided that during viral replication some ds-RNA is synthetized. Interestingly, in cell-free systems, these effector mechanisms are active against both cell and virus messenger RNA (mRNA), whereas in the intact cell they are capable of discriminating between cellular and viral mRNA. One possible mechanism of discrimination is the presence, in infected cells only, of ds-RNA. It is important, however, to note that recent evidence suggests that the antiviral state is not necessarily correlated with the induction of the 2′5′-oligo-A-synthetase or the 73,000-M_r protein kinase. In fact, in certain situations, cells may become resistant without expressing 2′5′-oligo-A- or 73,000-M_r protein kinase and vice versa.

Although IFNs were first recognized for their extraordinarily potent antiviral properties, it is now well established that they are a family of hormonelike molecules that exert a vast array of additional biologic effects that can be grouped into three main categories: inhibition of cell division, changes in plasma membrane and cytoskeleton physiology, and modulation of immune responsiveness.

Depending on cell type and experimental conditions, exposure to IFNs can result in either longer cell cycles (G1 and G2 normally being more affected than S) or the arrest of the cells predominantly in G1. Cellular RNA, DNA, and protein synthesis are only marginally affected by IFN treatment. An increase in actin fibers and fibronectin and a decrease in cell mobility has also been observed. Cell membrane changes observed in IFN-treated cells include decreased fluidity, inhibition of capping, and increased expression of surface markers, such as Fc receptors, β_2-microglobulin, and HLA antigens. The interplay of IFNs with immune responses is complex. For example, either the cytotoxicity of various effector cell types can be enhanced, or the replication of lymphocytes may be inhibited, or both, which may lead to suppression or enhancement of the various arms of the immunity.

The IFN system constitutes one of the first lines of defense against viruses. It becomes operative within a few hours of infection, and its activation during acute viral infections *in vivo* lasts from a few days to weeks. The evidence supporting a relationship between the IFN system and host ability to deal with viral infections is strong and may be summarized as follows:

1. In many viral infections there is a clear temporal correlation between IFN production and recovery.
2. Animals with a genetically defective IFN system are unusually vulnerable to viral infections.
3. In normal animals, inhibiting IFN production or action enhances the severity of infections.
4. The administration of IFN or IFN inducers protects against viral infections.

It is now believed that modulation of immune mechanisms contributes significantly to the overall antiviral activity of IFNs *in vivo*. The same is certainly true for the antitumor effect exerted by IFNs *in vivo*, as exemplified by the IFN-induced regression of experimental tumors resistant to the direct anticellular activity of IFN *in vitro*.

It should be borne in mind, however, that increasing evidence indicates that the three IFN types and several cloned subtypes may differ markedly in their relative efficacy as immunomodulatory, cell growth inhibitory, and antiviral agents. For a more comprehensive review, see Baron *et al.*[1]

2. THE INTERFERON IN PICORNAVIRUS INFECTIONS

2.1. *In Vivo* Experimental Models

Picornavirus infections provided a useful model for early studies on the role of IFN in viral infections, as reliable animal models of infection were available and their pathogenesis had already been partly elucidated. Pioneering studies during the 1960s[2] showed that, following infection of mice with encephalomyocarditis virus (EMCV), circulating IFN was present during viremia. Approximately 10^5 total units of IFN were found to be produced during this phase; the end of viremia was followed by the cessation of interferonemia. By contrast, administration of preformed IFN or IFN inducers[3–5] to mice infected with EMCV or with the very similar Columbia SK virus increased survival and delayed the appearance of paralytic symptoms, even when the virus was inoculated intracerebrally. These studies established that protection was more effective when IFN treatment was started before virus inoculation and was repeated daily. Further studies showed that mice infected with EMCV and treated with anti-IFN serum died early with symptoms of systemic infection, whereas in control mice the main cause of death was a later occurring encephalitis.[6] In mice treated with the antiserum, viral replication was detected earlier and was more prominent in all organs tested. Interestingly, IFN levels were also higher than in control animals, probably as a consequence of enhanced viral replication. However, the devastating effects of viral replication had already occurred at the time that high IFN levels were detected, and the damage was irreversible.

These results were confirmed and extended by reports[7–9] showing that treatment with IFN or IFN inducers is beneficial only when initiated prior to infection or during its early stages. Protection was associated with the early presence of IFN in the bloodstream, implying that the viremic phase might be prevented and target organs could not be reached by the virus. For instance, in mice infected with coxsackievirus B3 (CVB3) the IFN inducer polyinosinic–polycytidilic acid (poly I : C) prevented the development of myocardial lesions maximally when it was administered 6–12 hr prior to virus challenge, so that the IFN peak was concomitant with virus inoculation.[8] However, the levels of circulating IFN did not always reflect the extent of antiviral protection; it was suggested that some other concomitant antiviral mechanism, possibly evoked by IFN itself, was partly responsible for recovery from infection.[7]

The importance of IFN in the early phases of picornavirus infection was confirmed by experiments done by us in CVB3-infected mice. The administration of anti-$IFN_{\alpha/\beta}$ increased the percentage mortality and shortened average time to death postinfection (Fig. 1). The titers of circulating IFN were reduced initially as a result of antibody administration, but this was followed by a sharp rise 48 hr later. At this time, the levels of circulating IFN were higher than in virus-infected control mice (Fig. 2). These higher IFN levels were paralleled by an increase of viral content in the blood and various organs (data not shown). These results suggest that the outcome of CVB3 infection is strongly affected by the levels of circulating IFN available during the first hours of infection.

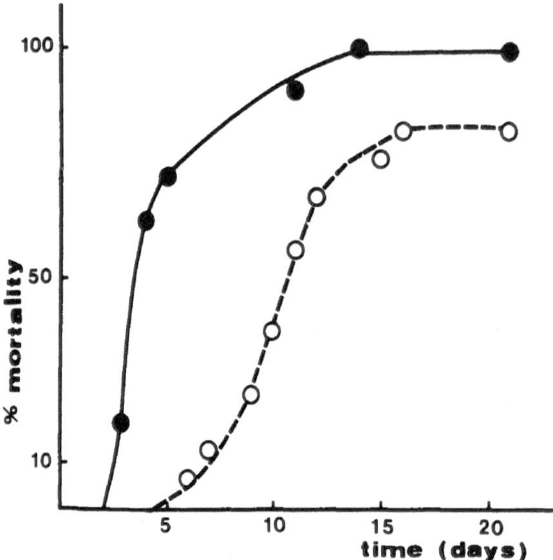

FIGURE 1. Effect of anti-IFN serum on the mortality of CVB3-infected mice. Two-month-old BALB/c mice were infected intraperitoneally with 10^7 $TCID_{50}$ CVB3 and simultaneously injected intravenously with 6000 units anti-interferon-α/β (IFN α/β antibody; ●) or control serum (○). Treatment with antiserum was repeated 48 hr later. The means of four separate experiments, using three to five mice per group, are shown. The average time to death postinfection was 10.7 and 5.8 for mice treated with control and anti-IFN serum, respectively ($p < 0.05$ in Student's t-test).

Probably the IFN present in the circulation after this crucial time, even if at high levels, fails to protect the host because viral spread has already occurred. CVB3-infected mice die with signs of diffuse viral replication in many organs.

Most studies discussed so far have been performed with mouse IFN $_{\alpha/\beta}$ or its inducers. Recently, it has been shown that IFN_γ is as potent as $IFN_{\alpha/\beta}$ in protecting mice against lethal EMCV infection.[10] This study emphasized that IFNs have a maximum effect when given by the same route as the virus, suggesting that the presence of IFN is especially important at the site of primary viral replication. This concept is in keeping with previous observations showing that the degree of IFN-mediated antiviral effect is strongly dependent on the concentration reached by IFN in the vicinity of the producing cells.[11] In this regard, it is important to mention that around producing cells, IFN levels are many times higher than can be reached by administering exogenous IFN.[12]

An interesting set of findings on the importance of the IFN system during picornavirus infections comes from studies using a mutant of mengovirus originally isolated by Simon *et al.*[13] by mutagenization of a wild-type strain that is poorly sensitive to the antiviral action of IFN.

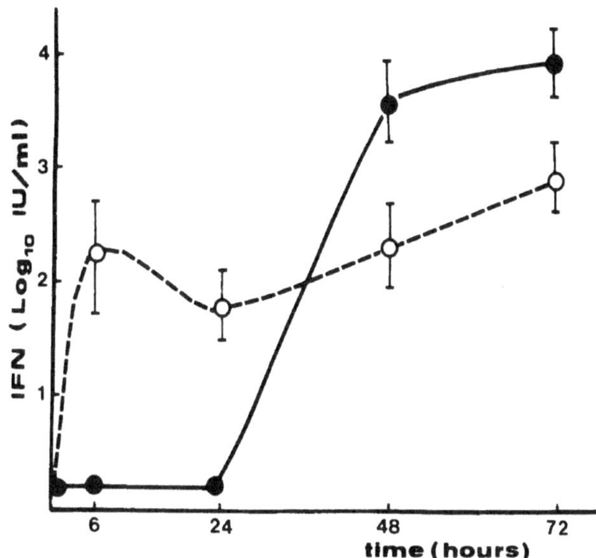

FIGURE 2. Effect of anti-IFN serum on the titers of IFN found in the serum of CVB3-infected mice. Mice were treated as in Fig. 1 (● refers to anti-IFN treated mice, ○ refers to control mice). At various times, blood samples were collected, and circulating IFN was titrated and characterized according to standard procedures.[1] The means ±SD of four separate experiments, using three to five mice per group, are shown. The statistical significance of the differences was evaluated by Student's *t*-test ($p < 0.05$ for all points).

This mutant (phenotype *Ifp*$^+$ as opposed to the wild-type *Ifp*$^-$) was more sensitive to IFN *in vitro,* a property that appeared to be genetically stable. Its sensitivity to IFN was proved to be a consequence of an enhanced capacity to induce IFN.[14] As a result of a priming effect elicited by treatment with exogenous IFN,[15] the cells were much more sensitive to the antiviral action of the IFN endogenously induced by *Ifp*$^+$; thus, this mutant was strongly inhibited. The different behavior of *Ifp*$^+$ and *Ifp*$^-$ viruses was not evident when the infected cells were unable to synthesize IFN.

Additional data suggest that a similar situation occurs with the vaccine strains of poliovirus as well as with the diabetogenic (D) and nondiabetogenic (B) variants of EMCV in mice.[16] The D variant replicates well in Langerhans islets, and the amount of serum IFN detected during infection is low. On the contrary, the B variant is capable of inducing higher circulating IFN levels and replicates poorly in islet cells. The ability to cause diabetes mellitus was found to be associated with the expression of an *Ifp*$^-$ phenotype by the D variant of EMCV, while the wild type counterpart behaved as *Ifp*$^+$. Furthermore, the neutralization of circulating IFN by anti-IFN antibody resulted in more viral antigen being present in pancreatic islets and in the development of diabetes after EMCV-B infection. Conversely, diabetes due to EMCV-D was prevented by exogenous IFN administration.[17] These results constitute strong evidence that the IFN system is involved in the pathogenesis of experimental EMCV-induced diabetes mellitus. They also suggest that the *Ifp* phenotype of the infecting virus may have important implications in naturally occurring picornavirus diseases. The severity of the infection could be modulated by the ratio between Ifp^+ and Ifp^- variants present in naturally infecting virus populations.

Although attractive, such hypothesis is contradicted by the observation that mice treated with either $IFN_{\alpha/\beta}$ or IFN_γ 4 days after EMCV-D challenge, i.e., before development of pancreatitis, virus replication in the brain and pancreas is enhanced, hyperglycemia is also enhanced, and signs of encephalomyocarditis become more evident.[18] Support for the idea that the presence of IFN during the days subsequent to the early phase of infection may adversely affect the course of infection comes from work on CVB3-infected mice.[19] Such detrimental effects could result from the immune-modulatory activity of IFN. For example, the presence of IFN at these stages of infection might delay the synthesis of antiviral antibody or enhance cell-mediated cytotoxicity.

2.2. Participation of the Immune System in the *In Vivo* Effects of Interferon

That immune mechanisms participate in the IFN-mediated control of picornavirus infection is substantiated by a number of findings. Pretreatment with the immunomodulating agent isoprinosine enhances the antiviral action of IFN against lethal EMCV infection in mice[20]; this effect was attributed to the stimulation of antiviral antibody responses or cell-mediated immunity. The antiviral effect of poly I : C-induced IFN (both exogenously administered and endogenously induced) was abolished by treatment with macrophage-toxic substances, such as silica or myocrisin in mice,[21] indicating that under these conditions macrophages are essential for IFN to be effective. Furthermore, low levels of endogenous IFN present in mice prior to viral infection have been shown to induce a constitutive antiviral state in peritoneal macrophages against EMCV infection,[22] suggesting that the capacity of macrophages to restrict EMCV replication stems from their interaction with pre-existing IFN.

Moreover, tunicamycin (TM) treatment of EMCV-infected mice strongly reduced the protection afforded by IFN or IFN inducers and increased virus titers in the brain.[23] It has been proposed that the inhibition of glycosilation of secretory and membrane-associated glycoproteins by this antibiotic prevents the normal interplay of IFN with the immune system and eventually reduces the pro-host action of IFN.

Additional evidence suggesting an important role of immune mechanisms in the protective effects of IFN includes data showing that treatment with the thymic factor thymostimulin significantly increases the survival rate of mengovirus-infected mice, probably through a nonspecific enhancement of endogenous IFN production,[24] and that an IFN inducer and CVB3-specific antibodies act synergistically in protecting CVB3-infected mice.[25] By contrast, the protective activity of IFN in mice infected with coxsackievirus A16 (CVA16) and enterovirus type 71 was not affected by depleting natural killer (NK) cells, macrophages, or T cells.[26] This might indicate that at least in certain picornavirus infections cell-mediated responses are less involved than humoral responses in the protective activity of IFN *in vivo*. This would be in keeping with the present understanding of how immune mechanisms contribute to recovery from such infections.[27]

It is also important to note that picornavirus infections can modulate the host's ability to respond to IFN inducers. For instance, it has been reported that peritoneal exudate cells and splenic lymphocytes harvested from mice infected with EMCV are hyporeactive to induction of IFN by Newcastle disease virus *in vitro*.[28,29] This effect was accompanied by a reduction in the proportion of B lymphocytes in the spleen, whereas T lymphocyte number and functions remained normal.[28]

2.3. Influence of Host Factors on Interferon Production

A possible interaction of hormonal factors with the IFN system in the pathogenesis of picornavirus infections is suggested by early data showing that cortisone-pretreated mice presented an enhanced susceptibility to CVB3 infection, higher viral titers in several organs, and increased IFN levels in the serum from the third day postinfection.[30] It is possible that cortisone induces a transient suppression of IFN production during the first hr postinfection; this in turn leads to enhanced viral replication and to a more generalized IFN induction.

Stress, probably through the release of cortisol, influences the interferonogenic response of mice to picornavirus infections. CVB3-infected mice forced to swim showed a significantly increased incidence and severity of heart lesions. It was shown that 24 hr postinfection, these mice exhibited no detectable levels of circulating IFN, whereas nonexercised controls did. Later, IFN levels were higher in the group that was most affected by infection, i.e., the exercised mice, probably reflecting the extensive virus replication that follows the early IFN depletion caused by the stressing exercise.[31] Again, these data call attention to the importance of the IFNs produced early after infection in determining the eventual outcome. The IFN produced at later stages appears incapable of coping with a virus that has already spread massively and of arresting the progression of pathogenetic events.

Interferon seems to be involved as well in the sex differences in susceptibility to enteroviruses, which have often been observed in humans and animals. Recent results in mice have emphasized the influence of gender in determining the extent of IFN production early after EMCV infection. Adult Swiss male mice proved more susceptible than females to this virus and produced less $IFN_{\alpha/\beta}$ during the first hour of infection. The gender did not influence the efficacy of exogenous IFN treatment.[32]

Further host factors that affect IFN response to picornavirus infections include age, food consumption, and tumor burdens.[33]

2.4. Interferon Induction and Sensitivity in Cell Cultures

Although in picornavirus-infected hosts circulating IFN is usually readily demonstrable, conflicting data exist on the ability of these naked viruses to induce IFN *in vitro*. For instance, hepatitis A virus was described as either capable[34] or incapable[35] of inducing IFN in cell cultures. Variations in the ability to induce IFN *in vitro* were also reported for enterovirus type 70 and CVA24, which are responsible for acute hemorrhagic conjunctivitis in humans.[36] CVB4 was found incapable of inducing IFN *in vitro* on BGMK, WISH, Hela, or WI38 cells.[37] Generally, in all systems tested, picornaviruses have been shown to be poor IFN inducers *in vitro*, at least 10–100-fold less potent than good inducers, such as Newcastle disease virus. For instance, despite the moderately high levels of IFN detected in CVB3-infected mice, in our hands CVB3 failed to induce detectable IFN in cultures of human or murine fibroblasts, leukocytes, and macrophages (F. Dianzani, M. R. Capiobanchi, and D. Matteuci, unpublished data).

Coxsackie B virus types 1–5 readily establish persistent infections in human lymphoid cell lines, but at no time is IFN detectable in persistently infected cultures. The addition to such cultures of a mixture of antibody to human IFN_{α}, IFN_{β}, and IFN_{γ} does not affect the virus cell equilibrium, excluding that undetectable levels of IFN play a role in the maintenance of the carrier state in this particular system.[38]

More general agreement exists on the *in vitro* sensitivity of picornaviruses to the antiviral action of IFN, although the MOI of challenge virus seems to affect the results markedly. For example, IFN cured human lymphoid cell lines and fetal heart cells persistently infected with CVB effectively.[39,40] Furthermore, IFN has been shown to act *in vitro* synergistically with antibodies or with the neutralizing activity found early in tears of patients affected with acute hemorrhagic conjunctivitis.[41] Variants displaying different sensitivity to IFN_{α} *in vitro* have been found in natural isolates of CVB4.[37] Interestingly, degrees of sensitivity to IFN and virulence *in vivo* of the variants correlated well, although no differences were observed in their tissue tropism.

Infection of animal cells by picornaviruses results in the rapid shutoff of cellular RNA synthesis followed by a more gradual decline of protein synthesis. This inhibition occurs early and precedes the onset of viral macromolecular synthesis. In cells pretreated with IFN, viral replication is inhibited, but the shutoff of host protein synthesis is not prevented. However, in IFN-treated cells, the shutoff is transient, and 24 hr postinfection, cellular protein synthesis returns to control values. The effect is not due to breakdown of host RNA, as cellular RNA is not drastically degraded and mRNA present in the cells at the time of infection maintains template activity.[42]

These results are consistent with the activation of ds-RNA-dependent nucleases through the 2′,5′oligo-A-synthetase system. Since the nuclease must be activated by ds RNA, destruction of RNA, including viral RNA, occurs only in the vicinity of the replication complex leaving unaltered most cellular RNA.[43] At the same time, certain initiation factors might be altered by infection even in IFN-treated cells, which might be responsible for the shutoff of cellular protein synthesis. Actinomycin D prevents the resumption of protein synthesis at 24 hr, suggesting that newly synthesized cellular RNAs are needed, possibly those, short-lived, that code for initiation factors.

2.5. The Interferons in Picornaviral Infections of Humans

Acute hemorrhagic conjunctivitis caused by enterovirus type 70 and CVA24, viral hepatitis A, and rhinoviral infections are among the picornavirus infections of humans studied in this respect. IFN_{β} was detected during the first days of illness in tear samples from patients with acute hemorrhagic conjunctivitis caused by CVA24. The appearence of IFN coincided with a decrease of virus titers in the tears suggesting that IFN inhibits viral replication in the eye. A CVA24-specific neutralizing activity was also detected in the tears within 24 hr from onset of clinical disease. This factor was synergistic with IFN in inhibiting CVA24 in tissue cultures. This leads to the conclusion that most likely the combined action of antibody and IFN contributes to reduce the length and severity of conjunctivitis.[41]

Early reports had indicated that patients with infectious hepatitis A do not produce IFN in the acute and convalescent stages of the disease.

However, a recent report has appeared suggesting that during the disease IFN may play some role in the elimination of virus-infected hepatic cells by enhancing cell- and antibody-mediated cytotoxicity.[35]

In 1973, Merigan *et al.*[44] showed that high doses of intranasally instilled human IFN_α protected volunteers against rhinovirus challenge. Large doses of IFN were required for protection, as explained by the rapid local clearance of topically applied IFN. Subsequent studies conducted at several centers have conclusively proven that both natural and recombinant IFN_α provide protection against colds experimentally induced with a range of rhinovirus serotypes and that protection depends not only on IFN dose but on timing and duration of treatment as well. IFN administration during the late incubation period of rhinovirus infection significantly alleviates but does not prevent clinical symptoms. Instead, postexposure prophylaxis of family contacts significantly reduces symptomatic cases, in the absence of manifestations of cumulative toxicity. However, treatment with high-dose IFN_α for more than 10 days causes adverse side effects, including nasal stuffiness, bloody mucus, mucosal erosions, and submucosal infiltrates of mononuclear cells.[45]

An interesting counterpart of such results is a report describing the impaired production of IFN_α by children displaying an undue susceptibility to respiratory viral infections.[46] This deficiency was demonstrated both directly in nasopharyngeal aspirates during the acute phases of infections and in short-term cultures of peripheral blood leukocyte stimulated with conventional IFN_α inducers. IFN_γ production stimulated by T-cell mitogens in lymphocyte cultures was instead normal. A causal relationship between IFN_α deficiency and failure to eradicate locally invasive viruses, such as rhinoviruses, from the upper respiratory tract was suggested.

3. CONCLUSIONS

Despite a poor ability to induce IFN *in vitro,* picornaviruses evoke substantial IFN responses *in vivo.* Thus, for example, the host responds to infection by these viruses with the early appearance of high-titer IFN in the blood. Such circulating IFN appears to be able to prevent or limit picornavirus spread to target organs, provided it is present early and in sufficient amounts to block viremia or to induce resistance in the endothelial cells responsible for anatomical barriers, such as the blood–brain barrier. In most models of systemic infection, the administration of anti-IFN antibody during this phase has been shown to cause a significant enhancement of viral replication and exacerbation of pathology. The enhanced late IFN production that may occur in anti-IFN antibody-treated hosts most likely is a consequence of the higher virus titers produced in target organs, and occurs at a time when virus-induced damage is no longer preventable.

In certain animal models, the administration of anti-IFN antibody at later stages of infection has resulted in an alleviation of picornavirus-induced pathogenesis and of exogenous IFN in an aggravation. This seems to be a reflection of the immunomodulating properties of IFN. It is now generally accepted that immune mechanisms are extensively involved in the amplification of IFN effects. Thus, if viral replication has not been successfully dealt with, it seems reasonable to assume that IFN may be responsible for either ameliorating or worsening virus-induced pathogenesis, depending on the immunomodulatory effects produced and on the role played by immune mechanisms in the genesis of tissue damage.

Interferon responses to picornaviruses are influenced by host factors known to affect the outcome of infection, including the hormonal situation. To what extent these factors mediate their effects through IFN remains to be determined. Also unknown is the molecular basis of the differences in IFN inducing ability or insensitivity to IFN presented by certain picornavirus variants. Such differences might contribute to determine the variations of virulence encountered in naturally circulating picornaviruses. Further studies along these lines seem warranted.

ACKNOWLEDGMENTS. The original experiments reported in this chapter were supported by Progetto Finalizzato Controllo delle Malattie da Infezione, CNR. Skillful secretarial help by Miss E. Di Francesco is greatly appreciated.

REFERENCES

1. Baron, S., Dianzani, F., and Stanton, J. (eds.), 1981–1982, The interferon system: A review to 1982, *Texas Rep. Biol. Med.* **41:**1–715.
2. Baron, S., Buckler, C. E., McCloskey, R. V., and Kirschstein, R. L., 1966, Role of interferon during viremia. I. Production of circulating interferon, *J. Immunol.* **96:**12–16.
3. Baron, S., Buckler, C. E., Friedman, R., and McCloskey, R., 1966, Role of interferon during viremia. II. Protective action of circulating interferon, *J. Immunol.* **96:**17–24.
4. Dianzani, F., Rita, G., Cantagalli, P., and Gagnoni, S., 1969, Effect of DEAE-dextran on interferon production and protective effect in mice treated with the double-stranded polynucleotide complex polyinosinic–polycytidylic acid, *J. Immunol.* **102:**24–27.
5. Gresser, I., Bourali, C., Thomas, M. T., and Falcoff, E., 1968, Effect of repeated inoculation of interferon preparations on infection of mice with encephalomyocarditis virus, *Proc. Sco. Exp. Biol. Med.* **127:**491–496.
6. Gresser, I., Tovey, M. G., Bandu, M. T., Maury, C., and Brouty-Boyé, D., 1976, Role of interferon in the pathogenesis of virus disease in mice as demonstrated by the use of

anti-interferon serum. I. Rapid evolution of encephalomyocarditis virus infection. *J. Exp. Med.* **144:**1305–1315.

7. Giron, D. J., Liu, R. Y., Hemphill, F. E., Pindak, F. F., and Schmidt, J. P., 1980, Role of interferon in the antiviral state elicited by selected interferon inducers, *Proc. Soc. Exp. Biol. Med.* **163:**146–150.
8. Norris, D., and Loh, P. C., 1973, Coxsackievirus myocarditis: Prophylaxis and therapy with an interferon stimulator, *Proc. Soc. Exp. Biol. Med.* **142:**133–136.
9. Olsen, G. A., Kern, E. R., Glasgow, L. A., and Overall, J. C. Jr., 1976, Effect of treatment with exogenous interferon polyinosinic acid–polycytidylic acid, or polyinosinic acid–polycytidylic acid–poly-L-lysine complex on encephalomyocarditis virus infections in mice, *Antimicrob. Agents Chemother.* **10:**668–676.
10. Shalaby, M. R., Hamilton, E. B., Benninger, A. H., and Marafino, B. J. Jr., 1985, *In vivo* antiviral activity of recombinant murine gamma interferon, *J. Interferon Res.* **5:**339–345.
11. Dianzani, F., Viano, I., Santiano, M., Zucca, M., and Baron, S., 1977, Effect of cell density on development of the antiviral state in interferon-producing cells: A possible model of *in vivo* conditions, *Proc. Soc. Exp. Biol. Med.* **155:**445–448.
12. Dianzani, F., Gullino, P., and Baron, S., 1978, Rapid activation of the interferon system in vivo, *Infect. Immun.* **20:**55–57.
13. Simon, E. H., Kung, S., Koh, T. T., and Brandman, P., 1976, Interferon-sensitive mutants of mengovirus. I. Isolation and biological characterization, *Virology* **69:**727–736.
14. Marcus, P. I., Guidon, P. T. Jr., and Sekellick, M. J., 1981, Interferon induction by viruses. VII. Mengovirus: "Interferon-sensitive" mutant phenotype attributed to interferon-inducing particle activity, *J. Interferon Res.* **1:**601–611.
15. Fleischmann, W. R., Jr., 1977, Priming of interferon action, *Tex. Rep. Biol. Med.* **35:**316–325.
16. Cohen, S. H., Bolton, V., and Jordan, G. W., 1983, Relationship of interferon-inducing particle phenotype to encephalomyocarditis virus-induced diabetes mellitus, *Infect. Immun.* **42:**605–611.
17. Yoon, J. W., Cha, C. Y., and Jordan, G. W., 1983, The role of interferon in virus-induced diabetes, *J. Infect. Dis.* **147:**155–159.
18. Gould, C. L., McMannama, K. G., Bigley, N. J., and Giron, D. J., 1985, Exacerbation of the pathogenesis of the diabetogenic variant of encephalomyocarditis virus in mice by interferon, *J. Interferon Res.* **5:**33–37.
19. Lutton, C. W., and Gauntt, G. J., 1985, Ameliorating effect of IFN-beta and anti-IFN-beta on coxsackievirus B3-induced myocarditis in mice, *J. Interferon Res.* **5:**137–146.
20. Chany, C., and Cerutti, I., 1977, Enhancement of antiviral protection against encephalomyocarditis virus by a combination of isoprinosine and interferon, *Arch. Virol.* **55:**225–231.
21. Stebbing, N., Dawson, K. M., and Lindley, I. J. D., 1978, Requirement for macrophages for interferon to be effective against encephalomyocarditis virus infection of mice, *Infect. Immun.* **19:**5–11.
22. Belardelli, F., Vignaux, F., Proietti, E., and Gresser, I., 1984, Infection of mice with antibody to interferon renders peritoneal macrophages permissive for vesicular stomatitis virus and encephalomyocarditis virus, *Proc. Natl. Acad. Sci. USA* **81:**602–606.
23. Maheshwari, R. K., Husain, M. M., Attallah, A. M., and Friedman, R. M., 1983, Tunicamycin treatment inhibits the antiviral activity of interferon in mice, *Infect. Immun.* **41:**61–66.
24. Klein, A. S., Fixler, R., and Shoham, J., 1984, Antiviral activity of a thymic factor in

experimental viral infections. I. Thymic hormonal effect on survival, interferon production and NK cell activity in mengo virus infected mice, *J. Immunol.* **132:**3159–3163.

25. Cho, C. T., Feng, K. K., McCarthy, V. P., and Lenahan, M. F., 1982, Role of antiviral antibodies in resistance against coxsackievirus B3 infection: interaction between preexisting antibodies and an interferon inducer, *Infect. Immun.* **37:**720–727.
26. Sasaki, O., Karaki, T., and Imanishi, J., 1986, Protective effect of interferon on infections with hand, foot and mouth disease virus in newborn mice, *J. Infect. Dis.* **153:**498–502.
27. Mims, C. A., and White, D. O., 1984, *Viral Pathogenesis and Immunology,* Blackwell Scientific Publications, Oxford.
28. Faden, H., Glasgow, L. A., and Overall, J. C. Jr., 1978, Variable effect of encephalomyocarditis virus on host defense mechanisms, *Infect. Immun.* **19:**94–100.
29. Stringfellow, D. A., Kern, E. R., Kelsey, D. K., and Glasgow, L. A., 1977, Suppressed response to interferon induction in mice infected with encephalomyocarditis virus, semliki Forest virus, influenza A2 virus, herpesvirus hominis type 2 or murine cytomegalovirus, *J. Infect. Dis.* **135:**540–551.
30. Rytel, M. W., 1969, Interferon response during coxsackie B-3 infection in mice. I. The effect of cortisone, *J. Infect. Dis.* **120:**379–382.
31. Reyes, M. P., and Lerner, A. M., 1976, Interferon and neutralizing antibody in sera of exercised mice with coxsackievirus B-3 myocarditis, *Proc. Soc. Exp. Biol. Med.* **151:**333–338.
32. Pozzetto, B., and Gresser, I., 1985, Role of sex and early interferon production in the susceptibility of mice to encephalomyocarditis virus, *J. Gen. Virol.* **66:**701–709.
33. Bendinelli, M., Ruschi, A., and Santopadre, G., 1965, Replicazione virale e produzione di interferone in topi a dieta carente infettati con virus mengo, *Riv. Ital. Igiene* **25:**191–204.
34. Kurane, I., Binn, L. N., Bancroft, W., and Ennis, F. A., 1985, Human lymphocyte responses to hepatitis A virus-infected cells: Interferon production analysis of infected cells, *J. Immunol.* **135:**2140–2144.
35. Vallbracht, A., Gabriel, P., Zahn, J., and Flehmig, B., 1985, Hepatitis A virus infection and the interferon system, *J. Infect. Dis.* **152:**211–213.
36. Langford, M. P., Villarreal, A. L., and Stanton, G. J., 1983, Antibody and interferon act synergistically to inhibit enterovirus, adenovirus and herpes simplex virus infection, *Infect. Immun.* **41:**214–218.
37. Jordan, G. W., and Bolton, V., 1985, Interferon-sensitive coxsackievirus variants in nature, *J. Interferon Res.* **5:**289–296.
38. Matteucci, D., Paglianti, M., Giangregorio, A. M., Capobianchi, M. R., Dianzani, F., and Bendinelli, M., 1985, Group B coxsackieviruses readily establish persistent infections in human lymphoid cell lines, *J. Virol.* **56:**651–654.
39. Kandolf, R., Canu, A., and Hofschneider, P. H., 1985, Coxsackievirus B3 can replicate in cultured human foetal heart cells and is inhibited by interferon, *J. Mol. Cell. Cardiol.* **17:**167–198.
40. Bendinelli, M., Matteucci, D., Conaldi, P. G., Giangregorio, A. M., Capobianchi, M. R., and Dianzani, F., 1988, Mechanisms of group B coxsackievirus persistence in human cells, *Eur. Heart J.* **7** (suppl. E):(in press).
41. Langford, M. P., Barber, J. C., Sklar, V. E. F., Clark, S. W. III, Patriarca, P. A., Onarato, I. M., Yin-Murphy, M., and Stanton, G. J., 1985, Virus-specific, early appearing neutralizing activity and interferon in tears of patients with acute hemorrhagic conjunctivitis, *Curr. Eye Res.* **4:**233–239.
42. Vaquero, C., Aujean-Rigaud, O., Sanceau, J., and Falcoff, E., 1981, Effect of inter-

feron on transient shut-off of cellular RNA and protein synthesis induced by mengo virus infection, *Antiviral Res.* **1:**123–134.
43. Nilsen, T. W., and Baglioni, C., 1979, Mechanisms for discrimination between viral and host mRNA in interferon-γ treated cells, *Proc. Natl. Acad. Sci. USA* **76:**2600–2604.
44. Merigan, T. C., Hall, T. S., Reed, S. E., and Tyrrel, D. A. J., 1973, Inhibition of respiratory virus infection by locally applied interferon, *Lancet* **1:**563–567.
45. Couch, R. B., 1984, The common cold: control?, *J. Infect. Dis.* **150:**167–173.
46. Isaacs, D., Clarke, J. R., Tyrrel, D. A. J., Webster, A. D. B., and Valman, H. B., 1981, Deficient production of leukocyte interferon (interferon-alpha) *in vitro* and *in vivo* in children with recurrent respiratory tract infections, *Lancet* **31:**950–952.

6

Interactions with the Immune System

MAURO BENDINELLI, PIER GIULIO CONALDI, and DONATELLA MATTEUCCI

1. INTRODUCTION

For many years, coxsackieviruses (CV) and other enteroviruses have been considered relatively unsophisticated pathogens. They were believed to produce acute infections only and to induce tissue damage and disease exclusively by causing lysis of infected target cells. This contention was based on consistent observations showing that, provided cells are susceptible to infection, CV dramatically damaged or destroyed them, owing to a rapid and virtually complete virus-mediated shutoff of protein synthesis followed by cell lysis in a few hours. Accordingly, necrotic lesions in target organs, such as central nervous system (CNS), pancreas, liver, and myocardium, were viewed solely as the result of local viral replication and cytopathic effect produced by the infecting CV, while immune responses were considered to play only a protective role.

As a result of clinical observations and far-reaching experimental data discussed in other chapters of this book, it is becoming increasingly apparent that CV infections do instead share many complexities classically associated with enveloped viruses. They can persist for a considerable time in infected patients, produce subtle changes of infected cells, cause chronic pathology, and trigger immunopathologic damage to infected and uninfected tissues. Thus, although many aspects of CV-in-

---

MAURO BENDINELLI, PIER GIULIO CONALDI, and DONATELLA MATTEUCCI • Institute of Epidemiology, Hygiene and Virology, University of Pisa, I-56100 Pisa, Italy.

duced pathogenesis remain to be understood, there is no doubt that the immune system exerts a far more complex role than originally believed.

This chapter reviews available information on the interplay established by CV with the immune system. Attention is especially focused on aspects that have recently emerged and may be relevant to the changing scenario of CV-induced pathogenesis. To place such data in perspective, a concise account of our present understanding of immune responses to CV and their role in resistance and tissue damage is also provided.

2. NONSPECIFIC MECHANISMS OF RESISTANCE TO COXSACKIEVIRUSES

Extensive data document the pivotal importance of the mononuclear phagocytic system in host defenses against viral infections in general. That cells of the monocyte–macrophage series are deeply involved in resistance to CV is suggested by a number of findings on experimentally infected mice. Early studies[1] evidenced a parallel between the decrease with age of susceptibility to CV and functional maturation of these cells and that infusion of peritoneal exudate cells (PEC) taken from uninfected syngenic adults protected suckling mice against the lethal effect of coxsackie B virus, type 3 (CVB3). Pretreatment of PEC donors with macrophage-activating substances enhanced protection, whereas macrophage-toxic or blocking treatments had opposite effects. It has also been suggested[2] that mononuclear inflammatory cells recruited into CVB-infected heart tissue are essential in the elimination of the infecting virus. Recently, ibuprofen, a nonsteroid anti-inflammatory drug that inhibits prostaglandin synthesis, was seen to worsen myocardial inflammation and necrosis during CVB3 infection of mice.[3] Since cells in the monocyte–macrophage lineage are a major source of prostaglandins, the result might imply that mediators released by such cells are key determinants in the containment and clearance of CV from infected target tissues, possibly through their influence on inflammatory response homeostasis. However, in these experiments, ibuprofen did not significantly affect virus persistence in the heart.

As expected, control of CV infection by macrophages is carried out in collaboration with the effectors of specific immunity. For example, passive protection of CVB3-infected suckling mice by adult PEC was enhanced by the concomitant administration of diluted antiviral antibody.[1] *In vitro* studies suggest, however, that macrophages can also exert an independent anti-CV activity, as peritoneal macrophages of nonimmune mice were shown to take up and inactivate extracellular CVB3.

This direct antiviral effect was increased by activation of macrophages and decreased by cyclophosphamide treatment of donor mice.[1,4]

The above findings indicate that macrophages are crucial in limiting CV dissemination in the mouse, but nothing is known of their relevance to natural infection of humans. Macrophage-mediated antiviral mechanisms were credited with the sharp reduction of CVB3 infectivity effected by freshly cultured human peripheral blood mononuclear cells,[5] but supporting evidence was not presented. It is also unknown as to whether antigen presentation and other accessory functions of these cells are required for eliciting CV-specific immune responses. That CV can reduce the ability of macrophages to collaborate with lymphocytes in the generation of antibody responses to heterologous antigens is discussed in Section 6.

Additional nonspecific defenses known to contribute to anti-CV resistance include interferon (IFN) (Chapter 5), restricted expression of cell receptors (Chapter 4), and barrier and clearance mechanisms operative in the gut.[6] Of special interest are recent findings[7] on natural killer (NK) cells in CV infections. NK cell-mediated cytotoxicity was activated during the early stages of CVB3 infection. Activity peaked at day three postinfection and then declined. Mice depleted of NK cells by inoculation of anti-asialo GM_1 antiserum early in the course of infection had CVB3 titers in heart tissues 11–670-fold higher than did noninoculated mice. Myocarditic lesions were also worsened in these mice. These results suggest that NK cells act to limit CV replication and subsequent pathology.

3. ANTIBODY RESPONSE TO COXSACKIEVIRUSES

As is the case with most viral infections, in naturally infected humans humoral responses to CV consist of the production of IgM, IgA, and IgG. Circulating IgM antibody appears shortly after initial infection and persists for 8–10 weeks. Homotypic IgG antibody appears later and persists for many years postinfection. Circulating IgA antibody also lasts several years postinfection.[8] CV-specific secretory IgA activity has been detected in the nasopharynx and intestines after infection with several serotypes.[9] It is well documented that IgG antibody titers in the circulation can be boosted by reinfection with the same CV serotype or by heterotypic responses elicited by other CV or related enteroviruses. It is unclear, however, whether such anamnestic responses suffice to explain the lifelong persistence exhibited by anti-CV antibody as well as by antibodies to other enteroviruses. Findings that CV infectivity and RNA can be demonstrated in the host for a considerable time after the acute phase of infection (Chapter 17) suggest a continuing antigenic stimulation as an alternative possibility.

Whereas the pattern depicted above is the accepted average pattern of humoral responses to CV, it is well recognized that in individual patients antibody responses can vary considerably. For example, following primary CVB infection, diversity has been documented in the maximum antibody titers achieved and in the rapidity of their decline.[10] In certain subjects with recurrent pericarditis, cardiac neurasthenia, myalgic enchephalitis, or orchitis anti-CVB IgM antibodies have been seen to persist for several months or years.[11] The reasons for such variations are unclear. Differences in clinical manifestations associated with CVB infection have been suggested to correlate with differences in antibody behavior. Genetic factors can also be involved. In the murine model, time of appearance and titer of neutralizing antibody as well as other parameters of CVB3 infection are under multigenic control involving both H-2-associated and non-H-2 genes.[12] Similarly, in humans a correlation between anti-CVB humoral reactions and HLA DR3 and DR4 genes was recently suggested,[13] but whether this is due to a higher susceptibility to CV of individuals with these haplotypes or involves genetic regulation of anti-CV antibody response remains to be determined. A further possibility is that CV modulate immune responsiveness and that such effect depends on the host's genetic makeup in humans as well as in mice (Section 7).

In mice, antibody response to CV appears to be at least partly thymus dependent (see Table I). For example, in *nu/nu* and *+/nu* NFR mice, neutralizing antibody titers did not differ greatly over the first month of infection but thereafter declined much faster in athymic than in euthymic mice. While IgM antibody production was comparable, IgG antibodies were markedly reduced in athymic animals.[15]

The CVB virion polypeptide containing the major epitopes is VP1, which bears both type- and group-specific immunogenic determinants. Antigenic moieties eliciting the production of neutralizing antibody have also been identified on VP2. Data indicate that laboratory and field strains of CV consist of populations of viral variants differing in antigenic and/or pathogenic properties (Chapter 10). Changes in antigenicity are most likely produced by a mechanism of antigenic drift based on the very high rate of evolution of RNA genomes. Selection with monoclonal antibodies during CVB replication *in vitro* led to the emergence of antigenic variants different from the parental strain and from one another.[19] Investigation of antibodies directed to individual epitopes and intratypic variants in infected patients may provide clues to many unresolved problems.

TABLE I
CVB3 Infection and Myocardial Lesions in T-Cell-Deficient Mice as Compared with Controls[a]

Type of deficiency and mouse strain	Virus in the heart: Titer	Virus in the heart: Persistence	Antibody response	Myocardial lesions: Incidence	Myocardial lesions: Tissue injury	Myocardial lesions: Cell infiltration	Lethality	Reference
Congenital thymus aplasia								
N:NIH(S) II*nu/nu*	Slight increase	Increase	Reduced duration (N)	Decrease	Decrease	Decrease	Increase	14
NFR *nu/nu*	Slight increase	Increase	Reduced duration (N) Decrease (EIFA)	Slight increase	No change	No change	NR	15
BALB/c *nu/nu*	No change	NR	Slight decrease (N)	Slight decrease	Decrease	Absent	No change	16
BALB/c *nu/nu*	NR	No change	NR	No change	No change	No change	Slight increase	17
Neonatal thymectomy BALB/c	No change	No change	No change (N)	NR	Decrease	Decrease	Decrease	18
Anti-thymocyte serum CD-1	No change	No change	Reduced duration (N)	NR	Decrease	Decrease	Increase	18

[a]N, neutralization; EIFA, enzyme immunofluorescence assay; NR, not reported.

3.1. Role in Resistance and Recovery

Clinical observations in individuals with congenital agammaglobulinemia prove that virus-specific antibodies are essential in the control of CV and related enteroviruses.[20,21] In humans, the mechanisms of antibody protection against CV infection are not established but, in analogy to poliovirus infection, it seems likely that a timely nasopharyngeal and gut secretory IgA response may be effective in limiting viral replication in mucous membranes and reducing chances of spread to internal tissues. Similarly, circulating neutralizing antibodies are most likely crucial in preventing the hematogenous spread of CV from the primary sites of replication to target organs and, when they develop too late to be successful in this respect, in facilitating recovery by terminating viremia and reducing the viral load in infected organs.

The murine model confirms the importance of humoral immunity in defense against CV infection. Susceptibility of adult mice to CVB is increased by treatments known to interfere with immune responsiveness including corticosteroids, cyclophosphamide, stress, and undernutrition. Investigation of the relative importance of circulating antibody and other mediators of antiviral defenses lead to the conclusion that humoral antibody is critical in limiting CVB3 infection.[2,4,22] Studies have shown that passive transfer of neutralizing antibodies is protective only before or within 24 hr postinfection, suggesting that antibodies act primarily by neutralizing virus, which would otherwise disseminate the infection.[23] In CVB3-infected mice, there was a critical fall of viremia that occurred between days 3 and 5 postinfection, depending on animal strain, concomitant with the appearance of detectable antibody in the circulation,[12] suggesting that antibody response is a major mechanism for clearing CV from the bloodstream.

It is of interest that neutralizing antibody synthesis in mice did not always correlate with suppression of CVB3 replication in infected tissues.[2,12] For example, cortisone administered before CVB3 infection lead to an abnormally high viral content in the heart, pancreas, and liver for more than 2 weeks. The extent of tissue necrosis and mortality was also increased. However, there was no evidence that neutralizing antibody response was impaired in steroid-treated mice. By contrast, a depression of mononuclear inflammatory cell immigration into the infected tissues was noted. It is therefore likely that inhibition of CV replication in infected tissues requires the concurrent activity of macrophages and other infiltrating inflammatory cells. The importance of the concomitant action of various effectors of antiviral defenses is also exemplified by findings showing that combined treatment of newborn mice with an IFN inducer and antiviral antibody resulted in synergistic protection against CVB3 challenge.[23]

There is satisfactory evidence that pre-existing antibodies afford a significant protection against reinfection with the homologous CV. However, protection against reinfection is not absolute. Among contacts of infected people, CV excretion was observed also in individuals with pre-existing antibody.[24] The importance of antigenic viral variants in reinfections has not been investigated. Whether the presence of neutralizing serum antibodies implies lifelong immunity against homologous CV-induced tissue damage and disease, as is the case for polioviruses, also remains to be determined. Mice vaccinated with an attenuated temperature-sensitive mutant of CVB3 could be reinfected with the homologous wild type but were protected against myocarditis even in the absence of assayable levels of neutralizing antibody. They however exhibited a more prompt and elevated antibody response than unvaccinated controls, as well as an altered T-cell response.[25]

3.2. Role in Tissue Damage

There is evidence that the humoral response also participates in the genesis of CV-induced disease. Early work[26] showed that CVB4-infected mice can develop immune complex nephritis. More recently, attention has been focused on the role of humoral immune mechanisms in the onset and evolution of CVB-induced pathology of heart tissue. Convincing evidence indicates that immunologic factors are involved in the genesis of heart damage in mice. Virus-specific antibodies that crossreacted with cardiac tissue and lysed vital cardiocytes in the presence of complement were detected in the serum of patients with post-CVB myocarditis.[27] Such antibodies might find an explanation in the existence of similar epitopes on viral and some normal host antigens (molecular mimicry), as suggested by the fact that one virus-specific monoclonal antibody has been found to react with the normal myocardium of several species.[28]

Heart-specific autoantibodies were also found in the sera of several strains of mice following infection with CVB3.[12] Although clearly related to infection, such antibodies did not crossreact with virion antigens. The observation that their development correlates with the ability of different strains of mice to mount an IgG antibody response to myocyte antigens and additional results discussed in other chapters indicate that these antibodies are most probably triggered by cardiac components rendered antigenic as a consequence of viral infection. Autoantibodies to kidney tubules, smooth muscle, and DNA were also present in the sera of CVB3-infected mice. These data demonstrate that a broad autoimmune response can be triggered by CV infection and add to the evidence discussed below that perturbation of immune functions is an important feature of CV infection.

4. CELL-MEDIATED IMMUNE RESPONSE TO COXSACKIEVIRUSES

Evidence concerning cell-mediated immune (CMI) responses of humans to CV is limited.[29] Investigation of the murine CVB model has, however, clearly documented that CV elicit CMI at least as demontrated by established *in vitro* assays. Spleen cells of adolescent mice challenged with CVB3 have repeatedly been reported to contain cytotoxic T lymphocytes (CTL) specifically reactive against target cells infected with the homologous virus. PEC from CVB3-infected or immunized mice have been seen to react specifically with cell lysates containing large quantities (at least 10^8 plaque-forming units) of CVB3 in macrophage migration inhibition assay (MIF).[30] Splenic T lymphocytes of mice exposed to attenuated or virulent CVB3 have been shown to respond mitotically to purified CVB3 particle antigen.[25] Comparable results were obtained by examining peripheral blood lymphocytes and PEC from CVB3-infected baboons,[31] thereby extending the demonstration of CV-specific CMI to a system that is phylogenetically closer to humans.

Cytotoxic T lymphocyte production has been particularly investigated in view of its suspected role in CVB-induced cardiac damage. These effector cells are generally measured by their ability to lyse neonatal fibroblasts and myocytes infected by the homologous CV in an *in vitro* ^{51}Cr-release assay. In the spleen of male mice, cytoxicity was detected on the third day postinfection, peaked by 1 week, and then declined to low levels by 2 weeks. Splenocytes obtained 3 and 5 days postinfection also exerted a low degree of cytotoxic activity for uninfected cells, but by 1 week postinfection there was little or no reactivity against these targets. While this early nonspecific response (probably related to NK cell activation) was equivalent in animals of either genders, the production of immunospecific CTL was considerably more frequent and stronger in males than in females.[32,33]

The defective CTL response of female mice has recently been attributed to active suppression due to the rapid development of suppressor cells in animals of this sex. Male lymph node cells cultured on monolayers of CVB3-infected target cells for 5 days generated significant cytolytic activity to myocyte targets. Such *in vitro* induction of CTL could be inhibited by coculture with T-cell-enriched spleen cells of immune females. Nonimmune female cells were not inhibitory.[34] A similar suppressive activity was tentatively identified in male mice infected with a substrain variant of CVB3 that induced little or no CTL activity and resulted in myocarditis.

4.1. Role in Resistance and Recovery

The multiplicity of host factors involved in both resistance and recovery from viral infections is well appreciated. Normally all these factors work together, so that any attempt to evaluate their relative importance by dissecting the immune system is artifactual and any conclusion an oversimplification. It is generally accepted, however, that, by and large, viral infections may be grouped according to whether their control is more dependent on humoral (e.g., poliomyelitis, yellow fever, dengue) or CMI mechanisms (e.g., herpes simplex, vaccinia, ectromelia). From what is known of the murine model, CV seem to belong to the first group. Comparison of CVB3 and CVB4 infections in T-cell-deficient and -sufficient mice has yielded somewhat contradictory results[35] (see Table I), but it would appear that the former group have no major defects in their ability to resist acute CVB challenge. Thus, the prevailing view is that CMI plays only a small part in the host's resistance to CV.

A role of T cells in resistance to CV is, however, likely. Macrophages and possibly other cellular components contribute to clear tissues from infecting CV. Whether such cells need to be activated by T-cell products, such as interleukin-2 (IL-2) and γ-interferon (IFN_{γ}), has not been established.[35] In mice rendered more susceptible to CVB3 by malnourishment, the severity of infection was reversed by the adoptive transfer of immune lymphoid cells but was not affected by the levels of humoral antibody in their blood.[22] Likewise, cyclosporin, an inhibitor of IL-2 production and other T-lymphocyte functions, given daily to CVB3-infected adult mice, resulted in augmented myocyte necrosis and mortality and in prolonged virus detectability.[36] In T-cell-deficient mice, CVB have been seen to persist longer than in fully immunocompetent animals, suggesting that intact T-cell function contributes significantly to terminate CV infections. It remains to be determined whether these effects are mediated by the already discussed thymus dependence of the antibody response to CVB or through CMI mechanisms.

4.2. Role in Tissue Damage

That CMI is involved in the genesis of CV-induced myocarditis in mice is suggested by several lines of evidence. Woodruff[32] and more recently Huber *et al.*[33] and others have provided convincing evidence that H-2 restricted CTL are implicated in the disease process (Chapter 7). T-cell-deprived mice frequently exhibited reduced cardiac pathology following CVB3 infection (Table I). A relationship has often been noted between development of CTL and severity of myocardiopathy. For example, cardiac necrosis induced by a cardiotropic variant of CVB3 (CVB3m) was greater in males than in females, which develop little or no CTL activity. Mortality, myofiber necrosis, and CTL development were reduced in males castrated prior to infection and returned to normal levels following treatment with testosterone. Similarly, treatment with the immunopotentiating drug levamisole increased the number of myocarditic lesions induced by CVB3m and stimulated spleen cells from these mice to greater cytotoxic activity against infected target cells.[37]

The nature of the antigen(s) serving as a target for the CTL in the CVB-infected heart is unclear. The question is a major one, because unenveloped viruses such as CV are not known to encode proteins that alter cell membranes and/or might be recognized by CTL and other cytotoxic cells. In the original work of Woodruff,[32] anti-CV antisera was unable to block immune spleen cell cytotoxicity. By hypertonic KCl extraction of heart tissue of CVB3m-infected mice and baboons, Paque *et al.*[30] obtained a novel antigen ($\sim$50,000 M_r), which reacted specifically with PEC of CVB3m-infected or immunized mice in the MIF assay but did not crossreact with CVB3 virion antigens. Recently, Lutton and Gauntt[38] demonstrated that CVB3-induced changes on the plasma membrane of murine neonatal skin fibroblasts, including the appearance of novel sugar residues. The relationship between these changes and vulnerability to cytotoxicity remains to be determined. However, these results might indicate that autoimmune processes triggered by CVB3 in mice encompass CMI mechanisms, as well as the humoral mediators discussed in Section 3.2.

To further complicate the issue, it was recently shown that mice infected with an amyocarditic variant of CVB3m acquired both new heart tissue antigen and specific CTL capable of recognizing such antigen and, nevertheless, developed no myocarditis. Thus, it would appear that such changes do not suffice for the induction of myocarditis. It seems likely that humoral and cellular factors act in concert in the production of CV-triggered tissue damage as well as in anti-CV resistance.[32,40] In any case, the clinical importance of CVB-induced cardiopathies encourages further effort in elucidating the underlying mechanisms. Investigation of immune pathways to CV-induced pathology might help developing rational means of management and prevention.

5. COXSACKIEVIRUS REPLICATION IN IMMUNOCOMPETENT CELLS

The encounter of viruses with immunocompetent cells may result in a wide spectrum of possible outcomes, which may account for differences in ultimate host susceptibility and viral pathogenicity.[41] Nevertheless, there is little information on CV interactions with such cells, especially in humans.

5.1. Lymphocytes

Inference from what is known of poliovirus infection has led to the assumption that primary multiplication of CV and related enteroviruses occurs in lymphatic tissues of the pharynx and gut. It is from these sites that CV are assumed to spread to target organs, such as liver, myocardium, pancreas, and CNS, giving rise to the varied clinical manifestations that can be associated with infection. Yet, while poliovirus has been shown to replicate in antigen-stimulated human lymphocytes, at least *in vitro,*[41] there is no direct evidence that CV infect lymphoid cells.

Apart from a concise report showing that freshly cultured human peripheral blood mononuclear cells shortened CVB3 survival *in vitro,*[5] the only hitherto published approach to this matter is an indirect one. A panel of cultured human lymphoid lines of B- and T-cell origin was challenged with the six CVB serotypes.[42] Most virus–cell combinations proved permissive but, on the whole, B-cell lines appeared somewhat more susceptible to CVB infection than did T-cell lines (Table II). In many cases, propagation of the infected cells led to the establishment of

TABLE II
Response of Human Lymphoid Cell Lines to Acute CVB Infection

	Infecting virus (viral yield[a]/viable cell density[b])					
Cell line (origin)	CVB1	CVB2	CVB3	CVB4	CVB5	CVB6
MOLT-4 (T)	+/+	+/+	+/+	±/+	±/+	±/−
CCRF CEM (T)	−/+	−/+	±/+	±/+	−/+	−/+
Ramos (B)	+/±	−/±	+/±	+/+	±/−	±/−
Jiyoye (B)	+/+	−/+	+/−	−/+	±/+	+/−
P3HR-1 (B)	+/−	±/±	±/−	−/−	+/−	±/−

[a]Viral yield (log $TCID_{50}$) at day 3 postinfection: +, >5.0; ±, 2.5–5.0; −, <2.5.
[b]Viable cell density at day 3 postinfection (percentage of uninfected cultures): +, >75%; ±, 25–75%; −, <25%.

long-term persistently infected cultures. This finding was interesting because cell permissiveness to CVB replication is usually associated with complete cytolysis of the cultures.

At least in the virus–cell combinations most thoroughly characterized, persistence was maintained by a carrier culture mechanism involving viral spread through the medium and replication in a minority of the cells at any given time. The properties of the persistently infected cultures (Table III) showed that known mechanisms of viral persistence in tissue cultures, such as IFN and defective interfering particles, did not appear to be involved in this system. Establishment and maintenance of persistence of CVB in lymphoid lines appears to be due to the presence and continuous replenishment of a small fraction of susceptible variants among a population of otherwise intrinsically resistant cells. Because viral receptors have been shown to be so important in initiating and maintaining CV infections (Chapter 4), susceptibility or resistance might involve differences at the adsorption-penetration step.

Whether these findings have any relevance to natural CV infections in man is presently unknown. They suggest, however, that lymphocytes might represent a chronic reservoir for CV rather than a substratum for rapid CV amplification. Evidence that CV can persist *in vivo* is limited, but the list of picornaviruses that have been shown to give chronic infections is steadily increasing. Since cells of lymphoid lines differ considerably from normal lymphocytes, these aspects warrant further investigation using freshly explanted human lymphocytes.

Even in mice, the susceptibility of lymphocytes to CV has not been thoroughly studied. Early investigators had detected high-titer CV in extracts of murine lymphoid organs and assumed that they resulted from local viral replication. Later studies[43] showed that the situation can be entirely different. Infectious center assay performed at various times postinfection detected less than one virus-producing cell per 10^6 thymus, spleen, and lymph node cells of CVB3-infected adult mice. Immunofluorescent staining of infected thymuses and spleens with two polyclonal anti-CVB3 sera gave substantially negative results and when the experiment was repeated using monoclonals, only a few scattered viral antigen-positive cells could be detected. The positive cells were present from a few hours postinfection; their number remained stable for several days. In addition, repeated attempts to grow CVB3 on primary cultures of mitogen-stimulated and -unstimulated murine lymphoid cells or to develop *in vivo* variants with increased lymphotropism gave consistently negative results. Since lymphoid cells of adult mice are refractory to CVB3, the bulk of viral infectivity found in extracts of lymphoid organs derives from either viremia or drainage of infected nonlymphoid tissues, or both.

TABLE III
Properties of Human Lymphoid Cell Lines Persistently Infected with CVB

Recently cloned virus establishes persistence as easily as uncloned virus.

Cultures produce virus for months or years without noticeable changes of morphology or proliferation.

During early passages, viral titers fluctuate irregularly without relation to cell viability, then stabilize.

At any given time, virus-producing cells range between 0.01 and 3%, with a tendency to decrease.

Culture fluids do not contain viral inhibitors.

Relative to input virus, chronic virus becomes progressively more temperature sensitive, more virulent for uninfected lymphoid cells, and attenuated for mice, and produces smaller plaques.

Virus cell equilibrium is not affected by anti-IFN antibodies, mitogens, interleukins, and so forth.

Infection can be cured with exogenous IFN or antiviral antibody.

Antiserum-cured cells show a slightly enhanced resistance to superinfection with the homologous CVB. Spontaneously cured cells become more resistant to heterologous CVB as well.

5.2. Macrophages

While there is evidence that human macrophages *in vitro* are permissive for polioviruses and some echoviruses and may represent early sites of virus amplification in infected individuals,[41] there are no studies of human macrophages as a possible target for CV replication. Murine macrophages appear not to support CV growth. Infectious center assays detected no virus-producing cells in PEC harvested from mice at various times after CVB3 infection.[4] Gauntt *et al.*[44] found that murine macrophages can take up and carry CVB3 for at least 24 hr but no evidence that myocarditic and amyocarditic substrains of CVB3 replicated in cultured lymph node macrophages. We obtained similar results with peritoneal macrophages (D. Matteucci and M. Bendinelli, unpublished data). Subsequent studies[45] have suggested an important role for endogenously produced IFN in maintaining the nonpermissive state of macrophages to a number of viruses, including one picornavirus. It is not known whether endogenous IFN is also responsible for the apparent resistance of murine macrophages to CV.

6. IMMUNODEPRESSION BY COXSACKIEVIRUSES

That viral infections can impair immunity is well known, but studies in this field with enteroviruses are few.[41] The first indication that CV can affect the functioning of the immune system came from studies showing that adult BALB/c mice infected with certain group A and B serotypes exhibited reduced antibody and CMI responses to unrelated antigens.[46] Subsequent investigations have focused on the antibody responsiveness of CVB3-infected mice.[47] Spleen cultures of infected mice produced fewer antibody-forming cells than did cultures from mock-infected animals. The impairment peaked 1 week postinfection and then remained unchanged over the second week. The reduced number of antibody-forming cells produced was not due to a shift in the kinetics of the response, could not be overcome by increasing antigen dosage, and was not sustained by decreased cell viability. While primary responses to thymus-dependent antigens were depressed by 50–75%, secondary responses and primary responses to thymus-independent antigens remained within the normal range.

Interestingly, starting at day 6 postinfection, a proportion of infected mice showed a progressive involution of lymphoid organs (Section 7), but immunodepression could not be attributed to such changes because cells from organs with normal weight and histology were also hyporesponsive.

The cellular basis of CVB3-induced immunosuppression was examined by conventional cellular immunology methods. The results demonstrated the existence of an accessory cell deficit (Table IV). Owing to the multiplicity and complexity of activities exerted by macrophages in the induction of antibody responses, it has not been possible to identify with certainty the exact macrophage function altered by CVB3 infection. An indication may be the failure of peritoneal macrophages, as compared with splenic macrophages, to restore the antibody responsiveness of infected cells. Since IA-positive cells represent 60–70% of splenic macrophages and only 10–15% of peritoneal macrophages, this finding might imply that CVB3 infection affects the antigen-dependent interaction between macrophages and T lymphocytes. Poliovirus-induced de-

TABLE IV
Evidence That Splenic Macrophages of CVB3-Infected Mice Are Impaired

Antibody reponses requiring little macrophage help are not depressed.
Adherent spleen cells of infected mice do not restore responsiveness of macrophage-depleted normal cells.
Macrophage-depleted cells of infected mice respond well when supplemented with normal adherent spleen cells.
Total spleen cells of infected mice respond well when supplemented with normal macrophage-enriched cells.
Electron microscopy shows that spleen macrophages contain phagocyted degenerating cells and are crammed with phagosomes.

TABLE V
Evidence That Spleen Cells of Infected Mice Contain Activated Nonspecific T Suppressor Cells

T-cell-enriched spleen cells of infected mice do not restore antibody responsiveness of normal T-cell-depleted spleen cells.
Total and T-cell-enriched spleen cells from infected mice suppress responsiveness of normal spleen cells in a dose-dependent manner.
T-cell-depleted spleen cells of infected mice do not suppress responsiveness of normal spleen cells.
Spleens contain enhanced proportions of $Lyt\text{-}1^{-}2^{+}$ cells.

pression of human lymphocyte mitogenesis has also been linked to a defect of antigen-presenting macrophages.[41] Suppression of mitogen-induced blastogenesis has been reported as well for CV.[37]

In addition, CVB3-infected spleens contained antigen-nonspecific suppressor cells, characterized as T lymphocytes (Ts) (Table V). However, there was no significant correlation between the degree of antibody hyporesponsiveness and the degree of suppressive activity detected in individual mice. Since normal macrophages restored the responsiveness of Ts-containing cultures, this might indicate that the macrophage defect is the direct cause of reduced responsiveness and Ts either are not important or act through the mediation of macrophages.

Coxsackievirus-infected patients can exhibit a transient mild granulocytopenia.[48] After exposure to CVB3 and other enteroviruses *in vitro,* human granulocytes have been shown to undergo substantial structural changes of the plasma membrane, increased adherence to endothelial cells, and lower responsiveness to chemotactic stimuli.[49] These changes might at least partly explain the scarcity of polymorphonuclear cells in CV-induced lesions.

We know of no evidence that suggests that CV suppress immunity in humans. There are, however, early observations that concomitant CV infections can lead to unusually severe poliovirus and protozoan infections[46]; recently, a case was reported of a CVB4-infected child in whom a fulminant pulmonary coccidioidomycosis developed and who mounted a defective CMI response to the latter pathogen.[50]

7. HISTOPATHOLOGY OF LYMPHOID ORGANS IN COXSACKIEVIRUS-INFECTED HOSTS

The histopathology of lymphoid tissues during CV infections of humans is little understood. The immune system of CV-infected mice is not only functionally altered but can also undergo a profound involution. In a systematic study of these aspects,[43] CVB3-infected mice exhib-

TABLE VI
Involution of Lymphoid Organs during CVB3 Infection of BALB/c Mice

Day post-infection	Mean weight (% of control)			
	Thymus	Spleen	Lymph nodes	Whole body
2	88	133	137	100
4	62	110	107	90
6	26	68	70	83
8	18	47	69	78
10	12	35	18	83
12	12	31	30	96

ited a dose- and time-dependent involution of the thymus, spleen, and lymph nodes. Susceptibility to systemic lymphatic atrophy was scarcely influenced by gender and age but appeared to be genetically determined. A correlation with severity of disease was observed in different animal strains. In susceptible mice, the thymus was the first organ to be affected, since a moderate degree of atrophy was already evident by 2 days postinfection, whereas in the spleen and lymph nodes the effect became noticeable at about 1 week postinfection (Table VI). Histologically, the thymus showed an early depletion of cortical thymocytes and loss of distinction at the corticomedullary junction, followed by cellular depletion of the medulla (Fig. 1). In the peripheral lymphoid organs, there was a generalized cellular depletion. Fibrosis was the eventual outcome, in either case. Counts of spleen cells showed a normal viability of residual cells and documented a parallel decline of the various lymphocyte classes and subclasses studied, except for an early temporary rise in Lyt-1^-2^+ cells. By electron microscopy, the only appreciable changes were in macrophages, which often contained phagocytized degenerating cells and were crammed with phagosomes. These effects could be reproduced (although with variable efficiency) with a number of different isolates and clones, but not with an attenuated strain of CVB3. A certain degree of thymus and spleen atrophy was also noted following CVB1 infection. CVB5 produced thymus atrophy only. A decrease of T cells has been noted in the lymph nodes of CVB3-infected mice.[37]

8. MECHANISMS AND SIGNIFICANCE OF DAMAGE TO THE IMMUNE SYSTEM

Alterations of lymphoid tissue architecture and cellularity have been described in several virus-infected hosts. These include mice in-

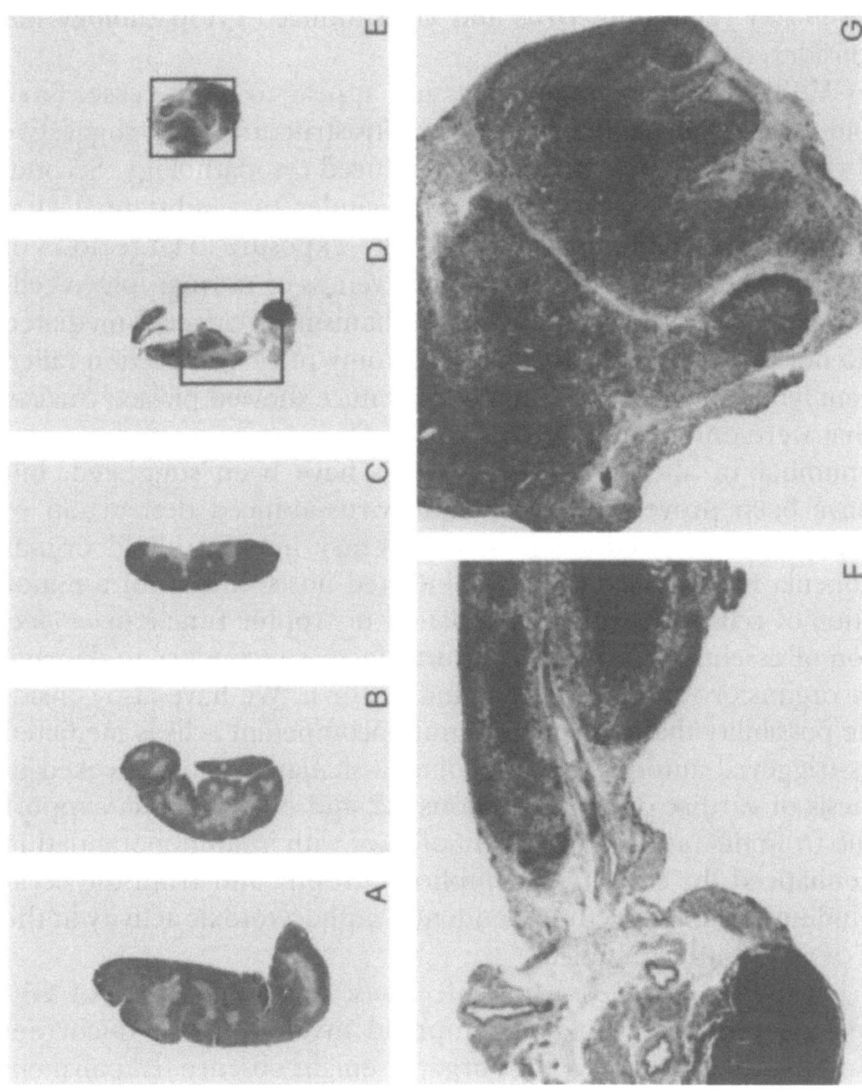

FIGURE 1. Representative thymuses of CVB3-infected mice at days 0 (**A**), 2 (**B**), 4 (**C**), 6 (**D,F**), and 8 (**E,G**); hematoxylin and eosin. (**A–E**) ×4.2; (**F,G**) ×50.

fected with lymphocytic coriomeningitis virus, Sendai virus, cytomegalovirus, and dengue virus; hamsters infected with selected arenaviruses; and humans infected with measles virus, immunodepressive retroviruses, and Lassa virus. It is also well appreciated that a variable degree of immunodepression is produced by most viral infections. Although the flurry of research in this field in recent years has not eliminated the uncertainties on the mechanisms involved, in many instances a direct effect of locally replicating virus and virus-induced cytopathology has been considered a major factor.[41]

In CVB3-infected mice, this does not appear to be the case. First, lymphoid tissues showed no histologic or ultrastructural signs suggestive of local viral replication or direct virus-induced cytopathology. Second, solid evidence (discussed in Section 5) excludes that substantial viral replication takes place in such organs. Finally, exposure to large doses of virus did not modify the antibody responsiveness of normal spleen cells *in vitro*. Data also indicate that classic mechanisms of adrenal-mediated stress do not play a major role. Adrenalectomy prior to infection failed to prevent lymphoid atrophy, and infected mice showed plasma cortisol levels that were only moderately increased.[43]

A number of alternative explanations have been suggested, but none have been proved. These include virus-induced destruction of lymphoid precursors, reduction of their entry into lymphoid organs (lymphopenia has been noted in CV-infected hosts[51]), lysis of a minor population of cells with immunoregulatory or trophic functions *in vivo*, depletion of essential nutrients or growth factors secondary to damage to other organs or vasal endothelia, and so forth. We have also considered the possibility that damage to immunocompetent cells is mediated by virus-triggered autoreactive phenomena similar to those invoked in the genesis of cardiac damage (Sections 3.2 and 4.2). Indirect support has come from the fact that treatment of mice with immunopotentiating agents enhanced the severity of lymphoid atrophy and from the occasional finding of complement-dependent lymphocytotoxic activity in the plasma of CVB3-infected mice.

As is the case for most viral infections, the implications of CV-induced immunodepression and lymphoid involution on concurrent CV-induced pathogenesis to other organs remain obscure. Impairment of immune functions is considered a mechanism by which viruses can dodge the immune defenses. However, since immunopathology is believed to play a significant role in CV and post-CV pathogenesis, the balance between detrimental and beneficial effects might be a positive one for the host. This calls for great attention in the therapeutic use of immunomodulatory agents in the treatment of CV-induced disease.

9. CONCLUDING REMARKS

This chapter discusses available information on the interplay CV establish with the host's immune system. Clearly, there is a huge disproportion between the clinical importance of these viruses and our knowledge of how the immune system behaves during the course of infection. The progress achieved in recent years is limited but nevertheless has opened new interesting fields to our vision and has made it evident that a better understanding of CV immunobiology could redirect current thinking of pathogenesis by these viruses.

ACKNOWLEDGMENTS. This work was supported by grants from the Italian National Research Council, Special Project Control of Infectious Diseases, and from the Ministry of Education.

REFERENCES

1. Rager-Zisman, B., and Allison, A. C., 1973, The role of antibody and host cells in the resistance of mice against infection by coxsackie B-3 virus, *J. Gen. Virol.* **19:**329–338.
2. Woodruff, J. F., 1979, Lack of correlation between neutralizing antibody production and suppression of coxsackievirus B-3 replication in target organs: Evidence for involvement of mononuclear inflammatory cells in host defences, *J. Immunol.* **123:**31–36.
3. Costanzo-Nordin, M. R., Reap, E. A., O'Connell, J. B., Robinson, J. A., and Scanlon, P. J., 1985, A nonsteroid anti-inflammatory drug exacerbates coxsackie B-3 murine myocarditis, *J. Am. Coll. Cardiol.* **6:**1078–1082.
4. Kabiri, M., and Hadaegh, M. D., 1977, Interaction of coxsackievirus B3 and peritoneal exudate cells of adult mice treated with cyclophosphamide, *J. Med. Virol.* **1:**183–191.
5. Denman, A. M., Rager-Zisman, B., Merigan, T. C., and Tyrrell, D. A. J., 1974, Replication or inactivation of different viruses by human lymphocyte preparations, *Infect. Immun.* **9:**373–376.
6. Loria, R. M., Kibrik, S., and Broitman, S. A., 1974, Peroral infection with group B coxsackievirus in the adult mouse: Protective functions of gut, *J. Infect. Dis.* **130:**539–543.
7. Godeny, E. K., and Gauntt, C. J., 1986, Involvement of natural killer cells in coxsackievirus B3-induced murine myocarditis, *J. Immunol.* **137:**1695–1702.
8. Hyöty, H., Huupponen, T., Kotola, L., and Leinikki, P., 1986, Humoral immunity against viral antigens in type 1 diabetes: Altered IgA-class immune response against coxsackie B4 virus, *Acta Pathol. Microbiol. Immunol. Scand. Sect. C* **94:**83–88.
9. Welliver, R. C., Drucker, M. M., and Ogra, P. L., 1982, Immunology of enteroviruses, in: *Comprehensive Immunology—Immunology of Human Infection.* Part II: *Viruses and Parasites; Immunodiagnosis and Prevention of Infectious Disease* (A. J. Nahmias and R. J. O'Reilly, eds.), pp. 185–203, Plenum Medical, New York.
10. Bell, E. J., and McCartney, R. A., 1984, A study of coxsackie B virus infections, 1972–1983, *J. Hyg. (Camb.)* **93:**197–203.
11. Tilzey, A. J., Signy, M., and Banatvala, J. E., 1986, Persistent coxsackie B virus specific IgM response in patients with recurrent pericarditis, *Lancet* **1:**1491–1492.
12. Wolfgram, L. J., Beisel, K. W., Herskowitz, A., and Rose, N. R., 1986, Variations in the

susceptibility to coxsackievirus B3-induced myocarditis among different strains of mice, *J. Immunol.* **136:**1846–1852.

13. Schernthaner, G., Banatvala, J. E., Scherbaum, W., Bryant, J., Borkenstein, M., Schober, E., and Mayr, W. R., 1985, Coxsackie-B-virus-specific IgM responses, complement-fixing islet-cell antibodies, HLA DR antigens, and C-peptide secretion in insulin-dependent diabetes mellitus, *Lancet* **2:**630–632.
14. Schnurr, D. P., Cao, Y., and Schmidt, N. J., 1984, Coxsackievirus B3 persistence and myocarditis in N:NIH(S) II nu/nu mice, *J. Gen. Virol.* **65:**1197–1201.
15. Schnurr, D. P., and Schmidt, N. J., 1984, Coxsackievirus B3 persistence and myocarditis in NFR nu/nu and +/nu mice, *Med. Microbiol. Immunol.* **173:**1–7.
16. Hashimoto, I., Tatsumi, M., and Nakagawa, M., 1983, The role of T lymphocytes in the pathogenesis of coxsackie virus B3 heart disease, *Br. J. Exp. Pathol.* **64:**497–504.
17. Robinson, J. A., O'Connell, J. B., Roeges, L. M., Major, E. O., and Gunnar, R. M., 1981, Coxsackie B3 myocarditis in athymic mice, *Proc. Soc. Exp. Biol. Med.* **166:**80–91.
18. Woodruff, J. F., and Woodruff, J. J., 1974, Involvement of T lymphocytes in the pathogenesis of coxsackie virus B_3 heart-disease, *J. Immunol.* **113:**1726–1734.
19. Cao, Y., Schnurr, D. P., and Schmidt, N. J., 1984, Monoclonal antibodies for study of antigenic variation in coxsackievirus type B4: Association of antigenic determinants with myocarditic properties of the virus, *J. Gen. Virol.* **65:**925–932.
20. Griffith, J. S., Katz, S. L., and Moore, M., 1977, Persistent enterovirus infections in agammaglobulinemia, in: *Microbiology—1977* (D. Schlessinger, ed.), pp. 488–493, American Society for Microbiology, Washington, D.C.
21. Johnson, J. P., Yolken, R. H., Goodman, D., Winkelstein, J. A., and Nagel, J. E., 1982, Prolonged excretion of group A coxsackievirus in an infant with agammaglobulinemia, *J. Infect. Dis.* **146:**712.
22. Woodruff, J. F, and Woodruff, J. J., 1971, Modification of severe coxsackievirus B3 infection in marasmic mice by transfer of immune lymphoid cells, *Proc. Nat. Acad. Sci. USA* **68:**2108–2111.
23. Cho, C. T., Feng, K. K., McCarthy, V. P., and Lenahan, M. F., 1982, Role of antiviral antibodies in resistance against coxsackievirus B3 infection: Interaction between preexisting antibodies and an interferon inducer, *Infect. Immun.* **37:**720–727.
24. Kogon, A., Spigland, I., Frothingham, T. E., Elveback, L., Williams, C., Hall, C. E., and Fox, J. P., 1969, The Virus Watch Program: A continuing surveillance of viral infections in metropolitan New York families. VII. Observations on viral excretion, seroimmunity, intrafamilial spread and illness association in coxsackie and echovirus infections, *Am. J. Epidemiol.* **89:**51–61.
25. Gauntt, C. J., Paque, R. E, Trousdale, M. D., Gudvangen, R. J., Barr, D. T., Lipotich, G. J., Nealon, T. J., and Duffey, P. S., 1983, Temperature-sensitive mutant of coxsackievirus B3 establishes resistance in neonatal mice that protects them during adolescence against coxsackievirus B3-induced myocarditis, *Infect. Immun.* **39:**851–864.
26. Sun, S.-C., Burch, G. E., Sohal, R. J., and Chu, K.-C., 1967, Coxsackie B4 viral nephritis in mice and its autoimmune-like phenomenon, *Proc. Soc. Exp. Biol. Med.* **126:**882–885.
27. Maisch, B., Trostel-Soeder, R., Stechemesser, F., Berg, P. A., and Kochsiek, K., 1982, Diagnostic relevance of humoral and cell-mediated immune reactions in patients with acute viral myocarditis, *Clin. Exp. Immunol.* **48:**533–545.
28. Saegusa, J., Prabhakar, B. S., Essani, K., McClintock, P. R., Fukuda, Y., Ferrans, V. J., and Notkins, A. L., 1986, Monoclonal antibody to coxsackievirus B4 reacts with myocardium, *J. Infect. Dis.* **153:**372–373.
29. Bruserad, G., Stevensen, M., and Thorsby, E., 1985, T lymphocyte responses to cox-

sackie B4 and mumps virus. II. Immunoregulation by HLA-DR3 and DR4 associated restriction elements, *Tissue Antigens* **26:**179–192.

30. Paque, R. E., Gauntt, C. J., Nealon, T. J., and Trousdale, M. D., 1978, Assessment of cell-mediated hypersensitivity against coxsackievirus B3 viral induced myocarditis utilizing hypertonic salt-extracts of cardiac tissue, *J. Immunol.* **120:**1672–1678.
31. Paque, R. E., Gauntt, C. J., and Nealon, T. J., 1981, Assessment of cell-mediated immunity against coxsackievirus B3-induced myocarditis in a primate model (*Papio papio*), *Infect. Immun.* **31:**470–479.
32. Woodruff, J. F., 1980, Viral myocarditis, *Am. J. Pathol.* **101:**426–479.
33. Huber, S. A., Job, L. P., and Auld, K. R., 1982, Influence of sex hormones on coxsackie B-3 virus infection in Balb/c mice, *Cell. Immunol.* **67:**173–189.
34. Job, L. P., Lyden, D. C., and Huber, S. A., 1986, Demonstration of suppressor cells in coxsackievirus group B, type 3 infected female Balb/c mice which prevent myocarditis, *Cell Immunol.* **98:**104–113.
35. Khatib, R., Khatib, G., Chason, J. L., and Lerner, A. M., 1983, Alterations in coxsackievirus B4 heart muscle disease in ICR Swiss mice by anti-thymocyte serum, *J. Gen. Virol.* **64:**231–236.
36. O'Connell, J. B., Reap, E. A., and Robinson, J. A., 1986, The effects of cyclosporine on acute murine coxsackie B3 myocarditis, *Circulation* **73:**353–359.
37. Gudvangen, R. J., Duffey, P. S., Paque, R. E, and Gauntt, C. J., 1983, Levamisole exacerbates coxsackievirus B3-induced murine myocarditis, *Infect. Immun.* **41:**1157–1165.
38. Lutton, C. W., and Gauntt, C. J., 1986, Coxsackievirus B3 infection alters plasma membrane of neonatal skin fibroblasts, *J. Virol.* **60:**294–296.
39. Lutton, C. W., Gudvangen, R. J., Nealon, T. J., Paque, R. E, and Gauntt, C. J., 1985, Cellular immune responses in mice challenged with an amyocarditic variant of coxsackievirus B3, *J. Med. Virol.* **17:**345–357.
40. Huber, S. A., and Lodge, P. A., 1986, Coxsackievirus B-3 myocarditis. Identification of different pathogenic mechanism in DBA/2 and BALB/c mice, *Am. J. Pathol.* **122:**284–291.
41. Friedman, H., Bendinelli, M., and Specter, S. (eds.), 1987, *Handbook on Viral Immunosuppression*, Dekker, New York, (in press).
42. Matteucci, D., Paglianti, M., Giangregorio, A. M., Capobianchi, M. R., Dianzani, F., and Bendinelli, M., 1985, Group B coxsackieviruses readily establish persistent infections in human lymphoid cell lines, *J. Virol.* **56:**651–654.
43. Matteucci, D., Toniolo, A., Conaldi, P. G., Basolo, F., Gori, Z., and Bendinelli, M., 1985, Systemic lymphoid atrophy in coxsackievirus B3-infected mice: Effects of virus and immunopotentiating agents, *J. Infect. Dis.* **151:**1100–1108.
44. Gauntt, C. J., Trousdale, M. D., LaBadie, D. R. L., Paque, R. E., and Nealon, T., 1979, Properties of coxsackievirus B3 variants which are amyocarditic or myocarditic for mice, *J. Med. Virol.* **3:**207–220.
45. Belardelli, F., Vignaux, F., Proietti, E., and Gresser, I., 1984, Injection of mice with antibody to interferon renders peritoneal macrophages permissive for vesicular stomatitis virus and encephalomyocarditis virus, *Proc. Natl. Acad. Sci. USA* **81:**602–606.
46. Bendinelli, M., Ruschi, A., Campa, M., and Toniolo, A., 1975, Depression of humoral and cell-mediated immune responses by coxsackieviruses in mice, *Experientia* **31:**1227–1229.
47. Bendinelli, M.. Matteucci, D., Toniolo, A., Patané, A. M., and Pistillo, M. P., 1982, Impairment of immunocompetent mouse spleen cell functions by infection with coxsackievirus B3, *J. Infect. Dis.* **146:**797–805.

48. Young, N. A., 1979, Coxsackievirus and echovirus, in: *Principles and Practice of Infectious Disease,* Vol. II (G. L. Mandell, R. G. Douglas, and J. E. Bennett, eds.), pp. 1104–1120, Wiley, New York.
49. Kirkpatrick, C. J., Bültmann, B. D., and Gruler, H., 1985, Interaction between enteroviruses and human endothelial cells *in vitro.* Alterations in the physical properties of endothelial cells plasma membrane and adhesion of human granulocytes, *Am. J. Pathol.* **118:**15–25.
50. Gururaj, V. J., Marsh, W. W., and Aiyar, S. R., 1984, Fulminant pulmonary coccidioidomycosis in association with coxsackie B_4 infection, *Clin. Pediatr.* **24:**406–408.
51. Gomez, M. P., Reyes, M. P., Smith, F., Ho, L. K., and Lerner, A. M., 1980, Coxsackievirus B3-positive mononuclear leukocytes in peripheral blood of Swiss and athymic mice during infection, *Proc. Soc. Exp. Biol. Med.* **165:**107–113.

7

The Role of Immune Mechanisms in Pathogenesis

SALLY ANN HUBER

1. INTRODUCTION

Although most coxsackievirus (CV) infections in humans are mild and frequently asymptomatic, in some individuals more serious manifestations develop, including meningitis, paralysis, encephalitis, type I (insulin-dependent) diabetes mellitus (IDDM), and myocarditis.[1–3] Association of particular viral infections with certain diseases usually depends on demonstration of rising antibody titers to the pathogen and a compatible clinical history. In both IDDM and myocarditis, approximately 50% of recently diagnosed patients demonstrate two- to fourfold rises in CV-specific antibodies compared with less than 20% of control patients.[2,3] Furthermore, several studies report dramatic increases in new diabetes and myocarditis cases coinciding with known CV epidemics and also demonstrate definite seasonality for these diseases (fall and winter) that correlate remarkably well with the seasonal distribution of picornavirus infections.[2,3] Despite the nearly convincing circumstantial evidence of CV etiology in myocarditis and diabetes, infectious virus is only rarely isolated from patients with active disease.[3]

2. ROLE OF VIRUS IN DISEASE PATHOGENESIS

Coxsackievirus can induce direct tissue injury by virus-mediated destruction of infected cells. Their basic replicative cycle consists of ad-

SALLY ANN HUBER • Department of Pathology, University of Vermont, Burlington, Vermont 05405.

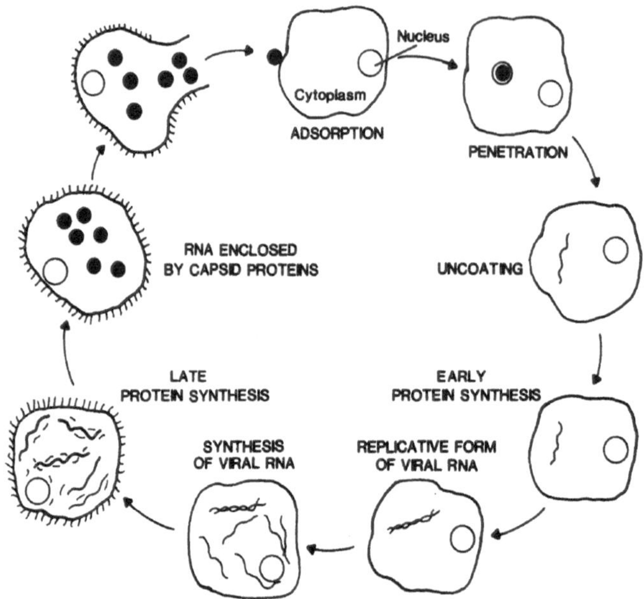

FIGURE 1. Schematic representation of coxsackievirus replication cycle.

sorption of the virus to specific receptors on the cell surface, followed by entry of the viral nucleic acid into the cell cytoplasm, replication of viral genome and proteins, and production of progeny virions. Frequently, infectious virions remain in the cytoplasm forming identifiable crystalline structures until rupture of the cell membrane brings release (Fig. 1). How rupture of the membrane occurs remains unclear, but it must be under viral control since the time required for cell lysis varies between picornaviruses.[4]

Although budding viruses insert well-defined viral antigens in the cell membrane, picornaviruses do not. Nonetheless, some virus-induced alterations in the membrane must occur. KCl extracts of cardiac tissue from CV-infected mice stimulate delayed hypersensitivity responses (migration inhibition factor release) by lymphocytes from the sensitized animals.[5] Furthermore, cytolytic T lymphocytes (CTL) from CV-infected mice can readily lyse infected but not uninfected cells[6] and distinguish between cells infected with CV and other viruses.[7] The novel antigen induced on the infected cell surface and recognized by the immune system has not been characterized; however, growing evidence supports the concept that the cell not the viral genome codes for this antigen. Hyperimmune antibody to the virion does not bind to CV-infected cells as determined by immunofluorescence and immunoperoxidase-staining procedures.[8] Thus, the antigen is probably not a capsid protein. More recently, Lutton and Gauntt[9] showed that certain lectins interact specifically with virus-infected cells, indicating that the new antigen has polysaccharide components. Since picornaviruses are not known to produce glycoproteins, these workers suggest that alterations in cell metabolism instigated by the infection cause the antigenic changes recognized by the immune system. An alternate explanation could be cell-mediated glycosylation of virus proteins; however, this possibility seems remote. Indeed, our own unpublished work confirms that of Gauntt and colleagues.

DNA-initiated mRNA synthesis and host cell protein production are equally inhibited in myocytes treated with either actinomycin D or CV. Furthermore, actinomycin D treatment without viral infection produces the identical antigenic changes on the myocytes as CV. Thus, the immune system must recognize either degradation products or neoantigens present on the infected cell that directly result from inhibition of cellular protein synthesis.

3. EVIDENCE FOR IMMUNE PATHOGENESIS IN PICORNAVIRUS INFECTIONS

Although both the growth of the virus in the tissue and the immunity triggered by the infection could cause cell damage, current evidence predominantly implicates the immune system as the major pathogenic mechanism in certain CV-associated diseases. Murine models of CV-induced myocarditis and encephalomyocarditis (EMC) virus-induced IDDM provide significant evidence for immune pathogenesis in picornavirus diseases. Virus rapidly appears in the heart and pancreas after infection, usually peaks within 3–4 days, and is quickly eliminated within 2–3 weeks.[10–12] Early alterations in these organs consist of either scattered hypereosinophilic myocytes[3,13,14] or degranulated β-cells[15,16] and correspond to periods of maximal virus concentrations in the tissue. Inflammation begins toward the end of the first week and may persist for days or weeks after infectious virus is no longer evident. Mononuclear cell infiltrates are intimately associated with tissue damage,[3,14,16] but whether these cells functioned primarily in eliminating virus, limiting cellular damage and protecting the surrounding uninfected cells, or in producing tissue injury has remained unclear. Numerous studies using T-lymphocyte-deficient mice give the strongest evidence supporting lymphoid cell involvement in pathogenesis[13,17–19] (Fig. 2). Animals deprived of functional T cells either by the use of congenital athymic (nude) animals, by administration of rabbit antimouse thymocyte serum or by thymectomy, irradiation,and bone marrow reconstitution, develop minimal inflammation after infection but no permanent injury or scarring (fibrosis). Maximal virus concentrations in the tissues of immuno-

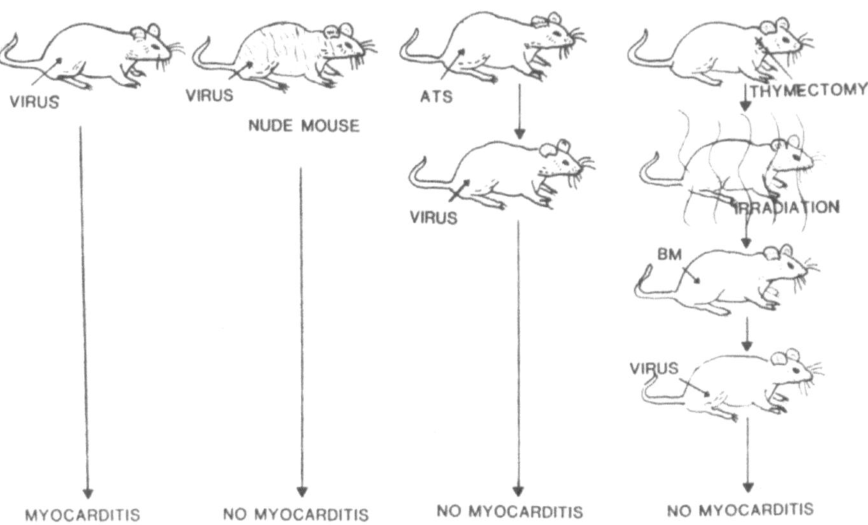

FIGURE 2. Experiments using T-lymphocyte-deficient mice demonstrate the importance of immunopathogenic mechanisms in CV-induced myocarditis.

logically intact and compromised mice remain equivalent and virus elimination occurs normally in both groups. These observations result in two conclusions. First, pathogenesis of picornavirus-induced diseases must be T-cell dependent, since little damage occurs in the absence of these lymphocytes. Second, T cells are not required for control of the viral infection. Indeed, natural killer (NK) cell, macrophage, and IgM virus neutralizing antibody primarily mediate virus elimination in picornavirus infection.[3]

Despite these studies, other investigators find myocarditis and diabetes unchanged in T-cell-deficient mice.[20,21] Although several factors may explain the reported discrepancies between these various groups, the most likely explanation is that multiple pathogenic mechanisms arise in individuals infected under different conditions. Only some mechanisms may involve T-cell-dependent immunity. Indeed, Craighead and co-workers[18] recently demonstrated a direct correlation between hyperglycemia and EMC virus concentrations in the pancreases of infected DBA/2 mice. T-cell depletion of these animals did not affect either the incidence or severity of diabetes. Furthermore, minimal insulitis is observed in immunologically intact DBA/2 islets, although extensive degranulation of the β-cells is noted. In a second strain, the pattern of disease differs strikingly. EMC virus infection of BALB/cBY mice produces severe hyperglycemia, but there is no correlation between glucose levels and virus concentrations in the pancreas. T-lymphocyte depletion of this strain results in marked reduction in diabetes susceptibility. Viral elimination occurs identically in T-lymphocyte-deficient and -sufficient mice of both strains, however, again demonstrating that T cells are not involved in control of viral infection. Thus, in this case, either virus or immune mechanisms could cause diabetes, and the particular pathogenic mechanism involved in the disease depends on the genetic background of the infected host.

4. MECHANISMS OF PATHOGENESIS

The immune system is complicated, and many factors triggered by viral infections could participate in tissue injury. Figure 3 illustrates the various potential mechanisms of cell lysis caused by infiltrating lymphoid cells in CV infections. Evidence for involvement of particular mechanisms is given below.

4.1. Cytolytic T Lymphocytes

Picornavirus-infected mice generate cytolytic T lymphocytes (CTL), which efficiently kill myocytes in animals developing myocarditis[3,6–8,13] and β-cells in animals developing diabetes.[22] CTL appear in peripheral lymphoid organs simultaneously with tissue injury in both models, and large numbers of these cells can be demonstrated in myocardial infiltrates using immunohistochemical staining procedures.[23] Furthermore, T-lymphocyte-depleted BALB/c mice demonstrate minimal cardiac in-

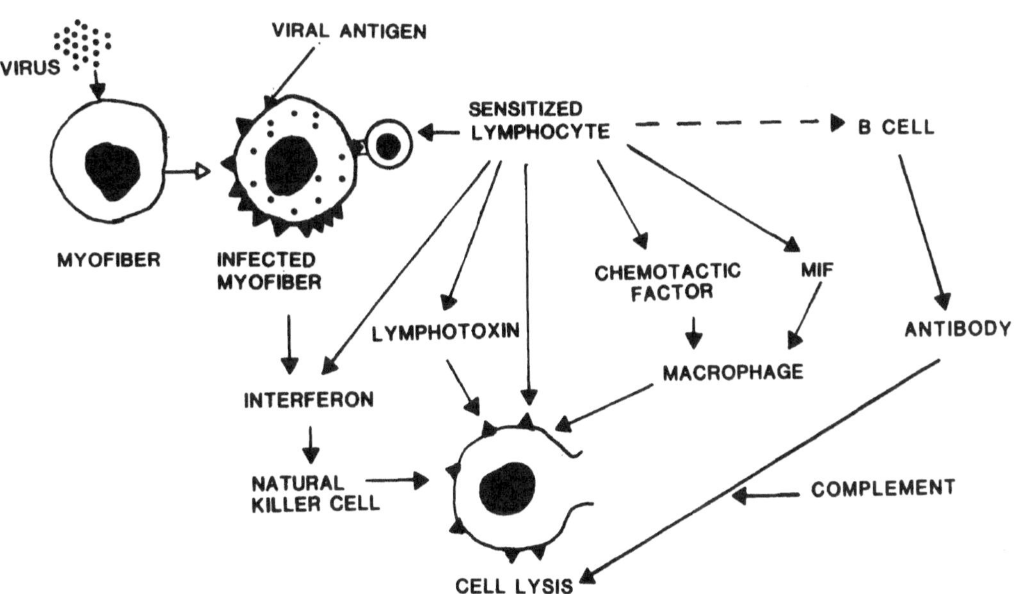

FIGURE 3. Possible mechanisms of myocyte lysis during CV infection.

jury after CV infections but, when adoptively transfused with highly purified populations of sensitized helper and cytolytic/suppressor cells, animals given cytolytic cells develop extensive myocardial lesions while mice given helper T (Th) cells remain protected.[24] Finally, BALB/c mice can be depleted of helper and cytolytic/suppressor cells *in vivo* using monoclonal antibodies to L3T4 and Lyt 2, the respective antigenic markers on the two subpopulations.[25,26] Only animals depleted of the cytolytic/suppressor cells fail to develop myocarditis.[27] Thus, multiple lines of evidence implicate cellular immunity as the predominant mediator of cardiac injury in BALB/c mice.

Two distinct CTL populations arise in CV-inoculated animals.[6] One CTL population lyses infected (virus-specific CTL) and the other uninfected (autoreactive CTL) myocytes. Virus-specific CTL must recognize antigens newly induced by the virus on the myocyte cell membrane, while the autoreactive CTL population recognizes antigens present on uninfected cardiocytes. What significance the two T-cell populations have in the pathogenesis of myocarditis remains obscure, but autoreactive CTL produce markedly more cardiac necrosis when transfused into infected T-lymphocyte-deficient mice than virus specific CTL. Second, autoreactive CTL can induce myocarditis in uninfected mice, while the virus-specific CTL cannot (J. Fohlman and S. Huber, unpublished observations). Thus, viral infection must initiate autoimmunity to tissue antigens and the autoimmune mediators cause the tissue injury seen in the disease.

What role does the virus-specific CTL play in CV-induced myocarditis? Recent circumstantial evidence from this laboratory indicates that virus-specific CTL might be required for autoreactive CTL generation. Some variants of CV fail to initiate myocarditis despite viral infection of the heart.[28] Animals infected with the nonmyocarditic CVB3o lack detectable autoreactive CTL[29] but have suppressor T cells that efficiently inhibit myocarditis development.[30] Since autoreactive CTL predominantly cause cardiac injury *in vivo,* but this suppressor cell primarily prevents virus-specific CTL responses *in vitro,* we believe that the virus-specific CTL may precede autoreactive CTL in myocarditis and may be required for sensitization of the latter cell. A proposed mechanism for this involvement is given in Section 5.

It is not completely understood how CTL kill target cells, although various mechanisms have been suggested. Lymphocytes may produce chemotactic, macrophage-activating, and migration-inhibiting factors that concentrate activated macrophages in the tissue in which phagocytosis and release of toxic molecules, including free radicals, induce target cell death.[5,31] Alternatively, the CTL may lyse cells directly.[32] Several hypotheses suggest that CTL either exude activated enzymes from their cell surface, which degrade target cell membranes,[33] open ionic channels in the target cell membrane resulting in osmotic lysis,[34] or produce toxic compounds called lymphotoxins which enter the target cell and inhibit cellular metabolism.[35]

4.2. T Helper Cells and Humoral Immunity

Th and B lymphocytes interact in the production of immunoglobulins—IgG, IgA, and IgE—antibody classes. In many autoimmune diseases, tissue injury presumably results from autoantibody-mediated destruction of target cells either through complement-dependent or antibody-dependent cell-mediated mechanisms. Animals experimentally infected with various viruses, such as reo, parainfluenza, and vaccinia, can develop autoantibodies; in the case of reovirus, these autoantibodies may be responsible for producing polyendocrine disease, including diabetes.[36] Until recently, little evidence existed for pathogenic autoantibodies in picornavirus-induced diseases. Heart-reactive antibodies (HRA) occur in many clinical myocarditis cases,[37] in which predisposing picornavirus infections are suspected; however, no definitive proof presently exists that these autoantibodies cause cardiac injury. HRA also arise in the sera of CV-infected BALB/c mice.[38] These factors probably develop incidentally to the disease process and are not directly involved in tissue damage. The HRA belong to the IgM class and occur equally in T-lymphocyte-deficient and immunologically intact mice, although only the latter animals develop cardiac lesions.

Rose and colleagues[39] were the first investigators to describe potentially pathogenic IgG autoantibodies in CV infections. These investigators have found that 87% of infected A-strain mice develop autoantibodies reacting to both skeletal and heart muscle, however, only the antibody specifically reacting to heart muscle correlates well with pathogenesis. DBA/2 mice give additional evidence for pathogenic humoral autoimmunity in CV-induced myocarditis.[40] Although only approximately 30–40% of DBA/2 mice show serologic evidence of IgG HRA within 7 days of infection, nearly all animals show IgG deposits in their hearts, while no deposits are seen in equally damaged BALB/c tissue. Furthermore, *in vivo* depletion of Th cells and the third component of complement (C3) using monoclonal antibody to L3T4[27] and cobra venom factor,[40] respectively, abrogates myocarditis in infected DBA/2 mice, but not in similarly treated BALB/c animals. Histologically, the lesions in HRA-induced myocarditis differ from those in CTL-induced injury.[39] Inflammation in the humorally mediated disease appears as a diffuse interstitial mononuclear cell infiltrate, while lesions in BALB/c mice are focal and show considerable myocyte necrosis.[3,6]

Host genetics play a substantial role in determining individual disease susceptibility and pathogenic mechanism in myocarditis. Many investigators[39–41] report wide variations in disease susceptibility between genetically defined inbred mouse strains. Although no estimate of the number of genes controlling myocarditis have been given, clearly this disease is regulated by multiple loci. Evidence suggests genes associated with the major histocompatibility complex (MHC) strongly influence the incidence and titers of HRA, while genes outside the MHC may control duration of viremia and viral infection in the animals.[39] Most importantly, association of CV-induced myocarditis with the MHC further implicates immunopathogenic mechanisms in tissue damage, since this complex represents the predominant immunoregulatory locus in mammalian species.

Little is known either about the antigens recognized by HRA or whether the antigens inducing humoral and cellular autoimmunity are the same. Although most investigators do not observe reactivity of either homologous or heterologous hyperimmune CV antibodies for cardiac tissue,[3,13] Notkins *et al.*[36] describe a monoclonal CVB4-neutralizing antibody, screened from 65 hibridoma clones, which also reacts to heart antigens but not to other tissues. This observation indicates cross-reactivity may exist at low frequency between certain CV and heart antigens. Investigators using polyclonal antisera may not observe crossreactivity simply because the relevant antibody exists as a minor contaminant of the virus neutralizing antibody pool. Whether such limited quantities of autoantibody could produce disease remains unknown. Nonetheless, the work of Notkins certainly raises the question of whether autoimmunity may occasionally result inadvertently from misguided immunity to viral antigens.

4.3. Interferon and Natural Killer Cells

α-, β-, and γ-interferon (IFN) are proteins derived from virus-infected cells and stimulated lymphocytes. These molecules have, among other attributes, antiviral and immunoregulatory properties.[42]

4.3.1. Antiviral Effects

Viral infection of a cell stimulates IFN mRNA transcription from cell genes. Translation of messenger RNA (mRNA) results in IFN, which then activates the production of antiviral proteins in the infected and adjacent uninfected cells.[42] These antiviral proteins interfere with translation of the viral RNA. Thus, IFNs do not directly interact with viruses but inhibit infection by metabolically preventing viral replication.

4.3.2. Natural Killer Cells

Interferons also cause natural killer (NK) cell proliferation and activation.[43] NK cells efficiently lyse virus infected cells, and augmentation of NK cell activity usually mirrors kinetics of virus elimination in infected animals.[45,46] Unlike T-lymphocyte immunity, NK cells react not to specific antigens but recognize a wide range of infected and neoplastic cells. No evidence exists linking NK cells to tissue injury in CV infections. Indeed, some studies indicate highly effective NK cell responses may be protective,[44,45] presumably because early elimination of the virus may preclude both virus-specific and autoreactive CTL generation and production of HRA.

4.3.3. Antigen Presentation

Initiation of immune responses requires presentation of the antigenic epitope in association with either class I or class II MHC antigens.[46] Macrophages, dendritic cells, and other members of the reticuloendothelial system (RES) normally express MHC antigens on their plasma membranes and function as the primary initiators of immune responsiveness. Many cells, including β-cells and cardiocytes, express little or no MHC antigens[47] and are therefore presumably poor inducers of immunity. However, IFN_{γ}, primarily produced by activated lymphocytes, causes enhanced MHC antigen expression on lymphoid and RES cells and induces new antigen expression on normally negative cells.[48] Thus, cells previously not involved in antigenic sensitization can now interact with T lymphocytes stimulating immune responses to molecules present on their cell surface. Where the presented antigens are exclusively associated with either virus or virus-infected cells, the resulting immunity should aid in virus elimination. However, when normal tissue antigens are presented, autoimmunity might occur.

4.4. Macrophages

Macrophages account for 50% or more of the mononuclear inflammatory cells in virus-induced myocarditis and diabetes[2,3,10,13,23] but, as with NK cells, macrophages may be more protective than damaging to tissue. T-lymphocyte-depleted mice display normal macrophage activity but no diabetes or myocarditis after infection. This observation might indicate that macrophages play no significant role in disease pathogenesis; however, without stimulated T lymphocytes, various macrophage chemotactic and activating factors would not be present to concentrate large numbers of phagocytic cells *in situ*. Therefore, while tissue injury remains T-cell dependent, lymphokine-activated macrophages may be directly responsible for cell destruction. At least one investigator (B. MacPherson, University of Vermont, personal communication) correlates elevated phagocytic activity and superoxide production in macrophages from EMC-infected mice with diabetes induction and administration of carageenan, silica, and trypan blue, all inhibitors of macrophages, prevents diabetes development in infected animals. Although circumstantial, this evidence could implicate macrophage in β-cell injury.

Better evidence exists for macrophage involvement in controlling viral infections. While virus-neutralizing antibody and IFN are usually credited with virus clearance,[3,42,49,50] macrophage may also be involved. High concentrations of hyperimmune serum prevent lethal CV infections in suckling mice but, at lower antibody concentrations, animal survival depends on transfer of macrophages with the serum.[49] Second, mice treated with cortisone show significant impairment of CV elimination, although virus-neutralizing antibody titers remain elevated in these animals, and IFN levels are not affected by this treatment.[50]

5. HYPOTHETICAL MODEL FOR PICORNAVIRUS-INDUCED AUTOIMMUNITY IN MYOCARDITIS

Figure 4 presents one interpretation of the events following CV infections that may lead to autoimmunity and myocarditis. A similar model could be developed for CV-induced diseases in a variety of infected tissues. Virus entering the body is disseminated into many tissues, including heart and peripheral lymphoid organs. The infection rapidly induces IFN_{α} and IFN_{β} and antiviral molecules in infected and adjacent uninfected cells.[42] IFN produced in the peripheral lymphoid organs activates NK cells and macrophages that spill into the lymphatics and blood. Small numbers of these effector cells may enter the heart, either killing infected myocytes or scavenging cell debris from the few myocytes destroyed by virus. In either case, the injury produced is minimal and undetected by visual histologic examination of tissue sections.[3] However, some macrophages loaded with myocyte debris probably return to the peripheral lymphoid organs, and preliminary sensitization of the immune system to virion and myocyte antigens begins. Soon (3–4 days), IgM virus-neutralizing antibodies arise and join with NK cells and macrophages in fighting the infection.[3,14,45] By day 5, virus-specific CTL and IgG virus-neutralizing antibodies appear.[3,13,14] The CTL lyse infected myocytes and produce various lymphokines, including macrophage chemotactic and activating factors resulting in additional mononuclear cell infiltrations in localized areas of the heart in which virus persists. Under the pressure of the host's antiviral responses, the infec-

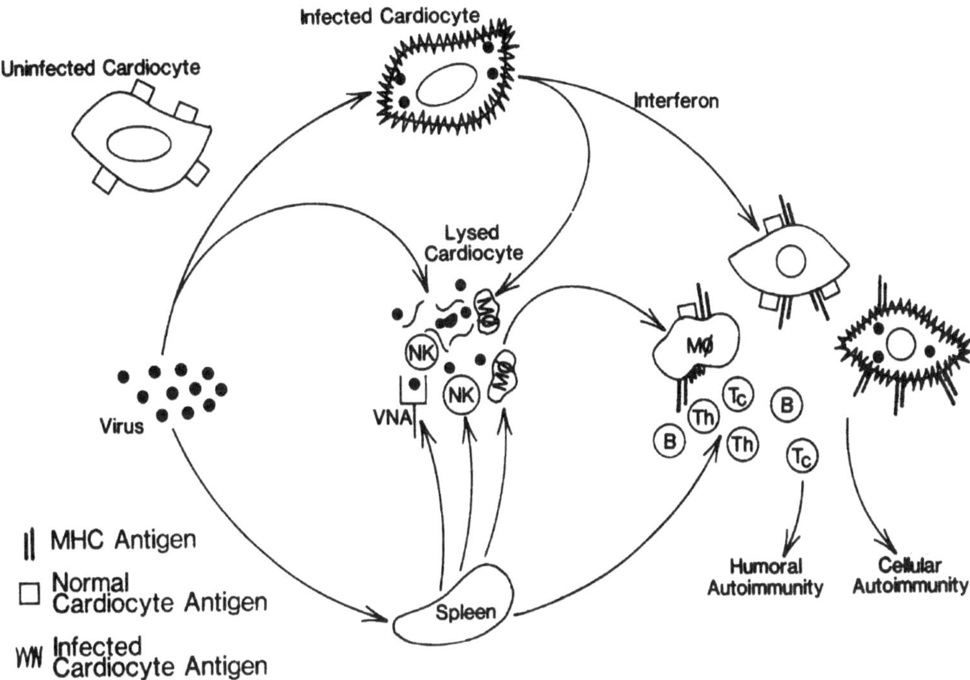

FIGURE 4. Schematic representation of a hypothetical model of CV-induced myocarditis. VNA, virus-neutralizing antibody; MØ, macrophage; NK, natural killer cell; B, B lymphocyte; Tc, cytolytic T lymphocyte; Th, T helper cell.

tion rapidly abates and within several more days all infectious virus disappears.

This juncture forms a critical point in myocarditis development. The above sequence of events probably occurs whenever virus infects the heart and explains the mild inflammatory cell infiltrate observed in myocarditis-resistant and T-lymphocyte-deficient animals postinfection. Normally, these cells disappear when virus is eliminated from the organ, leaving no detectable residual damage to the myocardium.[10,15,39,41] Whether more severe disease arises probably depends on the induction of autoimmunity to normal myocyte antigens. Autoimmune sensitization may arise if macrophages in the heart phagocytize myocyte debris and present relevant self-antigens to the immune system. Alternatively, the myocytes may also act as aberrant antigen-presenting cells if IFN_γ produced during infection by virus-immune lymphocytes induces MHC antigen expression on otherwise negative cardiocytes. Thus, the myocytes would present their own cell-surface antigens, resulting in autoimmunity. Circumstantial evidence for the latter hypothesis exists. Botazzo and colleagues[51] reported a case of IDDM occurring in a child soon after a known viral infection. β-cells in the islets of Langerhans showed dramatic increases in MHC antigen expression that may have triggered the presumably pathogenic autoimmune response in the disease.

In conclusion, while much work remains to be done in this area, we now recognize that picornavirus infections trigger a complex cascade of events in the host which ultimately lead to tissue damage. Various mechanisms may produce cell injury under different conditions. Certainly, cell death due to virus-mediated lysis occurs but, in many cases, tissue injury results mainly from immunopathogenic mechanisms. Precisely how the virus triggers immune and autoimmune responses remains enigmatic. Future work in this area should broaden our knowledge in both virology and immunology and ultimately lead to a better understanding of the pathogenesis of picornavirus-induced diseases.

REFERENCES

1. Davis, B. D., Dulbecco, R., Eisen, H. N., Ginsberg, H. S., and Wood, W. B., Jr. (eds.), 1969, *Microbiology*, Harper & Row, New York.
2. Huber, S. A., and MacPherson, B. R., 1984, Virus and insulin-dependent diabetes mellitus, in: *Immunology of Clinical and Experimental Diabetes* (S. Gupta, ed.), pp. 295–327, Plenum, New York.
3. Woodruff, J. F., 1980, Viral myocarditis: A review, *Am. J. Pathol.* **101:**425–484.
4. Rekosh, D. M. K., 1977, The molecular biology of picornaviruses, in: *The Molecular Biology of Animal Viruses* (D. P. Nayak, ed.), pp. 63–110, Dekker, New York.
5. Paque, R. E., Gauntt, C. J., Nealon, T. J., and Trousdale, M. D., 1978, Assessment of cell-mediated hypersensitivity against coxsackievirus B3 viral-induced myocarditis utilizing hypertonic salt extracts of cardiac tissue, *J. Immunol.* **120:**1672–1678.
6. Huber, S. A., and Lodge, P. A., 1984, Coxsackievirus B-3 myocarditis in Balb/c mice: Evidence for autoimmunity to myocyte antigens, *Am. J. Pathol.* **116:**21–29.
7. Huber, S. A., Job, L. P., and Woodruff, J. F., 1980, Lysis of infected myofibers by coxsackievirus B-3 immune T lymphocytes, *Am. J. Pathol.* **98:**681–694.
8. Wong, C. Y., Woodruff, J. J., and Woodruff, J. F., 1977, Generation of cytotoxic T lymphocytes during coxsackievirus B-3 infection. 1. Model and viral specificity, *J. Immunol.* **118:**1159–1164.
9. Lutton, C. W., and Gauntt, C. J., 1976, Coxsackievirus B3 infection alters plasma membranes of neonatal skin fibroblasts, *J. Virol.* **60:**294–296.
10. Woodruff, J. F., and Woodruff, J. J., 1974, Involvement of T lymphocytes in the pathogenesis of coxsackie virus B3 heart disease, *J. Immunol.* **113:**1726–1734.
11. Gatmaitan, B. G., Chason, J. L., and Lerner, A. M., 1970, Augmentation of the virulence of murine coxsackievirus B3 myocardiopathy by exercise, *J. Exp. Med.* **131:**1121–1136.
12. Adesanya, C. O., Goldberg, A. H., Plear, W. P. C., Thorn, K. A., Young, N. A., and Abelmann, W. H., 1976, Heart muscle after experimental viral myocarditis, *J. Clin. Invest.* **57:**569–575.
13. Woodruff, J. F., Wong, C. Y., and Woodruff, J. J., 1977, Cytotoxic T cells in coxsackieviral disease, in: *Immune Effector Mechanisms in Disease* (M. E. Weksler, S. O. Litwin, R. R. Riggio, and G. W. Siskind, eds.), pp. 207–237, Grune & Stratton, New York.
14. Huber, S. A., and Job, L. P., 1983, Cellular immune mechanisms in coxsackievirus group B, type 3 induced myocarditis in Balb/c mice, in: *Myocardial Injury* (J. J. Spitzer, ed.), pp. 491–508, Plenum, New York.

15. Yoon, J. W., Onodera, T., Jenson, A. B., and Notkins, A. L., 1978, Virus induced diabetes mellitus. XI. Replication of coxsackie B3 virus in human pancreatic beta cell cultures, *Diabetes* **27:**778–781.
16. Craighead, J. E., 1980, Experimental models of juvenile onset (insulin-dependent) diabetes mellitus, *Monog. Pathol.* **21:**166–176.
17. Buschard, K., Rygaard, J., and Lund, E., 1976, The inability of diabetogenic virus to induce diabetes mellitus in athymic (nude) mice, *Acta Pathol. Microbiol. Scand. [C]* **84:**299–303.
18. Huber, S. A., Babu, P. G., and Craighead, J. E., 1985, Genetic influences on the immunologic pathogenesis of encephalomyocarditis (EMC) virus-induced diabetes mellitus, *Diabetes* **34:**1186–1190.
19. Hashimoto, I., and Tomatsu, T., 1978, Myocardial changes after infection with coxsackie virus B3 in nude mice, *Br. J. Exp. Pathol.* **59:**13–20.
20. Khatib, R., Khatib, G., Chason, J. L., and Lerner, A. M., 1983, Alterations in coxsackievirus B4 heart muscle disease in ICR Swiss mice by anti-thymocyte serum, *J. Gen. Virol.* **64:**231–236.
21. Robinson, J. A., O'Connell, J. B., Roeges, L. M., Major, E. O., and Gunnar, R. M., 1981, Coxsackie B3 myocarditis in athymic mice, *Proc. Soc. Exp. Biol. Med.* **166:**80–91.
22. Babu, P. G., Huber, S. A., and Craighead, J. E., 1986, Contrasting features of T lymphocytes mediated diabetes in encephalomyocarditis virus infected Balb/cBy and Balb/c CUM mice, *Am. J. Pathol.* **124:**193–198.
23. Deguchi, H., Kitaura, Y., Morita, H., Kotaka, M., and Kawamura, K., 1985. Cell-mediated immunity in coxsackie B3 virus myocarditis in mice. In situ characterization by monoclonal antibody of mononuclear cell infiltrates, in: *Myocarditis and related disorders* (M. Sekiguchi, E. G. J. Olsen, and J. F. Goodwin, ed.), pp. 221–227, Springer-Verlag, Tokyo.
24. Guthrie, M., Lodge, P. A., and Huber, S. A., 1984, Cardiac injury in myocarditis induced by coxsackievirus group B, type 3 induced myocarditis in Balb/c mice is mediated by Lyt 2^+ cytolytic lymphocytes, *Cell. Immunol.* **88:**558–567.
25. Waldor, M. K., Sriram, S., Hardy, R., Herzenberg, L. A., Herzenberg, L. A., Lancer, L., Lim, M., and Steinman, L., 1985, Reversal of experimental allergic encephalomyelitis with monoclonal antibody to a T-cell subset marker, *Science* **227:**415–417.
26. Cantor, H., and Boyse, E. A., 1975, Functional subclasses of T lymphocytes bearing different Ly antigens. II. Cooperation between subclasses of Ly^+ cells in the generation of killer activity, *J. Exp. Med.* **141:**1390–1405.
27. Estrin, M., Herzum, M., Lodge, P. A., and Huber, S. A., 1987, Coxsackievirus B3-induced myocarditis: Autoimmunity is $L3T4^+$ T helper cell and IL-2 independent in Balb/c mice, *Am. J. Pathol.* **127:**335–341.
28. Gauntt, C. J., Trousdale, M. D., LaBadie, D. R. L., Paque, R. E., and Nealon, T., 1979, Properties of coxsackievirus B3 variants which are amyocarditic or myocarditic for mice, *J. Med. Virol.* **3:** 207–220.
29. Huber, S. A., and Job, L. P., 1983, Differences in cytolytic T cell response of BALB/c mice infected with myocarditic and non-myocarditic strains of coxsackievirus group B, type 3, *Infect. Immun.* **39:**1419–1427.
30. Estrin, M., Smith, C., and Huber, S. A., 1986, Antigen specific suppressor T cells prevent cardiac injury in Balb/c mice infected with a nonmyocarditic variant of coxsackievirus group B, type 3, *Am. J. Pathol.* **125:**578–584.
31. Chensue, S. W., Kunkel, S. L., Higashi, G. I., Ward, P. A., and Boros, D. L., 1983, Production of superoxide anion, prostaglandins and hydroxyeicosatetraenoic acids by macrophages from hypersensitivity-type (Schistosoma mansoni egg) and foreign body-type granulomas, *Infect. Immun.* **42:**1116–1125.

32. Berke, G., and Amos, D. B., 1973, Mechanisms of lymphocyte-mediated cytolysis. The LMC cycle and its role in transplantation immunity, *Transplant. Rev.* **17:**71–121.
33. Ferluga, J., and Allison, A. C., 1975, Cytotoxicity of isolated plasma membranes from lymph node cells, *Nature (Lond.)* **255:**708–710.
34. Ferluga, J., and Allison, A. C., 1974, Observations on the mechanism by which T-lymphocytes exert cytotoxic effects, *Nature (Lond.)* **250:**673–675.
35. Granger, G. A., and Williams, T. W., 1968, Lymphocyte cytotoxicity in vitro: Activation and release of a cytotoxic factor, *Nature (Lond.)* **218:**1253–1254.
36. Notkins, A. L., Onodera, T., and Prabhakar, B., 1984, Virus-induced autoimmunity, in: *Concepts in Viral Pathogenesis* (A. L. Notkins, and M. B. A. Oldstone, eds.), pp. 210–215, Springer-Verlag, New York.
37. Maisch, B., 1984, Diagnostic relevance of humoral and cell-mediated immune reactions in patients with acute myocarditis and congestive cardiomyopathy, in: *Cardiology* (E. I. Chazov, V. N. Smirnov, and R. G. Oganov, eds.), pp. 1327–1338, Plenum, London.
38. Huber, S. A., Lyden, D. C., and Lodge, P. A., 1985, Immunopathogenesis of experimental coxsackievirus induced myocarditis: Role of autoimmunity, *Herz* **10:**1–10.
39. Wolfgram, L. J., Beisel, K. W., Herskowitz, A., and Rose, N. R., 1986, Variations in the susceptibility to coxsackievirus B3-induced myocarditis among different strain of mice, *J. Immunol.* **136:**1846–1852.
40. Huber, S. A., and Lodge, P. A., 1986, Coxsackievirus B-3 myocarditis. Identification of different pathogenic mechanisms in DBA/2 and Balb/c mice, *Am. J. Pathol.* **122:**284–291.
41. Gauntt, C. J., Gomez, P. T., Duffey, P. S., Grant, J. A., Trent, D. W., Witherspoon, S. M., and Paque, R. E., 1984, Characterization and myocarditis capabilities of coxsackievirus B3 variants in selected mouse strains, *J. Virol.* **52:**598–605.
42. Stewart, W. E., 1979, *The Interferon System,* Springer-Verlag, New York.
43. Biron, C. A., and Welsh, R. M., 1982, Blastogenesis of natural killer cells during viral infections in vivo, *J. Immunol.* **129:**2788–2796.
44. Godney, E. K., and Gauntt, C. J., 1987, Murine Natural Killer cells limit coxsackievirus B3 replication, *J. Immunol.* **139:**913–918.
45. Huber, S. A., Job, L. P., and Woodruff, J. F., 1981, Sex related differences in the pattern of coxsackievirus B-3 induced immune spleen cell cytotoxicity against virus infected myofibers, *Infect. Immun.* **32:**68–73.
46. Klinkert, W. E. F., LaBadie, J. H., O'Brien, J. P., Beyer, C. F., and Bowers, W. E., 1980, Rat dendritic cells function as accessory cells and control the production of a soluble factor for mitogenic responses of T lymphocytes, *Proc. Natl. Acad. Sci. USA* **77:**5414–5423.
47. Leszczynsky, D., Renkonen, R., and Hayry, P., 1985, Turnover of dendritic cells in rat heart, *Scand. J. Immunol.* **22:**351–355.
48. Basham, T. Y., and Merigan. T. C., 1983, Recombinant interferon-γ increases HLA-DR synthesis and expression, *J. Immunol.* **130:**1492–1494.
49. Rager-Zisman, B., and Allison, A. C., 1973, The role of antibody and host cells in the resistance of mice against infection by coxsackie B-3 virus, *J. Gen. Virol.* **19:**329–338.
50. Woodruff, J. F., 1979, Lack of correlation between neutralizing antibody production and suppression of coxsackievirus B3 replication in target organs: Evidence for involvement of mononuclear inflammatory cells in host defense, *J. Immunol.* **123:**31–36.
51. Bottazzo, G. F., Dean, B. M., McNally, J. M., Mackay, E. H., Swift, P. G., and Gamble, D. R., 1985, In situ characterization of autoimmune phenomena and expression of HLA molecules in the pancreas in diabetic insulitis, *N. Engl. J. Med.* **313:**353–360.

8

Application of Monoclonal Antibodies to the Study of Coxsackieviruses

BELLUR S. PRABHAKAR

1. INTRODUCTION

Coxsackieviruses (CV) are highly prevalent pathogens with the ability to cause fatal diseases in humans. Although coxsackie B virus (CVB) infections are relatively benign in adults, they can be fatal in children and have been shown to be the cause of meningoencephalitis, myocarditis, hepatitis, and some cases of diabetes.

The fact that a given type of CVB can produce a variety of clinical syndromes suggested that there may be naturally occurring variants of the virus within each type that are biologically and antigenically distinct but indistinguishable using polyclonal reference serum. Since the available methods had serious limitations in detecting subtle differences between various isolates, monoclonal antibodies were produced and used to elucidate the antigenic differences among both clinical and laboratory isolates of coxsackievirus B4 (CVB4). The monoclonal antibodies provided a basis for the selection of CVB4 variants and were used to determine the frequency of antigenic variation. Monoclonal antibodies with different specificities provided new probes for analysis of the structural composition of CVB4 viruses. These antibodies are beginning to allow

---

BELLUR S. PRABHAKAR • Laboratory of Oral Medicine, National Institute of Dental Research, National Institutes of Health, Bethesda, Maryland 20892.

us to establish a correlation between the antigenic nature of the variants and their biologic activity.

2. GENERATION AND CHARACTERIZATION OF MONOCLONAL ANTIBODIES

Hybridomas that secrete CVB4-specific monoclonal antibodies were obtained by fusing myeloma cells with splenic lymphocytes from 4–6-week-old BALB/c mice previously immunized with purified CVB4.[1] The fused cells were plated at 2×10^5 cells/well into 96-well microtiter plates in 100 μl DMEM medium containing 15% fetal calf serum (FCS) and hypoxanthine, aminopterin, and thymidine.[2] When the hybridomas grew to confluency, the culture supernatants were initially screened against 100 tissue-culture infectious $dose_{50}$ ($TCID_{50}$) of the JVB strain of CVB4 (which was used for immunizing BALB/c mice) in a microneutralization assay.[1] The hybridomas that were positive for the secretion of anti-CVB4 antibodies were cloned by seeding a single cell per well using the limiting dilution method.[3] The clones were retested by the microneutralization assay and also by a liquid-phase radioimmunoassay (RIA) using labeled purified CVB4 virus.[1] In this way, more than 70 hybridomas producing anti-CVB4 monoclonal antibodies were isolated. Most of the hybridomas produced antibodies belonging to the IgG isotype.[1] The predominant subclass was IgG2a. Most of the antibodies that neutralized the virus also reacted well in the radioimmunoassay (RIA). However, some antibodies that neutralized poorly bound well to the virus in the RIA. In our studies, we used neutralization assay for initial screening and therefore obtained mostly neutralizing antibodies.

3. ANTIGENIC VARIANTS AMONG NATURALLY OCCURRING CLINICAL ISOLATES

In our studies, we obtained fresh clinical isolates from a number of different health departments.[1] Each of the isolates was grown once in our laboratory in monkey kidney cells (LLC-Mk2). One hundred $TCID_{50}$ of each virus was tested against a panel of 18 neutralizing monoclonal antibodies in a microneutralization assay. These studies led to several important findings (Fig. 1). On the basis of their reactivity, the panel of monoclonal antibodies recognized at least 15 different epitopes. Antibodies 86-3 and 204-4 gave similar reactivities; therefore, they are thought to recognize the same epitope. Similarly, 128-2, 208-2, and 288-3 recognized the same epitope. There is a high frequency of anti-

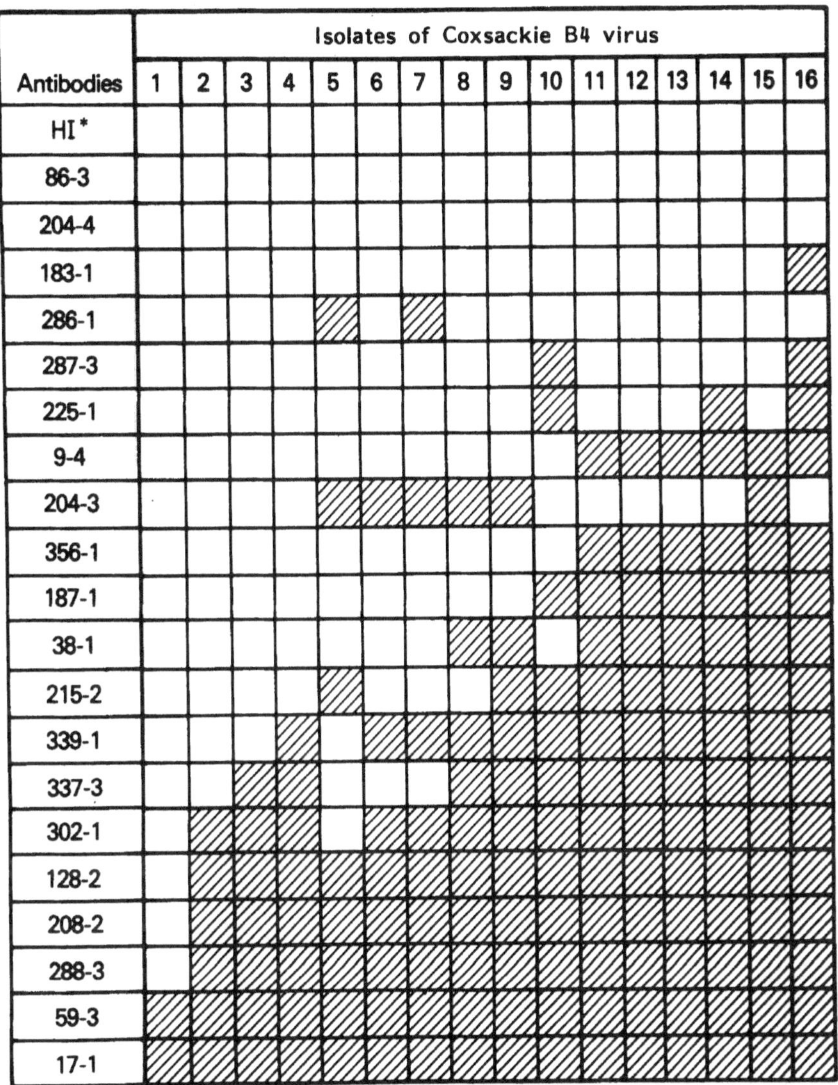

FIGURE 1. Antigenic variation among natural isolates. Open boxes show complete neutralization of the virus (100 $TCID_{50}$), hence no cytopathology, and hatched boxes show incomplete neutralization of the virus. Virus number 1 represents the prototype JVB strain. Viruses 2–16 represent the clinical isolates. (See Prabhakar *et al.*[1] for further details.)

genic variation among natural isolates; the natural isolates differed from the prototype virus (#1) in anywhere from 2 (isolate #2) to 12 (isolate #16) of 15 different epitopes being recognized by the panel of monoclonal antibodies; these isolates also showed considerable differences in the range of epitopes expressed. Some of the clinical isolates (i.e., #11–13) exhibited identical patterns of reactivity. However, all isolates were neutralized by some of the monoclonal antibodies and by the hyperimmune serum showing that these viruses also possess antigenic determi-

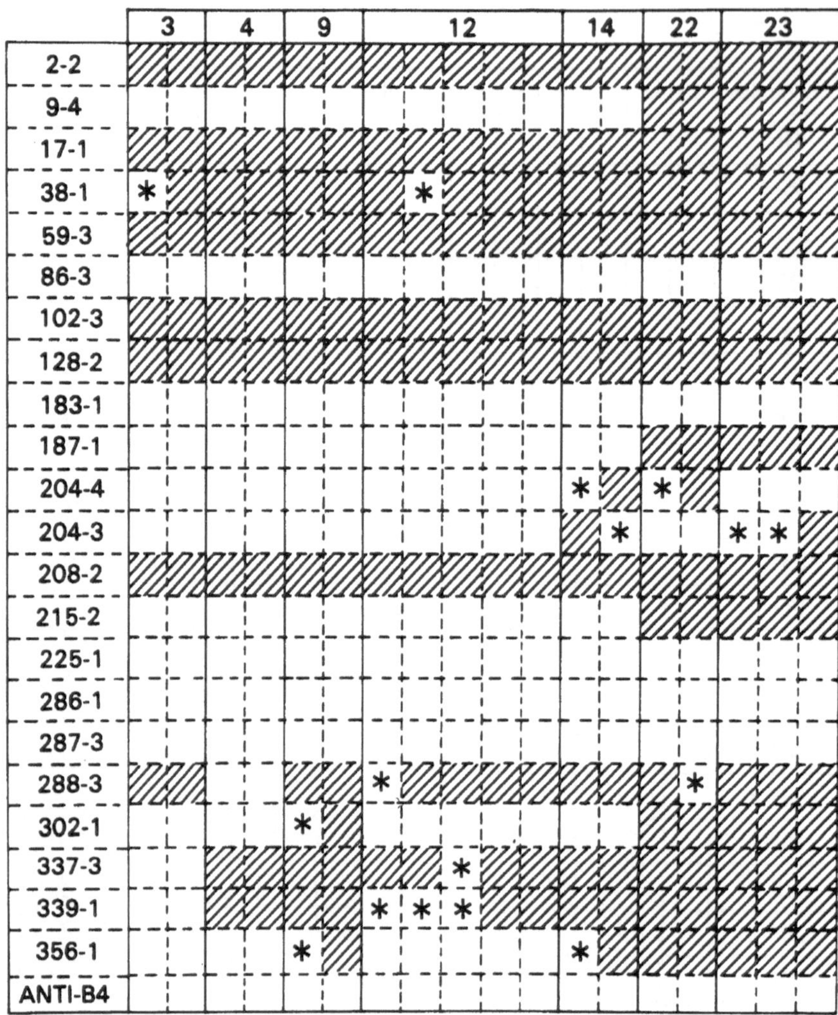

FIGURE 2. Antigenic analysis of isolates from the same patient. Numbers 3, 4, 9, 12, 14, 22, and 23 indicate different patients. Two to five isolates from each patient were tested for the expression of epitopes. Dissimilarities in the expression of epitopes are indicated by asterisks (*). Open boxes indicate neutralization and hatched boxes indicate lack of complete neutralization.

nants that are stable and common to all isolates. These, perhaps, are the determinants responsible for the type specificity.

In order to see whether CVB4 can undergo variation within a given individual during the course of an infection, we tested several isolates from the same patient obtained at different times postinfection. A known amount (100 $TCID_{50}$) of each of the isolates was tested with our panel of monoclonal antibodies using a microneutralization assay. The results illustrated in Fig. 2 showed that all the isolates obtained from a given individual showed similar patterns of reactivity. However, variation in certain epitopes was readily detectable. This might be due to the extremely high sensitivity of the microneutralization assay used for the detection of variants. Perhaps a high proportion of the virus particles in each of the isolates did express a given epitope, but a small proportion that did not express the epitope may have escaped neutralization and resulted in the cytopathic effect, giving the appearance that the isolate was negative for that epitope. Nevertheless, it is clear that new variants can arise *in vivo* and can be detected using monoclonal antibodies.

4. SELECTION OF ANTIGENIC VARIANTS

Our studies with clinical isolates suggested that the rate of mutation of neutralization epitopes on CVB4 might be high. In order to test this, we determined the frequency of antigenic variation among CVB4 isolates. This was done by titrating plaque-purified CVB4 isolates in the presence of different monoclonal antibodies obtained in the form of ascites fluid. The frequency of antigenic variation at a given epitope was determined by dividing the virus titer obtained in the absence of antibody by that obtained in the presence of a given monoclonal antibody. These studies revealed that the frequency of antigenic variants ranged from 10^{-4} to $10^{-5.8}$.[1] The frequencies found in this study were similar to the frequencies found with a number of other viruses.[4,5] Although such a high rate of mutation among CVB4 exists, it had not been detected earlier because hyperimmune polyclonal sera were used for typing. This high frequency of antigenic variation may also explain the surprisingly large number of antigenic variants that we found in natural isolates of CVB4 (Fig. 1).

5. CHARACTERIZATION OF ANTIGENIC VARIANTS

Variants selected in our laboratory using monoclonal antibodies were analyzed using two assays. In the first assay, 100 $TCID_{50}$ of each of the variants were tested against a panel of antibodies in a microneutralization assay. Absence of cytopathic effect (CPE) indicated complete neutralization of the virus. Presence of CPE indicated that a proportion of the virus inoculum resisted neutralization. In the second assay, 100 PFU of the virus were tested against the same panel of monoclonal antibodies in a plaque-reduction assay.[6] Table I compares these two assays, which used three variants (E, E1, and E2) and a panel of monoclonal antibodies. These studies demonstrated that even when 95% of the E2 virus was neutralized by antibody 204-4 (as determined by plaque-reduction assay), it was scored as non-neutralizable in the $TCID_{50}$ assay. These studies

TABLE I
Comparison of Neutralization by $TCID_{50}$ and Plaque-Reduction Assays

	Virus[c]					
	E		E1		E2	
Antibodies	$TCID_{50}$	Plaq	$TCID_{50}$	Plaq	$TCID_{50}$	Plaq
Imm. serum	+[a]	100[b]	+	100	+	100
183-1	+	100	+	100	+	100
86-1	+	100	+	100	+	100
204-4	+	98	−	84	−	95
215-2	+	96	−	78	−	80
356-1	−	83	−	69	−	56
339-1	−	71	−	73	−	34
Medium	−	0	−	0	−	0

[a]+, complete neutralization of the virus in the $TCID_{50}$ assay; −, cytopathology.
[b]Numbers indicate percentage plaque reduction.
[c]Virus E is a plaque-purified virus. E_1 and E_2 represent plaque variants of the parental E virus. See Prabhakar *et al.*[6] for further details.

showed that a small proportion of viruses lacking particular epitopes resulted in the CPE and gave the appearance that the entire virus population was resistant to neutralization. Furthermore, it showed that the $TCID_{50}$ assay is a particularly sensitive assay for detecting epitope negative variants.[6] Therefore, this highly sensitive assay was used to test a series of variants for their ability to interact with a panel of monoclonal antibodies. The variants, as expected, did not react with the antibodies used for their selection (e.g., variant 9-286 did not react with antibody 286-1) (Fig. 3). In addition, the variants also failed to react with a number of other antibodies with which the parental virus reacted (e.g., virus 9-286 failed to react with antibodies 102-3, 183-1, 187-1, 204-4). Moreover, the variants also revealed epitopes that were not present on the parental virus (e.g., virus 9-286 expressed an epitope detected by antibody 204-3 that was not present on the parental virus 9).[7]

Similar studies on the appearance and disappearance of epitopes were carried out using plaque-purified viruses and their progeny clones (without selection in the presence of monoclonal antibodies)[6] (Fig. 4). These studies showed that some progeny viruses were very stable and gave reactivity patterns similar to those found with the parental virus (e.g., A1, A2, and A3 reactivities were identical to that of parent A virus). However, certain other progeny viruses showed differences in the pattern of reactivity with monoclonal antibodies (e.g., B1, B2, and B3 differed from parental B virus). Certain epitopes present on the parental

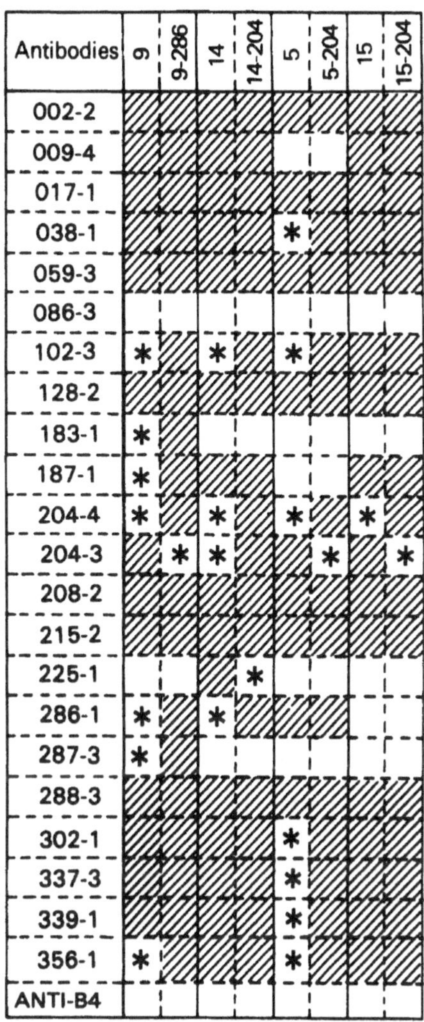

FIGURE 3. Analysis of variants selected in the presence of monoclonal antibodies. Viruses 9, 14, 5, and 15 represent parental plaque-purified viruses. Viruses 9–286, 14–204, 5–204, and 15–204 represent variants selected with antibodies 286 and 204, respectively. Asterisks (*) show epitopes that differ in the parental virus and the corresponding antigenic variants. Open boxes indicate complete neutralization. Hatched boxes show lack of complete neutralization. (See Prabhakar *et al.*[7] for further details.)

virus (e.g., epitope recognized by 339-1 and 38-1 on parental virus D) were missing on the progeny virus (D1 and D2). Conversely, certain epitopes that were missing on the parental virus (e.g., the epitope recognized by antibody 356-1 on B virus) were present on the progeny (e.g., B1, B2, and B3 virus). These and other studies demonstrated that, when variants are tested with a panel of antibodies directed against different

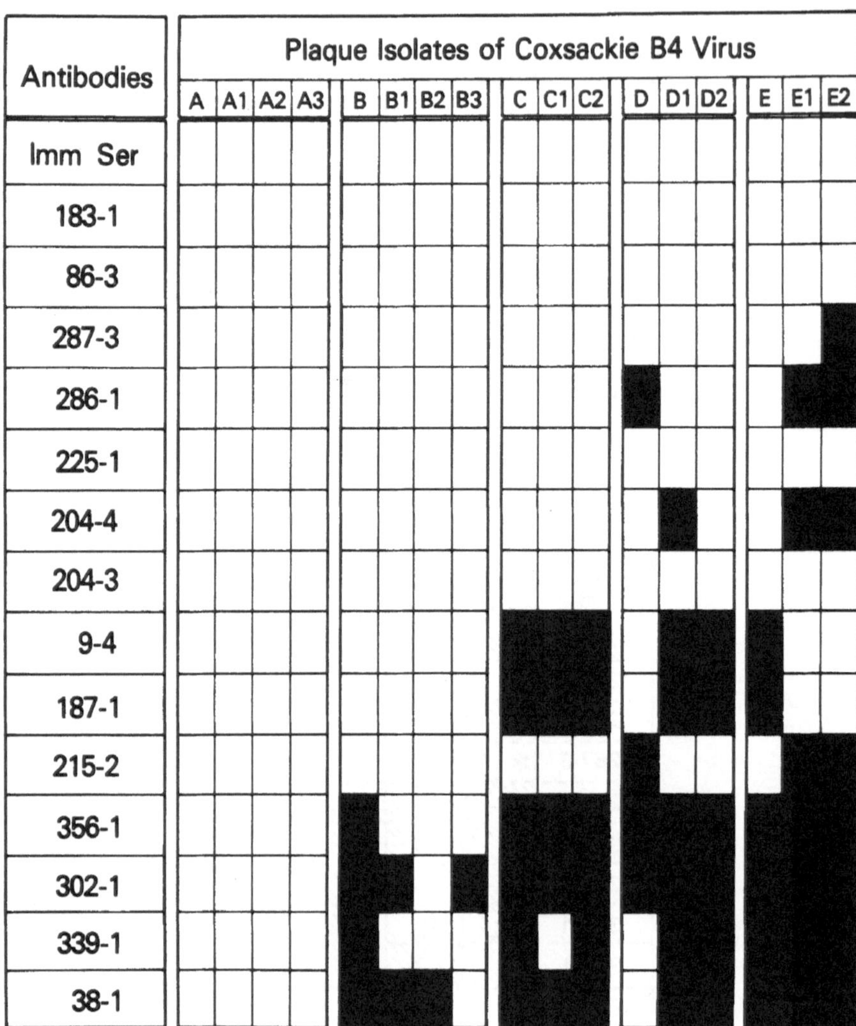

FIGURE 4. Antigenic variation among plaque isolates of CVB4. Viruses A, B, C, D, and E represent parental plaque-purified viruses. Viruses A1, A2, A3, B1, B2, B3, C1, C2, D1, D2, E1, and E2 represent plaque isolates of the corresponding parental viruses. Open boxes indicate neutralization and closed boxes indicate lack of complete neutralization. Note that A1, A2, and A3 displayed reactivity identical to that of the parental A virus. However, except for C2, all other plaque isolates differed from their parental virus. (See Prabhakar *et al.*[6] for further details.)

epitopes, changes in one or more epitopes are detected at frequencies greater than 10^{-2}.[6]

6. DEMONSTRATION OF CONSERVED AND NONCONSERVED EPITOPES ON CVB4

Although very high frequencies of antigenic variants were found with different laboratory and clinical isolates, we were also interested to see how well certain epitopes were conserved among natural isolates of CVB4. Forty-seven fresh clinical isolates were tested against a panel of 16 neutralizing monoclonal antibodies.[6] Six of the monoclonal antibodies (i.e., 183-1, 86-3, 287-3, 286-1, 225-1, and 204-4) neutralized 94–98% of the clinical isolates tested and were therefore considered antibodies recognizing highly conserved (HC) epitopes (Fig. 5). Six other monoclonal antibodies (i.e., 204-3, 9-4, 187-1, 215-2, 356-1, and 302-1) neutralized 51–79% of the isolates tested and were grouped as antibodies recognizing moderately conserved (MC) epitopes. The other four (i.e., 339-1, 38-1, 337-3, and 288-3) neutralized only 21% or less of the isolates and were grouped as those recognizing poorly conserved (PC) epitopes. Of the 47 clinical isolates tested, 22 possessed all the HC and, with very few exceptions, all the MC epitopes. Twelve of the isolates had most, but not all, of the HC and MC epitopes. By contrast, 13 of the isolates had most of the HC epitopes but few of the MC and PC epitopes. Despite the marked antigenic diversity seen in MC and PC epitopes, none of the clinical isolates escaped neutralization by the polyclonal anti-CVB4 antiserum in the $TCID_{50}$ assay. This is perhaps because serum consists of a mixture of monoclonal neutralizing antibodies that can interact with many different independent epitopes. Although most of the antigenic changes might be innocuous, certain changes, especially in HC epitopes, might result in alterations in biological properties.[6,8]

7. NEUTRALIZATION EPITOPE DIVERSITY AMONG LABORATORY ISOLATES

Two human isolates of CVB4, i.e., Edwards (Edw) and JVB, were compared to evaluate diversity/stability of neutralization epitopes and to search for strain specific determinants.[9] Polyclonal antisera raised against the two parental and several laboratory isolates failed to show any strain-specific determinants.[9] However, when a panel of monoclonal antibodies were used, we were able to demonstrate strain specific differences (Table II). Seven of the monoclonal antibodies (i.e., 86-3, 225-1, 183-1, 86-7, 286-5, 225-5, and 287-2) neutralized both viruses, suggesting that they were reacting with stable epitopes. Four of the monoclonal antibodies (i.e., 9-4, 187-8, 214-5, and 356-2) neutralized all the JVB viruses and none of the Edw viruses. Epitopes recognized by these four monoclonal antibodies were present on JVB isolates obtained either after multiple passages in culture (Table II) or after plaque purification of parental JVB virus (not shown) indicating that these epitopes are highly stable but are unique to JVB virus. These studies and further analysis using the plaque-reduction assay[9] demonstrated not only qualitative but also quantitative differences between parental Edw

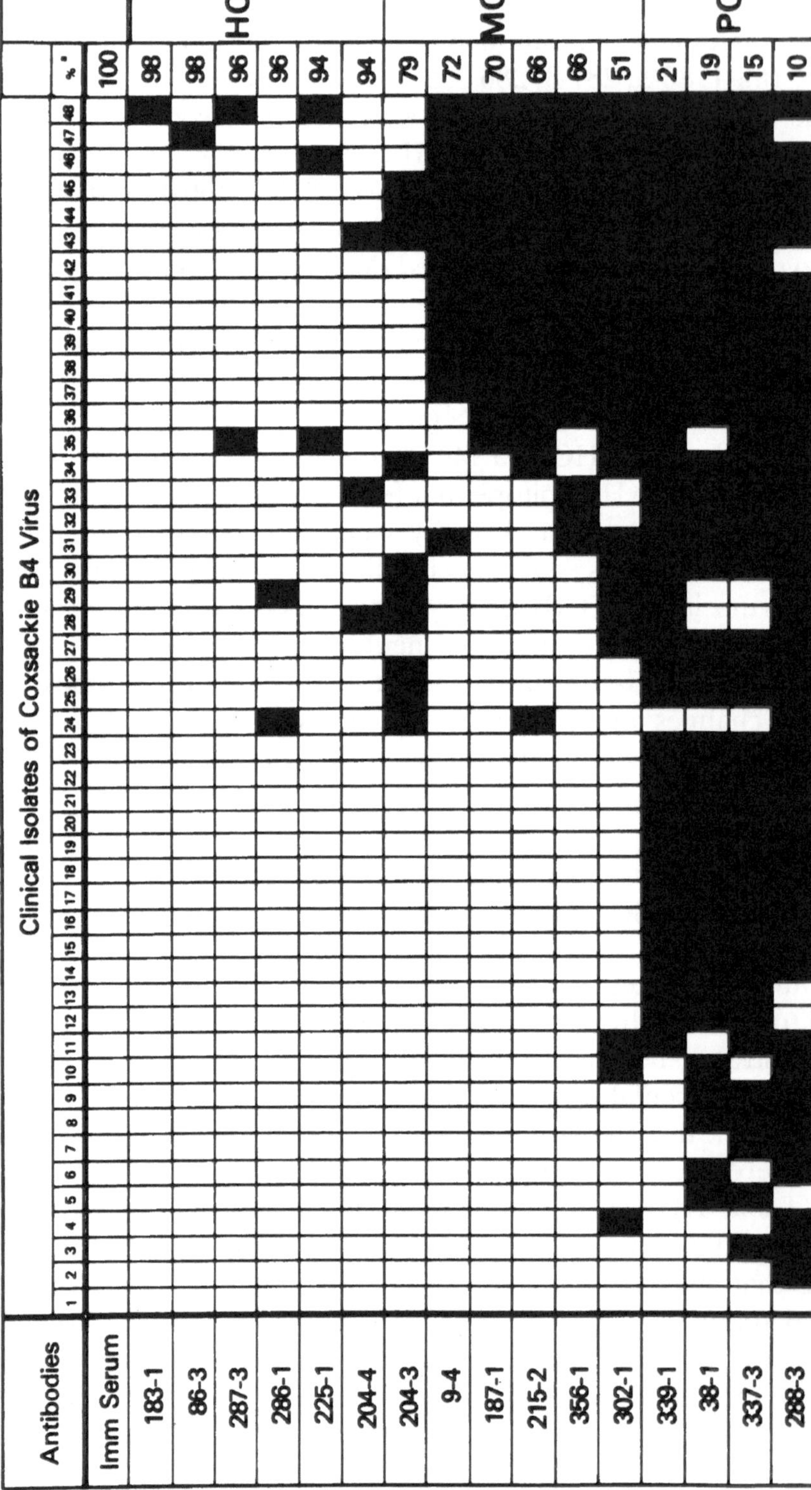

FIGURE 5. Delineation of conserved and nonconserved epitopes on clinical isolates of CVB4. Open boxes indicate complete neutralization and closed boxes indicate incomplete neutralization. Asterisk (*) Percentage isolates neutralized. HC, MC, and PC: Antibodies that react with highly, moderately and poorly conserved epitopes, respectively. (See Prabhakar *et al.*[6] for further details.)

TABLE II
Neutralization of Coxsackievirus B4 Isolates with Monoclonal Antibodies[a,b]

	Monoclonal antibodies										
	Conserved on both JVB and Edw							Conserved on JVB			
Virus	86-3	225-1	183-1	86-7	286-5	225-5	287-2	9-4	187-8	214-5	356-2
JVB	+	+	+	+	+	+	+	+	+	+	+
JVB-P1	+	+	+	+	+	+	+	+	+	+	+
JVB-P8	+	+	+	+	+	+	+	+	+	+	+
Edw	+	+	+	+	+	+	+	−	−	−	−
Edw-P1	+	+	+	+	+	+	+	−	−	−	−
Edw-P8	+	+	+	+	+	+	+	−	−	−	−

[a]JVB-P1 and Edw-P1 and JVB-P8 and Edw-P8 viruses represent first and eighth passages, respectively, in buffalo green monkey kidney cells. See Webb *et al.*[9] for further details.
[b]+, complete neutralization of the virus; −, incomplete neutralization of the virus.

and JVB viruses and their derivatives. In this context, it is interesting to note that the Edw isolate is highly pathogenic toward adult BALB/c mice (100% mortality), whereas JVB is nonpathogenic. This suggests that there might be a relationship between the presence or absence of certain antigenic determinants on CVB4 variants and their pathogenic potential.[9]

8. SELECTION OF ANTIGENIC VARIANTS WITH CHANGES IN THEIR BIOLOGIC PROPERTIES

A given type of CVB4 can produce a number of different diseases. It is not known whether this is due to antigenic variants with different tissue tropisms or whether it represents chance infection of given target tissues. Since these viruses exhibit very high frequency of spontaneous mutations ($> 10^{-2}$), it has been difficult to establish a link between a particular antigenic phenotype and a given disease.[6,8] Therefore, studies were initiated to determine whether variants with different biologic properties can be selected by growing the virus in the presence of monoclonal antibodies. More than half a dozen antigenic variants selected in the presence of different monoclonal antibodies (i.e., S0–S5 and V1, Table III) were tested for their ability to kill suckling mice. The animals infected with these viruses showed a mortality of 63–100% (S0–S5, Table III). These studies showed that antigenic variants selected in the presence of a monoclonal antibody may show partial attenuation. One of the variants selected with monoclonal antibody 356-1 (V1, Table III) was further selected by growing the virus in the presence of monoclonal

TABLE III
Death Due to Infection of Mice with CVB4 Variants[a]

Variant virus[b]	Selecting monoclonal antibody	D/T[c]	Percentage dead
CVB4	None	24/24	100
S0	93[d]	13/16	81
S1	285	8/8	100
S2	102	21/24	88
S3	86	14/22	64
S4	204	15/24	63
S5	183	20/24	83
V0[e]	None	39/40	98
V1	356	39/40	95
V2	356 + 183	10/40	25
V3	356 + 183 + 204	2/40	5

[a]See Prabhakar *et al.*[12] for further details.
[b]Variants were selected by growing plaque-purified virus in the presence of the indicated monoclonal neutralizing antibody.
[c]Number dead over number tested.
[d]Control ascites with no anti-CVB4 activity.
[e]CVB4 was plaque-purified twice to obtain V0.

antibody 183-4. This double variant (V2, Table III) when inoculated into suckling mice caused approximately 25% mortality showing substantial attenuation. Further selection of this double variant by growing the virus in the presence of monoclonal antibody 204-4 resulted in almost complete attenuation (V3, Table III), with only 5% of the animals showing mortality.[12] Together, our studies indicate that variants selected with a single monoclonal antibody show little or no attenuation, whereas variants selected with more than one antibody show much greater attenuation (V2 and V3, Table III). These and the studies with JVB and Edw viruses mentioned earlier strongly suggest that variants of CVB4 with different biologic properties do exist and can be selected from the same parental pool.[8,12]

9. CHARACTERIZATION OF VIRAL POLYPEPTIDES

Studies characterizing the viral polypeptides of CV have been limited. However, work carried out in Crowell's laboratory using polypeptides from CVB3 showed that polypeptide VP2 is responsible for the induction of neutralizing antibodies.[13] Our efforts to delineate functional domains of various CVB4 polypeptides using monoclonal antibodies have yielded equivocal results. When virus-infected cell extracts were immunoprecipitated with monoclonal antibodies and then subjected to polyacrylamide gel electrophoresis (PAGE), we found more than one polypeptide precipitated. All the monoclonal antibodies tested precipitated polypeptides VP1, VP2, and VP3. Some of the antibodies, in addition, have also precipitated the precursor polyprotein VP0.[12] There are perhaps several reasons for this. First, the epitope recognized by a given neutralizing monoclonal antibody is present on more than one polypeptide; therefore, the antibody binds to all the polypeptides. Second, although the antibody binds to an epitope present on one of the polypeptides, it can still precipitate all the polypeptides because of the strong association of the proteins. With the exception of one antibody (356-1), none of the 11 monoclonal antibodies tested thus far has reacted with viral polypeptides in Western blots. However, antibody 356-1 appears to bind primarily to VP1 (B. Prabhakar, unpublished observations). This suggests that at least one of the epitopes responsible for virus neutralization is present on VP1. Animals immunized with individual polypeptides failed to show any antiviral antibody response. However, when all four polypeptides were mixed prior to inoculation, a good antiviral antibody response was produced. This suggested that more than one viral polypeptide might be required to induce an effective immune response. Further detailed studies are required to understand the relationship between the structure of the virus and its function.

10. ANTI-IDIOTYPIC ANTIBODIES TO ANTI-CVB4 MONOCLONAL ANTIBODIES

Jerne's network hypothesis (1974) gave rise to the idea of using anti-idiotypic antibodies as probes for viral receptors and as potential immunogens for the induction of antiviral antibodies.[14] This is based on the premise that the antibody combining site (paratope) of an antibody will have structures complementary to the antigenic determinant it recognizes. This antibody (Ab1) can be used as an antigen to elicit an anti-idiotypic antibody response (Ab2). Some of the Ab2 antibodies will possess paratopes complementary to the paratope of Ab1. Therefore, some of the Ab2 paratopes have structures resembling the antigenic determinants recognized by Ab1. If Ab1 is an antiviral antibody that recognizes a structure on the virus responsible for induction of neutralizing antibodies or for attachment to cells, then Ab2 antibodies may resemble that viral structure on the virus. These Ab2 antibodies may therefore recognize the receptor or elicit an immune response against the virus. These types of experiments have been successful with several hormones[15–17] and viral antigens.[18–22]

Anti-idiotypic antibodies have been used successfully in experimental animals as vaccines against several viruses, including poliovirus,[22] rabies virus,[20] herpes virus,[18] and hepatitis virus.[19] Anti-idiotypic antibodies made against antireovirus antibodies have shown antireceptor activity.[23]

We generated polyclonal anti-idiotypic antibodies (anti-Ids) against three anti-CVB4 monoclonal antibodies with independent specificities.[24] These anti-Ids were tested for their ability to bind to CVB4 receptors and to elicit anti-CVB4 antibodies. The anti-Ids reacted specifically with the immunizing antibodies and inhibited their ability to bind to and neutralize CVB4. However, the anti-Ids exhibited no apparent effect when tested for their ability to inhibit CVB4 binding to receptor positive cells.[24,25] It was possible that the anti-Ids bound to the receptor but failed to inhibit virus attachment. Therefore, the ability of the anti-Ids to bind to the receptor positive cells (BGMK and HeLa) was tested. We found that none of the anti-Ids could bind to these cells. There are several possible explanations for this: (1) the monoclonal antibodies against which anti-Ids were made did not bind to the receptor attachment protein on the virus; (2) the anti-Ids may have bound to a region close to the paratopes of Ab1, preventing Ab1 binding to and neutralizing the virus without being truly complementary to the paratope of Ab1; (3) there might be more than one type of receptor for CVB4 attachment. This is unlikely because there was no effect of any of our anti-Ids on the initial rate of virus binding and the anti-Ids themselves did not bind to the cells.[24,25]

Although these results are in contrast to the results obtained with reovirus,[23] our results may reflect the general experience using a number of virus systems.[20,22] Recently published reports on the structure of two picornaviruses, human rhinovirus 14 and poliovirus type I, reveal that the receptor attachment protein might be buried in the 25-Å deep canyons present in the viral capsid protein.[26,27] Since the rim of the canyon is rather small, it is conceivable that our monoclonal antibodies were restricted from entering the canyon.[25] This might explain the lack of success in raising Ab2s with antireceptor activity.

In other experiments, mice immunized with anti-idiotypic antibodies showed a relatively small anti-anti-idiotypic antibody (Ab3) response.[24,25] When the same sera were tested for neutralizing antibodies, there was little, if any, increase in the antibody titer above background. This relatively poor response might be due to using a suboptimal dose of anti-Id for immunization. Alternatively, the paratopes of Ab2 antibodies may not have been truly complementary to the paratopes of Ab1. These and other studies[20,22] make it clear that not all the anti-Ids and anti-anti-Ids will be of the "internal image" (complementary to the paratopes) type and that not all "internal image" type Ab2 antibodies can bind the receptor or elicit antiviral antibodies.[24,25]

11. MOLECULAR MIMICRY AND AUTOIMMUNITY

Several studies have suggested that there might be a humoral or cellular autoimmune component to coxsackievirus-induced myocarditis. We wanted to see whether antibodies made against CVB4 were able to crossreact with normal myocardial tissue. We tested more than 60 anti-CVB4 monoclonal antibodies against frozen acetone-fixed sections of normal mouse heart and a variety of other tissues, using an indirect immunofluorescence assay.[28] One of the anti-CVB4 neutralizing antibodies, designated 356-1 (IgG2a), reacted brilliantly with the mouse myocardium and none of the other tissues tested, including smooth and skeletal muscles (Fig. 6). By immunoperoxidase staining of longitudinal sections of heart, this antibody was shown to react with the transverse striations of the myofibrils.

Our studies provide evidence that antiviral antibodies can sometimes react with normal tissues. In this connection it is interesting to note the recent studies by Srinivasappa *et al.*,[29] in which more than 600 monoclonal antibodies against 11 different viruses were screened against a panel of 14 different tissues. Approximately 3.5% of these antibodies reacted with cells in normal tissues, showing that the reactivity of antiviral antibodies with normal tissues is a common phenomenon. However, it is still not clear whether these crossreacting antibodies are deleterious. This will probably depend on the affinity of the interaction and on the location of the autoantigen. Further studies to understand the fine structure of the antigen to which antibody 356-1 binds and its effects on cardiac muscle should prove useful.

12. PERSPECTIVE

It is somewhat surprising that to date monoclonal antibodies against none of the coxsackieviruses (A or B) except CVB4 have been reported. Although CV form one of the largest groups of human pathogens, relatively little is known about their structure. Our studies have begun to demonstrate the usefulness of monoclonal antibodies in understanding the structure and biology of CVB4 viruses. Using monoclonal antibodies, several important observations have been made. All isolates of CVB4 including natural isolates, laboratory isolates derived from plaque-purified parental viruses, and antigenic variants selected using monoclonal

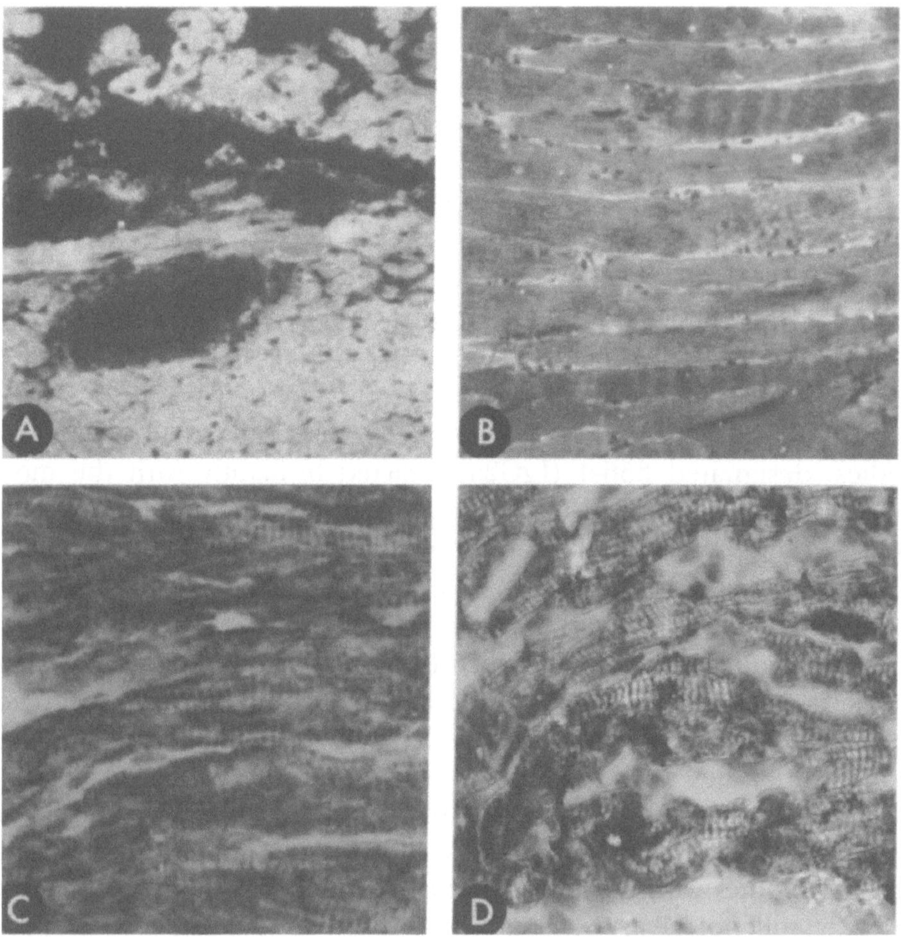

FIGURE 6. Reactivity of monoclonal anti-CVB4 antibody 346-1 with cardiac muscle. (**A**) Strong reactivity with atrium and ventricle of mouse. (**B**) Lack of specific reactivity with the skeletal muscle. (**C**) Peroxidase staining of mouse heart showing transverse striations. (**D**) The specimen in C viewed under polarized light microscope showing A bands in the myofibrils. (See Saegusa *et al.*[28] for further details.)

antibodies show multiple epitope changes. The epitopes on CVB4 viruses have been grouped into highly, moderately and poorly conserved epitopes. These antibodies have made it possible to select variants with changes in pathogenic potential. These antibodies have been used to delineate the structure and function of the viral proteins and to understand the similarities that exist between host proteins and the virus. It would have been very difficult, if not impossible, to make some of these observations without using monoclonal antibodies. Considering how useful these antibodies have been in the study of the biology of CVB4, it is certain that a similar approach would improve our understanding of other coxsackieviruses.

ACKNOWLEDGMENTS. The author is indebted to Dr. Abner Louis Notkins for his support and guidance and is grateful to Eloise Mange for excellent editorial assistance.

REFERENCES

1. Prabhakar, B. S., Haspel, M. V., McClintock, P. R., and Notkins, A. L., 1982, High frequency of antigenic variants among naturally occurring human coxsackie B4 virus isolates identified by monoclonal antibodies, *Nature (Lond.)* **300:**374–376.
2. Prabhakar, B. S., Haspel, M. V., and Notkins, A. L., 1984, Monoclonal antibody techniques applied to viruses, in: *Methods in Virology*, Vol. VII (C. Maramorosch and H. Koprowski, eds.), pp. 1–18, Academic, New York.
3. McKearn, T. J., 1980, Cloning of hybridoma cells by limiting dilution in fluid phase. *Monoclonal Antibodies: Hybridomas: A New Dimension in Biological Analyses* (R. H. Kennett, T. J. McKearn, and K. B. Bechtol, eds.), p. 374, Plenum, New York.
4. Holland, J. J., 1984, Continuum of change in RNA virus genome, in: *Concepts in Viral Pathogenesis* (A. L. Notkins and M. B. A. Oldstone, eds.), pp. 137–143, Springer-Verlag, New York.
5. Holland, J., Spindler, K., Horodyski, F., Grabau, E., Nichol, S., and Vandepol, S., 1982, Rapid evolution of DNA genomes, *Science* **215:**1577–1585.
6. Prabhakar, B. S., Menegus, M. A., and Notkins, A. L., 1985, Detection of conserved and non-conserved epitopes on Coxsackievirus B_4: Frequency of antigenic changes, *Virology* **146:**302–306.
7. Prabhakar, B. S., McClintock, P. R., Haspel, M. V., Wohlenberg, C., Menegus, M., and Notkins, A. L., 1983, Monoclonal antibodies as probes to study antigenic variations in coxsackie B viruses, *Igiene Mod.* **80:**1071–1079.
8. Prabhakar, B. S., and Notkins, A. L., 1984, Antigenic variants of viruses and their relevance to clinical disease, in: *Concepts in Viral Pathogenesis* (A. L. Notkins and M. B. A. Oldstone, eds.), pp. 158–162, Springer-Verlag, New York.
9. Webb, S. R., Kearse, K. P., Foulke, C. L., Hartig, P. C., and Prabhakar, B. S., 1986, Neutralization epitope diversity of Coxsackievirus B_4 isolates detected by monoclonal antibodies, *J. Med. Virol.* **20:**9–15.
10. Cao, Y., Schnurr, D. P., and Schmidt, N. J., 1984, Differing cardiotropic and myocarditic properties of group B type 4 coxsackievirus strains, *Arch. Virol.* **80:**119–130.
11. Cao, Y., Schnurr, D. P., and Schmidt, N. J., 1984, Monoclonal antibodies for study of antigenic variation in coxsackievirus type B4: Association of antigenic determinants with myocarditic properties of the virus, *J. Gen. Virol.* **65:**925–932.
12. Prabhakar, B. S., Srinivasappa, J., and Ray, U. R., 1987, Selection of Coxsackievirus B_4 variants with monoclonal antibodies results in attenuation, *J. Gen. Virol.* **68:**865–869.
13. Beatrice, S. T., Katze, M. G., Zajac, B. A., and Crowell, R. L., 1980, Induction of neutralizing antibodies by the coxsackievirus B3 virion polypeptide, VP2, *Virology* **104:**426–438.
14. Jerne, N. K., 1974, Towards a network theory of the immune system, *Ann. Immunol. Inst. Pasteur* **125C:**373–389.

15. Sege, K., and Peterson, P. A., 1978, Use of anti-idiotypic antibodies as cell-surface receptor probes, *Proc. Natl. Acad. Sci. USA* **75**:2443–2447.
16. Wassermann, N. H., Penn, A. S., Freimuth, P. I., Treptow, N., Wentzel, S., Cleveland, W. L., and Erlanger, B. F., 1982, Anti-idiotypic route to anti-acetylcholine receptor antibodies and experimental myasthenia gravis, *Proc. Natl. Acad. Sci. USA* **79**:4810–4814.
17. Schreiber, A. B., Couraud, P. O., Andre, C., Vray, B., and Strosberg, A. D., 1980, Anti-alprenolol anti-idiotypic antibodies bind to β-adrenergic receptors and modulate catecholamine-sensitive adenyl cyclase. *Proc. Natl. Acad. Sci. USA* **77**:7385–7389.
18. Kennedy, R. C., and Dreesman, G. R., 1985, Immunoglobulin idiotypes: Analysis of viral antigen-antibody systems, *Prog. Med. Virol.* **31**:168–182.
19. Kennedy, R. C., Melnick, J. L., and Dreesman, G. R., 1984. Antibody to hepatitis B virus induced by injecting antibodies to the idiotype, *Science* **223**:930–931.
20. Reagan, K. J., Wunner, W. H., Wiktor, T. J., and Koprowski, H., 1983, Anti-idiotypic antibodies induce neutralizing antibodies to rabies virus glycoprotein, *J. Virol.* **48**:660–666.
21. Ertl, H. C. J., and Finberg, R. W., 1984, Sendai virus-specific T-cell clones: Induction of cytolytic cells by an anti-idiotypic antibody directed against a helper T-cell clone, *Proc. Natl. Acad. Sci. USA* **81**:2850–2854.
22. Uytdehaag, F. G. C. M., and Osterhaus, A. D. M. E., 1985, Induction of neutralizing antibody in mice against poliovirus type II with monoclonal anti-idiotypic antibody, *J. Immunol.* **134**:1225–1229.
23. Nepom, J. T., Weiner, H. L., Dichter, M. A., Tardieu, M., Spriggs, D. R., Gramm, C. F., Powers, M. L., Fields, B. N., and Green, M. I., 1982, Identification of a hemagglutinin-specific idiotype associated with reovirus recognition shared by lymphoid and neural cells, *J. Exp. Med.* **155**:155–167.
24. McClintock, P. R., Prabhakar, B. S., and Notkins, A. L., 1986, Anti-idiotypic antibodies against anti-Coxsackievirus B_4 monoclonal antibodies, in: *Virus Attachment and Entry into Cells* (R. L. Crowell and K. Lonberg-Holm, eds.), pp. 36–43, American Society for Microbiology, Washington, D.C.
25. McClintock, P. R., Prabhakar, B. S., and Notkins, A. L., 1986, Anti-idiotypic antibodies to monoclonal antibodies that neutralize Coxsackievirus B_4 do not recognize viral receptors, *Virology* **150**:352–360.
26. Hogle, J. M., Chow, M., and Filman, D. J., 1985, Three dimensional structure of poliovirus at 2.9, A resolution, *Science* **229**:1358–1365.
27. Rossmann, M. G., Arnold, E., Erickson, J. W., Frankenberger, E. A., Griffith, J. P., Hecht, H-J., Johnson, J. E., Kamer, G., Luo, M., Mosser, A. G., Rueckert, R. R., Sherry, B., and Vriend, G., 1985, Structure of a human common cold virus and functional relationship to other picornaviruses, *Nature (Lond.)* **317**:145–153.
28. Saegusa, J., Prabhakar, B. S., Essani, K., McClintock, P. R., Fukuda, Y., Ferrans, V. J., and Notkins, A. L., 1986, Monoclonal antibody to Coxsackievirus B_4 reacts with myocardium. *J. Infect. Dis.* **153**:372–373.
29. Srinivasappa, J., Saegusa, J., Prabhakar, B. S., Gentry, M. K., Buchmeier, M. J., Wiktor, T. J., Koprowski, H., Oldstone, M. B. A., and Notins, A. L., 1986, Molecular mimicry: Frequency of reactivity of monoclonal antiviral antibodies with normal tissues, *J. Virol.* **57**:397–401.

9

Host Conditions Affecting the Course of Coxsackievirus Infections

ROGER M. LORIA

1. INTRODUCTION

A search of the literature showed that specific review articles on host conditions affecting viral infection or specifically group coxsackievirus (CV) infection could not be found in the major data bases. Consequently, the reader is referred to general reviews on group CV and/or enteroviruses for additional details.[1–8] The paucity of reviews in this area is indicative of a general trend that has emphasized the role of the infectious agent and minimized the importance of the interaction between the host and the infecting virus.

The terms host conditions or host factors are used throughout this review to connote physiologic, metabolic, genetic, and pathophysiologic conditions that influence host response to CV infection. The importance of host conditions in affecting the course of viral disease was accentuated in the hypothesis advanced by Huppert and Wild.[9] In their paper, these workers suggest that certain viral infections cause disease by a hit-and-run mechanism. Accordingly, viral infection would cause the initial damage or trigger an event(s), then be cleared from the host, and have no further role in the disease process. Host conditions would then be the

ROGER M. LORIA • Departments of Microbiology, Immunology, and Pathology, Virginia Commonwealth University; School of Basic Health Sciences, Medical College of Virginia, Richmond, Virginia 23298.

principal determinants of the disease process. Indeed, Loria *et al.*[10] suggested that CVB4 might be a trigger agent in the host with a genetic predisposition to diabetes mellitus; host conditions, in this case the genetics, would determine the type of response to infection.

The humoral or cellular immune responses to CV are considered only when pertinent to the understanding of the effects of host conditions on infection with CV. Several chapters in this volume review topics that can be regarded as specific host conditions. Consequently, several of these subjects are discussed only briefly to maintain continuity.

The coxsackieviruses, group A and B, are composed of coxsackieviruses A1–22 and 24 and coxsackieviruses B1–6. Furthermore, many of the newer isolates, enteroviruses 68–72, induce clinical syndromes similar to the group CV. The clinical syndromes associated with group CV are listed in Table I. The wide range of clinical syndromes caused by one group of viruses illustrates their opportunistic nature. It may also be an indication that other factors contribute to the heterogeneity of clinical syndromes that group CV will induce.

2. INFECTION: ROUTE, DOSE, SITE OF VIRUS LOCALIZATION

2.1. Mode or Route of Virus Inoculation

Host susceptibility to CV infection may vary considerably, whether virus inoculation is done by the intracerebral, intravenous, subcutaneous, intraperitoneal, or oral route.[1–13] Generally, it has been accepted that the natural route of CV infection is the fecal–oral route, identical to that for the poliovirus group. However Maximovich *et al.*[14] reported infecting 1–2-day-old mice with CVA5 and CVB4 by aerosols. Although considered intestinal viruses, the CV group also cause a considerable number of respiratory illnesses[15] in humans and can also infect the host by the respiratory route.[16,17] Relatively few investigators use the oral route for CV infection. Kaplan and Melnick[12] concluded that the oral infection of newborn mice with CV was 10,000 times less sensitive than the intraperitoneal route. However, these workers stated that this "apparent resistance may be due merely to the techniques employed and these animals may not have actively swallowed the full dose of virus." Studies on the ingestion of CVB5 by 6-week- and 4-month-old mice were undertaken by Gevaudan and Charrel,[18,19] who concluded that the intensity and duration of the viremic stage in mice are a function of the titer and the dose of virus inoculated. These investigators were unable to

TABLE I
Clinical Syndromes Associated with Coxsackieviral Infection[a]

Coxsackieviruses, group A, types 1–24[b]
Acute hemorrhagic conjunctivitis (type 24)
Acute lymphatic or nodular pharyngitis (type 10)
Aseptic meningitis (types 1–11, 14, 16–18, 22, 24)
Common cold (types 21, 24)
Encephalitis (types 2, 5–7, 9)
Epidemic myalgia (types 4, 6, 10)
Exanthem (types 4–7, 9, 16)
Hand-foot-and-mouth disease (types 5, 10, 16)
Hepatitis (types 4, 9)
Herpangina (types 1–6, 8, 10, 22)
Infantile diarrhea (types 18, 20–22, 24)
Paralysis (infrequently) (types 4, 7, 9)
Pneumonitis of infants (types 9, 16)
Upper respiratory illness (types 10, 21, 24)
Coxsackieviruses, group B, types 1–6
Aseptic meningitis (types 1–6)
Encephalitis (types 1–3, 5)
Epidemic myalgia (types 1–5)
Hepatitis (type 5)
Paralysis (infrequently) (types 2–5)
Pericarditis myocarditis (types 1–5)
Pleurodynia (types 1–5)
Rash (type 5)
Severe systemic infection in infants, meningoencephalitis and myocarditis (types 1–5)
Undifferentiated febrile illness (types 1–6)
Upper respiratory illness and pneumonia (types 4, 5)
Enterovirus, types 68–72[c]
Acute hemorrhagic conjunctivitis (type 70)
Hand-foot-and-mouth disease (type 71)
Hepatitis (type 72)
Meningoencephalitis (types 70, 71)
Paralysis (types 70, 71)
Pneumonia and bronchiolitis (type 68)

[a]Adapted from Kibrick[1] and Melnick.[2]

[b] Coxsackievirus 23 was never formally accepted as a new type, as it was found to be identical with the previously described echovirus 9.

[c]Since 1968, new enterovirus types have been assigned enterovirus type numbers rather than being subclassified as coxsackieviruses or echoviruses. The vernacular names of the previously identified enteroviruses have been retained.

demonstrate any significant neutralizing antibody levels, suggesting that only a subclinical infection was obtained.

Loria *et al.*[20] compared the results of neonate outbred CD-1 mice infected with CVB5 by intubation or by intraperitoneal injection. The results showed that the oral route and the parenteral route of infection were equally efficacious in establishing infection in the neonate. By contrast, the adult host possesses an effective mechanism against CV infections by the oral route. This type of host defense mechanism appears to have two major components, one of which is a barrier system that prevents penetration of CV from the intestinal lumen into the circulation. It was shown that in the young adult mouse the intestinal barrier can reduce the penetration of virus from the intestinal lumen into the circulation by at least 100,000-fold. In addition to the barrier system, the second defense is an active virus-clearance system which is located on the mucosal side of the intestinal tract. It functions by preventing the replication and clearance of virus adsorbed to the intestinal mucosa. This virus-clearance system is not operative in either the neonate or on the serosal side of the intestinal tract. Both the intestinal barrier and virus-clearance systems are host factors that affect CV infection. These systems are age dependent, and their development is associated with differentiation, maturation, and aging.[21,22]

2.2. Dose Effect

Although animals may be susceptible to infection by CV, high doses of CV may alter the course of the disease. This is particularly evident in newborn mice, in which high doses may cause rapid death without paralysis within 48 hr, most likely due to toxicity. However, a low inoculum, below the minimal infectious dose, may result in a nonproductive infection. Such small doses of virus may be inactivated prior to cellular infection, insufficient to form a primary focus,[16] or may be too small to generate sufficient virus progeny to sustain systemic virus spread to the particular target organs. Clearly, the minimal oral infectious dose in the neonate and adult may differ by as much as 100,000-fold.[21,22].

Finally, despite a scarcity of data, it is suggested that the route of infection, the size of the minimal infectious dose, and the site of virus localization are all conditions modulated by the host and determine in part the type of syndrome that may develop.

3. TEMPERATURE

The influence of environmental and body temperature on susceptibility to CV infection was described by Boring, showing that mice maintained at 4°C had a higher susceptibility to infection.[23] These observations were confirmed by Gevaudan *et al.*[24–26] Subsequently, Teisner and Haar[27] showed that newborn mice aged 1–6 days had a body temperature below 35.5°C and that there is a correlation between the change in body temperature and development of resistance to CVB1. Gevaudan and Charrel[26] reported that the elevation of body temperature is beneficial to the host infected with a temperature-sensitive CVB5, since it reduced mortality considerably. The effect of temperature on resistance to CV was studied in order to assess the role of interferon in CV infections,[28] since interferon production is temperature dependent. Mice maintained at 4°C succumbed to infection but had higher interferon titers than did surviving animals maintained at 25°C. Interferon does not appear to have a significant protective role in CV infections, even though CV variants with differing interferon sensitivity exist in nature.[29] Finally, adaptation to an ambient temperature lower or higher than the normal may function as a selective condition to select and limit various temperature-sensitive mutants present in the infectious dose.[26,30]

4. EFFECT OF AGE AND AGING ON COXSACKIEVIRUS INFECTION

Age of Infected Animal

Sigel[31] and Bang *et al.*[13] reviewed the factors involved in age-related susceptibility to viral infection by the intraperitoneal route. The striking age-related high mortality rates of newborn mice when infected by CVB is in contrast to the mild or inapparent infections in adults.[32] A pronounced decrease in susceptibility to viral infection is evident on the first day of life, with the nadir in susceptibility reached on the tenth day.[33] Similar results were obtained for CVB3 infection by Burnstein.[34] These observations are summarized in Fig. 1.[34a]

Some of the proposed explanations for age-specific CV susceptibility are as follows:

1. Delayed or impaired antibody formation (Dalldorf[35])
2. Development of immunologic tolerance (Hotchin[36])
3. Interferon tissue content (Heineberg *et al.*,[37] Rytel[38])
4. Increased adrenocortical activity (Dalldorf,[35] Rytel[38])
5. Sex hormones (Berkovich *et al.*[39,40]
6. The presence of natural tissue barriers like the blood–brain barrier (Sabin,[41] Siegel[31]) and the intestinal barrier (Loria *et al.*[21,22])
7. Availability of virus-receptor sites (Kunin,[42] Crowell[43])

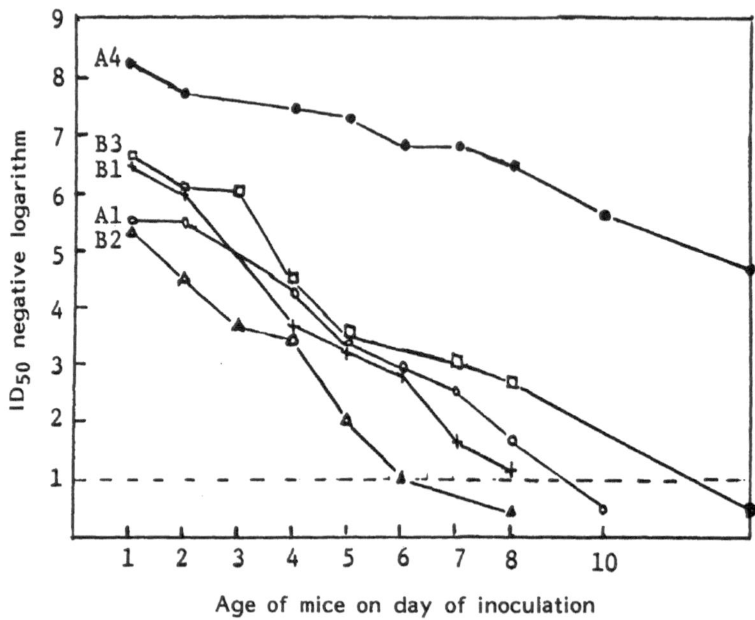

FIGURE 1. Increased resistance, with age, of newborn mice inoculated with five antigenically distinct coxsackieviruses. (Modified from Melnick and Curnen[34a].)

None of these factors alone is adequate to explain the age-related susceptibility of viral infection.

Tissue susceptibility to CV also varies with age, as reported by Grodums *et al.*[44,45] While the brain loses its viral susceptibility during the first 2 weeks of life, there is an increase in susceptibility of the heart and brown fat. Grodums and Dempster[46] allude to the tissue tropism and to the possible existence of receptors in their statement, "whichever components of the age factor are concerned, they operate very differently for brain, heart or brown fat tissue." It has been observed that a common pattern of reaction exists in the mouse brain for all types and virus strains tested. Within a period of 2 weeks, complete resistance to damage is asserted. The role of cellular receptors as determinants of viral tropism is reviewed in Chapter 4, this volume.

Khatib *et al.*[47] demonstrated that susceptibility to CVB1–4 was age dependent, species dependent, and differentially organ dependent. These investigators challenged Swiss-Webster mice aged 2 and 14 days and 3–5 months, by the intraperitoneal route. Immediate death occurred only in newborns (2-day-old) in which fulminant encephalitis developed. In the newborn group infected with CVB1 and B4, maximal virus titer in the heart was associated with necrotizing transmural myocarditis. However, in 14-day-old mice infected with CVB2 and CVB3, a nontransmural myocarditis developed. Maximal cardiac virus titer and histopathologic changes were not concurrent. It was suggested that the mechanism may be mediated by an immunopathologic process. Destruction of the exocrine pancreas was more evident in 3–5-month-old than in the younger animals, and pathologic changes coincided with maximal virus titers.

Hornberger and Plotkin[48] extended these observations to the cellular level by reporting that human fetus fibroblasts in culture are susceptible to CVB4 infection, while adult fibroblasts are not. Fibroblast from older fetuses are also resistant to CVB4 infection. These investigators concluded that this was an age-dependent phenomenon associated with reduced attachment and blockade of virus penetration into the cells. Morag *et al.*[49] investigated the prevalence of antibodies to CVA9, CVB1–4, and echovirus types 4, 7, and 9 and enterovirus 72, hepatitis A (HAV) in a homogeneous defined population. At the age of 2–4 years, 40–69% of individuals tested had antibodies to CV and echovirus, in contrast to 4% that had antibodies for HAV, indicating that the acquisition of antibodies and resistance to group CV and echovirus occurred earlier in life than to HAV. At age 5–17 years, 85% had antibodies to five of the following enteroviruses CVA9, CVB1–4, and echovirus 4, 7, and 9, but only 10% had antibodies to HAV. Prevalence of antibodies to CV and echovirus were found to be higher in females than in males by age 18 and older, but not in the group aged 18 years and younger. It is suggested that this may be associated with child care and close contact. The results also demonstrate the relative greater exposure to and specific age distribution of group CV as compared with hepatitis A virus, two agents that infect humans by the oral route.

Noah and Urquhart[50] reported that the age distribution of CVB infection shows a small but definite difference as compared with other enteroviruses in being relatively more common in the neonates and the younger children than CVA or echoviruses. Indeed, the age distribution of the cases associated with CVA, mostly CVA9, was similar to that of echovirus, except for a slight preponderance of cases in the 5–9 age group. Kaplan *et al.*[51] noted another type of age distribution. They reported the isolation of 602 positive cultures of CVB, 81 of these (13.45%) from infants under 3 months of age. Aseptic meningitis was the most frequent syndrome observed in 48 of 77 (57%). Three age-associated patterns of death were noted: a rapid pattern in patients 2–17 days old, a biphasic pattern at 8–24 days old, and a group with a progressive illness; 24 mothers had evidence of virus-like infection 10 days antepartum and 5 days postpartum.

Rosenberg *et al.*[52] state that the symptoms and the sequelae of CV infections are not uniform when viral spread occurs in nurseries.[53] In their opinion, premature infants are more affected, presumably because of an inadequate capacity to develop an inflammatory response. DeSa[54] reviewed 3196 autopsies and reported that 21 of 24 cases of isolated myocarditis occurred in infants under 1 year of age. In 16 of these infants, there were no antecedent signs prior to death; sometimes, a short clinical history of less than 24 hr was observed. Serologic and virologic data of CVB as the causative agent were available only in two cases, due to the lack of suspicion of viral myocarditis.

A study of 666 neonates, of which 586 infants were cultured at one to four weekly home visits revealed that the incidence of nonpolio enterovirus isolation was 12.8%. The overall prevalence of enterovirus excretion was 5.3% over the 1-month follow-up period. Risk of viral infection was associated only with lower socioeconomic status ($p < 0.0001$) and lack of breast-feeding ($p < 0.0001$). In late summer and early fall, enteroviral infections are common during the first year of life; of these, 79% of infections were asymptomatic, but 21% of all infants with positive viral cultures were readmitted to the hospital.[55]

Chamberlain *et al.*[56] performed a study on 49 children, 5–8 years old, who had a proven enterovirus CNS infection during the first year of life; 21 of these cases were CV infections. These investigators could not prove that CNS virus infection during the first year of life affected the later development of those children who recovered from the original illness.

King *et al.*[57] showed that the development of homotypic or heterotypic antibodies to CVB1–6 responses was age related; 81% (29 of 36) of children aged 6 months to 4 years had a homotypic response, whereas 77% (44 of 57) of persons aged 15 years had heterotypic responses.

5. FETAL AND NEONATAL INFECTIONS

5.1. Pregnancy as a Host Condition

The available evidence indicated that pregnancy is a host condition associated with greater susceptibility to infection by group CV agents.[58–60] Reyes *et al.*[61] isolated CVB5 from the cervix and throat of four women and from the rectum in three during the third trimester of pregnancy. Symptoms in the women were undifferentiated febrile illness or aseptic meningitis. The newborn infants had no symptoms of CV infection. By contrast, Baker and Philips[62] reported four mothers, with only one having any clinical symptoms of a flu-like syndrome, but all four babies had fulminant infection. CVA3 and CVB2 were isolated from two individuals and CVB4 from two other cases.

Kibrick and Benirschke[63] reported the first case of intrauterine infection with CVB3. Benirschke was unable to demonstrate any histopathologic changes in CVB3-infected placentas. These observations are in good agreement with experimental data.[64–66] CVA4 was also recovered from the meconium of a newborn infant with multiple defects and who died from bronchopneumonia.

Available data clearly illustrate that the group CV can pass through the placental barrier and cause *in utero* infection. When CVB has been shown to cause an intrauterine infection, it appears to occur late in pregnancy and overlaps with delivery.[52] Furthermore, evidence indicates that there is an age preference for CVB infections and that this age-dependent susceptibility is operating at the host, tissue, and cell levels.

5.1.1. *Teratogenicity*

Serologic studies by Brown[67–69] and Brown and Karunas,[70] reported a significant association between infection with group CVB1–5 during pregnancy and congenital heart disease, even though many of the infections in the mothers had been asymptomatic. Serum from 22,935 women, 630 mothers of infants with anomalies, and 1164 controls were examined. Retrospective serologic evidence of infection during first trimester and last 6 months of pregnancy with CVB1–5, CVA9, and echovirus 6 and 9 was sought.

A positive correlation between maternal infection and infant anomaly with CVB2–4, and CVA9 was found. The overall anomaly rate associated with first trimester infections with CVB4 was significantly higher than in controls. Maternal CVB2 infection throughout pregnancy, CVB4 during first trimester, and infection with at least one of the five CVB were all associated with urogenital anomalies when compared to controls. CVA9 infection was associated with digestive anomalies and CVB3 and CVB4 with cardiovascular defects.

Hanshaw *et al.*[71] in a study involving several hundred patients were unable to show a connection between CV and congenital heart disease. Hall and Reed[72] evaluated 350 patients in a study of congenital contracture of the joints, arthrogryposis, and could not document an effect of CV in humans. However, Drachman *et al.*[73] reported that CVA inoculation on day 7 of gestation resulted in viral myopathy and contracture of the joints in chicks. Koskimies *et al.*[74] performed a prospective study of 48,000 obstetrical patients in order to identify possible teratogenic factors. A study of 274 mothers of children with maldevelopment examined for possible viral infection showed seroconversion of CVB5 titers to be increased in this group. Neither maternal age nor previous abortion could be implicated as causative factors; however, consumption of drugs, especially analgesics, was higher in mothers with increased antibody titers to viruses. Koskimies *et al.*[74] conclude that no causal determination of any single factor could be established. Consequently, the teratogenecity of CV remains in question.

5.1.1.a. Experimental Data. Modlin and Crumpacker[76] used CVB1 to infect by the gastrointestinal (GI) route CD-1 outbred pregnant mice. Transplacental infection of the fetus occurred transiently during maternal infection. Pregnant mice infected in the third trimester had significantly higher geometric mean virus titers in the blood, heart, liver, and uterus, and infection persisted longer than in nonpregnant mice ($p = 0.04$). Based on the degree of histopathologic findings, Modlin and Crumpacker concluded that CVB1 infection in late gestation is more severe than in mice infected at earlier gestational age. Furthermore, pregnant mice infected during early gestation (7 days) had lower geometric mean virus titers in the blood ($p = 0.02$) and virus persistence was shorter than in pregnant mice infected after 14 days. Fourty-eight percent of all animals infected during the early stages of the pregnancy spontaneously aborted, while late gestation-infected mice had markedly lower rates of spontaneous abortions.

Lansdown[66] infected Theiler Swiss mice by the intramuscular or intraperitoneal route with CVB3 on the eighth day of gestation. He observed a 40.15% reduction in fetal weight in CVB3-infected fetuses, $p < 0.001$. Thymus weight was 61.11% lower than in uninfected controls. The ratio between the thymus weight and body weight in CVB3-infected fetuses was 35.06% lower. Maternal pancreatic exocrine insufficiency was also evident.[64] When this investigator[66] infected these animals on days 12–14 of gestation, fetal weight was reduced by 29.80%, $p < 0.001$. However, heart weight was increased by 27.08% (NS) and the heart weight to body weight ratio was augmented by 137.04%. As also observed by Modlin *et al.*,[76] survival of neonates born of mothers infected with virus was markedly lower. Macroscopically, CVB3-infected neonate hearts had prominent auricles and an abnormal ventricular form, with a bifid heart apex and an increased size of the right ventricle relative to the left. However, no major histopathological abnormalities were evident. Miranda *et al.*[77] reported that females that had recovered from CVB5 infection had a reduced fecundity, as evident from the reduced capacity to become pregnant and a reduction in mean litter size and greater perinatal mortality. Gauntt *et al.*[78] inoculated neonatal or 7-day-old CD-1 outbred mice with either of two temperature-sensitive mutants, ts1 and ts6, or the parent CVB3. These animals develop a forebrain anomaly (porencephaly or hydraencephaly), which is dependent on the age at time of inoculation and the viral variant used.

The age-dependent resistance to virus-directed tissue destruction was not caused by either increased neutralizing antibodies or interferon. The replication of ts6 mutant was approximately 100-fold higher in the neonatal brain than in the brain of 7-day-old animals. The inability of the neonatal mice to maintain a body temperature of 37°C and exhibiting body temperatures down to 35°C could favor the replication of *ts*6 over the ts1 mutant.

Other members of the group CV have also been implicated in fetal infections and/or passage through the placental barrier; these include CVA8,[54] CVB3,[79] and CVB4.[80] Flamm[81] showed that rabbits infected intravenously with either CVA1 or CVA9 had blastocyte infection in the early stages, while the amniotic fluid was infected in the later stages of pregnancy.

6. GENETICS

The influence of host genetics on the susceptibility to virus and on the nature of the interaction with the infectious virus has been documented by many investigators.[82–84] Host virus interaction, per se, is a complex phenomenon in which the host genetic component is polygenic in nature and the H-2 (MHC)-associated factor(s) are only one facet of this system. The host conditions are one aspect of this interaction that are considered in this section.

Medrano and Green[82] and Miller *et al.*[85] reported that cellular susceptibility to poliovirus was associated with the presence of cellular receptors coded by genetic information present on human chromosome 19. The cellular receptors of polio II, CVB3, and Echo II were also shown to be determined by chromosome 19 genetics.

6.1. Diabetes

The influence of host genetics has been evident in situations in which a particular mutation or host condition is associated with an increased susceptibility to CV infection. One such host condition is the genetic predisposition to diabetes mellitus. Loria *et al.*[10] reported that the diabetes mutation, db, on chromosome 4 in the inbred C57BL/KsJ mouse exerted a specific influence on the host response to CVB4 infection, as determined from the LD_{50} and the percentage of cumulative mortality response. This host influence was specific, since two other diabetes mutations—the yellow–obese, A^y/a, or the obese–diabetes, ob/ob on chromosome 2 and 6, respectively—did not have the same effect. Mice with the misty coat color mutation, m/m located 1 cen-

TABLE II
Pancreatic Titers of Coxsackievirus B4 Titers in Diabetic and Nondiabetic Mice[a]

Genotype	Phenotype	Infecting dose[b]	Pancreatic titer[c,d]
C57BL/KsJ db+/db+	Diabetes	1.04	9.56
C57BL/KsJ db+/+m	—	1.20	9.88
C57BL/KsJ +m/+m	Coat color	3.11	9.55
C57BL/6J ob/ob	Obese–diabetes	0.80	9.27
C57BL/6J +/+	Control	0.90	9.04
C57BL/6J A^y/a	Yellow–obese	2.65	9.80
C57BL/6J a/a	Control	2.14	9.24

[a]Mean log virus titers of all pancreatic tissues: 9.54 ± 0.1.
[b]Dose is in log PFU/animal.
[c]Titer of virus is expressed as log PFU/g tissue.
[d]Performed 3 days after intraperitoneal infection. All values are the mean of three independent titrations, each done in duplicate.

timorgan (cM) from the diabetes mutation db on the same chromosome, also had a different LD_{50} and cumulative mortality response than the diabetic mutant mice.

The virus concentrations in the pancreas of all genotypes and their inbred controls were essentially identical, 9.54 ± 0.1 log PFU/g tissue, irrespective of the differences in the inoculating doses (Table II). The results illustrate that the virus genotype determines the titers in the pancreas, while the host genetics determine the susceptibility and the rate of host–virus interaction.

The difference between the influence of host genetics on susceptibility to CVB4 and on the host response to the infection was further outlined by Cook *et al.*[86] The susceptibility to CVB4 in inbred mice strains SWR/J, DBA/2, and C57BL/6, as determined by LD_{50} values, was 6.5, >9.0, and 3.6, log PFU/animals, respectively. Even though CVB4 was most lethal to C57BL/6 mice, no significant pancreatic histopathologic change was evident. The most severe pancreatic pathology, i.e., acinar necrosis with acute interstitial inflammation and islet atrophy, was observed in the SWR/J mice, which displayed an intermediate susceptibility to virus-induced mortality. The most resistant mice to CVB4, DBA/2, displayed an acute and chronic interstitial pancreatic inflammation. Glucose tolerance was used to monitor pathophysiologic changes following CVB4 infection. The most significant increase in glucose tolerance was observed in SWR/J mice at 7 and 21 days postinfection; these animals had an intermediate susceptibility to virus-induced mortality. By contrast, evidence of chemical diabetes 21 days after CVB4 infection was present in the C57BL/6 mice that did not have any pancreatic histological abnormalities but that had the highest mortality. These findings illustrate that notwithstanding the infecting viral strain, host conditions, i.e., genetics, hormonal and immunologic, may have a major role in determining the host response to a particular virus infection.

Inbred diabetic mutant mice C57BL/KsJ db+/db+, when maintained on a restricted dietary intake to prevent overeating, do not develop spontaneous diabetes. Neutralization antibodies to CVB4 were not produced in these mice following infection with one-half a CVB4 LD_{50} dose. However, the same homozygous db+/db+ mice allowed to overeat develop overt diabetes and, when challenged with CVB4, produced neutralizing antibodies after a lag period.[87] These findings illustrate the influence of the host diabetic genotype and phenotypic expression on the course of CVB4 infection. Montgomery *et al.*[88] further showed that diabetic mutant mice were unable to produce adequate levels of virus-specific IgG but responded with a high level of nonspecific antibodies following viral challenge, suggesting a polyclonal activation by CVB4. The results from these experiments with diabetic mice are in essential agreement with the results from test on human diabetics. Banatvala *et al.*[89] reported that four IDDM patients, all of whom were Austrian, tested negative for CVB-specific IgM at the onset of the illness but were positive when next tested approximately 3 months after onset of IDDM. They further comment that they had frequently observed several patients with acute CVB in whom specific IgM responses did not appear until some weeks after the onset of symptoms. The specificity of the interaction between the human with a genetic predisposition to diabetes and CVB is also illustrated in a previous study by King *et al.*[57] These workers found a positive IgM response to CVB1–6, as measured by enzyme-linked immunosorbent assay in 39% (11 of 28) children 3–14 years of age, in whom IDDM developed during the same year. A homotypic response to CVB4 was evident in 45% (5 of 11) and in 9% (1 of 11) to CVB5. Islet cell cytoplasmic antibodies (IgG) and complement-fixing islet cell antibodies were detected only in six that were positive for CVB-specific IgM, suggesting that these antibodies do not crossreact. CVB-specific IgM responses were present in only 5.5% (16 of 290) of age-matched nondiabetic London children whose sera were collected during the same period. Banatvala *et al.*[89] confirmed and extended these observations. Recently diagnosed IDDM patients from England, Austria, and Australia and their matched controls were tested for their IgM responses to CVB1–5. In 30 percent (37 of 122), there was a positive IgM response at age <15, while only 6% (15 of 204) of their matched controls were positive ($p < 0.005$). Schernthaner *et al.*[90] extended these findings by showing that 96% of patients with positive CVB4-specific IgM response had at least one of the HLA genes, DR3, DR4 associated with genetic predisposition to diabetes. In an independent study of a CVB4 outbreak, Niklasson *et al.*[91] reported that the only individual to develop diabetes was found to have the HLA-DR phenotype 3,4 which is associated with IDDM. All 22 CVB4-positive individuals were tested for ICCA and islet surface antibodies and were found negative.

In their study, Frisk *et al.*[92] found evidence of an acute infection within 2 months in 54% (13 of 24) of newly diagnosed IDDM children. CVB-specific IgM responses were detected by reverse radioimmunoassay (RIA) in 67% (16 of 24) patients on the day of IDDM diagnosis. Age-matched controls (nondiabetic) during the same period did not have CV-specific antibodies.

Cardiovascular

Host genetics and host conditions may also have a significant role in determining the outcome of CV cardiovascular infection. Since this topic is discussed in detail in Chapter 7 this volume, only a brief statement is presented here. Huber and Lodge[93] isolated two populations of cytotoxic T lymphocytes (CTL) from male BALB/C mice inoculated with heart-adapted CVB3 (Nancy). One of these CTL populations reacts against uninfected myocytes and another against virus-infected myocytes. These observations indicated that host genetic factors, probably autoimmune in nature, mediated the CVB3 myocarditis. Indeed, Schnurr *et al.*[94] state that two mechanisms appear to be operative in causing CVB3 myocardiopathy; one requires normal thymic function and the other depends on viral persistence. Host factors determine which of the mechanisms is in operation.

7. NUTRITION

Severe malnutrition in the form of marasmus reduces the normal body weight by 60%, causing a drastic increase in susceptibility to CVB3. Graded undernutrition, which reduced body weight by only 34%, did not produce such an effect.[95,96] Restoration of marasmic mice to normal food intake on the day of infection resulted in no deaths 6 days later. Woodruff[95,96] showed that marasmus was associated with increased CVB3 titer in tissues, gross cardiac lesions, and massive hepatic necrosis. Starvation caused a marked depression of the mononuclear cell inflammatory response with severe atrophy of the lymphoid tissues and lymphocytopenia.

Feigin *et al.*[97] showed that infection with CVB3 in mice was associated with a decrease in the total concentration of blood amino acids, particularly the amino acids necessary for *de novo* synthesis of proteins. Loria *et al.*[98] infected mice with CVB5 perorally and measured intestinal absorption. *In vivo* intestinal perfusion 3 days postinfection was associated with an increase in glucose (26.5%) and leucine (17%) absorption. These results suggest that host nutrients and nutrient reserves are being taxed during viral infection. The host nutritional condition may contribute markedly in competition with the virus for specific nutrients.

Overnutrition in the form of excessive caloric intake or excessive intake of one particular nutrient can also be a form of malnutrition. Loria *et al.*[99] reported that dietary-induced hypercholesteremia in outbred or inbred mice was associated with a 100% increase in susceptibility to CVB5. Infection of hypercholesteremic animals was associated with leukopenia, severe fatty metamorphosis, and focal necrosis of the liver, as well as ileus, cardiomyolysis, and lack of inflammatory response.

The synergistic effects of hypercholesteremia and CVB5 infection were further documented by Campbell *et al.*,[100] showing increased replication and persistence of virus in tissues, particularly the aorta. There was a marked augmentation of CVB5-mediated cardiopathy, resulting in persistent cardiomyolysis, which was not evident in normocholesteremic virus-infected animals. Hypercholesteremia was also associated with pathologic changes in the aorta of CVB5-infected mice that became evident only months after acute infection. The increased susceptibility to CVB5 was shown to be associated with physiologic changes, which altered host resistance, particularly accumulation of intrahepatic cholesterol. Maximum susceptibility to CVB5 was shown to coincide with a 2.5-fold increase in the ratio of hepatic cholesterol to protein. This metabolic imbalance led to a reduced virus clearance rate from the blood and liver. These dietary conditions were shown to impair the host immune response independently, leading to increased susceptibility to bacterial infection and a reduced ability to reject tumor cells.[101] The decreased resistance to CVB5 in hypercholesteremic mice was primarily mediated by a defect in the nonspecific immune responses of macrophages and monocytes.[102]

8. EFFECT OF HORMONES ON INFECTION

Gelfand *et al.*[103] found a significantly higher enterovirus isolation rate among males than among females during population surveillance studies for poliovirus and coxsackievirus. Indeed, several investigators have demonstrated a relationship between sex hormones and susceptibility to CVB1 in male mice.[39,40] Ainbender *et al.*[104] reported that the female IgA response to poliovirus was significantly higher than that of males. However, Huber *et al.*[105] and Lynden and Huber[106] showed that administration of progesterone to castrated male and female mice prior to virus inoculation resulted in increased virus concentrations as well as the cellular and humoral CV-specific immune response, which in this case led to a more severe myocarditis. Ainbender *et al.* proposed that progesterone may directly increase viral replication or independently enhance both viral replication and T-cell responses.

Individual hormones such as cortisone were shown to increase the susceptibility of adult mice to CVB infection.[107] Rytel[38] suggested that cortisone-mediated susceptibility to CV infection was due to a decrease in the level of interferon mediated by this hormone. However, virus titers were increased in cortisone-treated mice and directly proportional to interferon titers. Consequently, it appears that the interferon response may be determined by the degree of viral replication, rather than limiting CV replication. Interferon may have only a limited role in CV infections.

Gatmaitan *et al.*[108] showed that when CVB3-infected mice were forced to swim in a warm pool, the virulence of CVB3 was drastically augmented. Viral replication was augmented 530-fold over control. Reyes *et al.*[109] reported that this severe exercise-induced stress may be mediated by host hormonal conditions. Using Swiss ICR mice infected with CVB3, they report that exercise is associated with the release of the catecholamines, epinephrine, and norepinephrine from the adrenal medulla into the plasma. Accordingly, these neurohormones regulate the intracellular concentrations of cyclic adenosine monophosphate (cAMP) and cyclic guanosine monophosphate (cGMP), which modulate humoral and cellular immune responses as well as inflammation, leading to the increase in susceptibility to CV infections. Recently, Jamal and Hansen[110] reported that abnormal single-fiber electromyography (SFEMG) was evident in 40 patients with postviral fatigue syndrome; 35 of the 40 patients had a CV infection. These observations, as well as the recent interest in postviral fatigue syndrome, may have considerable application.

9. OTHER AGENTS

Various agents have been shown to have a marked influence on the host response to group CV infection. Rezkalla and Khatib[111] showed that nonsteroidal anti-inflammatory agents, such as salicylates and indomethacin, may adversely affect the course of acute CVB3 murine myocarditis. The use of these anti-inflammatory agents for the relief of aches or fever during CVB3 myocarditis may essentially aggravate the disease process. Andrews *et al.*[112] illustrated that botulinum toxin increases the replication of CVA in the skeletal muscle of mice. The effects of the toxin on the muscle were similar to the effects of surgical denervation, suggesting that synaptic transmission may have a role in regulating the susceptibility of muscle to CVA. Treatment with gold salts was done experimentally in CVB3-infected mice and shown to increase mortality and viral replication markedly.[113] Other factors, such as psychological stress[114] and X-irradiation,[115,116] also may have a negative influence on the outcome of CV infection.

10. SUMMARY

Host conditions contribute to the outcome of infection and the evidence that they have a significant role in group coxsackievirus infection is abundant. Host conditions such as age, temperature, genetics, nutrition, hormones, stress, medication, toxins, and radiation all influence the outcome of CV infection. While the properties of the coxsackieviruses and their role in the mediation of infection and immunity need to be evaluated, the contribution of the host condition to the course and outcome of viral infection should not be overlooked.

REFERENCES

1. Kibrick, S., 1964, Current status of coxsackie and Echo viruses in human disease, *Prog. Med. Virol.* **6:**27–70.
2. Melnick, J. L., 1985, Enteroviruses: Polioviruses, coxsackieviruses, echoviruses, and newer enteroviruses, in: *Virology* (B. N. Fields, D. N. Knipe, R. M. Chanock, J. L. Melnick, R. M. Roizman, and R. E. Shope, eds.) pp. 739–794, Raven, New York.
3. Cherry, J. D., 1983, Enteroviruses, in: *Infectious Diseases of the Fetus and Newborn Infant* (J. S. Remington and J. D. Klein, eds.), pp. 290–334, WB Saunders, Philadelphia.
4. Mims, C. A., 1982, Hot and microbial factors influencing susceptibility, in: *The Pathogenesis of Infectious Disease,* 2nd ed. (C. A. Mims, ed.), pp. 237–258, Academic, New York.
5. Buddingh, J. G., 1965, Pathogenesis and pathology of viral infections, in: *Viral and Rickettsial Infection of Man,* 4th ed. (F. L. Horsfall and E. Tamm, eds.), pp. 339–355, JB Lippincott, Philadelphia.
6. Platt, H., 1965, Pathogenesis, in: *Viral and Rickettsial Infections of Animals* (A. O. Betts and C. J. York, eds.), pp. 167–210, Academic, New York.
7. Beveridge, W. I. B., 1965, Epidemiology of virus diseases, in: *Viral and Rickettsial Infections of Animals* (A. O. Betts and C. J. York, eds.), pp. 335–364, Academic, New York.
8. White, D. O., and Fenner, F. J., 1986, *Medical Virology,* pp. 119–145, Academic Press, New York.
9. Huppert, J., and Wild, T. F., 1984, Virus disease without a virus, *Ann. Virol.* **135:**327–333.

10. Loria, R. M., Montgomery, L. B., Corey, L. A., and Chinchilli, V., 1984, Coxsackie virus B4 infection in animals with diabetes mellitus genotype, *Arch. Virol.* **81**:251–262.
11. Johnson, T., 1954, Family infections by coxsackie viruses, *Arch. Ges. Virusforsch.* **5**:384–400.
12. Kaplan, A. S., and Melnick, J. L., 1951, Oral administration of coxsackie viruses to newborn and adult mice, *Proc. Soc. Exp. Biol. Med.* **76**:312–315.
13. Bang, F. B., and Luttrell, C. N., 1961, Factors in the pathogenesis of virus disease, in: *Advances in Virus Research,* Vol. 8 (K. N. Smith and M. A. Lauffer, eds.), pp. 199–244, Academic, New York.
14. Maximovich, N. A., and Suptel, E. A., 1966, Changes caused by and multiplication of virus coxsackie in tissue of susceptible animals, *Acta Virol.* **10**:425–429.
15. Hable, K. A., O'Connel, E. J., and Herrmann, E. C., 1970, Group B Coxsackieviruses as respiratory viruses, *Mayo Clin. Proc.* **45**:170–177.
16. Lycke, E., 1983, General aspects of pathogenesis of virus infections, in: *Textbook of Medical Virology* (E. Lycke and E. Norrby, eds.), pp. 112–124, Butterworths, London.
17. Magee, W. E., and Miller, O. V., 1970, Individual variability in antibody response of human volunteers to infection of the upper respiratory tract by coxsackie A-21 virus, *J. Infect. Dis.* **122**:127–138.
18. Gevaudan, P., and Charrel, J., 1963, Etude de la virémie expérimentale de la souris adulte après ingestion de virus C, *C. R. Soc. Biol.* **162**:2232–2237.
19. Gevaudan, P., and Charrel, J., 1964, Infection expérimentale de la souris adulte par ingestion de virus coxsackie. Chéminement du virus au cours des phases pré et postvirimiques, *C. R. Soc. Biol.* **158**:2387–2390.
20. Loria, R. M., Kibrick, S., and Broitman, S. A., 1974, Peroral infection with group B coxsackievirus in the newborn mouse, *J. Infect. Dis.* **130**:225–230.
21. Loria, R. M., Kibrick, S., and Broitman, S. A., 1974, Peroral infection with group B coxsackievirus in the adult mouse: Protective function of the gut, *J. Infect. Dis.* **130**:539–543.
22. Loria, R. M., Shadoff, N., Kibrick, S., and Broitman, S. A., 1976, Maturation of intestinal defenses against peroral infection with group B coxsackievirus in the mouse, *Infect. Immun.* **13**:1397–1401.
23. Boring, W. D., Zurhein, G. M., and Walker, D. L., 1956, Factors influencing host virus interactions. II. Alteration of coxsackievirus infection in adult mice by cold, *Proc. Soc. Exp. Biol. Med.* **93**:273–277.
24. Gevaudan, P., and Charrel, J., 1966, Influence de la temperature sur l'volution de l'infection expérimentale du souriceau nouveaux-né par le virus Coxsackie, *C. R. Soc. Biol.* **160**:139–141.
25. Gevaudan, P., Charrel, J., and Sautet, G., 1965, Psychrosensibilité de la souris male à l'infection par le virus coxsackie, *C. R. Soc. Biol.* **159**:1188–1191.
26. Gevaudan, P., and Charrel, J., 1966, Effet favorable de l'hypothermie relative sur l'evolution de la maladie du souriceau nouveau-né inoculé par un variant chaud de virus Coxsackie, *C. R. Soc. Biol.* **160**:820–826.
27. Teisner, B., and Haahr, S., 1974, Poikilothermia and susceptibility of suckling mice to coxsackie B1 virus, *Nature (Lond.)* **247**:568.
28. Haahr, S., and Teisner, B., 1973, The influence of different temperatures on mortality, virus multiplication and interferon production in adult mice infected with CVB1, *Arch. Ges. Virusforsch.* **42**:273–277.
29. Jordan, G. W., and Bolton, V., 1985, Interferon-sensitive coxsackievirus variant in nature, *J. Interferon Res.* **5**:289–296.
30. Gauntt, C. J., Paque, R. E., Trousdale, M. D., Gudvangen, R. J., Barr, D. T., Lipotich, G. J., Nealon, T. J., and Duffey, P. S., 1983, Temperature-sensitive mutant of cox-

sackievirus B3 establishes resistance in neonatal mice that protects them during adolescence against coxsackievirus B3-induced myocarditis, *Infect. Immun.* **39:**851–864.

31. Sigel, M. M., 1952, Influence of age on susceptibility to virus infection with particular reference to laboratory animals, *Annu. Rev. Microbiol.* **6:**247–280.
32. Gifford, R., and Dalldorf, G., 1951, The morbid anatomy of experimental coxsackie virus infection, *Am. J. Pathol.* **27:**1047–1064.
33. Howes, D. W., 1954, Studies of coxsackie viruses. I. Comparison of age-susceptibility relationships in mice, *Aust. J. Exp. Biol. Med. Sci.* **32:**253–264.
34. Burnstein, T., 1960, Relative susceptibility of young mice and hamsters to coxsackie B-3 virus, *Proc. Soc. Exp. Biol. Med.* **105:**306–308.

34a. Melnick, J. L., and Curnen, E. C., 1952, The coxsackie virus group, in: *Viral and Rickettsial Infections of Man,* 2 Ed., (T. M. Rivers, ed.), p. 342. J. P. Lippincott, Philadelphia.

35. Dalldorf, G., 1955, The coxsackie viruses, *Annu. Rev. Microbiol.* **9:**277–296.
36. Hotchin, J. E., Weigand, H., and Benson, L. M., 1960, Immunological tolerance and virus infection with lymphocytic choriomeningitis in mice, *Fed. Proc.* **19:**196 (abst).
37. Heineberg, H., Gold, E., and Robbins, F. C., 1964, Difference in interferon content in tissue of mice of various ages infected with coxsackie B-1 virus, *Proc. Soc. Exp. Biol. Med.* **115:**947–953.
38. Rytel, M. W., 1969, Interferon response during coxsackie B-3 infection in mice. I. The effect of cortisone, *J. Infect. Dis.* **120:**379–382.
39. Berkovich, S., and Ressel, M., 1965, Effect of gonadectomy on susceptibility of the adult mouse to coxsackie B1 virus infection, *Proc. Soc. Exp. Biol. Med.* **119:**690–694.
40. Berkovich, S., and Ressel, M., 1967, Effect of sex on susceptibility of adult mice to coxsackie B1 virus infection. *Arch. Ges. Virusforsch.* **22:**246–251.
41. Sabin, A. B., and Ward, R., 1941, The natural history of human poliomyelitis. I. Distribution of virus in the nervous and nonnervous tissues, *J. Exp. Med.* **73:**771–793.
42. Kunin, C. M., 1962, Virus–tissue union and the pathogenesis of enterovirus infections, *J. Immunol.* **88:**556–569.
43. Crowell, R. L., 1976, Comparative generic characteristics of picornavirus receptor interaction, in: *Cell Membrane Receptors for Viruses, Antigens and Antibodies, Polypeptides, Hormones, and Small Molecules.* (R. F. Beers and E. G. Basset, eds.), pp. 179–203, Raven Press, New York.
44. Grodums, E. I., and Dempster, G., 1959, The age factor in experimental coxsackie B-3 infection, *Can. J. Microbiol.* **5:**595–604.
45. Grodums, E. I., and Dempster, G., 1959, Myocarditis in experimental coxsackie B-3 infection, *Can. J. Microbiol.* **5:**605–615.
46. Grodums, E. I., and Dempster, G., 1962, Pathogenesis of coxsackie group B viruses in experimental infection, *Can. J. Microbiol.* **8:**105–113.
47. Khatib, R., Chason, J. L., Silberberg, B. K., and Lerner, M. A., 1980, Age-dependent pathogenicity of group B coxsackieviruses in Swiss-Webster mice: Infectivity for myocardium and pancreas, *J. Infect. Dis.* **141:**394–403.
48. Hornberger, E., and Plotkin, S., 1983, Human skin fibroblasts are nonpermissive to coxsackie B4 infection: An age dependent phenomenon, *Intervirology* **19:**195–200.
49. Morag, A., Margalit, M., Shuval, H. I., and Fattal, B., 1984, Acquisition of antibodies to various coxsackie and Echo viruses and hepatitis A virus in agricultural communal settlements in Israel, *J. Med. Virol.* **14:**39–47.
50. Noah, N. D., and Urquhart, A. M., 1980, Virus meningitis and encephalitis in 1979, *J. Infect.* **2:**379–383.
51. Kaplan, M. H., Klein, S. W., McPhee, J., and Harper, R. G., 1983, Group B cox-

sackievirus infections in infants younger than three months of age: A serious childhood illness, *Rev. Infect. Dis.* **5:**1019–1032.
52. Rosenberg, H. S., Kohl, S., and Vogler, C., 1981, Viral infections of the fetus and the neonate, *Monog. Pathol.* **22:**133–200.
53. Gear, J. H. S., and Measroch, V., 1973, Coxsackie virus infections of the newborn, *Prog. Med. Virol.* **15:**42–62.
54. deSa, D. J., 1986, Isolated myocarditis as a cause of sudden death in the first year of life, *Forensic Sci. Int.* **30:**113–117.
55. Jenista, J. A., Powell, K. R., and Menegus, M. A., 1984, Epidemiology of neonatal enterovirus infection. *J. Pediatr.* **104:**685–690.
56. Chamberlain, R. N., Christie, P. N., Holt, K. S., Huntley, R. M., Pollard, R., and Roche, M. C., 1983, A study of school children who had identified virus infections of the central nervous system during infancy, *Child Care Health Dev.* **9:**29–47.
57. King, M. L., Shaikh, A., Bidwell, D., Voller, A., and Banatvala, J. E., 1983, Coxsackie-B-virus-specific IgM responses in children with insulin-dependent (juvenile-onset; type I) diabetes mellitus, *Lancet* **1:**1397–1399.
58. Dalldorf, G., and Gilford, R., 1954, Susceptibility of gravid mice to coxsackie virus infection, *J. Exp. Med.* **99:**21–27.
59. Surjus, A., 1961, Effet du virus coxsackie B3 sur la souris gestante et sa transmission transplacentaire, *Ann. Inst. Pasteur* **100:**825–827.
60. Meyer, J., and Loffler, H., 1968, Infektin gravider Mause mit coxsackievirus, *Pathol. Microbiol.* **32:**139–140.
61. Reyes, M. P., Zalenski, D., Smith, F., Wilson, F. M., and Lerner, M. A., 1986, Coxsackievirus-positive cervices in women with febrile illnesses during the third trimester in pregnancy, *Am. J. Obstet. Gynecol.* **155:**159–161.
62. Baker, D. A., and Philips, C. A., 1980, Maternal and neonatal infection with coxsackievirus, *Obstet. Gynecol.* **55:**128–158.
63. Kibrick, S., and Benirscke, K., 1958, Severe generalized disease (encephalohepatomyocarditis) occurring in the newborn period due to infection with coxsackie virus group B, *Pediatrics* **22:**857–875.
64. Lansdown, A. B., 1977, Coxsackievirus B3 infection in pregnancy and its influence on fetal heart development, *Br. J. Exp. Pathol.* **58:**378–385.
65. Lansdown, A. B., 1978, Viral infections and disease of the heart, *Prog. Med. Virol.* **24:**70–113.
66. Lansdown, A. B., 1977, Histological observations on thymic development in fetal and newborn mammals subject to intrauterine growth retardation, *Biol. Neonate* **31:**252–259.
67. Brown, G. C., 1970, Maternal virus infection and congenital anomalies, *Arch. Environ. Health* **21:**362–365.
68. Brown, G. C., 1966, Recent advances in the viral aetiology of congenital anomalies, *Adv. Teratol.* **1:**55–80.
69. Brown, G. C., 1968, Coxsackie virus infections and heart disease, *Am. Heart J.* **75:**145–146.
70. Brown, G. C., and Karunas, R. S., 1972, Relationship of congenital anomalies and maternal infection with selected enteroviuses, *Am. J. Epidemiol.* **95:**207–217.
71. Hanshaw, J. B., and Dudgeaon, J. A., 1978, *Viral Diseases of the Fetus and Newborn,* WB Saunders, Philadelphia.
72. Hall, J. G., and Reed, S. D., 1982, Teratigens associated with congenital contractures in humans and in animals, *Teratology* **25:**173–191.
73. Drachman, D. B., Weiner, L. W., Price, D. L., and Chase, J., 1976, Experimental Arthrogryposis caused by viral myopathy, *Arch. Neurol.* **33:**362–367.

74. Koskimies, O., Lapinleimu, K., and Saxen, L., 1978, Infections and other maternal factors as risk indicators for congenital malformations: A case-control study with paired serum samples, *Pediatrics* **61:**832–837.
75. Taina, E., Hanninen, P., and Gronroos, M., 1985, Viral infections in pregnancy, *Acta Obstet. Gynecol. Scand.* **64:**167–73.
76. Modlin, J. F., and Crumpacker, C. S., 1982, Coxsackievirus B infection in pregnant mice and transplacental infection of the fetus, *Infect. Immun.* **37:**222–226.
77. Miranda, Q. R., Kirk, R. S., and Beswick, T. S. L., 1972, The long term effects of neonatal coxsackie B5 infection in mice: Reduced fecundity of recovered females, *J. Pathol.* **109:**183–193.
78. Gauntt, C. J., Huntington, H. W., Arizpe, H. M., Gutvangen, R. J., and Deshambo, R. M., 1984, Murine forebrain anomalies induced by coxsackievirus B3 variants, *J. Med. Virol.* **14:**341–355.
79. Soike, K., 1967, Coxsackie B3 virus infection in the pregnant mouse, *J. Infect. Dis.* **117:**203–208.
80. Selzer, G., 1969, Transplacental infection of the mouse fetus by coxsackie viruses, *Israel J. Med. Sci.* **5:**125–127.
81. Flamm, H., 1966, Some considerations concerning the pathogenesis of prenatal infection, in: *The Prevention of Mental Retardation Through Control of Infectious Diseases* (H. C. Eichenwald, ed.), pp. 79–87, U.S. Government Printing Office, Washington, D.C.
82. Medrano, I., and Green, H., 1973, Picornavirus receptors and picornavirus multiplication in human-mouse hybrid cell lines, *Virology* **54:**515–524.
83. Coullin, P., Bone, A., Relourcet, R., and Van-Cog, N., 1976, Permissivité de clone cellulaire hybride hommesouris à trois enterovirus: Polio II, coxsackie B3, et echo II: Role dû chromosome humain F.19, *Pathol. Biol. (Paris)* **24:**195–203.
84. Gerald, P. S., and Bruns, G. A., 1978, Genetic determinants of viral susceptibility, *Birth Defects* **14:**1–7.
85. Miller, D. A., Miller, O. J., Dev, V. G., Hashmi, S., Tantravahi, R., Medrano, L., and Green, H., 1974, Human chromosome 19 carries a poliovirus receptor gene, *Cell* **1:**167–174.
86. Cook, S. A., Loria, R. M., and Madge, G. E., 1982, Host factors in coxsackievirus B4 induced pancreopathy, *J. Lab. Invest.* **46:**377–383.
87. Loria, R. M., Montgomery, L. B., and Tuttle-Fuller, N., and Gregg, H. M., 1986, Influence of diabetes mellitus heredity is associated with impaired immunity to coxsackievirus B4, *Diabetes Res. Clin. Pract.* **2:**91–96.
88. Montgomery, L. B., and Loria, R. M., 1986, Humoral immunity in hereditary and overt diabetes mellitus, *Med. Virol.* **12:**255–268.
89. Banatvala, J. E., Bryant, J., Schernthaner, G., Borkenstein, M., Schoeber, E., Brown, D., DeSilve, L. M., Menser, M. A., and Silink, M., 1985, Coxsackie B, mumps rubella, and cytomegalovirus specific IgM responses in patients with juvenile-onset insulin-dependent diabetes mellitus in Britain, Austria, and Australia, *Lancet* **1:**1409–1411.
90. Schernthaner, G., Banatvala, J. E., Scherbaum, W.. Bryant, J., Borkstein, M., Schoeber, E., and Mayr, W. R., 1985, Coxsackie B virus specific IgM responses, complement fixing islet cell antibodies, HLA Dr antigens and C peptide secretion in insulin dependent diabetes mellitus, *Lancet* **2:**630–632.
91. Niklasson, B. S., Dobersen, M. J., Peters, C. J., Ennis, W. H., and Moller, E., 1985, An outbreak of coxsackievirus B infection followed by one case of diabetes mellitus, *Scand. J. Infect. Dis.* **17:**15–18.
92. Frisk, G., Fohlman, J., Kobbah, M., Ewald, U., Tuvemo, T., Diderholm, H., and Friman, G., 1985, High frequency of coxsackie-B-virus-specific IgM in children de-

veloping type I diabetes during a period of high diabetes morbidity, *J. Med. Virol.* **17**:219–227.

93. Huber, S. A., and Lodge, P. A., 1986, Coxsackievirus B-3 myocarditis. Identification of different pathogenic mechanisms in DBA/2 and Balb/c mice, *Am. J. Pathol.* **122**:284–91.
94. Schnurr, D. P., and Schmidt, N. J., 1984, Coxsackievirus B3 persistence and myocarditis in NFR nu/nu and +/nu mice, *Med. Microbiol. Immunol. (Berl.)* **173**:1–7.
95. Woodruff, J. F., 1970, The influence of quantitative post-weaning under nutrition on coxsackievirus B-3 infection of adult mice. I. Viral persistence and increased in severity of lesions, *J. Infect. Dis.* **121**:137–163.
96. Woodruff, J. F., 1970, The influence of quantitative postweaning under nutrition on coxsackievirus B-3 infection of adult mice. II. Alteration of host defenses mechanisms, *J. Infect. Dis.* **121**:164–181.
97. Feigin, R. D., Middelkamp, J. N., and Reed, C. A., 1972, Murine myocarditis due to coxsackie B3 virus: Blood amino acid, virologic and histopathologic correlates, *J. Infect. Dis.* **126**:574–584.
98. Loria, R. M., Kibrick, S., and Broitman, S. A., 1977, Pathophysiological aspects of coxsackievirus B intestinal infection, *Am. J. Clin. Nutr.* **30**:1876–1879.
99. Loria, R. M., Kibrick, S., and Madge, G., 1976, Infection in hypercholesteremic mice with coxsackievirus B, *J. Infect. Dis.* **133**:655–662.
100. Campbell, A. E., Loria, R. M., and Madge, G. E., Coxsackievirus B cardiopathy and angiopathy in the hypercholesteremic host, *Atherosclerosis* **31**:295–306.
101. Kos, W. L., Loria, R. M., Snodgrass, M. J., Cohen, D., Thorpe, T. G., and Kaplan, A. M., 1979, Inhibition of host resistance by nutritional hypercholesteremia, *Infect. Immun.* **26**:658–667.
102. Campbell, A. E., Loria, R. M., Madge, G. E., and Kaplan, A. M., Suppression of immunity of coxsackievirus by dietary hypercholesteremia, *Infect. Immun.* **37**:307–317.
103. Gelfand, H. M., 1961, The occurrence in nature of the coxsackie and ECHO viruses, *Prog. Med. Virol.* **3**:193–244.
104. Ainbender, E., Weisinger, R. B., Hevizy, M., and Hodes, H. L., 1968, Differences in the immunoglobulin classes of polio antibody in the serum of men and women, *J. Immunol.* **101**:92–98.
105. Huber, S. A., Job, L. P., and Woodruff, J. F., 1981, Sex-related differences in the pattern of coxsackievirus B-3-induced immune spleen cell cytotoxicity against virus-infected myofibers, *Infect. Immun.* **32**:8–73.
106. Lynden, D. L., and Huber, S. A., Aggravation of coxsackievirus, group B, type 3 induced myocarditis and increase in cellular immunity to myocyte antigens in pregnant Balb/c mice and animals treated with progesterone, *Cell. Immunol.* **87**:462–472.
107. Kilbourne, E. D., and Horsfall, F. L., Jr., 1951, Lethal infection with coxsackievirus of adult mice given cortisone, *Proc. Soc. Exp. Biol. Med.* **77**:135–138.
108. Gatmaitan, B. G., Chason, J. L., and Lerner, A. M., 1970, Augmentation of virulence of murine coxsackievirus B3 myocardiopathy by exercise, *J. Exp. Med.* **131**:121–136.
109. Reyes, M. P., Thomas, J. A., Smith, F. E., and Lerner, A. M., 1982, Elevated thymocytes norepinephrine and cyclic guanosine 3′5′ monophosphate in T-lymphocytes from exercised mice with coxsackievirus B3 myocarditis, *Biochem. Biophys. Res. Commun.* **109**:704–708.
110. Jamal, G. A., and Hansen, S., 1985, Electrophysiological studies in the post-viral fatigue syndrome, *J. Neurol. Neurosurg. Psychiatry* **48**:691–694.
111. Rezkalla, S., Khatib, G., and Khatib, R., 1986, Coxsackievirus B3 murine myocarditis:

Deleterious effect of nonsteroidal anti-inflammatory agents, *J. Lab. Clin. Med.* **107**:393–395.

112. Andrews, C. G., Drachman, D. B., Pestronck, A., and Narayan, O., 1984, Susceptibility of skeletal muscle to coxsackie A2 virus infection: Effects of Botulinum toxin and denervation, *Science* **223**:714–716.
113. Kabiri, M., Basiri, E., and Kadivar, D., 1978, Potentiation of coxsackievirus B3 infection in adult mice pretreated with gold salt, *J. Med. Virol.* **3**:125–136.
114. Friedman, S. B., Ader, R., and Glasgow, L. A., 1965, Effects of psychological stress in adult mice inoculated with coxsackie B viruses, *Psychosom. Med.* **27**:361–368.
115. Schneck, L., and Berkovich, S., 1965, Effects of X-irradiation on the susceptibility of neonatal rat brain and muscle to coxsackie B1 virus infection, *Proc. Soc. Exp. Biol. Med.* **118**:658–661.
116. Cheever, F. S., 1953, Multiplication of coxsackie virus in adult mice exposed to roentgen radiation, *J. Immun.* **71**:431–435.

10

The Possible Role of Viral Variants in Pathogenesis

CHARLES J. GAUNTT

1. INTRODUCTION

Many different kinds of coxsackievirus (CV) variants have been described and studied with the goal of trying to relate a set of characteristics of a given variant to its pathogenicity in cell cultures or in animal models. In reading about studies about a particular variant, one must remember that any virus stock represents a genotypically heterogeneous population from which phenotypic expression of a dominant variant(s) is observed and measured. Murine models of CV diseases offer exciting possibilities for determining the molecular basis of virulence in these viruses.

2. ORIGIN OF VARIANTS

2.1. Mutation Frequencies in Enteroviruses/Coxsackieviruses

Enterovirus variants arise as a result of mutation and/or recombination events coupled with positive or negative selection pressures operative in the immediate environment of the infected cell. Mutation rate per base in single-stranded RNA viruses has been estimated[1] to be $\sim 10^{-4}$. Prabhakar *et al.*[2] used neutralizing monoclonal antibodies pre-

---

CHARLES J. GAUNTT • Department of Microbiology, The University of Texas Health Science Center at San Antonio, San Antonio, Texas 78284.

pared against the prototype coxsackievirus B4 (CVB4) to assess mutation in capsid antigens of 15 naturally occurring CVB4 isolates and calculated a frequency of $10^{-4.0}$–$10^{-5.8}$. Subsequent microneutralization analyses of 47 clinical isolates of CVB4 with 18 monoclonal antibodies showed three classes of epitopes that were either highly, moderately, or poorly conserved.[3] Studies of several plaque isolates arising from five initial plaques of the JVB prototype revealed epitope appearance or disappearance at a frequency often exceeding 10^{-2}. This apparent high variation in the neutralizing epitope mosaic, likely due to point mutation(s), could not be detected in neutralization assays with hyperimmune antiserum.[3] Similar high rates of mutation were measured in foot-and-mouth disease virus (FMDV) serotypes O and C during serial low multiplicity passages in tissue culture.[4] T1 ribonuclease-oligonucleotide fingerprint mapping of genomic RNA from 34 individual FMDV clones showed each clone to differ in two to eight mutations from the average parental sequence.[4] Some variants showed a selective advantage, i.e., increased yield of infectious progeny.[4] These data are in contrast to the recent report of Parvin *et al.*,[5] who examined mutation rates in the VP1 gene of poliovirus type 1 propagated in tissue culture in absence of selective pressure. Direct nucleotide sequence analyses of VP1 in 105 clones from single plaques showed no mutations, a frequency of less than 2.1×10^{-6} mutations per nucleotide per infectious cycle.[5] Based on an estimate of about five infectious cycles per poliovirus plaque, Parvin *et al.*[5] calculated that a plaque would contain fewer than 8% mutant virions. Although the two different approaches cannot be directly compared, the 100–10,000-fold difference between mutation rates estimated for enteroviruses by the latter work versus the former publications may have several explanations. The difference could be due to (1) number of virus passages and number of replication cycles within each passage, (2) size of nucleotide sequence under study, (3) selective pressures exerted by the host cell, as illustrated by the influence of unknown host cell factor(s) on the production of defective-interfering particles[6]; or perhaps (4) extent of heterogeneity in the initial virus population. The documented emergence of variants, including CVB4 variants,[2,3] must be taken into account during production of large quantities of virus for subsequent study or for production of vaccine virus.

2.2. Recombination in Enteroviruses

Variants of enteroviruses also arise by genetic recombination, at a frequency approximately 1% of the frequency of base substitutions,[7] or 10^{-6}. Recombination has not been demonstrated for the CV. Naturally occurring poliovirus recombinants have been isolated that contain genomic segments from two[8] or three[9] different serotypes. Infectious recombinant viruses have been generated in tissue cultures infected with FMDV[10] or polioviruses.[11–13] In all recombinant enteroviruses, the capsid protein encoding sequence remains intact during any exchange of genetic material, preserving the serotype designation. Recombinant CV with enhanced replication advantages may explain the origin of some of the explosive epidemics due to these viruses.[14,15]

2.3. Variants and Virulence—The Point of it All

Variants of both group A (CVA) and CVB have been described since the discovery of these viruses nearly 40 years ago.[16] The majority of studies have centered on CVB, likely because of their greater ease of propagation in tissue culture. Excellent CVB-animal models of myocarditis[17–20] and diabetes[21] focus attention on understanding mechanisms of disease and defining virulence. Molecular cloning of the genomes of several CVB now offers the possibility of describing virulence in recombinant viruses in terms of a nucleotide sequence(s), secondary structure of the RNA genome, and/or tertiary structure of viral proteins. Seven recombinant type 1 poliovirus constructs between virulent Mahoney and attenuated Sabin strains suggested that attenuation to neurovirulence for monkeys is multiregional on the genome, including the 5′-terminal nontranslated region.[13]

3. TYPES OF COXSACKIEVIRUS GROUP B VARIANTS

3.1. Antigenic Variants

Intratypic variants of CVB have been studied for years.[16] Many variants were antigenically different from the homologous prototype strain in neutralization tests.[22–24] The gold standard assay for identification of new antigenic variants, the kinetic neutralization test, is still usefully applied to isolation of new CVB variants. This test was recently used to show that clinical isolates of CVB4 collected over a 15-year period were antigenically quite diverse, whereas echovirus 30 isolates collected over a 23-year period were antigenically quite homogeneous.[26] Thus, the CVB appear to be evolving. Is the generation of intratypic antigenic variants open-ended or within a definable number range in which variants evolve under mutation and selection pressures at random? Or is there some temporal cyclical scheme defined by sets of acceptable or forbidden (lethal) amino acid substitutions (M. A. Pallansch, unpublished data)? Neutralization immunogens on the surface of intratypic antigenic CVB variants need to be defined by monovalent antisera or monoclonal antibodies to specific peptides, as has been done for poliovirus type 1[27] and human rhinovirus type 14,[28] respectively. This information would be invaluable in determining the potential value of vaccine candidate strains of CVB in the future.

3.2. Receptor Variants

The initial event in CV infections is the attachment to specific receptors on the cell surface.[29] Host range and organ and tissue specificity are determined by this interaction and thus cell receptors are initial determinants of pathogenicity. All six CVB serotypes (prototypes) bind to the same receptor on HeLa cell membranes.[29] Demonstration of additional receptors for CVB serotypes, and thus an extended host range, was demonstrated by Reagan *et al.*[30] Human rhabdomyosarcoma (RD) cells are refractory to replication of CVB due to lack of receptors. These researchers isolated variants of CVB1, CVB3, CVB5, and CVB6 after only two or three undilute passages in RD cells. Each intratypic CVB-RD population agglutinated human erythocytes and produced small plaques on HeLa cells. These variants were present at $<1/10^5$–$<1/10^7$ plaque-forming units (PFU) in the original serotype stocks. The capsid polypeptide VP1 of a CVB3-RD variant exhibited an altered electrophoretic mobility in SDS-polyacrylamide gels. CVB3-RD attached equally well to either HeLa or RD cells, whereas parental CVB3 attached only to HeLa cells. The CVB3-RD variant is avirulent (amyocarditic) in mice, whereas the parent CVB3 is myocarditic (P. S. Whitter, B. J. Landau, B. Goldberg, M. Schultz, and R. L. Crowell, unpublished data). These data demonstrate how a cell population within a human organ or tissue might select and amplify a variant(s) with subsequent pathological consequences.

3.3. Interferon-Sensitive Variants

Interferon (IFN)-sensitive stable plaque variants were isolated from a clinical isolate of CVB4 by selection in cells treated with IFN.[31] IFN-sensitive variants were less virulent (mortality) than were IFN-insensitive variants, although tissue tropism was not different. Reduced virulence in IFN-sensitive variants occurred regardless of whether variants were isolated from original virus stock or from stock after seven passages in mouse pancreas tissue *in vivo*. None of the variants was diabetogenic in mice. The IFN-sensitive variants did not induce IFN in four different cell lines, suggesting that sensitivity was not due to IFN-inducing particles in the stocks, as has been demonstrated for stocks of IFN-sensitive variants of mengovirus[32] and encephalomyocarditis virus.[33] IFN-sensitive variants could not be obtained from clinical isolates of CVB3 or CVB5. Thus IFN-sensitive CVB4 variants exist in nature, and this property can modulate virulence in mice.

3.4. Temperature-Sensitive Mutants

Conditional lethal mutants of the temperature-sensitive (*ts*) type have been isolated from a myocarditic $CVB3_m$ parent strain.[34] Eight mutants were isolated following mutagenesis of parent stock in presence of 5-fluorouracil (5-FU), and two were spontaneous mutants. Complementation analyses were performed with neutral red dye-tagged *ts* mutants and permitted assignment of nine *ts* mutants into three non-overlapping complementation groups.[34] The *ts* events in three prototypes at 39.5°C were found to be in (1) assembly (*ts*11, group I), (2) a capsid polypeptide defect that prevents proper configuration at nonpermissive temperature during capsid construction (*ts*5, group II), and (3) inability to synthesize viral RNA (*ts*1, group III).[35] Temperature-sensitive variants of many viruses are prevalent in nature, and certain *ts* variants are pathogenic in animal models, although diseases caused were less severe than diseases caused by the parental wild-type virus.[36] Many of the diseases produced in animals by *ts* variants involved the central nervous system (CNS).[36] None of the CVB3 *ts* mutants induced myocarditis in adolescent mice.[37] Prototypes of the three groups varied in lethality for neonatal CD-1 mice; *ts*11 and *ts*1 were lethal for about one half the neonates, whereas *ts*5 was lethal in >90% of neonates.[37] Some mice that survived inoculation with *ts*1 had brain damage. Subsequent studies with *ts*1 and *ts*6 mutants (group III) showed that intracranial inoculation induced hydranencephaly or porencephaly, depending on age of the infant mouse at the time of inoculation and genotype of the virus variant.[38] Subsequent to inoculation as neonates, adolescent survivors of *ts*1 had porencephaly and hydranencephaly with similar frequencies (35%), whereas most (70%) of the *ts*6 survivors developed hydranencephaly. The parent $CVB3_m$ was uniformly fatal in neonates, but inoculation of 7-day-old infants produced some hydranencephaly in survivors. Sections of brains from *ts*6-inoculated hydranencephalic mice revealed severe meningeal reactions, necrotizing encephalitis, and liquifactive necrosis in the cerebrum; no pathology was found in the pons, medulla, or cerebellum.[38] Intraperitoneal inoculation or peroral feeding of *ts*6 to gravid mice resulted in approximately 4% of offspring with porencephaly.[39] The forebrain anomalies histologically resemble those found in infants. Four of 28 newborn infants with severe anatomical defects of the brain, including hydranencephaly, had antibody to at least one CVB in ventricular fluids, and two of the infants did not have the same antibody in their serum.[40] Virus was not recovered from ventricular fluids or sera of any infant.

3.5. Variant Virus Particles with New Properties

Purified preparations of prototype (Van Barscholten) CVB4 particles internally contain a protein kinase that is capable of phosphorylating capsid proteins VP1, VP3, and VP4.[41] An anomalous virus particle, termed membrane-bound virions (MBV), contains nearly 20-fold more enzymatic activity than do virions. The activity is cyclic nucleotide independent, requires Mg^{2+} or Mn^{2+} ions, is catalytic over pH 6.0–8.0 (optimum at pH 8.0), and phosphorylates serine residues. Several preparations of both virions and MBV of the E2 diabetogenic variant of the CVB4 Edwards isolate lack this serine kinase under various experimental conditions. Future studies will ascertain whether this enzyme might serve as a marker covariant with diabetogenic potential. Other enteroviruses, including FMDV[42] and poliovirus,[43] contain protein kinases internally in virus particles. Likely of host origin, this enzyme(s) may play a role in early events during replication.

3.6. Defective-Interfering and Other Aberrant Particles

Defective-interfering particles (DIP) of poliovirus[44,45] and mengovirus[46] have been described. DIP can considerably influence the pathogenicity of many virus populations in animal models, and DIP can cause new diseases not induced by virion preparations largely deficient in DIP.[47,48] Although DIP are generally detected during undilute passage of viruses in tissue culture, DIP have been produced in animal tissues during some infections.[48] DIP of the type described for poliovirus and mengovirus, in which 13–16% or 4–6% of genomes are deleted, respectively, have not been reported for the CVB. More than 70 serial undilute passages of $CVB3_m$,[34,37] prototype *ts* mutants of the three complementation groups derived from $CVB3_m$, myocarditic and amyocarditic revertants of each of the three *ts* mutant prototypes and an avirulent CVB3 ($CVB3_o$)[49] did not generate DIP in HeLa cells at 34° or 39.5°C (C. J. Gauntt, unpublished data). These same HeLa cells generated detectable poliovirus type 1 DIP during undilute passages 42–45, and the DIP had similar characteristics to DIP described by Cole *et al.*[44] Recent publications by Chatterjee and associates[41,50,51] report the existence of a new population of MBV arising during infection of HeLa cells with CVB. MBV fulfill several criteria described for enterovirus DIP.

The CVB5 MBV are polygonal, slightly smaller than virions, sediment around 107 S, exhibit a buoyant density of ~1.30 g/ml in CsCl gradients, contain 35 S genomes and only three capsid proteins and at least seven additional proteins (some of host origin) and are infectious.[50] MBV can interfere with replication of CVB5 virions in HeLa cells co-infected with MBV and virions. The MBV of CVB4 have many similar properties; in addition, the 35 S genomic RNA is linked to two VPg of different charge and is distinguishable from virion 35S RNA in one oligonucleotide by oligonucleotide fingerprint mapping.[51] The CVB4 MBV are 75- to >200-fold less infectious than virions. CVB4 MBV inhibit host protein synthesis less severely than do virions and partially block the extent of inhibition by virions. The MBV of CVB4 also interfere with replication of CVB4 virions. Additional studies of the diabetogenic capacity of MBV and of CVB4 MBV–virion mixtures in mice may provide information on the pronounced variation in CVB4 isolates for expression of diabetogenic potential.

4. COXSACKIEVIRUS GROUP B VARIANTS AND ANIMAL MODELS OF DISEASES

4.1. Myocarditic Variants Derived in Tissue Cultures

The capacity of CVB for induction of acute and chronic myocarditis in several animal models is well documented.[8,17,19,20,52] Genotype of the challenge virus is one of the variables that influence induction and severity of disease. Studies of two CVB3 Nancy prototype variants—one an amyocarditic tissue culture-adapted variant ($CVB3_o$) and the other a myocarditic mouse passage-adapted variant ($CVB3_m$)—were assayed for properties covariant with myocarditis in a murine model.[49,53] These two strains could not be differentiated by several virological or *in vivo* properties. Innumerable passages of $CVB3_o$ in tissue culture were likely sufficient for attenuation because a myocarditic variant could not be recovered following many mouse heart tissue passages *in vivo*.[53] Cardiac tissues from mice inoculated with $CVB3_m$, but not $CVB3_o$, contained new extractable antigens that elicited production of cell migration-inhibition factor (MIF) by peritoneal lymphoid cells from $CVB3_m$-inoculated mice in a cell-migration-inhibition assay.[49] This virus-induced neoantigen(s) was not found in normal heart tissue or liver or splenic tissues of $CVB3_m$-inoculated myocarditic mice and does not contain detectable CVB3 antigens or induce neutralizing antibody to CVB3.[54,55] Exclusion chromatography estimated the size of neoantigen(s) to be ~50,000 M_r.[55] In studies by Huber and associates,[56,57] comparison of these two variants *in vivo* showed that virus-specific cytotoxic T lymphocytes isolated from the spleen of inoculated animals preferentially lysed target cells (neonatal myofibers) infected with the homologous variant. Thus these variants induced noncrossreactive virus-induced antigens on the surface of target cells. The differential myocarditic capacities of these two strains may also partly be due to induction of suppressor T cells. Administration of cyclophophamide, a drug that is thought to preferentially eliminate suppressor T cells,[58,59] prior to $CVB3_o$ inoculation resulted in a significantly increased number of myocarditic lesions when compared with inoculation of $CVB3_o$ alone.[54] In another study, it was shown that low-dose cyclophosphamide treatment of $CVB3_o$-inoculated mice induced sufficient $Lyt2^+$ (suppressor T) lymphocytes, which upon transfer to mice prior to $CVB3_m$ inoculation ameliorated myocarditis.[60] Thus these two immunologically indistinguishable CVB3 variants express *in vivo* properties that are readily differentiated by the cell-mediated immune system.

Studies of two revertants of *ts* mutants of $CVB3_m$[37] that belong to the same complementation group showed differences from the parent virus in myocarditic properties and in their genomes.[61] The *ts*1R variant produced little to no myocarditis in any of nine mouse strains examined, whereas *ts*10R could be differentiated from $CVB3_m$ on the basis of the severity of myocarditis induced in mice of selected $H\text{-}2^b$ and $H\text{-}2^k$ haplotypes and in the male versus female myocarditis responses of two additional inbred strains. These three variants could not be differentiated by several virologic and biochemical properties. Differences in the three variant genomes were detected by T1 ribonuclease oligonucleotide fingerprint maps that showed two to six RNA sequence changes. Comparative studies of $CVB3_m$ and an amyocarditic revertant of a *ts* mutant derived from $CVB3_m$[37] (*ts*5R) showed that infection of mouse heart tissues resulted in induction of an extractable neoantigen(s) that could induce production of MIF *in vitro*.[62] Mice inoculated with *ts*5R or $CVB3_m$ produced cytotoxic T lymphocytes (CTL) that could specifically lyse fibroblast target cells infected with either virus. In low-dose cyclophosphamide-treated mice, *ts*5R induced myocarditis, suggesting that *ts*5R induction of T-suppressor cells may have played a role in amelioration of disease. Thus, viral induction of new antigen(s) on the surface of target cells and viral induction of specific CTL that can recognize these new antigenic changes do not always result in induction of myocarditis. Recently, we showed that a nonlytic infection of murine neonatal skin or adolescent mouse cardiac fibroblast cultures by $CVB3_m$ resulted in alteration of the plasma membrane of infected cells, detectable by binding of the lectin *Ulex europaeus* agglutinin I.[63] Maximum binding of lectin occurred at 72 hr postinoculation. Fibroblasts infected with *ts*5R showed weak binding of lectin at 72 hr postinoculation.[62] Thus, quantitative levels of virus-induced surface change may be another explanation for virulence expression. Both $CVB3_m$ and *ts*5R activate natural killer (NK) cells in infected mice; NK cells provide some protection against CVB3-induced myocarditis by limiting viral replication in heart tissue.[64]

CVB4 variants have been examined for markers covariant with neurovirulence[65] and cardiovirulence[66] in mice. Passage of the prototype JBV strain in the presence of guanidine hydrochloride resulted in the selection of guanidine-resistant mutants that were aneurovirulent in newborn mice.[65] In a second study, an aneurovirulent guanidine-resistant mutant was also found to be amyocarditic in suckling HAM/ICR and Swiss ND-4 mice, whereas the prototype was myocarditic in the former mice and amyocarditic in the latter mice.[66] Guanidine resistance was not a stable covariant marker for the amyocarditic property. Both a wild-type CVB4 isolated from a human with myocarditis and the derived guanidine-resistant variant isolated from the wild-type population caused myocarditis in both strains of suckling mice.[66]

Monoclonal antibodies produced against prototype (JVB) and field (Mil) strains of CVB4 were used to obtain nine antigenic variants no longer susceptible to the selecting antibody.[67] These variants were sorted into five distinct groups based on reactivity with 13 different monoclonal antibodies. The antigenic variants differed from parent virus and from each other in greater or lesser cardiotropic and/or myocarditic properties, suggesting that antibodies in a host could select for variants with greater or lesser pathogenic potential than the predominant variant in a population. Recent studies of the JVB and Edwards CVB4 strains and several intratypic variants of these strains with monoclonal antibodies directed against JVB showed that both quantitative and qualitative differences exist in the expression of epitopes on these strains.[68] These preliminary studies suggest it may be possible to differentiate viruses into avirulent/virulent groups on the basis of antigenic structure and specific surface characteristics (epitopes) of a variant may be an expression of the virulence potential of a given genotype.

4.2. Myocarditic Naturally Occurring Isolates

Naturally occurring isolates of CVB are being examined for properties covariant with myocarditis capability. Evolution of poliovirus vaccine strains during passage in humans is well documented at the nucleotide sequence level.[8,69] Analyses of genome sequences are beginning to be applied to human isolates of CVB, which are bountiful[14] (24% of >18,000 enteroviruses in 1970–1979).

A CVB3 variant, isolated from a patient with severe diarrhea and fever (SK-74), was compared with T-70, a variant isolated from a healthy child, with a prototype Nancy strain virus, and with a mouse-passaged variant of the Nancy strain virus (PMH) relative to myocarditis-inducing capabilities in mice.[70,71] Variants SK-74 and PMH caused high mortality (40–85%) in mice, heart tissues of these mice contained high titers of virus, and both variants induced significant myocarditis. All four variants induced lesions in the pancreas, although lesions induced by T-70 were less marked.

Studies of eight naturally occurring human isolates of CVB4 suggested that viral replication in heart tissues appeared to be requisite for development of myocarditis although some relatively amyocarditic variants replicated in heart tissues.[72] Only one of two isolates from patients with myocarditis produced myocarditis in mice. Four of eight isolates caused myocarditis at high (80–90% of animals) or intermediate (33–66%) frequencies following intraperitoneal inoculation.

Preliminary studies of eight naturally occurring isolates of CVB3 for myocarditic properties in adolescent mice have begun at the genome sequence level (S. M. Tracy, M. A. Pallansch, and C. J. Gauntt, unpublished data). In comparison to the myocarditic Nancy strain variant ($CVB3_m$), two isolates are myocarditic and the remaining six are amyocarditic. Of many biologic properties examined *in vitro* and *in vivo,* the myocarditic isolates induce a significantly higher viremia on days 1 and 2 postinoculation than the amyocarditic isolates. Restriction fragments representing >98% of a Nancy prototype genome were used in Southern-blot analyses with near full-length complementary DNA (cDNA) copies of the genomes of all eight isolates. The isolates showed considerable nucleotide sequence divergence from the prototype myocarditic strain, and cDNA of some isolates had little to no homology by hybridization with two to five restriction fragments.

4.3. Diabetes Mellitus Variants Derived in Tissue Culture

The CVB are thought to be etiologic agents for some cases of pancreatic disease in humans and studies with CVB, particularly CVB4, in mice support this hypothesis.[73,74] Although CVB4-mouse models of virus-induced diabetes show virus-induced β-cell damage and insulin-dependent hyperglycemia, a major problem is variable expression in induction of disease in inbred animals using a single virus stock. Intratypic variants may be one explanation for this variability.[75] Three stable plaque variants of the CVB4 Edwards isolate and the parent virus replicated to similar levels in the pancreas of adolescent mice at 3 days postinoculation. However, immunoperoxidase assays showed that the plaque variants (particularly variant E2) accumulated far more viral antigens within islet cells, compared with the parental virus. These data suggest that selection of stable diabetogenic variants from a heterogeneous stock is possible. A subsequent comparative study[76] of the diabetogenic potential of the prototype (Van Barscholten) CVB4 with the E2 variant confirmed variability in induction of hyperglycemia.[75] At 8 weeks postinoculation, approximately 78% of the E2 variant-infected mice were hyperglycemic, compared with only 9% of the prototype virus-infected mice.[76] Insulin synthesis in E2 variant-infected mice was well below the normal mouse level at this time. Two-dimensional gel electrophoresis showed reductions to undetectable levels of many cellular proteins in beta cells of mice infected with the E2 but not the prototype variant.[76] Subsequently, some cellular proteins reappeared in virus-negative β-cells of E2 variant-inoculated mice. Infections with the E2 variant thus induced long term alterations in β-cells, suggesting functional impairment was a factor in virus-induced hyperglycemia. This innovative study offers encouragement that a biochemical understanding of CVB4-induced murine diabetes mellitus may soon be at hand.

4.4. Diabetes Mellitus Naturally Occurring Isolates

Naturally occurring clinical isolates of CVB have been examined for their potential to cause a diabetes-like disease in mice.[77,78] In one study of CVB4 isolates,[77] seven of 12 isolates induced hyperglycemia in mice within 2–4 days postinoculation. Isolates were obtained from patients with upper respiratory illness and aseptic meningitis. The type of specimen from which the isolate was obtained appeared to be a factor. All three fecal isolates and four of six throat swab isolates induced hyperglycemia, whereas the three cerebrospinal fluid isolates did not. Variants obtained during passage of CVB4 in cultured murine β-cells are more diabetogenic for mice than are variants obtained after passage in murine fibroblast cultures.[78] In another study, 37 clinical isolates of CVB1, CVB3, CVB4, and CVB5 were assessed for altering glucose metabolism in mice.[79] Twelve isolates resulted in minor abnormal glucose indices in one or more of six mice, but virulence was not correlated with source of the specimen or type of illness in the patient. Sequential infection of mice with CVB3, CVB4, and CVB5 resulted in only 25 percent of mice with abnormal glucose indices. Inoculation of autoimmune New Zealand (NZB × NZW) F1 male mice with CVB3, CVB4, or CVB5 resulted in only a transient elevation in blood glucose levels and acute acinar pancreatitis.[79] These investigators concluded that the diabetogenic potential of field strains of CVB is limited and that undefined genetic and/or other host factors must be operative to convert animal β-cell damage to clinically overt disease. Recent studies[80] with reference strains of CVB3 and CVB4 passaged in monkey β-cell cultures prior to inoculation into Patas monkeys also suggested that unknown host factors are essential for viral induction of even a transient diabeteslike disease.

5. COXSACKIEVIRUS GROUP A VARIANTS

Intratypic variants of CVA have been reported for years, including variants expressing drug resistance, resistance to agar inhibitors and antigenic variants.[16] Clinical isolates of CVA7 have been differentiated from each other upon the basis of lesion formation in brown fat of cotton rats.[81]

The oligonucleotide fingerprint mapping technique was applied to eight CVA10 isolates obtained in 1978 or 1981–1982 during outbreaks of hand-foot-and-mouth disease or herpangina in a limited area of Japan.[82] The oligonucleotide maps of the four 1978 isolates showed that they were related to each other by 85–93% in their large T1 oligonucleotides, and their map was clearly distinct from the maps of the 1981–1982 isolates. The four 1981–1982 isolates shared only 17–34% of their large oligonucleotides. These data suggested to the authors that one variant likely entered the area in 1978 and underwent genomic changes and selection during human passage. Oligonucleotide maps of the 1981–1982 isolates suggested that at least two and perhaps more independently evolving CVA10 isolates were brought into the area.

6. UNACCOUNTED-FOR COXSACKIEVIRUS VARIANTS

Replication of CV is generally accompanied by lysis of the host cell.[16,83] Numerous exceptions to this generalization have been reported for CVB. CVB or CVA variants that could establish persistent infections in the absence of cytopathology might be associated with diseases different than those known to be caused by the prototype viruses. The possibility that cells persistently infected by some variants might establish a quasi *cos*-like cell system in humans and serve as repositories of CV genome information for subsequent complementation or genetic recombination events must be considered. Indeed, such a *cos*-like cell culture has been established under laboratory manipulation with a deletion mutant of poliovirus.[84]

An isolate of CVB2 replicated in 20 subcultures of human fetal diploid cells without cytopathology and this variant established a persistent infection.[85] Organ cultures of human fetal aorta supported replication of CVB5 without detectable gross degenerative changes and released virus for ≤12 weeks.[86] Cultures of CVB3-inoculated murine neonatal skin fibroblasts initially exhibited some cytopathology but surviving cells became persistently infected and released virus throughout >60 subcultures.[87] Rat insulinoma cells infected with CVB4 resulted in a persistent infection that lasted for 70 days in absence of any cytopathology.[88] Most insulinoma cells were infected, but insulin synthesis in infected cells was not altered in comparison with uninfected cells. Of considerable interest is the recent report of CVB establishing persistent infections of cultured human lymphoid cell lines in absence of cytopathology.[89] This exciting avenue of research must be extended to include the use of freshly isolated cultures of human leukocytes. Mice inoculated with CVB3 contain virus-positive mononuclear leukocytes in their blood for a short period after challenge.[90] In addition, $CVB3_m$ infects murine neonatal skin fibroblasts and replication occurs in absence of cytopathology,[62,63] with some cultures containing virus for a week or more (C. J. Gauntt, unpublished data). Both CVB3 and CVB4 strains can replicate in cultured fetal baboon aortic smooth muscle cells in the absence of cytopathology at the light microscopic level, although electron microscopy showed nonspecific ultrastructural cytopathologic alterations in some cells.[91] Thus, the CVB can replicate in certain cells in absence of cytopathology and are capable of establishing persistent infections in some cell lines that can be subcultured many times. We need to know if CVB persistence is associated with some cases of human diseases.

7. FUTURE GOALS

The goal is to understand the molecular basis of virulence in picornaviruses. Comparative nucleotide sequence analyses and secondary/tertiary RNA genome analyses must be performed on stable virulent and avirulent laboratory-derived quasi-isogenic variants. Similar studies of clinical isolates could be of value provided the researcher can confine the studies to a defined area of the genome; otherwise, multiple mutations unrelated per se to virulence and/or alterations in the genome introduced by genetic recombination events may obscure findings relevant to understanding the basis of virulence. Storage of variant genomes as cDNA may be a more stable repository of this information for future studies.

Additional animal models of disease need to be developed for studies of CV virulence and for a subsequent understanding of the mechanisms of disease(s) induction. In addition to established models of myocarditis and diabetes mellitus, models for polymyositis in mice have been developed.[92,93] Responses to CV in animals are nearly as numerous as one finds in humans infected by these viruses.[16] Adolescent and adult rodent models of diseases are the obvious choice. Gravid mice could also provide excellent models for studies of congenital defects induced by the CVs. The CVB can cross the placenta to infect fetuses,[94] and certain variants have been shown to cause forebrain anomalies in fetuses.[39]

Molecular mimicry[95] describes the sharing of epitopes by viruses and normal cells, a concept proved several times by reactions of monoclonal antibodies prepared to several different viruses with various normal tissues.[95–97] Molecular mimicry is a common phenomenon: of more than 600 monoclonal antibodies to 11 different viruses, approximately 3.5% reacted with specific cells in the organs of normal uninfected mice.[98] One of 66 neutralizing monoclonal antibodies against CVB4 was recently found to react specifically with transverse striations (A bands) of myofibrils in normal mouse hearts.[99,100] Infection of some strains of mice with CVB3 results in production of autoantibodies which react with normal murine heart tissues.[41,101,102] Some patients with acute perimyocarditis produce antibodies that can lyse human myocardial cells cultured *in vitro*.[103] The importance of these antibodies as a pathologic component in chronic diseases in humans could be assessed in currently available rodent models of CV diseases.

Atomic resolution of the three dimensional structures of poliovirus type 1[104] and human rhinovirus type 14[105] have been achieved at 2.9–3.0 Å. These studies described major surface features of the particles and identified neutralizing immunogenic regions on external protrusions. Similar analyses of CVB variants could provide some understanding of the molecular and atomic structure bases for differences in neutralization epitopes, receptors, and pathogenicity in animal models.

Monoclonal antibodies have been produced that react specifically with cell receptors and that prevent attachment of the CVB,[106,107] the polioviruses,[108] and human rhinoviruses.[109] Additional specific antibody probes should be generated to provide useful information on the host range of variant CVB viruses and the possible separation of variants into virulent/avirulent groups and to identify the binding site (the cleft on the icosahedral face?[105]) on virus particles.

ACKNOWLEDGMENTS. I thank Richard Colonno and Mark Pallansch for their critical readings of an early draft of this manuscript. Bellur Prabhakar, Nando Chatterjee, Ji-Won Yoon, and Sally Huber receive a special thanks for sharing with me several preprints of their soon to be published work. Sincere appreciation is extended to Mary Devadoss for bringing order to, and typing, this manuscript. Work in the author's laboratory is partially supported by grants from the National Heart, Lung, and Blood Institute (HL 21047) and the ERACE Foundation, Los Angeles.

REFERENCES

1. Holland, J., Spindler, K., Horodyski, F., Grabau, B., Nichol, S., and Vandepol, S., 1982, Rapid evolution of RNA genomes, *Science* **215:**1577–1585.
2. Prabhakar, B. S., Haspel, M. V., McClintock, P. R., and Notkins, A. L., 1982, High frequency of antigenic variants among naturally occurring human coxsackie B4 virus isolates identified by monoclonal antibodies. *Nature* **300:**374–376.
3. Prabhakar, B. S., Menegus, M. A., and Notkins, A. L., 1985, Detection of conserved and nonconserved epitopes on coxsackievirus B4; Frequency of antigenic change, *Virology* **146:**302–306.
4. Sobrino, F., Dávilla, M., Ortin, J., and Domingo, E., 1983, Multiple genetic variants arise in the course of replication of foot-and-mouth disease virus in cell culture, *Virology* **128:**310–318.
5. Parvin, J. D., Moscona, A., Pan, W. T., Leider, J. M., and Palese, P., 1986, Measurement of the mutation rates of animal viruses: Influenza A virus and poliovirus type 1, *J. Virol.* **59:**377–383.
6. Huang, A. S., 1973, Defective interferring particles, *Annu. Rev. Microbiol.* **27:**101–117.
7. Rueckert, R. R., 1986, Picornaviruses and their replication, in: *Fundamental Virology* (B. N. Fields, D. M. Knipe, R. M. Chanock, J. L. Melnick, B. Roizman, and R. E. Shope, eds.), pp. 357–390, Raven, New York.
8. Minor, P. D., John, A., Ferguson, M., and Icenogle, J. P., 1986, Antigenic and molecular evolution of the vaccine strain of type 3 poliovirus during the period of excretion by a primary vaccinee, *J. Gen. Virol.* **67:**693–706.
9. Kew, O., and Nottay, B., 1984, Evolution of the oral polio vaccine strains in humans occurs by both mutation and intramolecular recombination, in: *Modern Approaches to Vaccines* (R. Channock and R. Lerner, eds.), pp. 357–362, Cold Spring Harbor, New York.
10. King, A. M. Q., McCahon, D., Slade, W. R., and Newman, J. W. I., 1982, Recombination in RNA, *Cell* **29:**921–928.
11. Romanova, L. I., Tolskaya, E. A., Kolesnikova, M. S., and Agol, V. I., 1980, Biochemical evidence for intertypic genetic recombinations of polioviruses, *FEBS Lett.* **118:**109–112.
12. Tokskaya, E. A., Romanova, L. A., Kolesnikova, M. S., and Agol, V. I., 1983, Intertypic recombination in poliovirus: Genetic and biochemical studies, *Virology* **124:**121–132.
13. Omata, T., Kohara, M., Kuge, S., Komatsu, T., Abe, S., Semler, B. L., Kameda, A., Itoh, H., Arita, M., Wimmer, E., and Nomoto, A., 1986, Genetic analysis of the attenuation phenotype of poliovirus type 1, *J. Virol.* **58:**348–358.
14. Moore, M., Kaplan, M. H., McPhee, J., Bregman, D. J., and Klein, S. W., 1984, Epidemiologic, clinical and laboratory features of Coxsackie B1–B5 infections in the United States, 1970–79, *Public Health Rep.* **99:**515–522.
15. Schoub, B. D., Johnson, S., McAnerney, J. M., Dos Santos, I. L., and Klaassen, K. I., 1985, Epidemic coxsackie B virus infection in Johannesburg, South Africa, *J. Hyg. (Camb.)* **95:**447–455.
16. Crowell, R. L., and Landau, B. J., 1979, Picornaviridae: Enteroviruses-Cox-

sackieviruses, in: *CRC Handbook Series in Clinical Laboratory Science,* Virology and Rickettsiology, volume 1, part 1 (G. D. Hsuing and R. Green, eds.), pp. 131–155, CRC, Ft. Lauderdale, Florida.

17. Lerner, A. M., and Wilson, F. M., 1973, Virus myocardiopathy, *Prog. Med. Virol.* **15:**63–91.
18. Woodruff, J. F., 1980, Viral myocarditis. A review, *Am. J. Pathol.* **101:**425–483.
19. Herskowitz, A., Traystman, M. D., and Beisel, K. W., 1986, Murine viral myocarditis—New insights into mechanisms of disease, *Heart Failure* **2:**86–91.
20. McManus, B. M., Gauntt, C. J., and Cassling, R. S., 1986, Immunopathologic basis of myocardial injury, *Cardiovasc. Clin.* **18**(No. 2)**:**163–184.
21. Yoon, J. W., 1983, Viruses in the pathogenesis of type 1 diabetes in: *Current Problems in Clinical Biochemistry. Diabetes and Immunology: Pathogenesis and Immunotheraphy* (H. Kolb, G. Schernthaner, and F. A. Gries, eds.), (H. Aebi, M. Berger, and V. C. Dubach, eds.), pp. 11–38, Hans Huber, Bern, Switzerland.
22. Choppin, P. W., and Eggers, H. J., 1962, Heterogeneity of coxsackie B4 virus: Two kinds of particles which differ in antibody sensitivity, growth rate, and plaque size, *Virology* **18:**470–476.
23. Richter, F. A., MacPherson, L. W., Campbell, J. B., and Labzoffsky, N. A., 1972, Studies on intratypic variants of coxsackie B1 virus, *Arch. Ges. Virusforsch.* **38:**77–84.
24. Brown, F., and Wild, F., 1974, Variation in the coxsackievirus type B5 and its possible role in the etiology of swine vesicular disease, *Intervirology* **3:**125–128.
25. Harris, T. J. R., Doel, T. R., and Brown, F., 1977, Molecular aspects of the antigenic variation of swine vesicular disease and coxsackie B5 viruses, *J. Gen. Virol.* **35:**299–315.
26. Ash, P., Leong, W. A., Kennett, M. L., and Schnagl, R. D., 1985, Neutralization kinetic analysis of echovirus 30 and coxsackievirus B4 strains revealed little antigenic variation amongst the echovirus strains, *Aust. J. Exp. Biol. Med.* **63:**219–221.
27. Chow, M., Yabrov, R., Bittle, J., Hogle, J., and Baltimore, D., 1985, Synthetic peptides from four separate regions of the poliovirus type 1 capsid protein VP1 induce neutralizing antibodies, *Proc. Natl. Acad. Sci. USA* **82:**910–914.
28. Sherry, B., Mosser, A. G., Colonno, R. J., and Rueckert, R. R., 1986, Use of monoclonal antibodies to identify four neutralization immunogens on a common cold picornavirus, human rhinovirus 14, *J. Virol.* **57:**246–257.
29. Crowell, R. L., and Landau, B. J., 1983, Receptors in the initiation of picornavirus infections, in: *Comprehensive Virology.* Vol. 18 (H. Fraenkel-Conrat and R. R. Wagner, eds.), pp. 1–42, Plenum, New York.
30. Reagan, K. J., Goldberg, B., and Crowell, R. L., 1984, Altered receptor specificity of coxsackievirus B3 after growth in rhabdomyosarcoma cells, *J. Virol.* **49:**635–640.
31. Jordan, G. W., and Bolton, V., 1985, Interferon-sensitive coxsackievirus variants in nature, *J. Interferon Res.* **5:**289–296.
32. Marcus, P. I., Guidon, P. I., and Sekellick, M. L., 1981, Interferon induction by viruses. VII. Mengovirus: interferon-sensitive mutant phenotype attributed to interferon-inducing particle activity, *J. Interferon Res.* **1:**601–611.
33. Cohen, S. H., Bolton, V., and Jordan, G. W., 1983, Relationship of the interferon inducing particle phenotype of encephalomyocarditis virus to virus-induced diabetes mellitus, *Infect. Immun.* **42:**602–611.
34. Trousdale, M. D., Paque, R. E., and Gauntt, C. J., 1977, Isolation of coxsackievirus B3 temperature-sensitive mutants and their assignment to complementation groups, *Biochem. Biophys. Res. Commun.* **76:**368–375.
35. Gauntt, C. J., Trousdale, M. D., Lee, J. C., and Paque, R. E., 1983, Preliminary characterization of coxsackievirus B3 temperature-sensitive mutants, *J. Virol.* **45:**1037–1047.

36. Richman, D. D., and Murphy, B. R., 1979, The association of the temperature-sensitive phenotype with viral attenuation in animals and humans: Implications for the development and use of live virus vaccines, *Rev. Infect. Dis.* **1**:413–433.
37. Trousdale, M. D., Paque, R. E., Nealon, T., and Gauntt, C. J., 1979, Assessment of coxsackievirus B3 *ts* mutants for induction of myocarditis in a murine model, *Infect. Immun.* **23**:486–495.
38. Gauntt, C. J., Jones, D. C., Huntington, H. W., Gudvangen, R. J., and DeShambo, R. M., 1984, Murine forebrain anomalies induced by coxsackievirus B3 variants, *J. Med. Virol.* **14**:341–355.
39. Marlin, A. E., Huntington, W. H., Arizpe, H. M., Gudvangen, R. J., Brans, Y. W., and Gauntt, C. J., 1985, Coxsackievirus group B and hydranencephaly, *Concepts Pediatr. Neurosurg.* **6**:147–160.
40. Gauntt, C. J., Gudvangen, R. J., Brans, Y. W., and Marlin, A. E., 1985, Coxsackievirus group B antibodies in the ventricular fluid of infants with severe anatomic defects in the central nervous system, *Pediatrics* **76**:64–68.
41. Chatterjee, N. K., and Nejman, C., 1986, Protein kinase in nondiabetogenic coxsackievirus B4, *J. Med. Virol.* **19**:353–365.
42. Grubman, M. J., Baxt, B., LaTorre, J. L., and Bachrach, H. L., 1981, Identification of a protein kinase activity in purified foot-and-mouth disease virus, *J. Virol.* **39**:455–462.
43. Scharli, C. E., and Koch, G., 1984, Protein kinase activity in purified poliovirus particles and empty viral capsid preparations, *J. Gen. Virol.* **65**:129–139.
44. Cole, C. N., Smoler, D., Wimmer, E., and Baltimore, D., 1971, Defective interfering particles of poliovirus, *J. Virol.* **7**:478–485.
45. Nomoto, A., Jacobson, A., Lee, Y. F., Dunn, J., and Wimmer, E., 1979, Defective intefering particles of poliovirus: Mapping of the deletion and evidence that the deletions in the genome of DI (1), (2) and (3) are located in the same region, *J. Mol. Biol.* **128**:179–196.
46. McClure, M. A., Holland, J. J., and Perrault, J., 1980, Generation of defective interfering particles in picornaviruses, *Virology* **100**:408–418.
47. Huang, A. S., and Baltimore, D., 1977, Defective interfering animal viruses, in: *Comprehensive Virology*, Vol. 10 (H. Fraenkel-Conrat and R. R. Wagner, eds.), pp. 73–116, Plenum, New York.
48. Holland, J. J., Kennedy, S. I., Semler, B. L., Jones, C. J., Roux, L., and Grabau, E. A., 1980, Defective interfering RNA viruses and the host-cell response, in: *Comprehensive Virology*, Vol. 16 (H. Fraenkel-Conrat and R. R. Wagner, eds.), pp. 137–192, Plenum, New York.
49. Gauntt, C. J., Trousdale, M. D., LaBadie, D. R. L., Paque, R. E., and Nealon, T., 1979, Properties of coxsackievirus B3 variants which are amyocarditic or myocarditic for mice, *J. Med. Virol.* **3**:207–220.
50. Chatterjee, N. K., Samsonoff, W. A., and Tuchowski, C., 1983, Isolation and characterization of a membrane-bound population of group B coxsackieviruss, *J. Virol.* **45**:832–841.
51. Chatterjee, N. K., and Nejman, C., 1985, Membrane-bound virions of coxsackievirus B4: Cellular localization, analysis of the genomic RNA, genome-linked protein, and effect on host macromolecular synthesis, *Arch. Virol.* **84**:105–118.
52. Lerner, A. M., and Reyes, M. P., 1985, Coxsackievirus myocarditis—With special reference to acute and chronic effects, *Prog. Cardiovascular Dis.* **27**:373–394.
53. Roesing, T. G., Landau, B. J., and Crowell, R. L., 1979, Limited persistence of viral antigen in coxsackievirus B3-induced heart disease in mice, *Proc. Soc. Exp. Biol. Med.* **160**:382–386.
54. Paque, R. E., Gauntt, C. J., Nealon, T. J., and Trousdale, M. D., 1978, Assessment of

cell-mediated hypersensitivity against coxsackievirus B3 viral-induced myocarditis utilizing hypertonic salt extracts of cardiac tissue, *J. Immunol.* **120:**1672–1678.

55. Paque, R. E., Straus, D. C., Nealon, T. J., and Gauntt, C. J., 1979, Fractionation and immunologic assessment of KCl-extracted cardiac antigens in coxsackievirus B3 viral-induced myocarditis, *J. Immunol.* **123:**358–364.
56. Huber, S. A., and Job, L. P., 1983, Differences in cytolytic T cell response of BALB/c mice infected with myocarditic and non-myocarditic strains of coxsackievirus group B, type 3, *Infect. Immun.* **39:**1419–1427.
57. Huber, S. A., and Lodge, P. A., 1984, Coxsackievirus B-3 myocarditis in BALB/c mice. Evidence for autoimmunity to myocyte antigens, *Am. J. Pathol.* **116:**21–29.
58. Sym, S. N., Miller, S. D., and Claman, H. N., 1977, Immune suppression with supraoptimal doses of antigen in contact sensitivity. I. Demonstration of suppressor cells and their sensitivity to cyclophosphamide, *J. Immunol.* **119:**240–244.
59. Rollinghoff, M., Starzinski-Powitz, A., Pfizenmaier, K., and Wagner, H., 1977, Cyclophosphamide-sensitive T lymphocytes suppress the in vivo generation of antigen-specific cytotoxic T lymphocytes, *J. Exp. Med.* **145:**455–459.
60. Estrin, M., Smith, C., and Huber, S., 1986, Antigen specific suppressor T cells present cardiac injury in Balb/c mice infected with a nonmyocarditic variant of coxsackievirus group B, type 3, *Am. J. Pathol.* **125:**578–584.
61. Gauntt, C. J., Gomez, P. T., Duffey, P. S., Grant, J. A., Trent, D. W., Witherspoon, S. M., and Paque, R. E., 1984, Characterization and myocarditic capabilities of coxsackievirus B3 variants in selected murine strains, *J. Virol.* **52:**598–605.
62. Lutton, C. W., Gudvangen, R. J., Nealon, T. J., Paque, R. E., and Gauntt, C. J., 1985, Cellular immune responses in mice challenged with an amyocarditic variant of coxsackievirus B3, *J. Med. Virol.* **17:**345–357.
63. Lutton, C. W., and Gauntt, C. J., 1986, Coxsackievirus B3 infection alters plasma membrane of neonatal skin fibroblasts, *J. Virol.* **60:**294–296.
64. Godeny, E. K., and Gauntt, C. J., 1986, Involvement of natural killer cells in coxsackievirus B3-induced murine myocarditis, *J. Immunol.* **137:**1695–1702.
65. Jimes, S., and Jamison, R. M., 1983, Coxsackievirus B4: *In vitro* genetic markers and virulence, *Arch. Virol.* **77:**1–11.
66. Jimes, S., Jamison, R. M., and Grafton, W. D., 1984, Coxsackievirus B4: *In vitro* genetic markers and cardiovirulence, *Arch. Virol.* **81:**345–351.
67. Cao, Y., Schnurr, D. P., and Schmidt, N. J., 1984, Monoclonal antibodies for study of antigenic variation in coxsackievirus type B4: Association of antigenic determinants with myocarditic properties of the virus, *J. Gen. Virol.* **65:**925–932.
68. Webb, S. R., Kearse, K. P., Foulke, C. L., Hartig, P. C., and Prabhakar, B. S., 1986, Neutralization epitope diversity of coxsackievirus B4 isolates detected by monoclonal antibodies, *J. Med. Virol.* **20:**9–15.
69. Kew, O. M., Notlay, B. K., Hatch, N. H., Nakano, J. H., and Obijeski, J. F., 1981, Multiple genetic changes can occur in the oral poliovaccines upon replication in humans, *J. Gen. Virol.* **56:**337–347.
70. Komatsu, T., Hashimoto, I., and Kohara, T., 1983, Variation in virulence of coxsackie virus B3 in the heart of mice. I. Comparison of mortality and virus growth in the heart and other organs, *Microbiol. Immunol.* **27:**265–272.
71. Hashimoto, I., Komatsu, T., and Kohara, T., 1983, Variation in virulence of coxsackie virus B3 in the heart of mice. II. Pathological comparison, *Microbiol. Immunol.* **27:**335–345.
72. Cao, Y., Schnurr, D. P., and Schmidt, N. J., 1984, Differing cardiotropic and myocarditic properties of group B type 4 coxsackievirus strains, *Arch. Virol.* **80:**119–130.

73. Notkins, A. L., and Yoon, J. W., 1984, Virus-induced diabetes mellitus, in: *Concepts in Viral Pathogenesis* (A. L. Notkins and M. B. A. Oldstone, eds.), pp. 241–247, Springer-Verlag, New York.
74. Barrett-Conner, E., 1985, Is insulin-dependent diabetes mellitus caused by coxsackievirus B infection? A review of the epidemiologic evidence, *Rev. Infect. Dis.* **7**:207–215.
75. Hartig, P. C., Madge, G. E., and Webb, S. R., 1983, Diversity within a human isolate of coxsackie B4: Relationship to viral-induced diabetes, *J. Med. Virol.* **11**:23–30.
76. Chatterjee, N. K., Haley, T. M., and Nejan, C., 1985, Functional alterations in pancreatic β cells as a factor in virus-induced hyperglycemia in mice, *J. Biol. Chem.* **260**:12786–12791.
77. Kuno, S., Itakagi, A., Yamazaki, I., Katsumoto, T., and Kurimura, T., 1984, Pathogenicity of newly isolated coxsackievirus B4 for mouse pancreas, *Arch. Virol.* **28**:433–436.
78. Jordan, G. W., Bolton, V., and Schmidt, N. J., 1985, Diabetogenic potential of coxsackie B viruses in nature, *Arch. Virol.* **86**:213–221.
79. Yoon, J. W., Onodera, T., and Notkins, A. L., 1978, Virus-induced diabetes mellitus. XV. Beta cell damage and insulin-dependent hyperglycemia in mice infected with coxsackie virus B4, *J. Exp. Med.* **148**:1068–1080.
80. Yoon, J-W., London, W. T., Curfman, B. L., Brown, R. L., and Notkins, A. L., 1986, Coxsackie virus B4 produces transient diabetes in nonhuman primates, *Diabetes* **35**:712–716.
81. Grist, N. R., and Roberts, G. B. S., 1966, Coxsackie A 7 virus infections of rodents, *Arch. Ges. Virusforsch.* **19**:454–463.
82. Kamahora, T., Itagaki, A., Hattori, N., Tsuchie, H., and Kurimura, T., 1985, Oligonucleotide fingerprint analysis of coxsackievirus A10 isolated in Japan, *J. Gen. Virol.* **66**:2627–2634.
83. Dagan, R., and Menegus, M. A., 1986, A combination of four cell types for rapid detection of enteroviruses in clinical specimens, *J. Med. Virol.* **19**:219–228.
84. Sarnow, P., Bernstein, H. D., and Baltimore, D., 1986, A poliovirus temperature-sensitive RNA synthesis mutant located in a noncoding region of the genome, *Proc. Natl. Acad. Sci. USA* **83**:571–575.
85. Maverakis, N. H., Schmidt, N. J., Riggs, J. L., and Lennette, E. H., 1973, Carrier cultures of human fetal diploid cells infected with coxsackievirus type B2, *Arch. Ges. Virusforsch.* **43**:289–297.
86. Blacklow, N. R., Rose, F. B., and Whalen, R. A., 1975, Organ culture of human aorta: Prolonged survival with support of viral replication, *J. Infect. Dis.* **131**:575–578.
87. Schnurr, D. P., and Schmidt, N. J., 1984, Persistent infection of mouse fibroblasts with coxsackievirus, *Arch. Virol.* **81**:91–101.
88. Frank, J. A., Jr., Schmidt, E. V., Smith, R. E., and Wilfert, C. M., 1986, Persistent infection of rat insulinoma cells with coxsackie B4 virus, *Arch. Virol.* **87**:143–150.
89. Matteucci, D., Paglianti, M., Giangregorio, A. M., Capobianchi, M. R., Dianzani, F., and Bendinelli, M., 1985, Group B coxsackieviruses readily establish persistent infections in human lymphoid cell lines, *J. Virol.* **56**:651–654.
90. Gomez, M. P., Reyes, M. P., Smith, F., Ho, L. K., and Lerner, A. M., 1980, Coxsackievirus 3-positive mononuclear leukocytes in peripheral blood of Swiss and athymic mice during infection, *Proc. Soc. Exp. Biol. Med.* **165**:107–113.
91. Godeny, E. K., Sprague, E. A., Schwartz, C. J., and Gauntt, C. J., 1986, Coxsackievirus group B replication in cultured fetal baboon aortic smooth muscle cells, *J. Med. Virol.* **20**:135–149.

92. Strongwater, S. L., Dorovini-Zis, K., Ball, R. D., and Schintzer, T. J., 1984, A murine model of polymyositis induced by coxsackievirus B1 (Tucson strain), *Arthritis Rheum.* **27**:433–442.
93. Ytterberg, S. R., 1985, Coxsackievirus B-1 induced murine polymyositis requires active muscle infection by live virus to initiate disease, *Arthritis Rheum.* **28**:69 (abst.).
94. Modlin, J. F., and Crumpacker, C. S., 1982, Coxsackievirus B infection in pregnant mice and transplacental infection of the fetus, *Infect. Immun.* **37**:222–226.
95. Fujinami, R. S., Oldstone, M. B. A., Wrobewska, Z., Frankel, M. E., and Koprowski, H., 1983, Molecular mimicry in virus infection: Crossreaction of measles virus phosphoprotein or of herpes simplex virus protein with human intermediate filaments, *Proc. Natl. Acad. Sci. USA* **80**:2346–2350.
96. Gould, E. A., Chanas, A. C., Buckley, A., and Clegg, C. S., 1983, Monoclonal immunoglobulin M antibody to Japanese encephalitis virus that can react with a nuclear antigen in mammalian cells, *Infect. Immun.* **41**:774–779.
97. Haynes, B. F., Robert-Guroff, M., Metzgar, R. S., Franchini, G., Kalyanaraman, V. S., Parker, T. J., and Gallo, R. C., 1983, Monoclonal antibody against human T cell leukemia virus p19 defines a human thymic epithelial antigen acquired during ontogeny, *J. Exp. Med.* **157**:907–920.
98. Srinivasappa, J., Saegusa, J., Praghakar, B. S., Gentry, M. K., Buchmeier, M. J., Wiktor, T. J., Koprowski. H., Oldstone, M. B. A., and Notkins, A. L., 1986, Molecular mimicry: Frequency of reactivity of monoclonal antiviral antibodies with normal tissues, *J. Virol.* **57**:397–401.
99. Notkins, A. L., Onodera, T., and Prabhakar, B., 1984, Virus-induced autoimmunity, in: *Concepts in Viral Pathogenesis* (A. L. Notkins and M. B. A. Oldstone, eds.), pp. 210–215, Springer-Verlag, New York.
100. Saegusa, J., Prabhakar, B. S. Essani, K., McClintock, P. R., Fukuda, Y., Ferrans, V. J., and Notkins, A. L., 1986, Monoclonal antibody to coxsackievirus B4 reacts with myocardium, *J. Infect, Dis.* **153**:372–373.
101. Wolfgram, L. J., Beisel, K. W., and Rose, N. R., 1985, Heart-specific autoantibodies following murine coxsackievirus B3 myocarditis, *J. Exp. Med.* **161**:1112–1121.
102. Huber, S. A., and Lodge, P. A., 1986, Coxsackievirus B-3 myocarditis. Identification of different pathogenic mechanisms in DBA/2 and BALB/c mice, *Am. J. Pathol.* **122**:284–291.
103. Maisch, B., 1984, Diagnostic relevance of humoral and cell mediated immune reactions in patients with acute myocarditis and congestive cardiomyopathy, in: *Cardiology* (E. I. Chazov, V. N. Smirnov, and R. G. Oganov, eds.), pp. 1327–1338, Plenum, New York.
104. Hogle, J. M., Chow, M., and Filman, D. J., 1985, Three-dimensional structure of poliovirus at 2.9 Å resolution, *Science* **229**:1359–1365.
105. Rossmann, M. G., Arnold, E., Erickson, J. W., Frankenberger, E. A., Griffith, J. P., Hecht, H-J., Johnson, J. E., Kamer, G., Luo, M., Mosser, A. G., Rueckert, R. R., Sherry, B., and Vriend, G., 1985, Structure of a human common cold virus and functional relationship to other picornaviruses, *Nature (Lond.)* **317**:145–153.
106. Campbell, B. A., and Cords, C. E., 1983, Monoclonal antibodies that inhibit attachment of group B coxsackievirus, *J. Virol.* **48**:561–564.
107. Crowell, R. L., Field, A. K., Schleif, W. A., Long, W. L., Colonno, R. J., Mapoles, J. E., and Emini, E. A., 1986, Monoclonal antibody that inhibits infection of HeLa and rhabdomyosarcoma cells by selected enteroviruses through receptor blockade, *J. Virol.* **57**:438–445.
108. Minor, P. D., Pipkin, P. A., Hockley, D., Schild, G. C., and Almond, J. W., 1984,

Monoclonal antibodies which block cellular receptors of poliovirus, *Virus Res.* **1**:203–212.

109. Colonno, R. J., Callahan, P. L., and Long, W. J., 1986, Isolation of a monoclonal antibody that blocks attachment of the major group of human rhinoviruses, *J. Virol.* **57**:7–12.

11

Persistent Infections

DAVID P. SCHNURR and NATHALIE J. SCHMIDT

1. INTRODUCTION

Experimental persistent viral infections *in vitro* in cell cultures and *in vivo* in animal models are of considerable practical and theoretical interest in exploring possible mechanisms for viral persistence and pathogenesis in the human host. Human enteroviruses, including the group B coxsackievirus (CVB), are important pathogens generally thought of as being highly lytic viruses that are unlikely to establish persistent infections. Reports of CVB persistence in HeLa cells appeared during the 1960s[1,2] and in human fetal diploid (HFD) cells in 1973[3]; and more recently, a number of laboratories have observed CVB persistence both *in vitro* in various types of cell cultures[4–7] and *in vivo* in the mouse model.[8–10] These persistent infections could be useful as models for chronic disease, as they include infection of human heart fibroblasts,[5] rat pancreatic cells,[4] human lymphocytes,[6] and mouse skin fibroblasts[7] *in vitro* and mouse heart,[8,9] spleen,[10] and pancreatic[10] cells *in vivo*.

2. *IN VITRO* PERSISTENT INFECTIONS

Group B coxsackievirus persistent infections of HeLa, HFD, lymphoid, pancreatic, and fibroblastic cells have been characterized with

DAVID P. SCHNURR • Center for Advanced Medical Technology, San Francisco State University, San Francisco, California 94132; Viral and Rickettsial Disease Laboratory, Division of Laboratories, California State Department of Health Services, Berkeley, California 94704. NATHALIE J. SCHMIDT • Viral and Rickettsial Disease Laboratory, Division of Laboratories, California State Department of Health Services, Berkeley, California 94704.

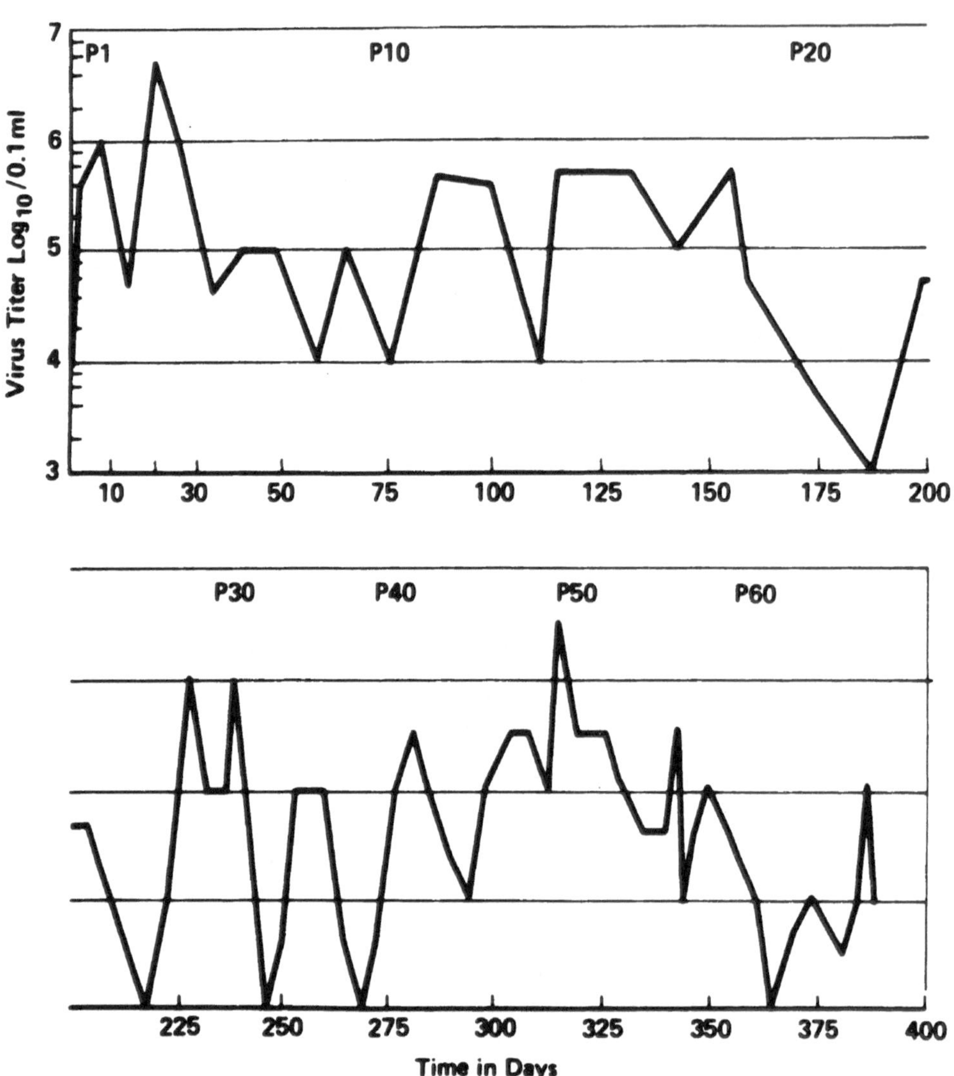

FIGURE 1. $TCID_{50}$ of CVB3 in supernatant fluids of persistently infected NFR fibroblasts sampled at the time of subculture. P1 = subculture #1, etc. (From Schnurr and Schmidt,[7] by permission of publisher.)

respect to cell destruction (CPE), virus production, number of infected cells, and resistance to superinfection. CPE was observed following infection of NFR mouse fibroblasts with CVB3 or CVB4,[7] of human lymphocytes with CVB types 1–6,[6] of HeLa cells with CVB3 or CVB5,[1,2] and of human heart cells with CVB3 but not when HFD cells were infected with CVB3[3] or when mouse pancreatic cells were infected with CVB4.[4]

Subculture of HeLa, HFD, heart cells, or lymphoid cells or continued maintenance of monolayers of pancreatic cells resulted in cultures that continued to release infectious virus. HeLa cells continued to release infectious virus for up to 3 years,[2] while cultures of human lymphocytes and murine fibroblasts did so for up to 2 years (Fig. 1).

Spontaneous cure of persistently infected cultures occasionally occurred after short-term serial subculture, e.g., after five passages of CVB3 in mouse fibroblasts[7] or after long-term persistence, e.g., after 14 months of CVB3 persistence (D. P. Schnurr, unpublished data) or at 4.5 and 8.5 months in lymphoid cells infected with CVB2 and CVB5 respectively.[6] Persistently infected HFD cells continued to release CVB until the cells could no longer be subcultured. The persistently infected HFD cells could be subcultured only 37 times, as compared with 47 subcultures for uninfected controls.[3] CVB persistent infection could be terminated by the addition of specific antibody[2,6] in some instances. However, in other cases in the absence of antibody, infectious virus was released for as long as the cultures were maintained.[1–7]

When tested, it appeared that nearly every cell was susceptible to virus during the acute infection, as greater than 90% of pancreatic cells or mouse fibroblasts were positive for virus by indirect immunofluorescence or infectious center assay[7] (Table I). In some cells, long-term virus persistence resulted in cultures in which few cells continued to produce infectious centers or viral antigen (0.1% in lymphoid cells at 12 months,[6] 5.9% in mouse cells at 12 months,[7] or less than 1% in HFD at 32, 35, and 38 passages.[3]) By contrast, 50% of mouse pancreatic cells were positive for CVB antigen at 60 days postinfection.[4]

Interference with homologous (Schnurr, unpublished) and heterologous CVB occurred in CVB persistently infected cells,[1,2,6] but there was no interference with replication of other heterologous viruses, such as poliovirus or vesicular stomatitis virus.[1,2,7] Interference in HeLa cells persistently infected with CVB3 or CVB5 was due to a block in absorption or penetration of homologous or heterologous CVB.[1,2] RNA extracted from CVB1 or CVB5 could undergo one round of multiplication

TABLE I
Percentage of NFR Fibroblasts Producing Infectious Centers after Infection with Various Doses of CBV3(M)[a]

	Cells producing infectious centers at days postinfection (%)			
MOI[b]	1	2	4	7
10	100	70	100	100
1	7	11	100	100
0.1	0	0.5	100	100
0.01	0	0	4	100

[a]M, myocarditic strain of CVB3.
[b]MOI, multiplicity of infection.

in CVB3 persistently infected cells, demonstrating that the block resulting in interference occurred prior to uncoating of viral RNA in the cells and that subsequent steps in viral multiplication were not inhibited.[2]

2.1. Mechanisms of Persistence

Persistent infections *in vitro* may be broadly grouped as carrier cultures or as steady-state infections. Carrier cultures are characterized by the death of all infected cells, so that the persistent infection is maintained by the extracellular spread of the virus to other susceptible cells. In this type of persistent infection, infected cells cannot be cloned and persistence can be cured by the addition of specific antiviral antibody. Factors in the medium, such as antibody or interferon (IFN), may protect most cells in the culture from infection by progeny virus released from infected cells. The progeny virus may be less lytic and may include defective interfering (DI) particles or temperature-sensitive (*ts*) mutants, which can interfere with the lytic parental virus. Carrier cultures are usually associated with lytic viruses.

In a steady-state infection, most cells are infected, but the infected cells grow while allowing the virus to replicate. In this case, the virus is normally a nonlytic virus. The infection cannot be cured by the addition of specific antiviral antibodies, and infected cells can be cloned. Viral mutants such as DI or *ts* mutants arise, but they are not crucial to the maintenance of the persistent infection.

Coxsackie B virus persistent infections are best categorized as carrier cultures. Not all cells in the culture are infected, and the persistent infection can be terminated by the addition of specific antibody. This type of persistent infection can be initiated and/or maintained by viral factors such as defective interfering (DI) particles or *ts* mutants, by antiviral agents such as antibody or IFN, or by the inherent resistance of most cells in the culture to the virus.

Coxsackie B virus persistent infection of HeLa cells[1,2] required the continuous presence of human sera that probably contained low levels of antibody to CVB. However, the other CVB-persistent infections described did not require addition of human sera or any other apparent antiviral factor. In those cases, the most likely explanation for persistent infection is that most of the cells in the culture are resistant to infection, and that susceptible cells are continually regenerated. This hypothesis can be tested by obtaining clones of cells and determining their resistance or susceptibility to CVB. Eight of nine clones of previously uninfected lymphocytes were completely lysed by CVB3, suggesting that most cells were susceptible to CVB.[6] However, susceptible cells might have a higher cloning efficiency, or persistent infection might select for resistant cells. Cloning of CVB persistently infected human lymphocytes was not possible.[6]

Factors important in the initiation of a persistent infection might be quite different from those that maintain persistence, and it is possible that viral mutants or IFN do help stabilize the culture.

Human lymphoid[6] or mouse fibroblastic cells[7] persistently infected with CVB showed no evidence of IFN-like activity. Our laboratory has tested supernatant fluids from CVB persistently infected cells and observed no reduction in yield of progeny or plaque numbers from mixed infections with standard virus, suggesting a minimal or no role for extracellular antiviral factors.

Viral mutants such as DI particles or *ts* virions frequently have been observed to be present in persistent infections of cell culture.[11] DI particles have been described for picornaviruses.[12] However, there is no evidence to implicate them as factors in CVB-persistent infections. Frank *et al.*[4] were unable to detect any virus RNA from CVB-persistent infection at a density other than that of the standard virus. These workers concluded that there was no evidence for DI particles.

Temperature-sensitive virus was present in CVB-persistent infection of lymphoid cells.[6] However, Matteucci *et al.*[6] believed that these mutant virions were not critical to the persistent infection because these mutants appeared after establishment of persistence. Persistent infection *in vitro* with another picornavirus, echovirus type 6, did not result in the generation of *ts* mutants.[13]

Coxsackie B virus-persistent infections may require addition of antiviral factors such as antibody to prevent complete lysis of the culture, as occurs with HeLa cells. However, in many other cases, such as mouse or human fibroblasts, human lymphocytes, and rat pancreatic cells, persistent infection in the absence of antibody is possible. Since more than 90% of the cells are positive for infectious virus and viral antigen during acute infection,[7] the cells must cure themselves or a small fraction of resistant cells must continually provide new susceptible cells to maintain the culture. Experiments aimed at determining host-cell factors in persistent CVB infections could provide important information.

Host factors are a function of the type of cell that is infected. In comparing Buffalo green monkey kidney (BGMK) to NFR mouse cells as hosts for CVB, we have observed features of the mouse cells that may contribute to the ease of establishing persistent infection in the mouse cells. The stability of the cell membrane differs between the two cell types. In preliminary experiments for the measurement of protein synthesis, we observed that mouse cells, which can be persistently infected, were more resistant to solubilization by detergent (1% NP-40 + 0.5% SDS) than were BGMK cells, which are easily solubilized.

TABLE II
Growth of CVB3 in NFR Mouse and BGMK Cells

Hours after infection	BGMK		NFR mouse	
	Supernatant	Supernatant + cell lysate	Supernatant	Supernatant + cell lysate
2	3.5[a]	4.5	4.0	5.0
4	4.0	3.5	3.5	5.5
6	4.0	6.5	3.5	5.5
24	8.5[b]	8.5	6.5	8.5
48	—	—	8.5	8.5
96	—	—	7.0	8.0

[a] $TCID_{50}$/ml.
[b] All cells lysed by virus.

In comparing yields of infectious CVB3 from mouse cells with those from BGMK cells during acute infection, it was seen that maximum titers of infectivity were achieved more rapidly in BGMK cells but that equivalent titers were ultimately reached in mouse cells. CPE and accompanying cell lysis occurred in both cell types during acute infection, but some mouse cells could be subcultured. These cells either survived acute infection or were not infected during the acute phase.

Viral infectivity remained cell associated in mouse cells for a longer period of time than in BGMK cells. Ninety percent of infectivity was cell associated in mouse cells by 4 full days postinfection, whereas all BGMK cells were lysed by 24 hr postinfection (Table II). The extended duration of cell association of infectious virus seen in mouse cells could be related to membrane stability and might be a factor in the establishment of persistent infection.

Although CVB-persistent infections are generally carrier cultures, it is possible that steady-state infections might occur as well. CVB4-persistent infection of rat insulinoma cells was not sufficiently characterized to determine the mechanism. However, because most cells became infected and remained infected, it is possible that this was a steady-state infection. Steady-state infections *in vitro* have also been described for echovirus type 6[13] and hepatitis A virus (HAV).[14] Since HAV is a nonlytic virus *in vitro,* this is not surprising. However, echovirus type 6 is usually a highly lytic virus.

Persistent infection with echovirus type 6 was established by infecting cells at a high multiplicity of infection (MOI), cultivating the survivors, challenging with echovirus type 6, and cloning and challenging with the same virus a third time. These cells were maintained in culture for 3 years and released infectious virus continuously without cellular destruction. At passages 40 and 67, more than 80% of the cells were positive for echovirus type 6 antigen when tested by indirect immunofluorescence. In addition, persistently infected cells produced 35–38% infectious centers, and persistent infection could not be terminated by addition of homologous viral antibody. There was no evidence of *ts* mutants or IFN in the cultures. However, persistently infected cells produced noninfectious defective virions at a ratio of 2×10^6 particles per plaque-forming unit (PFU). This excess of defective particles may have included DI particles, but they were not tested for. Repeated infection at a high MOI was used to initiate this persistent infection, and this would provide ideal conditions for generation of DI particles.[15]

On several occasions, we have cultivated cells from hearts of mice persistently infected with CVB3.[9] These cells were maintained *in vitro* for 90 days with continuous release of infectious virus. The cells did not form complete cell sheets but instead were present throughout the culture vessel in scattered islands. Several attempts at subculture failed, either because the cell density was too low or because the cells did not divide. Because of the low cell density and the failure of subculture, studies on expression of viral antigens, infectious centers, or interference were not possible. Despite the lack of further characterization, these persistently infected cells might best be termed as steady state, since there was no replenishment of susceptible cells, yet the cells continued to release infectious virus for up to 90 days. Such continuous release of virus in heart tissue could have important implications for chronic heart disease.

2.2. Evolution of Virus and Host Cells during CVB3-Persistent Infection

During the course of a persistent infection, both the host cell and the virus exert selective pressure on each other, and the resultant coevolution may contribute to a more stable host–parasite relationship. Even though standard lytic CVB can initiate persistent infections in certain cell types without the participation of additional factors, such as antiviral antibody, IFN, or DI particles, considerable evolution may occur during persistence.

We have investigated a number of characteristics of CVB3 from persistent infections and have compared the virus from a persistent infection to the standard parental virus (CVB3–S) used to initiate the infection (manuscript in preparation). A number of phenotypic changes occurred during persistence (Table III) that reflect rapid evolution of the virus.

TABLE III
Properties of CVB3–S and CVB3 from Persistent Infections

	Virus strain	
Property	CVB3–S	CVB3 from persistent infections
Plaque morphology in BGMK cells at 37°C	3–4 mm	1 mm in 4 days
Virus titer in BGMK cells 24 hr postinfection	$10^{6.1}$ PFU/0.1 ml	$10^{4.5}$ PFU/0.1 ml
Neutralization by polyclonal anti-CVB3	Indistinguishable[a]	
ActD sensitivity	Low	High
Degree of cell association	Low	High
50% inhibition of protein synthesis in BGMK cells	1.5 hr Postadsorption	3.5 hr Postadsorption
50% inhibition of protein synthesis in mouse cells	3 hr Postadsorption	3 hr Postadsorption

[a]Neutralized to the same degree.

Many investigators have observed a gradual increase in the number of small plaque variants recovered from persistently infected cells.[11] Persistent infection with CVB3 in mouse cells was also characterized by the appearance of small plaque variants (Fig. 2). CVB3–S produced 90% large plaques, while the percentage of small plaques during persistent infection increased from 12% at passage 2 to greater than 90% at passage 41 (Table IV). A similar change in plaque morphology during CVB-persistent infection was observed for persistent infection of lymphocytic cells.[6]

Parental CVB3 has a number of features that suggest that it grows

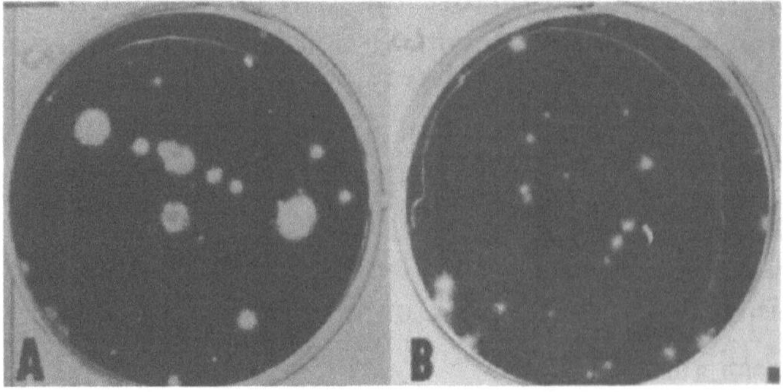

FIGURE 2. CVB3 plaques on BGMK cells: (**A**) large plaques produced by standard virus and (**B**) small plaques produced by virus from persistent infection.

TABLE IV
Evolution of Plaque Morphology and Actinomycin D Sensitivity

Source of virus	Fraction of small plaque forming virus (%)	ActD sensitivity[a]
Passage 2	12.5	51.0
Passage 11	28.1	0.2
Passage 21	40.2	0.07
Passage 31	73.1	0.53
Passage 41	93.3	0.64
Passage 90	91.8	2.81
Standard	10.0	42.50

[a]PFU in presence of 2.0 μg/ml ActD divided by PFU in absence of ActD.

more efficiently than virus from persistently infected cells. In BGMK cells at comparable MOIs, CVB3–S produced more than 90% more infectious progeny than did CVB3 from persistently infected cells at 24 hr postinfection.

The effect of CVB3–S and CVB3 from persistently infected cells on shutdown of host cell protein synthesis was tested on BGMK and mouse cells. Protein synthesis was measured by incorporation of [^{3}H]leucine into acid-precipitable material. As shown in Fig. 3, the shutdown of host cell protein synthesis in mouse cells had very similar kinetics for infections with both CVB3–S and CVB3 from persistent infections, with a major effect being seen by 4 hr postadsorption. By contrast, CVB3–S caused a more rapid shutdown in BGMK cells than did CVB3 from persistently infected cells. A major shutdown occurred by 2 hr postadsorption with CVB3–S as opposed to 4 hr for an equivalent shutdown by CVB3 from persistently infected cells. During persistent infection in mouse cells, CVB had become attenuated for its effect on protein synthesis of BGMK cells.

Increased actinomycin D sensitivity has been reported to occur for some strains of poliovirus and to be a marker for attenuation.[16] We tested the effect of actinomycin D on various passage levels of CVB3 from persistently infected cells and CVB3–S in mouse cells to determine whether increased sensitivity might have evolved during persistent infection. Considerably greater sensitivity was observed for CVB3 from persistently infected cells than for CVB3–S (Table V). In the presence of actinomycin D, passage 11 of CVB3 from persistently infected cells produced 0.2% of the infectious progeny produced in its absence, indicating that sensitivity to actinomycin D evolved very rapidly during persistence (Table IV).

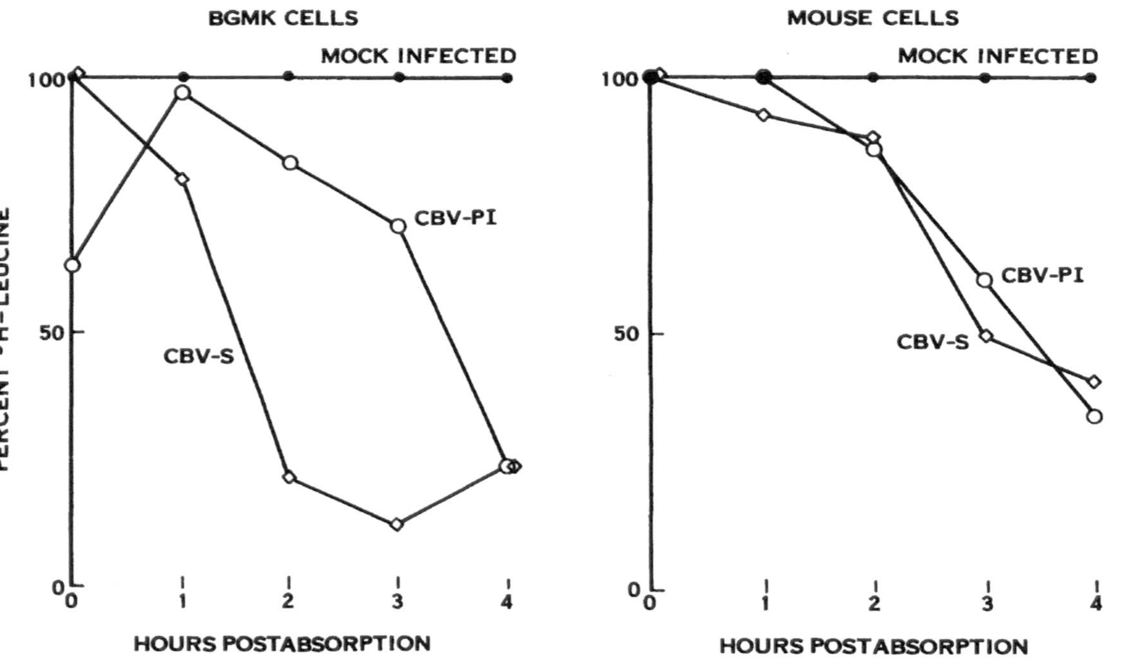

FIGURE 3. Inhibition of protein synthesis by CVB3–S and CVB3-PI in BGMK and mouse cells. Replicate cultures of BGMK or mouse cells were infected with CVB3–S or CVB3-PI at MOIs of between 10 and 20. Virus was absorbed for 1 hr, and fresh medium was added. At the indicated times media containing 1 uCi/ml ^{3}H leucine was added to 3 wells for 15 minutes. Wells were harvested and acid precipitable CPMs were determined. ●—● = mock infected, ○—○ = CVB3-PI, ◇—◇ = CVB3–S.

TABLE V
Replication of CVB3–S and CVB3 from Persistent Infections in the Presence of Actinomycin D[a]

Hours after infection[b]	Presence (+) or absence (−) of actinomycin D			
	CVB3–S[b]		CVB3 from persistent infections	
	+	−	+	−
2	2.5[c]	2.5[c]	0[c]	0[c]
3	4.5	3.5	0	0
4	5.5	5.5	0	3.0
5	5.5	5.0	0	3.5
6	4.5	4.5	0	4.5
24	4.0	5.0	3.0	6.5

[a]ActD concentration of 2.0 μg/ml.
[b]MOI 1–10.
[c]Log_{10} $TCID_{50}$/0.1 ml of supernatant.

It was of interest to determine whether the action of actinomycin D occurred on host cell metabolism or directly on the virus. To differentiate these possibilities, the effect of actinomycin D on yields of infectious progeny of CVB3 from persistently infected cell virus was compared under three different conditions, added (1) 2 hr prior to infection, remaining present throughout the duration of the experiment; (2) 2 hr before the infection, with cultures switched to medium without actinomycin D at the time of infection; and (3) at the time of infection. As compared with the control without actinomycin D, the addition of actinomycin D 2 hr before infection resulted in continuous suppression of viral multiplication for up to 5 days (as long as actinomycin D was present throughout the time of culture sampling) (Table VI). When actinomycin D was added 2 hr before infection and withdrawn at the time of infection, there was suppression in titers of viral infectious progeny for up to 27 hr postinfection, but by 30 hrs the titer of infectious virus was 80% of that of untreated control cultures. When actinomycin D was added at the time of infection, significant suppression of progeny was not apparent until 30 hr postinfection. Since it required 27–30 hr after withdrawal of actinomycin D until there was a significant increase in titer of progeny virus, we believe that actinomycin D blocked some aspect of cell metabolism, which is a prerequisite for completion of viral replication. Because host cell transcription is a major target for actinomycin D, we hypothesize that messenger RNA (mRNA) coding for a host protein required for viral replication had been blocked. Such host cell factors

TABLE VI
Effect of Actinomycin D Switch at the Time of Infection on Viral Replication in Previously Uninfected Mouse Cells

Actinomycin D	Hours after infection							
	1	3	6	24	27	30	48	120
None	2.5×10^{3}[a]	3.3×10^{3}	4.3×10^{4}	4.0×10^{6}	8.3×10^{7}	4.5×10^{7}	6.5×10^{7}	7.0×10^{7}
2.0 μg/ml added 2 hr before infection	1.6[b]	0.82	0.21	0.04	0.05	0.07	0.10	0.02
2.0 μg/ml added 2 hr before infection and removed at time of infection	1.6	0.76	0.19	0.04	0.05	0.82	0.43	0.46
2.0 μg/ml added at time of infection	3.2	1.1	1.9	0.9	0.64	0.07	0.15	0.05

[a]PFU/0.1 ml supernatant fluid.
[b]Ratio of titer to that in absence of actinomycin D.

have been described for poliovirus.[17,18] Under conditions in which limiting amounts of this protein are present, viral replication proceeds slowly or is blocked.

If actinomycin D acts directly on the virus, it should be equally effective in suppressing viral replication whether it is added at the time of infection or 2 hr prior to infection. The results showed that when actinomycin was added at the time of infection, maximum suppression of yield of infectious progeny from infections initiated with CVB3 from persistently infected cells did not occur until 30 hr postinfection, strongly suggesting that actinomycin acts on host cell metabolism rather than on the virus itself. The results of these experiments support the hypothesis that the effect of actinomycin D on yields of infectious progeny from infections initiated with virus obtained from persistently infected cells results from the effect of actinomycin D on host cell metabolism.

Parental CVB3 and CVB3 from persistently infected cells were indistinguishable when compared for neutralization by polyclonal CVB-3 (Nancy) immune serum or by the electrophoretic mobility of VP1–4 in polyacrylamide gels.

During the course of persistent infection, the cells may also evolve. For instance, a strain of our mouse cells spontaneously ceased production of infectious CVB after 90 passages. This strain of "cured" cells could be reinfected by homologous virus (CVB3), but no longer exhibited CPE even on acute infection, and produced 90% fewer infectious progeny than did control (previously uninfected) cells. Despite the lack of cytopathology and reduced viral titers, every cell was positive for viral antigen by avidin–biotin–peroxidase staining at 2 days after reinfection.

Thus, during persistent infection of mouse cells, the virus evolved toward less efficient growth, while the cells evolved toward greater resistance. In some cases, this co-evolution could result in spontaneous cure of the persistent infection.

3. *IN VIVO* PERSISTENT INFECTIONS

Coxsackie B virus persistence has not been reported in humans. However, persistent infections with other enteroviruses occur in immunodeficient humans,[19] and CVB persistence has been observed in murine models.[8–10] In all these cases except one,[10] enteroviral persistence was limited to immunodeficient or immunosuppressed hosts.

Wilfert *et al.*[19] described four cases of enteroviral persistent infection of the central nervous system (CNS) in agammaglobulinemic humans. Two of the cases involved persistence of echovirus type 9, and two involved echovirus type 30. The longest period of viral persistence occurred with echovirus type 30, which was isolated from cerebrospinal fluid (CSF) nine times over a 23-month period. During this period, there were also four negative isolation attempts. These patients were treated with pooled human serum globulin. It was noted that the passively acquired antibody protected the patients from viremia and distant visceral infection, but was unable to eliminate virus from the CNS. In these patients, there was no evidence of cellular immunodeficiency.

We have observed CVB3 persistence in two different strains of mice, NFR[8] and N : NIH(S)-II.[9] NFR athymic, nude, and euthymic mice were infected with CVB3 and assayed for infectious virus from spleen, pancreas, and heart at various times. Virus could be isolated from the hearts of nude mice for up to 28 days postinfection and on a single occasion from the spleen and pancreas at 21 days postinfection. In no case was virus isolated from euthymic mice after 7 days. We have also studied the persistence of CVB3 in athymic and euthymic N : NIH(S)-II mice. These outbred mice have an X-linked immunodeficiency gene that results in defective T-independent B lymphocytes. Virus could be isolated from the hearts of two of six euthymic mice at 14 days postinfection and from athymic mice for up to 94 days postinfection (Table VII). Heart cells from infected mice could be cultivated *in vitro* with the continuous release of CVB for up to 90 days. These cells did not divide; thus, it appears that some cells were capable of long-term survival while infected.

Coxsackie B virus type 1 persistence has been reported in BALB/c mice up to 180 days postinfection.[10] Although the number of isolation attempts was not indicated, virus was isolated from spleens of normal mice at 21 and 180 days. Virus was also isolated from blood, spleen and pancreas of immunosuppressed (cyclophosphamide treated) mice at 21 days and from pancreases at 180 days.

Infectious CVB cannot be isolated from immunocompetent mice beyond a maximum of about 14 days.[20] The relative role of cell-mediated immunity and antibody in termination of enteroviral infections is not entirely clear. In athymic mice, neutralizing antibody was present at 7 days p.i., although at lower titer than in euthymic mice.[8] Thus, although antibody may be important in limiting the spread of virus or preventing reinfection, direct or indirect participation of T lymphocytes may be required for termination of infection.

4. CONSEQUENCES OF PERSISTENT INFECTION

It is clear that persistent viral infections may have important medical consequences.[21] *In vitro* studies of persistent infections are useful, although ultimately *in vivo* verification of *in vitro* studies is required.

TABLE VII

Virus Recovery from Hearts of N : NIH(S)-II +/*nu* and *nu*/*nu* Mice Infected with CVB3(M)[a]

Mice	Age (weeks) at inoculation	Days postinoculation	No. of mice	Isolation results[b]	No. positive tested
+/*nu*	4	7	6	4.5, 3.5(3), 2.5, 2.0	6/6
	4	14	6	2.5, 2.0	2/6
	4	21	6		0/6
	4	28	5		0/5
nu/*nu*	4–5	7	4	5.0(2), 3.0	3/4
	9	7	8	4.5(2), 4.0(3), 3.5(2)	7/8
	12	7	2	4.0, 3.5	2/2
	9	14	6	≥5.5(2), 4.0, 3 not tested	2/3
	9	21	6	≥5.0, ≥4.0, 3 not tested	2/3
	4	28	5	Co-cultivation, 3 pos.	3/5
	9	28	8	5.0, 4.5, 3.5, 3 not tested	3/5
	12	28	2	Co-cultivation, 1 pos.	1/2
	4	56	1	Co-cultivation, 1 pos.	1/1
	9	56	3	Co-cultivation, 1 pos.	1/3
	4–5	94	3	≥4.5(2)	2/3

[a] M, myocarditic strain of CVB3.

[b] Log_{10} $TCID_{50}$ per heart; pos., positive. Each value indicates the virus recovery from the heart of one mouse; values in parentheses indicate the number of mice.

In vitro persistent infections with CVB have been established in heart, pancreatic and lymphoid cells, all of which could serve as important targets *in vivo*. Pathogenesis could result from direct destruction of cells, a host response against infected cells, or alteration of cellular function without cell destruction.

4.1. Pancreatic Cells

Persistent infection with lymphocytic choriomeningitis virus (LCMV) in mice has been suggested as a case in which persistent viral infection causes alteration of cellular function without destroying the target cell.[21,22] In this model, infected mice are hyperglycemic and have abnormal glucose tolerance levels. Because there appears to be normal pancreatic structure and β-cells, altered function due to LCMV infection has been implicated.

Persistent infection of rat insulinoma cells by CVB4 did not result in a decrease in insulin production in infected cells as compared with uninfected control cells.[4] Thus, for CVB4-persistent infection of insulin-producing cells, the results of an *in vitro* study did not support altered function without cell destruction as an explanation for what was observed in LCMV-infected mice *in vivo*.

Lymphocytic choriomeningitis virus or another virus such as CVB may serve an initiator function by modifying β-cells so that they become a target for autoimmune destruction or may act by some other mechanism.

4.2. Lymphoid Cells

In some cases, CVB-persistent infection of lymphocytes was easily established *in vitro*[6] and there is a report of CVB isolation from spleen cells 180 days postinfection.[10] Infection of lymphoid cells *in vivo* might result in a persistent infection or alter the function of cells so that a pathologic process such as autoreactivity could result. There are several reports of CVB3 causing generalized lymphoid involution, including splenic atrophy in mice.[23,24] However, in this case, the lymphoid cells could not be persistently infected *in vitro*. Thus splenic atrophy was probably not mediated by direct viral damage to lymphoid cells.

4.3. Heart Cells

Results of *in vitro* studies of human[5] and rat[25] heart cells have shown that although myocytes are infected and lysed by CVB, there are cells that do survive the infection. In the study with human heart cells, myocytes were lysed within 24 hr of infection and were then replaced by fibroblasts within 5 days. These cultures continued to release high titers of infectious CVB for the duration of the experiment (15 days). Although the persistent infection was not maintained for a longer period of time, it is possible that the cells could have been subcultured. Infectious virus was not tested for in the study using rat heart cells. However, by 48 hr postinfection myocytes exhibited CPE and decreased beating. Following the death of the myocytes, there was regrowth of surviving interstitial cells replacing the myocytes. Because infectious virus was not tested for, it is impossible to determine whether the cells were persistently infected.

In vivo studies of CVB-persistent infection are limited to those of NFR and N : NIH(S)-II mice. The hearts of 15 of 25 N : NIH(S)-II athymic mice harvested between 14 and 94 days postinfection yielded infectious virus (Table VII). These mice suffered significant mortality with increasing time postinfection. Of the 64 *nu/nu* mice infected, 14 died before they could be harvested, 1 died between 14 and 21 days, 5 died between 21 and 28 days, 3 died between 28 and 50 days, and 5 died between 50 and 94 days. The mortality was not due to aging, as only 1 of 16 uninoculated control *nu/nu* mice held from 21 to 94 days died. Although the mice that died before harvest were not autopsied, two moribund *nu/nu* mice harvested at 94 days postinfection had infectious virus in their hearts together with extensive myocarditis, and thus a chronic viral infection would seem to be a possible explanation for the mortality. Thus, an important consequence of persistent CVB virus in heart tissue is a high rate of mortality.

The mechanism of viral-induced heart disease is not entirely understood. In the BALB/c mouse model, there is considerable evidence that host-immune mechanisms, especially T lymphocytes, are important to the pathology.[20] Myofibers from BALB/c mice are relatively resistant to CVB induced cell death.[26] This finding is in contrast to the high degree of sensitivity of human and rat myocytes to CVB.[5,25] Genetic differences in the host cell as well as the strain of CVB may account for these contrasting results. During persistent infection of heart tissue, both immunologic and nonimmunologic tissue injury is possible. The continuing presence of infectious virus that could infect myocytes might destroy these cells directly (rat or human) or modify the cells so they are recognized and destroyed by host T cells (BALB/c mice). In N : NIH(S)-II athymic mice, CVB3 persistent infection of the heart was associated with increased mortality over the passage of time. Destruction of myofibers by virus could weaken the heart, resulting in dilated cardiomyopathy, and ultimately death. The ability to culture persistently infected heart cells (which do not divide or lyse) from infected N : NIH(S)-II athymic mice suggests a mechanism whereby infectious virus could be continually available to infect and lyse increasing numbers of myocytes.

In vivo CVB-persistent infections of human heart tissue have not been described. A recent study utilized CVB cDNA as a hybridization probe for detection of CVB-like nucleic acid and sequences in human myocardial biopsy specimens.[27] Positive hybridization signals were present in 9 of 17 patients with histologic evidence of active or healing myocarditis, whereas negative results were found in four heart specimens from patients without evidence of viral myocarditis. Several of the positive specimens were from patients with late healing or congestive cardiomyopathy. Neither of these was associated with acute viral infection, and it was suggested that CVB nucleic acid has persisted in the hearts of these patients. If confirmed, this would be clear evidence of CVB persisting in human heart and would provide a direct association between persistent infection and chronic heart disease.

5. DISCUSSION

It has become clear that CVB-persistent infections can be established both *in vitro* and *in vivo*. The mechanism(s) for the establishment and maintenance of these persistent infections are now being investigated. Mechanisms that have been proposed include modulation of lytic virus by viral mutants such as DI particles or *ts* virions, presence of IFN or antiviral antibody and/or cultures composed of a majority of resistant cells.

The CVB-persistent *in vitro* infection of mouse cells studied in our laboratory does not require viral mutants to initiate or maintain it, does not require viral antibody for maintenance, and shows no intefering activity in the supernatant fluid. Persistent infections were established on multiple occasions by inoculating uninfected mouse cells with the standard CVB and subculturing the cells at 3–7 days. The virus that initiated persistent infections was highly lytic in BGMK cells, destroying all cells during the acute infection. Thus, the establishment of the persistent infection was a reproducible event between the virus and the particular type of host cell. A better understanding of the mechanism of persistence requires study of host cell metabolism.

Since one strain of our mouse cells went from 100% of the cells being CVB antigen positive during acute infection to a very low percentage being antigen positive during persistence; because there was no evidence of cell destruction, we hypothesize that the cells can cure themselves of virus. One way in which the cells might regulate viral replication is by limiting production of a host cell product required for viral replication. In that case, the addition of an inhibitor of transcription such as actinomycin D might even further suppress viral multiplication. We have found that actinomycin D does suppress the multiplication of CVB3 from persistently infected cells to a much greater degree than it inhibits CVB3–S. The action of actinomycin D was virus specific and not host cell specific, as CVB3 from persistently infected cells were sensitive to actinomycin D in both BGMK and mouse cells, while CVB3–S was not susceptible to either cell type. Factors manifested as increased actinomycin D sensitivity could function in maintaining the persistent infection, but they would not account for the ease of establishment of persistence, because actinomycin D sensitivity was not measurable until persistence was established.

The stability of the cell membrane could be an important factor in the ability of infected cells to survive viral infection without a[illegible] ells lysing. In preliminary experiments desiged to measure inhibition ot protein synthesis, we observed that mouse cells that can be persistently infected were more resistant to solubilization by detergent than were the membranes of BGMK cells, which are easily solubilized. The stability of mouse cell membranes might also account for the greater degree of cell association of CVB3 in mouse cells as compared with BGMK cells, which could further relate to ease of establishment of persistent infection. The interaction of virus and host cells is complex. An understanding of the difference in CPE, cell lysis, and viral growth in BGMK as compared with NFR mouse cells will provide insight into the mechanisms of CVB persistence.

Coxsackie B virus-persistent infections may be important in the pathogenesis of certain chronic human disorders. *In vitro* and *in vivo* studies of CVB-persistent infection have focused on lymphoid, pancreatic, and heart infections. The evidence shows that CVB persistence does occur in mouse hearts and results in increased mortality. The mechanism may be by direct destruction of host cells or by the host-immune response. Whichever is the case, this is an exciting area that requires further investigation.

6. CONCLUSION

Experimental CVB-persistent infections can be established more readily than was once realized. The host cell is an important factor in determining whether a persistent infection will occur. Persistent infection of certain host cells *in vitro* is easily established. *In vivo* the host-immune response appears to be a major factor in determining whether persistence can occur. Thus far most evidence for CVB-persistent infections in *in vivo* mouse models comes from experiments using immunodeficient or immunocompromised mice. The only evidence for persistent infection in humans comes from recently reported nucleic acid hybridization studies on biopsied tissues; these findings need to be confirmed by other workers. CVB-persistent infections may have a role in certain chronic conditions such as diabetes or chronic heart disease.

ACKNOWLEDGMENTS. Parts of these studies were done by San Francisco State University graduate students Shigeo Yagi and Greg Harrowe. Support was provided by the California Affiliate of the American Heart Association.

REFERENCES

1. Crowell, R. L., 1963, Specific viral interference in HeLa cell cultures chronically infected with coxsackievirus B5 virus, *J. Bacteriol.* **86:**517–526.
2. Crowell, R. L., and Syverton, J. T., 1961, The mammalian-cell-virus relationship. VI. Sustained infection of HeLa cells by coxsackie B-3 virus and effect on superinfection, *J. Exp. Med.* **113:**419–435.
3. Maverakis, N. H., Schmidt, N. J., Riggs, J. L., and Lennette, E. H., 1973, Carrier cultures of human fetal diploid cells infected with coxsackievirus type B2, *Arch. Ges. Virusforsch.* **43:**289–297.
4. Frank, J. A., Schmidt, E. V., Smith, R. E., and Wilfert, C. M., 1986, Persistent infection of rat insulinoma cells with coxsackie B4 virus, *Arch. Virol.* **87:**143–150.
5. Kandolf, R., and Hofschneider, P. H., 1984, Effect of interferon on the replication of coxsackie B3 virus in cultured human fetal heart cells, in: *Viral Heart Disease* (H.-D. Bolte ed.), pp. 57–63, Springer-Verlag, New York.
6. Matteucci, D., Paglianti, M., Giangregorio, A. M., Capobianchi, M. R., Dianjani, F., and Bendinelli, M., 1985, Group B coxsackievirus readily establish persistent infections in human lymphoid cell lines, *J. Virol.* **56:**651–654.
7. Schnurr, D. P., and Schmidt, N. J., 1984, Persistent infection of mouse fibroblasts with coxsackievirus, *Arch. Virol.* **81:**91–101.
8. Schnurr, D. P., and Schmidt, N. J., 1984, Coxsackievirus B3 persistence and myocarditis in NFR nu/nu and PL/nu mice, *Med. Microbiol. Immunol.* **173:**1–7.
9. Schnurr, D. P., Cao, Y., and Schmidt, N. J., 1984, Coxsackievirus B3 persistence and myocarditis in N:NIH(S) II nu/nu and +/nu mice, *J. Gen. Virol.* **65:**1197–1201.
10. Bocharov, E. V., and Shalaurova, B. V., 1984, Persistence of coxsackie B1 virus in BALB/c mice, *Acta Virol.* **28:**345.
11. Youngner, J. S., and Preble, O. T., 1980, Viral persistence: Evolution of viral populations, in: *Comprehensive Virology*, Vol. 19 (H. Frankel-Conrat and R. R. Wagner eds.), pp. 73–135, Plenum Publishing Corp., New York.
12. Cole, C. N., Smoler, D., Wimmer, E., and Baltimore, D., 1971, Defective interfering particles of poliovirus. I. Isolation and physical properties, *J. Virol.* **7:**478–485.
13. Gibson, J. P., and Righthand, F., 1985, Persistence of echovirus 6 in cloned human cells, *J. Virol.* **54:**219–223.
14. Vallbracht, A., Hofmann, L., Wurster, K. G., and Flehmig, B., 1984, Persistent infection of human fibroblasts by hepatitis A virus, *J. Gen. Virol.* **65:**609–615.
15. Huang, A. S., 1973, Defective interfering viruses, *Ann. Rev. Microbiol.* **27:**101–117.

16. Schaffer, F. L., and Gorden, M., 1966, Differential inhibitory effects of actinomycin D among strains of poliovirus, *J. Bacteriol.* **91:**2309–2316.
17. Morrow, C. D., Gibbons, G. F., and Dasgupta, A., 1985, The host protein required for in vitro replication of poliovirus is a protein kinase that phosphorylates eucaryotic initiation factor-2, *Cell* **40:**913–921.
18. Dasgupta, A., Zabel, P., and Baltimore, D., 1980, Dependence of the activity of poliovirus replicase DNA or host cell protein, *Cell* **19:**423–429.
19. Wilfert, C. M., Buckley, R., Rosen, F. S., Whisnent, J., Oxman, M. N., Griffith, J. F., Katz, S. L., and Moore, M., 1977, Persistent enterovirus infections in agammaglobulinemia, in: *Microbiology-1977.* (D. S. Schlessinger ed.), pp. 488–493, American Society for Microbiology, Washington, D.C.
20. Woodruff, J. F., 1981, Viral myocarditis. *Amer. J. Pathol.* **101:**425–483.
21. Southern, P., and Oldstone, M. B. A., 1986, Medical consequences of persistent infection, *New Eng. J. Med.* **314:**359–367.
22. Oldstone, M. B. A., Southern, P., Rodriguez, M., and Rampert, M., 1984, Virus persists in B cells of islets of Langerhans and is associated with chemical manifestations of diabetes, *Science* **224:**1440–1443.
23. Bendinelli, M., Matteucci, D., Toniolo, A., Patane, A. M., and Pistillo, M. P., 1982, Impairment of immunocompetent mouse spleen cell functions by infection with coxsackievirus B-3, *J. Infec. Dis.* **146:**797–805.
24. Matteucci, D., Toniolo, A., Conaldi, P. G., Basolo, F., Gori, Z., and Bendinelli, M., 1985, Systemic lymphoid atrophy in coxsackie B-3 infected mice: Effects of virus and immunopotentiating agents, *J. Infec. Dis.* **151:**1100–1108.
25. Ying-Zhen, Y., and Dyke, J. W., 1985, Coxsackie B-2 virus infection in rat beating heart cell culture, *J. Virol. Meth.* **12:**217–224.
26. Huber, S. A., Job, L. P., and Woodruff, J. F., 1980, Lysis of infected myofibers by coxsackievirus B-3 immune T lymphocytes, *Amer. J. Pathol.* **98:**681–694.
27. Bowles, N. E., Olsen, E. G. J., Richardson, P. J., and Archard, L. C., 1986, Detection of coxsackie-B-virus-specific RNA sequences in myocardial biopsy samples for patients with myocarditis and dialated cardiomyopathy, *Lancet* **1:**1120–1123.

12

New Approaches to Laboratory Diagnosis

RÜDIGER DÖRRIES

1. INTRODUCTION

Classic laboratory techniques to diagnose coxsackievirus (CV)-induced diseases include isolation and typing of the causative agent from body fluids and/or demonstration of virus-specific antibodies in serum or cerebrospinal fluid (CSF) specimens.[1,2] These techniques generally are powerful in the diagnosis of viral infections; however, in the case of CV, they are hampered by many difficulties.

Isolation of these viruses requires either mice or tissue cultures, but unfortunately no single cell line susceptible to all CV serotypes is available, making routine isolation both cumbersome and time consuming. Usually, serodiagnosis of CV infections by demonstration of neutralizing, complement-fixing, or hemagglutinating antibodies is faster than are attempts to isolate the virus, but difficulties intrinsic to the CV group again result in considerable complications due to the high number of serotypes. Anamnestic reactions often give rise to remarkable titers of heterotypic antibodies, thereby reducing severely the significance of these assays. Although neutralizing antibodies are fairly homotypic, application of this test system is time consuming and expensive due to the necessity of keeping different tissue culture systems to comply with the restricted tissue specificity of certain virus types. Complement-fixing (CF) antibodies are often heterotypic or not detectable at all, and hemagglutination inhibition is impaired severely by the fact that only a minority of CV do agglutinate red blood cells (RBCs) of a restricted range of species.

RÜDIGER DÖRRIES • Institute for Virology and Immunobiology, University of Würzburg, D-8700 Würzburg, Federal Republic of Germany.

TABLE I
Methods to Detect Coxsackieviruses

Technique	Advantages	Disadvantages
EIA	Rapid, easy automation, evaluation by eye, marker enzyme very stable	Insensitive, restricted serotype specificity, broad panel of viral antigens and labeled antibodies necessary
Hybridization		
Dot, slot	Good sensitivity, very specific, access to diagnosis of myocardial complications	Radioactive probing, time consuming, difficult to quantitate
In situ	Excellent sensitivity, very specific, access to diagnosis of myocardial complications	
Isolation	Very sensitive, excellent characterization of isolates by monoclonal antibodies, conclusive for acute infection	Time consuming, laborious tissue culture, complex typing procedure

TABLE II
Methods to Detect Coxsackievirus-Specific Antibodies

Technique	Advantages	Disadvantages
Neutralization	Type specific, clinically relevant antibody, determination of IgM after FPLC of serum	Laborious tissue culture, time consuming, large number of infectious serotypes necessary
ISPRIA	Excellent sensitivity, rapid, only secondary antibodies to be labeled	Radioactive probing, rapid decline of antibody marker, see also ISPEIA
ISPEIA	Very sensitive, rapid, only secondary antibodies to be labeled, marker enzyme very stable, easy automation, evaluation by eye	Impairment of diagnosis by (1) crossreactive IgG and IgM; (2) persistent IgM, RF interference, competition by CV-specific IgG, large amounts of purified virus necessary
MACEIA	Very sensitive, less virus needed than in ISPEIA, no RF interference, rapid, marker enzyme very stable, easy automation, evaluation by eye	Impairment of diagnosis by (1) crossreactive IgM; (2) persistent IgM, competition of non-CV-specific IgM, broad panel of labeled, CV-specific antibodies needed, difficult adaptation of the system for detection of IgG

Since this situation has proved unsatisfactory for the clinician as well as for the laboratory, much effort has been put into the improvement of the laboratory diagnosis of CV infections in recent years (Tables I and II). This chapter discusses a number of newly developed approaches. Although some of these new techniques have not yet entered the routine diagnostic work, they may contribute significantly to a better and faster diagnosis of CV infections in the near future.

2. IMPROVEMENTS OF ESTABLISHED TECHNIQUES

2.1. Virus Isolation

Isolation of CV from body fluids or excretions has been hampered, especially in the case of group A coxsackieviruses (CVA), by the fact that these viruses grow scarcely and in a limited range of tissue-cultured cells or cannot be propagated at all *in vitro*. However, consequent search for CV-permissive cell types during the late 1970s led to proposal of the combinated use of three cell types that permit the propagation of all CVA and group B coxsackieviruses (CVB), except CVA1, CVA19, and CVA22.[3,4] Since two of these cells, RD cells and Vero cells, are permanent lines and the third, human fetal diploid kidney cells (HFDK), has a semipermanent character, their handling and maintenance are relatively easy. The introduction of this procedure in the diagnostic laboratory has greatly facilitated the isolation not only of CV, but also of other members of the enterovirus family, such as echovirus and poliovirus.[4] More recently, guinea pig embryo cells have been found to represent a substitution for isolation and propagation of certain CVA as sensitive and rapid as the classic inoculation of suckling mice.[5]

2.2. Neutralizing Antibodies

Multiple problems are associated with the determination of homotypic antibody responses in whole sera, due to a high proportion of anamnestic heterotypic IgG antibody responses in patients who have been exposed to other serotypes of CV before contact with a new one. This resulted in efforts to fractionate sera physicochemically in order to obtain IgG free IgM-containing aliquots of the serum. On the basis of early observations from different laboratories,[6–8] it was assumed that the demonstration of specific antibodies of the IgM class would indicate a current or recent infection without having to examine a second serum specimen collected within the following 2 weeks, thus speeding up the diagnosis considerably.

Although sucrose-gradient centrifugation has been shown to be a feasible and reliable method of obtaining enterovirus-specific IgM fractions,[7,9,10] in terms of speed and practicality, this technique has not been approved in many laboratories.

Another separation technique that has been proposed to detect CV-neutralizing IgM antibodies is the adsorption of IgG by protein A from whole sera. Initial successful application seemed to favor this simple and rapid separation procedure,[11] but increasing knowledge on the binding properties of protein A for different immunoglobulin iso- and subtypes demonstrated the limitations of this procedure. It has been shown that protein A does not bind IgG exclusively but, under certain circumstances, binds IgM as well. Moreover, lack of IgG_3 binding results in a significant contamination of the IgM fraction by this IgG subtype.[12] Since IgG_3 may have an important role in certain viral infections,[13] one has to be extremely careful in interpreting the neutralizing capacity of sera fractionated by protein A.

Application of liquid chromatography, either on the basis of size exclusion or ion exchange, will result also in fractionated serum specimens containing IgM with negligible IgG contamination.[14,15] However, the same criticism as for sucrose-gradient centrifugation applied to this approach; until recent years, most complications of the technique were overcome by the use of pressurized chromatography systems and the introduction of separation matrices with superb features concerning the homogeneity of the bead size and pressure resistance. The usefulness of these so-called high-pressure liquid chromatography (HPLC) or fast protein liquid chromatography (FPLC) systems has been shown in a variety of viral infections in which IgM fractions free of agent-specific IgG could be demonstrated by the use of either size exclusion[16] or ion-exchange chromatography.[17] The ease of setting up a system with automated sample application and programmed elution conditions makes these systems convenient for routine work. The highly reproducible separation of IgG and IgA from IgM antibodies within 6–10 min per serum specimen by this technique[17] facilitates the diagnosis of acute CV infections by the detection of a neutralizing IgM fraction in a single serum specimen.

3. NEW TECHNICAL APPROACHES

3.1. New Tools and Reagents to Detect the Virus

3.1.1. Enzyme Immunoassays to Detect Coxsackievirus-Specific Antigens in Clinical Specimens

The feasibility of using the highly specific and sensitive enzyme immunoassays (EIA) to detect virus-specific antigens in clinical specimens, such as blood, throat washes, cerebrospinal fluid (CSF), and stool has been repeatedly demonstrated for a broad pattern of viruses.[18] However, detection of CV-specific antigens by EIA has been described rarely. Based on a detection system for rotaviruses in stools, Yolken and Torsch[19,20] developed an EIA to detect CVA and CVB antigens in tissue culture supernatants as well as in clinical stool specimens. Monkey antibodies, monospecific for individual serotypes, were coated to polyvinyl microtiter wells and served as capture antibody for viral antigens present in the test specimen. After the nonbound material was washed off, captured viral antigens were detected indirectly by subsequent incubation with mouse antisera monospecific for the corresponding viral antigen and alkaline phosphatase (AP)-labeled goat antibodies specific for mouse IgG. AP-catalyzed hydrolysis of *p*-nitrophenylphosphate resulted in a yellow reaction product measurable at 405 nm in a microcolorimeter. To control the specificity of the system, wells coated with monkey preimmune sera were assayed in parallel. The sensitivity was tested by diluting viral antigens used in CF assays. CVB were detectable down to 10^{-2}–10^{-4} CF units, meaning that EIA is about 1000-fold more sensitive than the detection of antigen by CF. However, compared with virus isolation in tissue culture, this detection limit is rather insensitive since, at least for CVB, 0.01 CF units correspond to 1000 $TCID_{50}$/ml. By contrast, in tissue cultures infected with roughly one half the 24 CVA serotypes, viral antigens were detectable by EIA before a cytopathic effect (CPE) was noticeable, thereby increasing the speed of virus isolation procedures. Moreover, CVA and CVB serotype-specific antigens could be identified unequivocally in 8 of 11 stool specimens and in 5 of 11 rectal swab specimens, respectively. Problems that have not been solved completely within the context of these systems are those associated with crossreacting capture antibodies coupled to the solid phase and the low sensitivity when assaying clinical specimens directly. However, since the homotypic reaction of the detecting antibody is usually much stronger than its heterotypic reaction, and in view of the early detection of viral antigens in tissue culture before a CPE is noticeable, this type of assay is a promising base for further developments. Sensitivity and specificity can be improved by the introduction of the biotin–avidin detection system and monoclonal antibodies.

3.1.2. Nucleic Acid Hybridization to Detect Viral RNA in Tissue Specimens

As a consequence of rapid developments in gene technology in recent years, cloning and recombinant DNA technology have been introduced into the diagnosis of viral diseases. Especially in cases in which either the isolation of the etiologic agent is not feasible or serodiagnosis is unable to differentiate between related virus types, the detection of virus-specific sequences by dot or slot blot, sandwich, or *in situ* nucleic acid hybridization has led to the successful identification of viral genomic information in postmortem or biopsy material.[21] Although diagnostic application of these techniques has been reported repeatedly, especially in infections caused by DNA viruses, little information is available with respect to the CV group. The methodologic strategy to detect enteroviral information is as follows: genomic viral RNA is transcribed into a single-stranded complementary strand of DNA (ss-cDNA) by the enzyme reverse transcriptase. With the use of the enzymes DNA polymerase I and T4 DNA ligase, the ss-cDNA is transformed into a double-stranded DNA (ds-cDNA) copy of the viral RNA. Molecular cloning of this cDNA into bacterial plasmids results in a recombinant DNA that, once labeled with radioactive or biotinylated nucleotides, can be used as a probe to detect viral RNA in tissue specimens.

3.1.2.a. Dot- or Slot-Blot Hybridization. In this technique, a total deproteinized nucleic acid extract from the tissue specimen is usually immobilized on nitrocellulose paper and hybridized with the labeled DNA probe. Hyypiä *et al.*[22] used this technique to detect enterovirus-specific RNA. The cDNA was prepared from CVB3 RNA (strain Nancy) and cloned in the bacterial plasmid vector *Escherichia coli* pBR 322. The insert included 4.300 base pairs (bp) from the 3′ end of the CVB3 genome, which is known to encode the RNA-dependent RNA polymerase. Hybridization assays to nucleic acid extracts from infected tissue cultures revealed strong positive signals in the case of CVA9, CVB3, CVB4, echovirus 17, and poliovirus 3. The result was not surprising, since the probe had been prepared with a part of the genome expected to be highly conserved within the enterovirus group, and the probe was classified as group rather than serotype specific. Detection of enterovirus-specific genomic information was equally successful in extracts of tissue cultures inoculated with wild-type strains of enteroviruses from clinical stool specimens. Although a positive signal was not detected before the appearance of CPE, the group specificity of the labeled cDNA facilitated identification of enterovirus-containing stool specimens. No signals could be detected in cell extracts infected with herpes simplex virus 1 (HSV-1) adenovirus 2, Nebraska calf diarrhea virus, or measles virus. However, a positive reading was obtained rarely in attempts to detect virus-specific RNA directly in stool specimens. Only 1 of 8 specimens positive by virus-isolation assays gave a positive signal in the dot hybridization, demonstrating the limited sensitivity of the assay as compared with classic isolation procedures.

Recently, Bowles *et al.*[23] were able to detect CVB2 RNA in endomyocardial biopsy material in 9 of 17 patients with active or healed myocarditis or cardiomyopathy of unknown origin. Comparable to the work reported by Hyypiä *et al.*,[22] the labeled DNA probe was complementary to a 1.6-kb part extending from the 3′ end of the viral RNA and encoding the virus-specific RNA-dependent RNA polymerase. Evaluation of the hybridization signal with respect to signals achieved on known amounts of purified CVB2 RNA permitted an estimate of 10–100 CVB2 copies per cell equivalent detectable in biopsy material. No CVB RNA was detected in four patients with unrelated diseases. Interestingly, viral information was not only detected in patients in the acute stage of the disease, but also in late biopsies from patients with healed myocarditis, indicating a persistent infection of the myocardium by CVB. Although persistency generally is not associated with enteroviruses, these findings would fit perfectly with serologic data revealing persistence of CVB-specific IgM over years in myocardial complications (see Section 3.2.1.c).

The successful detection of genomic information in myocardial tissue of 53% of diseased patients shows the importance of this assay system in the diagnosis of myocarditis, since myocarditis in the context of a CVB infection is diagnosed mainly by serology and very rarely by virus isolation or demonstration of viral antigens in tissue specimens.

A significant improvement of this assay would be the detection of CV type-specific RNA. A promising step has been published by Tracy,[24] who used subgenomic fragments of the cloned CVB3 genome to characterize RNA homologies within the CVB group. Although the noncoding 5′-terminal end of CVB3 has shown to be well conserved within the CVB group, a 3.2–3.5-kb stretch from the 5′ end covering approximately 45% of the coding region could be identified as unique for CVB3 virus, thus offering the possibility of type-specific detection of CVB RNA in clinical specimens.

3.1.2.b. In Situ Hybridization. In contrast to slot or dot procedures, *in situ* hybridization detects viral information in the intact fixed tissue, permitting the direct identification of the virus-infected target cells and an estimate of the proportion of cells carrying the viral genome within a section. The technique is slow because the hybridized tissue specimen has to be exposed to a photographic emulsion sensitive for radioactive emission. Mostly ^{35}S-labeled cDNA is used for this technique, since the half-life of the isotope and the fairly good sensitivity of the coating emulsion for β-emissions permit exposure times up to weeks, making this technique very sensitive. By contrast, the long duration usually excludes this type of assay from diagnostic use in acute CVB infections. However, in prolonged CVB associated diseases, such as cardiac complications, the sensitive detection of viral RNA in biopsy material might give valuable information about target cell types. A promising experimental approach was reported by Kandolf *et al.*,[25] who evidenced

CVB3-infected cells of the myocardium in mice. On the basis of a complete CVB3 cDNA copy,[26] an *in situ* hybridization assay was set up, calculated to detect fewer than 10 RNA copies per infected cell. Extensive sequence homologies among different enterovirus types permit the sensitive detection of CVB, CVA, and echovirus-specific RNA using such a probe.

Future improvements of this system could involve the combination of DNA–RNA hybridization with immunohistology, opening the possibility to study interactions between the presence of genomic RNA, viral protein expression, and inflammatory immune response. Moreover, as in the case of dot-blot hybridization, subgenomic DNA probes should detect serotype specific RNA in tissue specimens, making for a better understanding of tissue tropism and other biological properties of individual CV types.

3.1.2.c. Factors Limiting the Use of Nucleic Acid Hybridization. Although nucleic acid hybridization technology is used in some laboratories for diagnostic purposes, several disadvantages prevent widespread application of these assays as diagnostic tools in routine work. First, there is a rapid decline of sensitivity due to the use of short-lived isotopes such as ^{32}P or ^{35}S to label the cDNA probe. Substitution of these markers with long-lived isotopes, e.g., ^{14}C or ^{3}H, improves the stability of the indicator cDNA, but the weak signal generated by these labels demands a long exposure time (up to several weeks) of the hybridized sample to a sensitive film. A real improvement seems to be the introduction of biotinylated nucleotides for labeling the cDNA. Hybridized specimens can then be detected with avidin-labeled enzymes, comparable to a solid-phase EIA. Although this label is stable, there is general agreement that so far the sensitivity of biotinylated cDNA probes is not as high as that of the radioactive-labeled counterparts. This is a significant limitation for this type of assay in CV diagnosis, especially in view of the lower overall sensitivity of hybridization in detecting enteroviruses as compared with classic isolation techniques.

Further problems can be caused by unspecific hybridization of the cDNA. Cross-hybridizing sequences in cDNA either of cellular or bacterial origin have shown to be the base of many false positive results. To avoid hybridization of cDNA to bacterial sequences in tissue due to vector-derived DNA sequences or detection of cellular DNA caused by cloning of cell-derived sequences, the virus specificity of the probe should be controlled rigorously.

In summary, RNA–DNA hybridization techniques, although not usable for routine CVB diagnosis at present, might become important tools for identification of the infecting serotype if their sensitivity will be enhanced and the cDNA probes will be labeled by nonisotopic markers of sufficient stability.

3.1.3. *Monoclonal Antibodies to Characterize Virus Isolates*

Enteroviruses have long been thought to be extremely stable with respect to their antigenicity. However, recently the use of monoclonal antibodies has demonstrated the existence of a remarkable number of variants among individual serotypes.

The use of neutralizing monoclonal antibodies to characterize antigenic variability of CVB4 isolates was described by Prabhakar *et al.*[27] Fresh clinical isolates from 15 cases were analyzed by neutralization with either a hyperimmune polyclonal reference serum or with a panel of 18 different monoclonals, raised against the CVB4 prototype strain JVB. All isolates were neutralized by the hyperimmune serum but showed different patterns of susceptibility to neutralization by the monoclonal antibodies. Usually the monoclonal antibodies that showed the highest neutralizing titers for the prototype strain were able to neutralize a broad spectrum of isolates, suggesting that they were detecting a fairly conserved epitope of CVB4 virus. By cross-neutralization experiments using the prototype virus, four clinical isolates, and four different monoclonal antibodies, the frequency of mutation in CVB4 virus was calculated to be 10^{-4}–$10^{-5.8}$. This high rate of mutation and number of variants may account for persistence of CVB4 in the human population as well as for the vast array of different diseases induced by CVB4.[27] This assumption was strengthened by data from the same laboratory[28] showing the presence of three different groups of epitopes on clinical isolates of CVB4, as defined by neutralizing monoclonal antibodies: highly conserved, moderately conserved, and poorly conserved. Passage in tissue culture changed the epitope composition of plaque-purified viral isolates at a rate of 10^{-2} for each epitope, even in the absence of selective pressures. Although the overall antigenic composition of the isolates was stable in the sense that the hyperimmune serum was able to neutralize all variants, these data may help explain subtle changes in biologic properties and the tissue tropism of CVB4.

Comparable data concerning antigenic variants of CVB4 have been published by Cao *et al.*[29] By cross-neutralization, five functional epitopes involved in neutralization could be identified on CVB4 variants selected on the basis of the presence of neutralizing monoclonal antibodies in the tissue culture fluid. Interestingly, these variants differed not only in their fine antigenic composition but also in their tissue-specific pathogenicity as tested in a mouse model. Compared with the parental strain, three types of variants were detected: one causing severe necrotic lesions in the myocardium, another producing less histopathology, and the third replicating poorly in the heart, strongly supporting the idea that variation in neutralizing epitopes may be closely related to the tissue tropism of CVB4.

In summary, the use of CV-specific monoclonal antibodies not only offers promising approaches to obtain greater insight into the antigenic variability of a group of viruses commonly considered very stable, but might also help establish procedures for the type-specific detection of viral antigens or virus-specific antibodies.

3.2. New Tools and Reagents to Detect the Antibody Response

3.2.1. Solid-Phase Immunoassays

Attempts to characterize the course of CV infections by the demonstration of virus-specific antibodies in serum and CSF have focused primarily on the detection of virus-specific IgM antibodies,[30] since their presence usually indicates the acute stage of the infection. Based on the availability of anti-IgM (μ-chain) antibodies of good quality, solid-phase assay systems were developed, permitting the detection of virus-specific IgM in unfractionated sera with high sensitivity and specificity within a working day.

Two principles have been found practicable to detect CV-specific IgM antibodies: the indirect solid-phase immunosassy (ISPIA), either as radioimmunoassay (ISPRIA) or as enzyme immunoassay (ISPEIA) and the so-called IgM antibody-capture immunoassay (MACIA), again either operating with radiolabeled (MACRIA) or with enzyme-labeled reagents (MACEIA). Generally, in ISPIA a solid support coated with the viral antigen is incubated with the serum specimen. Virus-specific IgM bound to the antigen are detected by subsequent incubation with labeled anti-IgM. By contrast, in MACIA, the anti-IgM is coated to the solid phase. Upon incubation with the serum specimen, the IgM are captured by the solid-phase-coupled anti-IgM, irrespective of their antigen specificity. Virus-specific IgM are evidenced by incubation with either labeled viral antigens or unlabeled viral antigens, followed by labeled virus-specific antibodies.

3.2.1.a. ISPIA to Detect Virus-Specific IgM. ISPIA to detect CV specific IgM has been applied using heavy-chain specific anti-IgM antibodies labeled with ^{125}I[31] or AP[32] as second antibody. ISPRIA was shown to be highly sensitive. During the acute stage of the infection, titers up to 1 : 131,072 were detectable in individual sera. Secondary infections by the same or related serotypes could be detected by an increase of virus-specific IgG in the absence of detectable IgM[31]. Using this assay, an average increase of 300-fold in sensitivity could be obtained as compared with neutralization. To avoid the complicated handling procedures of radiolabeled compounds and the short half-life of ^{125}I, the method was later modified by the introduction of an enzyme-labeled (AP) indicator antibody (ISPEIA).[32] Using this assay, the problems raised by heterotypic IgM responses in acute CVB infections were analyzed carefully. In general, three types of IgM responses could be identified: (1) in rare cases, a strictly homotypic response directed to the etiologic agent only; (2) in most instances, a positive response for other CVB serotypes, but at significantly lower titers than for the isolated serotype; and (3) a heterotypic response for CVB1 to 5.[32] Whereas the first two patterns did permit identification of the infecting serotype, no decision could be reached under this respect in the latter case. Moreover, some patients showed crossreactive IgM even for non-CVB enteroviruses. This finding limits the value of this assay severely, since a positive result may indicate a current infection with one out of some 60–70 possible enterovirus serotypes.

3.2.1.b. MACIA to Detect Virus-Specific IgM. Considerable work has been dedicated to the development of antibody capture assays to detect CV-specific IgM, since this method avoids false-positive results due to rheumatoid factor (RF) and requires less viral antigen than ISPEIA.

In a MACEIA described in 1980 by El-Hagrassy and Banatvala,[33] pooled antigens of CVB1–5 and pooled enzyme-labeled anti-CVB antibodies were used to characterize the IgM response in patients suffering from different CVB-associated diseases. The results were in good agreement with those reported more recently[34] based on the same technique. An average of approximately 85% of patients with acute CVB infection showed CVB group-specific IgM antibody titers that persisted for 6–8 weeks from clinical onset. Comparison by Chan and Hammond[34] of MACEIA and microneutralization test (NT) using sucrose-gradient fractionated sera from 19 patients with confirmed CVB4 isolation showed the superior sensitivity of MACEIA, since only 68% of acute CVB infections were deteced by NT. By contrast, in a recent study involving 273 sera, McCartney *et al.*[35] reported equal sensitivity of NT and MACEIA. However, the parameters considered by these workers indicative of a recent CVB infection were different (static titers equal or higher than 1 : 512 or a fourfold rise/fall in titers of unfractionated sera), and the serologic results were usually not validated with virus isolations, rendering direct comparison of the results rather difficult. Finally, in a group of 45 patients with aseptic meningitis, Bell *et al.*[36] reported 67% positive readings with MACEIA and only 22% of successful virus isolations, stressing again the excellent speed and sensitivity of this technique.

These results and the fact that the NT is still time consuming and cumbersome seem to favor MACEIA as the preferable assay in acute CVB infections. However, as in ISPEIA and MACRIA, a group-reactive IgM is detected routinely in MACEIA as well, again limiting its value for identifying the etiologic serotype.

3.2.1.c. Factors Limiting the Use of Solid-Phase Immunoassays. In both ISPIA and MACIA, the user is faced with several technical complications which have to be taken into account before application to routine work.[37]

Competition of antibodies can occur, lowering the sensitivity of the assay. In ISPIA, the presence of large amounts of virus-specific IgG in acute-phase sera can compete with virus-specific IgM for the limited number of viral epitopes[38] and in MACIA, nonvirus-specific IgM can compete with the usually lower amount of virus-specific IgM for the limited binding capacity of the solid-phase bound anti-IgM antibodies.

In ISPIA, RF, usually an IgM autoantibody produced in inflammatory diseases and directed against autologous IgG complexed with the appropriate antigen, may bind to virus-specific IgG attached to the viral solid-phase antigens. As a consequence, RF may produce false-positive results.[39] This complication, generally not observed in MACIA, can be overcome by preabsorbing the serum specimen with Latex-particle-bound anti-IgG, which can remove competing virus-specific IgG and RF in a single-step procedure.

In MACIA, labeled antigen or virus-specific antibody is necessary for each individual serotype, a tedious undertaking especially in the case of CVA. Moreover, calibration of the detecting antigens to a comparable sensitivity is more complex. In ISPIA, only the labeled anti-IgM needs to be standardized.

Apart from these technical complications, identification of the etiologic agent with these assays can be extremely difficult or even impossible due to the rise of heterotypic antibodies, which in some cases may completely obscure the homotypic response. This seems to prevent a correct diagnosis, especially in immunoassays for CV-specific IgG, since even significant increases of neutralizing antibody titers can escape detection.[40] However, heterotypic antibodies have also been detected in assays detecting IgM.[32,35,41,42] The situation is complicated even more by the fact that CVB-reactive IgM has been identified also in acute CVA, echovirus, and hepatitis A virus infections.[32,34,35,42] However, under circumstances that are poorly defined, homotypic or at least type-predominant IgM responses are mounted. As shown by Banatvala *et al.*,[43] the probability of detecting homotypic responses increases significantly with decreasing age of the patient in cases of juvenile onset of diabetes. Children up to 4 years of age responded homotypically in 81% of the cases, whereas 77% of children above 15 years of age raised a heterotypic response. As argued by Chan *et al.*,[34] these findings may indicate that the number or types of CVB or other enteroviruses to which the patient has been previously exposed will determine largely the type specificity of the IgM response of a current CVB infection.

Further fine investigation of crossreactivity amongst virus-specific IgM antibodies from patients with acute CV infection by Western blot analysis[32,44] demonstrated that heterotypic IgM antibodies interact predominantly with the virus protein VP1, pointing to a possible location of the group-specific epitopes on this structural protein. Interestingly, IgM antibodies of homotypic nature do also recognize VP1 preferentially, locating serotype-specific epitopes on the same structural protein. Data from experimental immunization protocols in animals suggest that CVB group specific determinants are located in cryptic sites of the virion. Using highly purified virus, captured by monospecific antibodies coated onto the bottom of a microtiterplate, Katze and Crowell[45] were able to demonstrate highly type-specific antibody responses in rabbits. By contrast, the use of urea-disrupted virus instead of intact particles showed a high proportion of heterotypic antibodies in these sera, a finding supporting the assumption that group-reactive determinants are usually not accessible for antibodies on the intact virion. It has to be expected, however, that incomplete particles are presented to the host's immune system during the course of most natural infections. Therefore, type-specific diagnosis in cases with activation of heterotypic IgM clones would only be possible by the exclusion of these antibodies from the reaction in immunoassays. This would be possible, either by using highly purified, intact virions as antigen or, even more desirable, a recombinant VP1 protein lacking the group-reactive parts. Therefore, the best homotypic IgM determinations are likely to be achieved by the antibody-capture technique, using highly purified viral antigens, since direct coating of viral particles to microtiter wells by incubation in alkaline low molar carbonate buffer, as usually performed in ISPIA, may damage the virus and expose group-reactive determinants.

In addition to the problem of crossreactivity, the diagnostic validity of specific IgM determination is hampered by the possible prolonged persistence of this antibody class. Normally, after acute infections, CVB-specific IgM is detectable for 6–8 weeks but, as demonstrated by Eggers and Mertens[46] in apparently healthy subjects recovered from CVB1-induced aseptic meningitis, it can persist for up to 6 months. CVB-specific IgM can persist for even longer periods in patients suffering from cardiac complications in association with CVB infections. Tilzey *et al.*[47] documented the persistence of heterotypic CVB-specific IgM for up to 5 years after the onset of clinical symptoms and seroconversion. These results might be explained either by viral persistence, which to date has never been recognized as a property of human picornaviruses, or by repeated induction of crossreactive IgM by infections with other enteroviruses.

In conclusion, enzyme immunoassays to detect virus-specific IgM are not yet capable of identifying the specific etiologic serotype in most acute CVB infections but can evaluate rapidly and with high sensitivity an acute enterovirus infection. Persistence of heterotypic IgM over years may however lead to false conclusions. Careful evaluation of clinical symptoms and later confirmation by NT may help reduce possible diagnostic errors.

4. CONCLUDING REMARKS

This chapter briefly reviews new and prospective technologies for the laboratory diagnosis of CV infections. It is clear that all modern approaches that have been exploited to improve the diagnosis of other viral diseases have also been more or less successfully applied to CV (Tables I and II). However, all attempts, aimed to improve the detection of either infectious virus, viral antigens, and viral genomic material or virus-specific antibodies, are still hampered by a core problem: the high number of different serotypes showing homology over extended parts of the genome makes it difficult to set up assay systems that permit a rapid, sensitive, and serotype-specific diagnosis. However, most probably, DNA recombinant technology should solve this problem in the near future. Cloned subgenomic cDNA of CVB3 is available today to detect RNA sequences unique for CVB3, and experiments are in progress to set up a serotype-specific hybridization assay using these as probes. Cloning of type-specific subgenomic information into expression vectors should also result in the production of antigens specific to each individual serotype, permitting more reliable serotype-specific IgM antibody determinations.

Besides improved specificity due to recombinant gene products, sensitivity should increase significantly if modern serologic assays, such as time-resolved fluorescence[48] or monoclonal antibodies specific for RNA–DNA hybrids,[49] will be introduced into CV diagnosis. The delay in the appearance of state-of-the-art technology in CV diagnosis is mainly attributable to the high number of serotypes. Eventually, however, progress in these technologies should offer a variety of rapid, precise, and reliable diagnostic tools.

REFERENCES

1. Melnick, J. L., Wenner, H. A., and Philips, C. A., 1979, Enteroviruses, in: *Diagnostic Procedures for Viral, Rickettsial and Chlamydial Infections*, 5th ed. (E. H. Lennette and N. J. Schmidt, eds.), pp. 471–534, American Public Health Association, Washington, D.C.

2. Melnick, J. L., 1982, Enteroviruses, in: *Viral Infections of Humans*, 2nd ed. (A. S. Evans, ed.), pp. 187–251, Plenum, New York.
3. Schmidt, N. J., Ho, H. H., and Lennette, E. H. 1975, Propagation and isolation of group A coxsackieviruses in RD cells, *J. Clin. Microbiol.* **2:**183–185.
4. Wecker, I., and ter Meulen V., 1977, RD cells in the laboratory diagnosis of enteroviruses, *Med. Microbiol. Immunol.* **163:**233–240.
5. Landry, M. L., Madore, H. P., Fong, C. K. Y., and Hsiung, G. D., 1981, Use of guinea pig embryo cell cultures for isolation and propagation of group A coxsackie viruses, *J. Clin. Microbiol.* **13:**588–593.
6. Schluederberg, A., 1965, Immune globulins in human viral infections, *Nature (London)* **205:**1232–1233.
7. Schmidt, N. J., Lennette, E. H., and Dennis, J., 1968, Characterization of antibodies produced in natural and experimental coxsackievirus infections, *J. Immunol.* **100:**99–106.
8. Schmidt, N. J., Magoffin, R. L., and Lennette, E. H., 1973. Association of group B coxsackieviruses with cases of pericarditis, myocarditis, or pleurodynia by demonstration of immunoglobulin M antibody, *Infect. Immun.* **8:**341–348.
9. Svehag, S. E., and Mandel, B., 1963, The formation and properties of poliovirus neutralizing antibody, *J. Exp. Med.* **119:**1–19.
10. Mertens, T., Hager, H., and Eggers, H. J., 1982, Epidemiology of an outbreak in a maternity unit of infections with an antigenic variant of echovirus 11, *J. Med. Virol.* **9:**81–91.
11. Reiner, M., and Wecker, E., 1981, A modified absorption-reduction method to detect virus-specific hemagglutinin inhibiting and neutralizing IgM antibodies, *Med. Microbiol. Immunol.* **169:**237–245.
12. Richman, D. D., 1983, The use of staphylococcal protein A in diagnostic virology, *Curr. Top. Microbiol. Immunol.* **104:**159–176.
13. Beck, O. E., 1981, Distribution of virus antibody activity among human IgG subclasses, *Clin. Exp. Immunol.* **43:**626–632.
14. Fahey, J. L., and Terry, E. W., 1978, Ion exchange chromatography and gel filtration, in: *Handbook of Experimental Immunology*, 3rd ed. (D. M. Weir, ed.), pp. 8.1–8.16, Blackwell Scientific Publications, Oxford.
15. Johnson, R. B., and Libby, R., 1980, Separation of immunoglobulin M (IgM) essentially free of IgG from serum for use in systems requiring assay of IgM-type antibodies without interference from rheumatoid factor. *J. Clin. Microbiol.* **12:**451–454.
16. Sann, G., Schneider, G., Loeke, S., and Doerr, H. W., 1983, Rapid fractionation of serum immunoglobulins by high pressure liquid gel permeation chromatography. Applications to routine serologic procedures, *J. Immunol. Methods* **59:**121–127.
17. Sampson, I. A., Hodgen, A. N., and Arthur, I. H., 1984, The separation of IgM from human serum by FPLC, *J. Immunol. Methods* **69:**9–15.
18. Yolken, R. H., 1982, Enzyme immunoassay for the detection of infectious antigens in body fluids, *Rev. Infect. Dis.* **4:**35–68.
19. Yolken, R. H., and Torsch, V., 1980, Enzyme-linked immunosorbent assay for the detection and identification of coxsackie B antigen in tissue cultures and clinical specimens, *J. Med. Virol.* **6:**45–52.
20. Yolken, R. H., and Torsch, V. M., 1981, Enzyme-linked immunosorbent assay for detection and identification of coxsackieviruses A, *Infect. Immun.* **31:**742–750.
21. Kulski, J. K., and Norval, M., 1985, Nucleic acid probes in diagnosis of viral diseases of man, *Arch, Virol.* **83:**3–15.
22. Hyypiä, T., Stålhandske, P., Vainionpää, R., and Pettersson, U., 1984, Detection of enteroviruses by spot hybridization, *J. Clin. Microbiol.* **19:**436–438.

23. Bowles, N. E., Richardson, P. J., Olsen, E. G. J., and Archard, L. C., 1986, Detection of coxsackie-B-virus-specific RNA sequences in myocardial biopsy samples from patients with myocarditis and dilated cardiomyopathy, *Lancet* **1:**1120–1123.
24. Tracy, S., 1984, A comparison of genomic homologies among the coxsackievirus B group: Use of fragments of the cloned coxsackievirus B3 genome as probes, *J. Gen. Virol.* **65:**2167–2172.
25. Kandolf, R., Ameis, D., Kirschner, P., Canu, A., and Hofschneider, P. H., 1986, In situ detection of viral genomes in myocardial cells by nucleic acid hybridization: A novel approach to the diagnosis of viral heart disease, presented at the *International Symposium on Inflammatory Heart Disease, May 15–17, Würzburg, Federal Republic of Germany.*
26. Kandolf, R., and Hofschneider, P. H., 1985, Molecular cloning of the genome of a cardiotropic coxsackie B3 virus: Full-length reverse-transcribed recombinant c-DNA generates infectious virus in mammalian cells, *Proc. Natl. Acad. Sci. USA* **82:**4818–4822.
27. Prabhakar, B. S., Haspel, M. V., McClintock, P. R., and Notkins, A. L., 1982, High frequency of antigenic variants among naturally occurring human coxsackie B4 virus isolates identified by monoclonal antibodies, *Nature (Lond.)* **300:**374–376.
28. Prabhakar, B. S., Menegus, M. A., and Notkins, A. L., 1985. Detection of conserved and nonconserved epitopes on coxsackievirus B4: Frequency of antigenic change, *Virology* **146:**302–306.
29. Cao, Y., Schnurr, D. P., and Schmidt, N. J., 1984, Monoclonal antibodies for study of antigenic variation in coxsackievirus type B4: Association of antigenic determinants with myocarditic properties of the virus, *J. Gen. Virol.* **65:**925–935.
30. Pattison, J. R., 1983, Tests for coxsackie B virus-specific IgM, *J. Hyg. (Lond.)* **90:**327–332.
31. Dörries, R., and ter Meulen, V., 1980, Detection of enterovirus specific IgG and IgM antibodies in humans by an indirect solid phase radioimmunoassay, *Med. Microbiol. Immunol.* **168:**159–171.
32. Dörries, R., and ter Meulen V., 1983, Specificity of IgM antibodies in acute human coxsackievirus B infections, analysed by indirect solid phase enzyme immunoassay and immunoblot technique, *J. Gen. Virol.* **64:**159–167.
33. El-Hagrassy, M. M. O., and Banatvala, J. E., 1980, Coxsackie-B-virus-specific IgM responses in patients with cardiac and other diseases, *Lancet* **2:**1160–1162.
34. Chan, D., and Hammond, G. W., 1985, Comparison of serodiagnosis of group B coxsackievirus infections by an immunoglobulin M capture enzyme immunoassay versus microneutralization, *J. Clin. Microbiol.* **21:**830–834.
35. McCartney, R. A., Banatvala, J. E., and Bell, E. J., 1986, Routine use of μ-antibody-capture ELISA for the serological diagnosis of coxsackie B virus infections, *J. Med. Virol.* **19:**205–212.
36. Bell, E. J., McCarney, R. A., Basquill, D., and Chaudhuri, A. K. R., 1986, μ-Antibody capture ELISA for the rapid diagnosis of enterovirus infections in patients with aseptic meningitis, *J. Med. Virol.* **19:**213–217.
37. Meurman, O., 1983, Detection of antiviral IgM antibodies and its problems—A review, *Curr. Top. Microbiol. Immunol.* **104:**101–131.
38. Ziegler, D. W., 1979, Determination of IgM antibodies in diagnostic virology, in: *Diagnosis of Viral Infecfions* (D. A. Lenette, S. Pecter, and K. D. Thompson, eds.), pp. 63–71, University Park Press, Baltimore.
39. Salonen, E.-M., Vaheri, A., Suni, J., and Wager, O., 1980, Rheumatoid factor in acute viral infections: Interference with determination of IgM, IgG and IgA antibodies in an enzyme immunoassay, *J. Infect. Dis.* **142:**250–255.
40. Hannington, G., Booth, J. C., Wiblin, C. N., and Stern, H., 1983, Indirect enzyme-

linked immunosorbent assay (ELISA) for detection of IgG antibodies against coxsackie B viruses, *J. Med. Microbiol.* **16:**459–465.

41. Pugh, S. F., 1984, Heterotypic reactions in a radioimmunoassay for coxsackie B virus specific IgM, *J. Clin. Pathol.* **37:**433–439.
42. Morgan-Capner, P., and McSorley, C., 1983, Antibody capture radioimmunoassay (MACRIA) for coxsackievirus B4 and B5-specific IgM, *J. Hyg. (Lond.)* **90:**333–349.
43. Banatvala, J. E., Bryant, J., Schernthaner, G., Borkenstein, M., Schober, E., Brown, D., De Silva, L. M., Menser, M. A., and Silink, M., 1985, Coxsackie B, mumps, rubella and cytomegalovirus specific IgM responses in patients with juvenile-onset insulin-dependent diabetes mellitus in Britain, Austria and Australia, *Lancet* **1:**1409–1412.
44. Mertens, Th., Pika, U., and Eggers, H. J., 1983, Cross antigenicity among enteroviruses as revealed by immunoblot technique, *Virology* **129:**431–442.
45. Katze, M. G., and Crowell, R. L., 1980, Immunological studies of the group B coxsackieviruses by the sandwich enzyme-linked immunosorbent assay (ELISA) and immunoprecipitation, *J. Gen. Virol.* **50:**357–367.
46. Eggers, H. J., and Mertens, T., 1986, Persistence of coxsackie B virus-specific IgM, *Lancet* **2:**284.
47. Tilzey, A. T., Signy, M., and Banatvala, J. E., 1986, Persistent coxsackie B virus specific IgM response in patients with recurrent pericarditis, *Lancet* **1:**1491–1492.
48. Halonen, P., Meurman, O., Lövgren, T., Hemmilä, I., and Soini, E., 1983, Detection of viral antigens by time-resolved fluoroimmunoassay *Curr. Top. Microbiol. Immunol.* **104:**133–146.
49. Boguslawski, S. J., Smith, D. E., Michalak, M. A., Mickelson, K. E., Yehle, C. O., Patterson, W. L., and Carrico, R. J., 1986, Characterization of monoclonal antibody to DNA–RNA and its application to immunodetection of hybrids, *J. Immunol. Methods* **89:**123–130.

13

General Pathogenicity and Epidemiology

NORMAN R. GRIST and DANIEL REID

1. GENERAL PATHOGENICITY OF COXSACKIEVIRUSES

Coxsackieviruses (CV) were distinguished from other enteroviruses by their ability to infect and cause disease in immature mice. This particular pathogenicity for rodents corresponds to a difference in the pattern of diseases produced in the normal human host, compared with those caused by other enteroviruses. Most of the properties of CV, however, do resemble those of other enteroviruses and there is considerable overlap between the pathogenetic behavior of all types of enteroviruses and also of their epidemiologic behavior.

1.1. Coxsackievirus A

Group A coxsackieviruses (CVA) typically cause severe generalized myositis and flaccid paralysis of skeletal muscles in newborn mice. Except for types A9 and A23 (alias echovirus type 9), most are difficult or as yet impossible to isolate in cell cultures, although many can be adapted to grow in cell cultures. Although they can cause myositis in the human host, affecting both skeletal muscle (Bornholm disease, epidemic myalgia) and sometimes cardiac muscle (acute myocarditis), they are less prominent causes of these diseases than are group B coxsackieviruses

NORMAN R. GRIST • Department of Infectious Diseases, University of Glasgow, Glasgow G12 8QQ, Scotland; Communicable Diseases Unit, Ruchill Hospital, Glasgow G20 9NB, Scotland. DANIEL REID • Communicable Diseases Unit, Ruchill Hospital, Glasgow G20 9NB, Scotland.

(CVB). More often they cause mild or inapparent infections, pharyngeal or oral lesions, acute neurologic diseases, or conjunctivitis.

Coxsackievirus A7

This anomalous virus attracted attention by causing outbreaks of paralytic disease in the U.S.S.R. and Scotland and occasional paralytic illnesses in South Africa and elsewhere.[1–5] Paralytic disease resembling that caused by polioviruses is produced by its inoculation into monkeys.[2,6] Fatal paralytic disease is produced in both newborn and adult cotton rats, merions, and hamsters, and in newborn but not adult mice, lemmings, and hooded rats; rabbits and guinea pigs are irregularly susceptible as newborn but not adult animals.[2,7] Experimentally infected adult rodents develop early transient viremia; the highest infectivity titers develop first in the brown fat, and then to lower titers in brain later[7] but without obvious signs of illness or histologic changes, except in cotton rats. The brown fat of cotton rats developed conspicuous necrosis both histologically and to the naked eye in animals inoculated with strains isolated from the Scottish outbreak, but not with the Russian AB 1V strain.[7,8] Lutynski and Chlap[9] reported scattered atrophic muscle lesions in infected adult rodents, which they suggested might be secondary to neurologic damage rather than the direct result of infection of muscle fibers. Another curious property of CVA7 is the specific hemagglutinin found in extracts of the carcasses of infected newborn mice, although undetected in cell cultures.[10,11] This is a useful marker for identifying these isolates, not reported for any other enterovirus, which provides a convenient hemagglutination inhibition test for diagnostic or serosurvey purposes.[12]

1.2. Coxsackievirus B

Typically, CVB cause spastic paralysis, tremors and incoordination, postural abnormalities, encephalitis, and stunting of the growth of newborn mice. These animals develop brown fat necrosis, which may be conspicuous to the naked eye on dissection, as well as encephalitis, often slight and patchy myositis, and sometimes pancreatitis. Cell cultures are sensitive and convenient for the isolation and propagation of CVB, although a few strains are most sensitively detected by inoculating newborn mice. Human infections produce a wide range of effects, often mild or inapparent. Paradoxically, in view of their relative myotropism in suckling mice, CVB more often affect skeletal and cardiac muscle than CVA, also causing visceral damage (e.g., to pancreas) as well as respiratory and neurologic disorders.

2. EPIDEMIOLOGIC OBSERVATIONS

In general, CV tend to spread as typical enteroviruses, mainly by fecal–oral routes and sometimes by respiratory routes or contagion. Epidemiology is dealt with more comprehensively in Chapter 21, but certain international and Scottish data are summarized below.

2.1. International Data

Surveillance of viral activity at international level has been carried out for many years by reports submitted to the World Health Organization (WHO). For enteroviruses, the data have been summarized periodically[13,14]; more recent information for the period 1975–1983 has been provided by the Virus Unit of WHO (Tables I and II). The data make it possible to compare and contrast the pattern of infection in different countries, to study international trends, and to observe the effects of vaccine policies against poliomyelitis in different countries. Limitations of this type of surveillance are that reporting may be incomplete, virologic tests may not be available in some countries, there is inevitably some selection of cases for testing, and reporting by country may be insensitive to local outbreaks, as the averaging effect may blunt the virologic picture. Despite these problems, it is possible to obtain useful information and for the main profiles of associated disease to emerge from the analysis.

During 1975–1983, WHO received 58,955 reports of enterovirus infections (Table I). Of these, 35.2% were CV infections, 9.8% of group A and 25.3% of group B (Table I). The commonest CV reported was type CVB4, followed by type CVB5, CVB2, CVB3, and CVA9. The commonest associated diseases concerned the nervous system, types CVB5, CVA9, and CVB4 predominating in meningitis and encephalitis. The next most frequently associated diseases involved the gastrointestinal (GI) and respiratory tracts, the same five types of virus predominating. As with other enteroviruses, CV infections were much more commonly found in children, especially in the 1–4-year age group (Table II).

2.2. Scottish Data

Although sharing some of the limitations of international surveillance, national collection of data provides more uniformity of reporting and sharper reflection of trends in the more limited geographic area covered, which might otherwise be masked by global or continental totals. In Scotland, laboratory reports of viral infections are received by the Communicable Diseases (Scotland) Unit.

TABLE I
Global Surveillance of Viral Diseases: Enteroviral Infections by Clinical Conditions: 1975–1983

	Respiratory			Central nervous system			Cardio-vascular	Muscle/joints	R.E. system and glands	Gastro-intestinal	Hepatitis	Skin and mucosae	Eye	Other and unspec.	Total
	Upper	Lower	Other	Men/enc.	Paral.	Other									
Coxsackievirus A															
Not typed	51	22	4	65	10	61	8	8	9	114	4	136	0	91	583
1	0	0	0	6	11	1	3	0	0	15	0	2	0	37	75
2	34	5	1	6	2	5	3	0	0	24	0	18	0	21	119
3	0	4	0	2	0	0	2	0	0	2	0	2	1	3	16
4	6	11	1	15	7	9	4	2	0	64	1	14	0	41	175
5	19	5	0	10	1	2	3	1	0	20	0	33	0	14	108
6	5	2	0	5	0	1	2	0	0	17	0	6	0	22	60
7	13	14	1	68	2	11	2	2	2	37	1	20	0	43	216
8	1	2	0	3	4	0	3	0	0	6	0	3	0	20	42
9	201	163	39	1049	15	127	18	15	19	485	12	161	5	498	2087
10	26	5	0	9	2	11	1	1	0	29	1	91	0	25	201
11	1	1	0	2	2	1	0	0	0	2	0	2	0	10	21
12	1	0	0	0	0	0	0	0	0	0	0	0	0	1	2
13	2	0	0	5	2	1	0	1	1	11	0	0	0	14	37
14	1	0	0	1	0	0	1	0	0	7	0	2	0	5	17
15	0	3	0	3	0	3	0	0	0	8	0	3	0	8	28
16	41	20	8	41	3	7	6	4	7	117	2	759	2	80	1097
17	1	3	0	1	1	0	0	0	0	2	0	1	0	8	17
18	0	1	0	5	1	0	0	0	0	4	0	1	1	5	18
19	0	1	0	0	0	0	0	0	0	0	0	0	0	0	1
20	0	3	0	4	2	1	0	0	1	4	0	1	0	9	25
21	6	9	0	7	0	1	0	0	0	13	0	3	0	12	51
22	0	1	0	1	2	0	0	0	0	0	0	0	0	4	8
24	0	3	0	9	1	0	1	1	1	15	0	4	21	7	63
Total	409	278	54	1317	68	242	57	35	40	996	21	1262	30	978	5767
Coxsackievirus B															
Not typed	5	17	7	26	3	4	45	17	12	18	0	4	0	62	220
1	266	176	50	292	12	113	65	49	29	381	5	58	6	438	1940
2	312	276	62	564	21	156	130	57	26	541	8	70	6	662	2891
3	267	211	70	648	17	152	96	56	19	591	9	62	4	644	2846
4	327	278	74	790	29	225	169	70	31	743	8	75	10	861	3690
5	203	182	42	1070	27	109	77	48	30	563	9	84	7	559	3010
6	20	29	6	81	3	22	14	5	2	84	2	7	0	72	347
Total	1400	1169	311	3471	112	781	596	302	149	2921	41	360	33	3298	14,944
Echoviruses total	2047	2104	469	14815	178	1675	266	223	188	7763	140	976	72	7275	38191
Polioviruses total	8	1	0	2	10	1	1	0	1	4	0	2	0	13	43

TABLE II
Global Surveillance of Viral Diseases: Enteroviral Infections by Age: 1975–1983

	<1 year	1–4 years	5–14 years	15–24 years	25+ years	Unknown	Total
Coxsackievirus A							
Not typed	112	286	94	16	47	28	583
1	15	39	9	3	3	6	75
2	33	71	10	3	0	2	119
3	3	7	2	2	1	1	16
4	36	100	16	1	10	12	175
5	16	62	17	5	5	3	108
6	12	41	3	0	1	3	60
7	42	79	39	14	31	11	216
8	13	21	3	0	0	5	42
9	573	845	741	192	353	103	2807
10	36	121	24	4	12	4	201
11	1	16	1	0	2	1	21
12	0	2	0	0	0	0	2
13	10	21	2	2	0	2	37
14	4	9	4	0	0	0	17
15	8	13	4	0	1	2	28
16	103	575	178	45	130	66	1097
17	3	9	1	0	0	4	17
18	3	7	3	3	1	1	18
19	0	1	0	0	0	0	1
20	10	12	0	0	1	2	25
21	12	19	5	9	3	3	51
22	2	2	2	0	0	2	8
24	9	19	10	3	21	1	63
Total	1056	2377	1168	302	622	262	5787
Coxsackievirus B							
Not typed	17	33	25	23	98	14	210
1	532	667	368	84	212	77	1940
2	750	1033	523	107	354	124	2891
3	743	1012	495	138	330	128	2846
4	890	1310	669	171	485	165	3690
5	674	891	596	193	502	154	3010
6	93	99	68	21	55	11	347
Total	3699	5045	2744	737	2036	673	14934
Echoviruses							
Total	10,704	9825	9047	2679	4278	1658	38191
Polioviruses							
Total	13	24	2	1	1	2	43

TABLE III
Coxsackievirus Reports to Communicable Diseases (Scotland) Unit: 1971–1985

	1971	1972	1973	1974	1975	1976	1977	1978	1979	1980	1981	1982	1983	1984	1985	Total
Coxsackievirus A																
2	2	2	1	1	0	1	0	0	0	1	2	6	0	9	3	28
3	0	3	0	0	0	0	0	0	0	0	0	0	0	0	0	3
4	2	3	5	0	2	0	0	0	0	0	4	0	0	1	5	22
5	0	2	1	0	0	0	0	0	0	7	2	2	0	0	10	24
6	0	1	0	0	0	0	0	0	0	1	4	0	5	6	0	17
7	0	0	1	0	0	0	0	0	0	0	0	0	0	0	1	2
8	6	0	0	0	0	0	0	0	0	0	5	1	0	0	0	12
9	39	23	24	17	21	30	50	0	33	10	39	15	34	16	12	363
10	1	2	5	0	1	0	0	0	7	5	0	8	7	0	0	36
14	0	0	0	0	0	0	0	0	0	0	5	2	5	5	0	17
16	16	0	10	3	3	15	2	3	0	40	5	2	5	5	0	109
21	0	0	0	0	0	0	0	0	0	0	0	0	1	0	0	1
Other and untyped A	1	7	6	11	13	9	8	4	0	12	0	1	18	1	0	91
Coxsackievirus B																
1	22	17	19	10	9	58	32	35	16	27	32	29	68	53	59	486
2	54	62	73	23	41	32	17	74	48	39	117	65	131	165	188	1129
3	45	62	80	23	31	36	47	92	41	49	46	47	50	78	121	848
4	46	73	60	44	144	58	57	59	126	93	147	195	245	328	511	2186
5	44	10	19	22	6	35	48	34	20	64	31	3	9	64	81	490
6	1	1	1	1	1	0	0	7	4	4	4	8	5	4	0	41
Type B uncertain[a]	6	2	7	14	26	18	32	58	46	87	77	38	109	131	247	898

[a]Serologic diagnosis: rising or high titer to more than one CVB type.

Table III summarizes the reports for the period 1971–1985. It includes both virus isolations and also, for group B viruses, diagnoses based on serologic studies. Where no virus was isolated but a significant neutralizing antibody response (fourfold or greater rising titer, or standing titers of 512 or greater) was detected to more than one type of CVB, entries in Table III are made against the heading "B type uncertain." Group B infections were much more commonly reported, partly because these viruses are relatively easier to isolate in cell cultures and also because of the frequent use of serological tests for diagnosis, especially in cardiac illnesses and in recent years in cases of suspected myalgic encephalomyelitis. Of group A infections, type A9 and A16 were most often reported, both detectable in cell cultures. For the period 1971–1980, the clinical classification of the patients in whom these infections were diagnosed is available and is summarized in Table IV. This reemphasizes the scale of interest and testing for CVB infections in cardiac and neurologic cases. Varying incidence of individual virus types from year to year is reflected in Table III, most clearly with CVA16, the commonest cause of hand-foot-and-mouth disease, which shows a 2–4-year cycle with increased activity in 1971, 1973, 1976, and 1980. Increased prevalence during the summer and fall months is shown for CVA16 and most of the other types of virus (Table V). The usual male predominance of symptomatic enterovirus infections is shown in Table VI, except for the "B type uncertain" group, which consists entirely of serologically diagnosed cases which are largely older females with suspected cardiac or myalgic encephalopathic symptoms. Except for this group, Table VI also shows the preponderance of infections in the childhood age groups, shown well by the CVA infections diagnosed by virus isolation.

2.3. Data from the Regional Virus Laboratory, Glasgow

The varying incidence of eight selected enterovirus infections, including CVA7 and CVB5, as reflected by virus isolations at this laboratory serving a population of more than one million for the period 1957–1976 were tabulated in 1978.[13] Additional and more recent data are summarized below.

2.3.1. *Coxsackievirus A Infections*

Because the use of newborn mice for virus isolation is expensive and inconvenient, it is less often practiced now that cell-culture techniques are widely available. Nevertheless, important enterovirus infections undetectable by cell cultures may be revealed by the use of mice. Since

TABLE IV

Clinical Classification of Coxsackievirus Infections Reported to CDS Unit, 1971–1980

	Respiratory			Central nervous system							Skin and		Other and	
	Upper	Lower	Other	Men/enc.	Paral.	Other	Cardiovascular	Muscle/joints	R.E. system and glands	Gastrointestinal	mucosae	Eye	unspec.	Total
Coxsackievirus A														
2	0	0	0	3	1	0	1	0	1	1	0	0	1	8
3	0	1	0	0	0	1	0	0	0	0	0	0	1	3
4	3	2	1	1	0	0	2	0	0	1	1	0	1	12
5	3	0	0	1	0	2	1	0	0	3	0	0	0	10
6	1	0	0	0	0	0	0	0	0	1	0	0	0	2
7	0	0	0	1	0	0	0	0	0	0	0	0	0	1
8	1	0	0	0	0	1	0	0	0	0	0	0	4	6
9	41	18	4	90	2	13	3	2	2	39	11	0	22	247
10	4	3	0	3	0	4	0	1	0	2	1	0	3	21
16	0	0	1	0	0	0	0	0	0	45	42	0	4	92
Other A	0	2	0	0	0	0	0	0	0	3	0	0	0	5
Untyped A	3	2	1	4	0	11	0	0	1	25	14	0	4	65
Coxsackievirus B														
1	22	27	6	17	1	9	77	18	4	14	3	0	47	245
2	32	22	7	35	1	16	165	31	5	29	12	0	108	463
3	31	37	10	54	0	25	135	35	16	54	9	0	100	506
4	65	42	14	59	1	28	200	62	23	76	8	1	181	760
5	30	20	7	53	1	17	44	16	4	50	14	0	46	302
6	1	0	0	2	0	1	9	4	0	1	0	0	2	20
Type B uncertain[a]	13	24	5	22	3	3	113	39	6	3	5	0	60	296

[a]Serologic diagnosis: rising or high titer to more than one CVB type.

TABLE V
Coxsackievirus Infections, by Four-Week Periods, in Scotland: 1971–1985

	1–4	5–8	9–12	13–16	17–20	21–24	25–28	29–32	33–36	37–40	41–44	45–48	49–52	53	Total
A2	0	2	4	1	0	2	1	0	0	4	5	8	1	0	28
A4	1	0	6	0	0	4	1	2	1	0	4	3	0	0	22
A5	3	0	1	1	4	0	1	4	0	2	3	3	2	0	24
A6	5	1	0	0	2	1	0	0	1	0	2	4	1	0	17
A8	0	1	0	0	0	0	0	1	5	4	1	0	0	0	12
A9	15	24	13	9	10	18	27	27	39	30	79	59	13	0	362
A10	6	1	3	0	3	2	1	1	1	3	5	3	7	0	36
A16	14	2	12	2	1	0	3	13	9	7	22	10	14	0	109
Untyped and other type A	2	0	5	2	6	3	5	21	10	10	12	12	10	0	98
B1	26	42	30	22	23	35	43	40	34	68	46	47	30	0	486
B2	50	81	81	58	74	62	73	70	126	145	109	110	89	1	1129
B3	42	57	41	58	41	51	79	78	74	90	80	73	84	0	848
B4	111	110	164	132	145	149	151	148	212	247	231	220	165	1	2186
B5	15	29	20	18	18	21	55	62	58	48	48	52	44	2	490
B6	4	2	1	3	0	2	2	1	1	7	7	5	6	0	41
Type B uncertain[a]	49	60	61	56	75	69	64	56	86	101	77	74	67	3	898

[a]Serologic diagnosis: rising or high titer to more than one CVB type.

TABLE VI
Laboratory Evidence of Coxsackievirus Infections in Scotland, by Age and Sex: 1971–1985

Virus type	<1			1–4			5–14			15–24			>25			Unknown			Total		
	M	F	U	M	F	U	M	F	U	M	F	U	M	F	U	M	F	U	M	F	U
A2	2	2	0	10	5	3	5	0	0	0	0	0	0	0	0	0	1	0	17	8	3
A4	2	0	0	11	3	0	2	3	0	0	0	0	0	1	0	0	0	0	15	7	0
A5	4	1	0	9	5	0	1	0	0	1	1	0	1	0	0	0	0	1	16	7	1
A6	2	2	1	7	1	1	2	1	0	0	0	0	0	0	0	0	0	0	11	4	2
A8	2	1	0	5	3	0	0	1	0	0	0	0	0	0	0	0	0	0	7	5	0
A9	33	15	3	59	39	2	51	29	0	12	11	0	38	27	0	7	3	1	200	124	6
A10	2	6	0	14	7	0	1	0	0	0	1	0	5	0	0	0	0	0	22	14	0
A16	5	4	0	29	23	1	15	6	0	3	4	0	8	6	0	4	1	0	64	44	1
Untyped and other type A	5	6	1	24	16	5	12	9	2	2	2	0	3	5	0	3	0	2	50	38	10
B1	29	15	0	33	22	0	14	16	0	19	13	0	163	130	0	9	21	2	267	217	2
B2	46	34	1	71	38	2	49	38	0	43	50	0	401	284	3	32	36	1	642	480	7
B3	51	31	1	66	38	0	54	30	1	55	58	0	231	192	1	24	14	1	481	363	4
B4	57	36	4	99	77	5	83	77	0	113	132	0	727	677	3	45	48	4	1124	1047	16
B5	40	32	5	68	33	2	34	19	1	29	20	0	100	83	3	15	6	0	286	193	11
B6	5	3	0	6	2	0	2	1	0	3	0	0	10	8	0	0	1	0	26	15	0
Type B uncertain[a]	1	1	2	5	4	0	18	25	1	46	61	1	285	329	11	23	27	0	378	347	15

[a]Serologic diagnosis: rising or high titer to more than one CVB type.

TABLE VII
Annual Incidence of 185 Coxsackievirus Infections Identified by Inoculation of Newborn Mice: 1956–1973[a]

	Type A															
Year	1	2	3	4	5	6	7	8	10	14	16	18	20	23	?[b]	Type B[c]
1956	0	1	0	0	0	0	1	0	0	0	0	0	0	0	0	0
1957	0	0	0	0	0	0	0	0	0	0	0	0	0	0	0	0
1958	0	0	0	0	0	0	0	0	0	0	0	0	0	0	0	0
1959	1	2	0	2	1	0	37	2	0	0	0	0	0	0	2	1
1960	0	0	0	1	2	0	0	0	1	0	0	0	0	0	0	1
1961	0	0	1	1	0	0	1	0	0	0	0	0	1	0	0	0
1962	0	0	0	0	2	4	0	0	0	0	0	0	0	0	3	0
1963	1	0	0	3	0	1	15	0	3	3	0	1	0	0	1	0
1964	0	6	0	0	0	0	0	6	0	0	0	0	0	1	0	1
1965	1	0	0	1	0	0	0	0	0	0	0	0	0	0	0	0
1966	0	0	0	1	0	0	0	0	1	0	0	0	0	0	0	0
1967	1	1	0	5	3	3	9	1	0	0	0	0	0	0	0	4
1968	0	1	0	1	1	0	1	2	1	0	0	0	0	0	1	0
1969	0	0	0	1	0	0	6	0	2	0	0	0	0	0	0	0
1970	0	0	0	2	0	0	0	0	2	0	1	0	0	0	0	2
1971	0	1	0	0	0	0	0	4	0	0	0	0	0	0	2	0
1972	0	0	0	0	0	0	0	0	2	0	0	0	0	0	1	0
1973	0	0	0	2	1	0	1	0	0	0	4	0	0	0	2	0
Total	4	12	1	20	10	8	71	15	12	3	5	1	1	1	12	9

[a]Excluding specimens positive in cell culture.
[b]Type A myositis in mouselets but not antigenically typed.
[c]2 type 1, 4 type 2, 1 type 3, 1 type 4, 1 type 5.

information in this area is scarce, Table VII displays the viruses isolated, mainly from feces, over the period 1956–1973, in suckling mice inoculated with specimens that were not cytopathogenic for cell cultures.[15] Nine of the 185 isolates were group B viruses that had not been detected in cell cultures. The remainder produced the signs and the gross and microscopic features characteristic of group A viruses of 14 types; 12 of them were unidentified by the 24 prototype antisera.

Different group A viruses predominated in different years. The preponderance of CVA7 infections reflected interest in these poliolike viruses that were active mainly at 4-year intervals from 1959 to 1967 and have not been detected since 1973.[5] Six additional cases of nonparalytic CVA7 infection were diagnosed serologically (see Grist and Bell[5] and Table 3 therein). Table VII excludes the type CVA9 infections and most CVA16 isolates, which were mainly detected in cell cultures, as well as type CVA23 (echovirus 9) infections, except for one isolate that was

TABLE VIII
Sex Distribution of Coxsackievirus Infections Detected in Newborn Mice: 1956–1973

Sex	Coxsackievirus type							Total
	A2	A4	A7	A8	A10	Other A types	Group B	
Male	8	7	47	8	4	29	5	108
Female	4	13	22	7	7	15	3	71
Unknown	0	0	2	0	1	2	1	6
Total	12	20	71	15	12	46	9	185

undetected by cell culture but that paralyzed newborn mice. The expected seasonal distribution was found with 62% of the isolations in the months June through August.[15] There was a 3:2 male predominance (Table VIII), and most isolations were from children, 65% under 5 years of age (Table IX).

Table X shows the diagnoses associated with these infections. CVA7 caused most of the paralytic and meningeal illnesses, but not encephalitis. Next most frequent were respiratory and mucocutaneous syndromes, including six cases of hand-foot-and-mouth disease, five of these with CVA16 infection. Three of five cardiac illnesses were associated with CVA4 infection.[16] Allowing for the additional isolations of CVA9 and CVA16 in cell culture, the distribution of type A viruses in this survey is similar to that reported from hospitals in the rest of Britain, 1958–1969.[17] Except for CVA7, the six commonest type A viruses in the Glasgow series correspond to the six commonest types in the survey of normal children in the London area, 1958–1969.[18]

TABLE IX
Age Distribution of Coxsackievirus Infections Detected in Newborn Mice: 1956–1973

Age (years)	Coxsackievirus type							Total
	A2	A4	A7	A8	A10	Other A types	Group B	
<1	1	5	8	2	1	11	1	29
1–4	3	9	41	10	7	18	4	92
5–14	6	3	21	3	2	13	2	50
15–24	2	2	1	0	0	0	1	6
25+	0	1	0	0	1	1	0	3
Unknown	0	0	0	0	1	3	1	5
Total	12	20	71	15	12	46	9	185

TABLE X
Diagnosis of 185 Patients with Coxsackievirus Infections Detected by Mouselet Inoculation: 1957–1973

	Type A																
Diagnosis	1	2	3	4	5	6	7	8	10	14	16	18	20	23	?	Type B	Totals
Paralysis	0	3	1	3	0	2	9	0	1	0	0	0	1	0	2	1	23
Meningitis	1	6	0	6	4	4	61	4	6	3	0	1	0	0	4	5	105
Encephalitis	1	2	0	0	0	0	0	0	1	0	0	0	0	0	2	0	6
Respiratory	0	0	0	3	4	2	1	1	0	0	0	0	0	0	1	2	14
Herpangina	0	1	0	0	0	0	0	0	2	0	0	0	0	0	0	0	3
Hand-foot-and-mouth disease	0	0	0	1	0	0	0	0	0	0	5	0	0	0	0	0	6
Other skin rash	0	0	0	0	1	0	0	1	0	0	0	0	0	1	0	1	4
Diarrhea	1	0	0	0	1	0	0	0	0	0	0	0	0	0	0	0	2
Cardiac	1	0	0	3	0	0	0	1	0	0	0	0	0	0	0	0	5
Other diseases[a]	0	0	0	2	0	0	0	3	1	0	0	0	0	0	1	0	7
Well	0	0	0	2	0	0	0	5	1	0	0	0	0	0	2	0	10

[a] A4, systemic infantile infection, acute hip pain; A8, polyneuritis, febrile convulsion, acute hip pain; A10, fever and encephalopathy and multiple infections; ?, poliomyelitis contact with otitis media.

The 1959 outbreak of CVA7 infection provided the opportunity to test the significance of the relationship of this virus to neurological illnesses. Every fecal sample received from any type of illness during the peak period of the outbreak was inoculated into newborn mice. Type A7 virus was isolated from five of six paralytic illnesses, from 18/57 cases of nonbacterial meningitis (including 18 of the 20 from which no other enterovirus was isolated) and from only one of the 60 other patients, a young woman with fever and headache but with normal cerebrospinal fluid.

2.3.2. *Coxsackievirus B Infections*

Group B coxsackievirus infections diagnosed at the Regional Virus Laboratory during the period 1972–1983 were recently analysed by Bell and McCartney.[19] Type B6 virus was not isolated but 123 isolations of the other five types were made, mainly from children with respiratory or meningeal illnesses. Serologic tests for specific neutralizing antibodies in 8477 patients showed significantly rising or falling titers in 0.9%, static titers of 512 or more in 13%, and static titers of 256 in a further 13%, fluctuating from year to year. Most of these tests were undertaken in cases of suspected heart disease, of which 12% showed titers of these categories (33% in definite myopericarditis), and also from 1980 cases of suspected myalgic encephalomyelitis, of which 41% well-documented cases gave seropositive results by these criteria.

3. ROLE OF COXSACKIEVIRUSES IN VARIOUS DISEASE SYNDROMES

The main relationships of CV to human diseases have been referred to above and in Tables I, IV, and X. Other contributors deal with neurologic disorders, the study of which by Dalldorf first led to discovery of these viruses, as well as mucocutaneous syndromes, neonatal and cardiac infections, and the recent studies of chronic infections and diabetes. Additional syndromes are discussed below.

3.1. Gastrointestinal Disorders

One would expect enteroviruses, primarily infecting cells of the alimentary tract, to be significant causes of diarrhea or other gastrointestinal disturbances. In practice, this appears not to be so.[20,21]

3.1.1. Diarrhea

Enteroviruses, mainly echoviruses, have often been isolated from the feces of children with diarrhea, sometimes in outbreaks, but also from the feces of unaffected children with similar or lower frequency. No individual enterovirus has shown a consistent relationship to diarrhea, and it may be that these enteroviruses more often act as indicators of fecal pollution than as intestinal pathogens, the diarrhea being caused by other undetected viruses, such as rotaviruses and astroviruses.[20]

3.1.2. Hepatitis

Acute hepatitis commonly accompanies generalized severe CV infection, usually in neonates or occasionally in older children. Particles like enteroviral virions were visualized within the liver cells of a 28-year-old man with fatal myopericarditis, myositis, and hepatitis, who showed rising antibody titers to CVB2 and CVB5.[22] IgG and IgM antibodies to CVB were found in a 16-year-old woman with severe myocarditis and fulminant hepatic failure.[23]

3.1.3. Pancreatitis

Acute pancreatitis is a common feature of CVB infection of mice. It is also found as part of the multisystem disease in human neonates infected by CVB but is rarely symptomatic in older persons, although damage to islet cells may well result directly or indirectly in insulin-dependent diabetes mellitus, especially diabetes of juvenile onset. Acute pancreatitis in adults is rare, but cases were reported by Murphy and Simmul[24] and Ursing.[25] Concurrent cytomegalovirus (CMV) and CVB infections were reported in a woman in whom pancreatitis and other symptoms developed after receiving a heart–lung transplant.[26] Raised CVB3 and CVB4 antibody titers were found more frequently in cases of acute or recurrent pancreatitis than in controls,[27] and Nakao[28] reported raised amylase levels, suggesting pancreatic damage in 31% of CVB5 and 23% of CVA9 infections, but not in echovirus 4 or 6 infections.

3.2. Acute Respiratory Infections

Group A and B coxsackieviruses have often been associated with acute respiratory infections, usually of the upper but sometimes of the middle or even lower tracts, mainly in young children.[13,21] Clinically, these illnesses are individually indistinguishable from those caused by rhinoviruses and many other viruses, but during outbreaks there may be associated characteristic symptoms of myalgia or meningitis in the same patients or in associates. Pharyngitis and fever are common features. Outbreaks may affect infants in nurseries, institutions, or camps. Lower respiratory tract infection can occasionally be severe or fatal in the youngest children affected in outbreaks, but otherwise these infections are usually minor and self-limiting. The eyes are occasionally involved, concurrently or alone, for instance, acute, sometimes hemorrhagic, conjunctivitis due to CVA24.[29,30]

3.3. Diseases of Muscle

In addition to their effects on heart muscle in both mice and humans, discussed by others below, CV can cause painful acute myositis manifest as Bornholm disease, epidemic myalgia, or pleurodynia, devil's grippe, and other synonyms. Bornholm disease may be sporadic or may occur in association with epidemics together with nonspecific fevers, meningitis, or cardiac manifestations of the infection during the widespread epidemics of CVB5 infection in Europe.[31] Group B viruses are mainly involved, but group A types 4, 6, 9, 10, and 23 (echo 9) have also been implicated.[13,21] CVA9 was isolated from muscle from an infant with congenital myositis[32] and from that of an adult with chronic myositis.[33] Particles with the morphology of picornaviruses have been found in the myocytes of patients with polymyositis,[34] sometimes aggregated in crystalline arrays, which in one case reacted strongly with CVA9 antiserum and weakly with CVB2 antiserum.[35] Crystalline aggregates thought to be of CV were also reported in the muscle cells of two patients with fatal dermatomyositis.[36]

Exercise undertaken during the course of a CV infection may provoke acute myositis in humans as in mice.[37] Both primary cytopathogenic effects of the virus and immunopathologic processes may play a role in some of these conditions.

3.4. Miscellaneous Disease Associations

It is not surprising that, especially in children, such highly prevalent viruses as those of the enterovirus group, including CV, have been found in chance association with many types of illness not necessarily caused by these viruses. For instance, they have been rarely associated with acute or chronic glomerulonephritis and the hemolytic–uremic syndrome in children,[13,21] the latter now being attributable to enterotoxin-producing *Escherichia coli.* Orchitis is an occasional feature of Bornholm disease and is probably due to the corresponding CV. Infection has rarely been reported in arthritis, and serologic evidence of recent CVB infection was found in a girl with Still's disease and the hemophagocytic syndrome.[38] CVB3 was isolated from two children with opsoclonus–myoclonus.[39] An enterovirus with CVA properties in suckling mice was isolated during an outbreak of acute infectious lymphocytosis in children.[40]

Hand-foot-and-mouth disease caused by CVA16 infection has caused abortion in early pregnancy.[41,42] Fever was the commonest clinical feature associated with CVB infection in the 10-year study by Moore *et al.*,[43] who particularly warn against facile interpretation of the significance of association between enterovirus infection and a particular disease syndrome in the individual patient.[21]

4. CONCLUSIONS

One might have expected that such long-known viruses as those of the CV group, of such interest during the years of research and development of controls for acute poliomyelitis, would by now have declined in interest and importance as mainly causes of minor self-limiting acute infections in a world of improving living conditions and reduced spread of fecal pollution. In the event, fecal–oral transmission of many infectious agents is still significant in the polluted modern world, and CV not only cause a significant proportion of acute meningeal and other neurologic infections, especially in children, but are growing in interest and possible importance in relationship to cardiac and perhaps chronic and immune-modulated diseases in older adult groups. Improved socioeconomic and living conditions reduce fecal–oral transmission of enteric infections, postpone the average age of infection, and build up relatively nonimmune populations in which epidemic spread can periodically take place, under appropriate conditions. In the case of polioviruses, this necessitated control by the use of vaccines. The result of these changes in the case of CV and echoviruses may already be emerging and merits careful surveillance especially in the fields of adult and neonatal infections.

ACKNOWLEDGMENTS. We are grateful to the late Dr. F. Assaad, recently director of the Division of Communicable Diseases of the World Health Organization, for his interest and collaboration over many years and for the data presented in Tables I and II. We thank Mrs. Anne Carey for help in compiling the data in Tables III and VI, and Mrs. Nora Wilson for assistance with the typing and compilation of this chapter.

REFERENCES

1. Chumakov, M. P., Voroshilova, M. K., Zhevandrova, V. I., Mironova, L. L., Itzelis, F. I., and Robinson, I. A., 1956, Isolation and study of immunologic type IV poliomyelitis virus, *Prob. Virol.* **1**:16–19.
2. Voroshilova, M. K., and Chumakov, M. P., 1959, Poliomyelitis-like properties of AB-IV-Coxsackie A7 group of viruses, *Prog. Med. Virol.* **2**:106–170.
3. Grist, N. R., 1962, Type A7 coxsackie (type 4 poliomyelitis) virus infection in Scotland, *J. Hyg. (Camb.)* **60**:323–332.
4. Gear, J. H. S., 1984, Nonpolio causes of polio-like paralytic syndromes, *Rev. Infect. Dis.* **6**:S379–S384.
5. Grist, N. R., and Bell, E. J., 1984, Paralytic poliomyelitis and non-polio enteroviruses: Studies in Scotland, *Rev. Infect. Dis.* **6**:S385–S386.
6. Grist, N. R., and Roberts, G. B. S., 1962, Histological studies of coxsackie A7 poliomyelitis in man and monkeys, *J. Pathol.* **84**:39–44.
7. Grist, N. R., and Roberts, G. B. S., 1966, Coxsackie A7 virus infections of rodents, *Arch. Virusforsch.* **19**:454–463.
8. Frolova, M. P., and Savinov, A. P., 1960, The histopathology of experimental infections in rodents caused by the Karaganda strain of ABIV (Type 4 of poliomyelitis virus), *Vopr. Virusol.* **5**:309–315.
9. Lutynski, R., and Chlap, Z., 1966, Experimental infection with coxsackie A-7 virus. II. Course of the infection and morphology of the lesions in different species of rodents, *Acta Med. Pol.* **7**:71–82.
10. Grist, N. R., 1960, Isolation of coxsackie A7 virus in Scotland, *Lancet* **1**:1054–1055.
11. Williamson, J. D., and Grist, N. R., 1965, Studies on the haemagglutinin present in coxsackie A7-infected suckling mouse tissue, *J. Gen. Microbiol.* **41**:283–291.
12. Grist, N. R., 1965, Further studies of coxsackie A7 virus infection in the West of Scotland, *Lancet* **2**:261–263.
13. Grist, N. R., Bell, E. J., and Assaad, F., 1978, Enteroviruses in human disease, *Prog. Med. Virol.* **24**:114–157.

14. Grist, N. R., Bell, E. J., and Reid, D., 1975, The epidemiology of enteroviruses, *Scott. Med. J.* **20:**27–31.
15. Grist, N. R., 1983, Coxsackie viruses 1956–1973: The diagnostic value of suckling mice, *Comm. Dis. Scott. Wkly. Rep.* **83**(44):vii–ix.
16. Grist, N. R., and Bell, E. J., 1969, Coxsackie viruses and the heart, *Am. Heart J.* **77:**295–300.
17. Public Health Laboratory Service, 1970, Coxsackie A infections, *Br. Med. J.* **2:**674–675.
18. Gamble, D. R., 1962, Isolation of coxsackie viruses from normal children aged 0–5 years, *Br. Med. J.* **1:**16–18.
19. Bell, E. J., and McCartney, R. A., 1984, A study of coxsackie B virus infections, 1972–1983, *J. Hyg. (Camb.)* **93:**197–203.
20. Grist, N. R., and Madeley, C. R., 1977, Gastro-entérites à virus, in: *Encyclopédie Médico-Chirurgicale* Maladies Infectieuses. Fasc. 9055 A-10, Julian Prelat, Paris.
21. Moore, M., and Morens, D. M., 1984, Enteroviruses, including polioviruses, in: *Textbook of Human Virology* (R. B. Belshe, ed.), pp. 407–483, PSB, Littleton, Massachusetts.
22. Gregor, G. R., Geller, S. A., Walker, G. F., and Campomanes, B. A., 1975, Coxsackie hepatitis in an adult, with ultrastructural demonstration of the virus, *Mount Sinai J. Med.* **42:**575–580.
23. Read, R. B., Ede, R. J., Morgan-Capner, P., Moscoso, G., Portmann, B., and Williams, R., 1985, Myocarditis and fulminant hepatic failure from coxsackievirus B infection, *Postgrad. Med. J.* **61:**749–752.
24. Murphy, A. M., and Simmul, R., 1964, Coxsackie B4 virus infections in New South Wales during 1962, *Med. J. Aust.* **2:**443–445.
25. Ursing, B., 1973, Acute pancreatitis in coxsackie B infection, *Br. Med. J.* **3:**524–525.
26. Wreghitt, T. G., Taylor, C. E. D., Banatvala, J. E., Bryant, J., and Wallwork, J., 1986, Concurrent cytomegalovirus and coxsackie B virus infections in a heart–lung transplant recipient, *J. Infect.* **13:**51–54.
27. Capner, P., Lendrum, R., Jeffries, D. J., and Walker, G., 1975, Viral antibody studies in pancreatic disease, *Gut* **16:**866–870.
28. Nakao, T., 1971, Coxsackie viruses and diabetes, *Lancet* **2:**1423.
29. Lim, K. A., and Yin-Murphy, M., 1977, The aetiologic agents of epidemic conjunctivitis, *Singapore Med. J.* **18:**41–43.
30. Mirkovic, R. R., Schmidt, N. J., Yin-Murphy, M., and Melnick, J. L., 1974, Enterovirus etiology of the 1970 Singapore epidemic of acute conjunctivitis, *Intervirology* **4:**119–127.
31. Public Health Laboratory Service, 1967, Coxsackie B5 virus infections during 1965, *Br. Med. J.* **4:**575–577.
32. Tang, T. T., Sedmak, G. U., Siegesmund, K. A., and McCreadie, S. R., 1975, Chronic myopathy associated with Coxsackievirus type A9—a combined microscopical and virus isolation study, *N. Engl. J. Med.* **292:**608–611.
33. Freudenberg, E., Roulet, F., and Nicole, R., 1952, Kongenitale Infektion mit coxsackie-virus, *Ann. Paediatr.* **178:**150–164.
34. Mastaglia, F. L., and Walton, J. N., 1970, Coxsackie virus-like particles in skeletal muscle from a case of polymyositis, *J. Neurol. Sci.* **11:**593–599.
35. Gyorkey, F., Cabral, G. A., Gyorkey, P. K., Uribe-Botero, G., Dreesman, G. R., and Melnick, J. L., 1978, Coxsackievirus aggregates in muscle cells of a polymyositis patient, *Intervirology* **10:**69–77.
36. Chou, S. M., and Gutmann, L., 1970. Picornavirus-like crystals in subacute polymyositis, *Neurology (NY)* **20:**205–213.

37. Jones, J. V., 1975, Muscle pain induced by exercise in coxsackie pericarditis, *Br. Med. J.* **2:**100.
38. Heaton, D. C., and Moller, P. W., 1985, Still's disease associated with coxsackie infection and haemophagocytic syndrome, *Ann. Rheum. Dis.* **44:**341–344.
39. Kuban, K. C., Ephros, M. A., Freeman, R. L., Laffell, L. B., and Bresnan, M. J., 1983, Syndrome of opsoclonus–myoclonus caused by coxsackie B3 infection, *Ann. Neurol.* **13:**69–71.
40. Horwitz, M. S., and Moore, G. T., 1968, Acute infectious lymphocytosis. An etiologic and epidemiologic study of an outbreak, *N. Engl. J. Med.* **279:**399–404.
41. Ogilvie, M. M., and Tearne, C. F., 1980, Spontaneous abortion after hand-foot-and-mouth disease caused by coxsackie virus A16, *Br. Med. J.* **281:**1527–1528.
42. Walker, I. R., 1985, Abortion in early pregnancy, *Med. J. Aust.* **142:**489.
43. Moore, M., Kaplan, M. H., McPhee, J., Bregman, D. J., and Klein, S. W., 1984, Epidemiologic, clinical and laboratory features of coxsackie B1-5 infections in the United States, 1970–1979, *Public Health Rep.* **99:**515–522.

14

Coxsackievirus Infection in Children under Three Months of Age

MARK H. KAPLAN

1. INTRODUCTION

Coxsackievirus (CV) infection in infants under 3 months of age results in significant morbidity and mortality. In this age group coxsackie B (CVB) infection, which is more virulent than coxsackie A (CVA), leads to an estimated hospitalization of approximately 50–364 infants per/100,000 live births/year, with a mortality of 3.9 per 100,000 live births each year.[1] This contrasts with an estimate by Nahmias of the attack rate of *Herpes simplex* infection of the newborn of 3.3–29 cases per 100,000 live births each year with a mortality of 60%.[2] Unfortunately, there is little current epidemiologic monitoring of this infection in the newborn, and there is no uniform reporting system for fatal CVB infection or nosocomial infection within the United States.[3]

The ability of the CV to survive in water and in enteric secretions allows this virus to be most efficiently transmitted during the summer months.[4,5] In most studies of CV infection in the child under 3 months of age, infection has occurred in the months of June through September.[1,3,5] Evidence from our laboratory indicates that the peak transmission occurs when people tend to go swimming, which in more temperate climates is generally in the months of August and September.

MARK H. KAPLAN • Division of Infectious Disease and Immunology, North Shore University Hospital, Manhasset, New York 11030; Cornell University Medical College, New York, New York 10021.

All CVB strains can cause neonatal infection, but strains B2 and B4 are the most common, whereas B1 and B6 strains appear to be the most virulent.[1]

Coxsackie B virus can produce a wide variety of infections in children under 3 months of age. These infections can best be understood by dividing them into mild and rarely fatal infections acquired just after birth either from a source in a nursery or from some sick household member, and those that are generally severe and often fatal that are acquired transplacentally from an infected mother just before or at the time of birth.

2. MILD COXSACKIE B INFECTION ACQUIRED FROM A SICK HOUSEHOLD MEMBER OR NOSOCOMIALLY

It has been speculated that after virus is ingested, infection is initiated within tonsillar cells or within Peyer patches in the small intestine. Initial infection may result in tonsillitis and upper respiratory infection and/or in mild gastroenteritis. In the newborn infant, this may cause feeding disorders or diarrhea and vomiting. Following initial infection, there is then a stage of viremia. In children under 3 months of age, viremia is striking. Primary viral culture of blood often reveals high titers of circulating virus.[6] Dagan *et al.* demonstrated that an enterovirus could be isolated from the blood of 44 infected children. Serum revealed virus in 35% of samples taken and mononuclear cell cultures revealed virus in 32% of samples. CVB4 was the most common virus in blood culture. Children under 3 months of age were more likely to have a positive blood culture than were older children. Viremic children were more likely to have meningitis or septicemia-like viral infection. During the viremic phase, children may be afebrile. Viremia may not necessarily produce symptomatic illness.[7] Jahn and Cherry have described a newborn with a CVB1 infection in whom heavy enteroviral infection occurred in spite of a mild febrile illness. Virus was detected in concentrations of 1200 tissue culture infectious doses ($TCID_{50}$) from the throat, 12,000 $TCID_{50}$ from the rectal swab, and 320 $TCID_{50}$ from the blood. This child did quite well with his infection. Following viremia, children may develop meningoencephalitis, worsening tonsillitis, gastroenteritis, myositis, and rarely myocarditis. Involvement of the lung and liver in nontransplacentally acquired infection does not appear to occur.

The first phase of clinical illness often occurring after ingestion of virus may last 2–3 days, the viremic phase 2–3 days, and the last phase of illness 2–3 days. The most common symptom that children have is fever. Fever may be constant in over one-half of children but rarely lasts more than 6 days. About 30% of children have a diphasic illness with 2–3 days of fever, 2–3 days of no fever, and then another 2–3 days of fever. Some children will have fever only in the initial phase of infection or during viremia. Others will have fever only during the latter days of infection.[1] The presence of fever in children under three months of age presents a particularly difficult problem to the pediatrician. In this age group, bacterial meningitis and bacteremia can be occult. As a result, pediatricians will usually admit these children to the hospital for a full sepsis evaluation, to include blood culture and spinal tap. This practice results in the hospitalization of many babies with CBV infections for extensive evaluations of a usually self-limited infection. In one prospective study of newborns, 24% of neonates from whom CVB was isolated within the first 3 months of life ended up being hospitalized for evaluation of fever.[8] We found that children under 3 months of age were usually hospitalized for an average of 7.7 days.[1] More than two thirds of hospitalized children received parenteral antibiotics for this viral disorder. Some children were hospitalized for as long as 2 weeks, because they were felt to have possible bacterial meningitis.

In hospitalized children under 3 months of age, viral meningitis is the most common manifestation of CVB infection.[1] In almost all cases, children lack clinical features of meningitis, such as stiff neck and Kernig or Brudzinski sign. More than two thirds, however, have abnormal spinal taps with >10 cells/mm^3. Cell counts in spinal fluid were greater than 10 in more than 85% of children. It is unusual to have cell counts >1000 cells/mm^3. Although viral meningitis is often associated with lymphocytosis in the spinal fluid, more than 50% of children had predominantly polymorphonuclear cell response in the initial spinal fluid. We found that more than two thirds of children had spinal fluid glucose of <30 with blood sugar to spinal fluid sugars of 3:1. Sugars of <20 did not occur. Concentration of spinal fluid protein was generally elevated but was rarely >130 mg/dl. This is in contrast to the spinal fluid protein reported by MacCracken and Mize in neonatal meningitis, which generally was ≥520 mg/dl.[9]

In spite of the fact that meningitis is a mild condition, there has been great concern that serious long-term neurologic sequelae can occur as a result of viral infection of the brain of the child under 3 months of age.[10–15] In two separate studies, children were shown to have comparable intellect but had reduced receptive language function development.[10,14] Ocular abnormalities have developed in one child.[10] In another child, spasticity and below-normal intelligence occurred.[11] A third child developed spastic quadriplegia and seizures.[12] A fourth had delayed language development, hypotonicity, and microcephaly.[13] These serious sequelae are probably exceptional consequences of this infection.

Adequate age-matched prospective studies addressing this crucial issue have not yet been done. Clearly, this type of study is vital, because the subtle impact on language development and other neurologic parameters may be an underestimated consequence of this viral disorder, particularly in early life.

Some children may present with diarrhea, but CV-induced gastroenteritis has not been reported to be seriously dehydrating. We were unable to document any cases with dehydration of greater than 5%.

While myocarditis in intrapartum-acquired CV infection is the rule, postnatally acquired CV myocarditis appears to be rare. When it does occur, it is generally in older children and adolescents.

3. INTRAPARTUM (CONGENITAL) FULMINANT COXSACKIE B INFECTION OF THE NEWBORN

The most serious consequence of CVB in early life appears when infection occurs during the intrapartum period. There has not been any adequately done prospective study to assess the frequency of this type of infection. When infection occurs during the intrapartum period, it is severe and often fatal. The child at greatest risk for overwhelming infection is one who is born to a mother who has just developed infection within 10 days prior to delivery and/or who develops clinical illness within 7 days after delivery. Figure 1 depicts the timing of maternal illness in relationship to the onset of fatal infection in the neonate.[1] Not all mothers of babies born with overwhelming infection experience clinical illness. In the reported fatal cases, only two thirds of mothers reported clinical illness during the critical intrapartum period. Maternal illness is often mild and may produce only fever without other symptoms. Some mothers may develop pleurisy, and some may experience viral meningitis. It appears that children experiencing overwhelming infection are born during a period of maternal viremia that results in transplacental transmission of virus. In one case, virus was isolated from amniotic fluid at the time of delivery, confirming the transplacental route of transmission of this fulminant infection.[1] There is really inadequate information about the role played by maternal antibody in protecting the newborn from overwhelming infection. It has been postulated that children born during maternal viremia do not get any maternal antibody to the virus or may lose the antiviral effect from the release of interferons or other antiviral proteins by the placenta at a time when placental viremia is occurring. Several animal models of transplacental viral infection have been developed in order to understand the role of maternal antibody in protecting the neonate from overwhelming infection.[15,16]

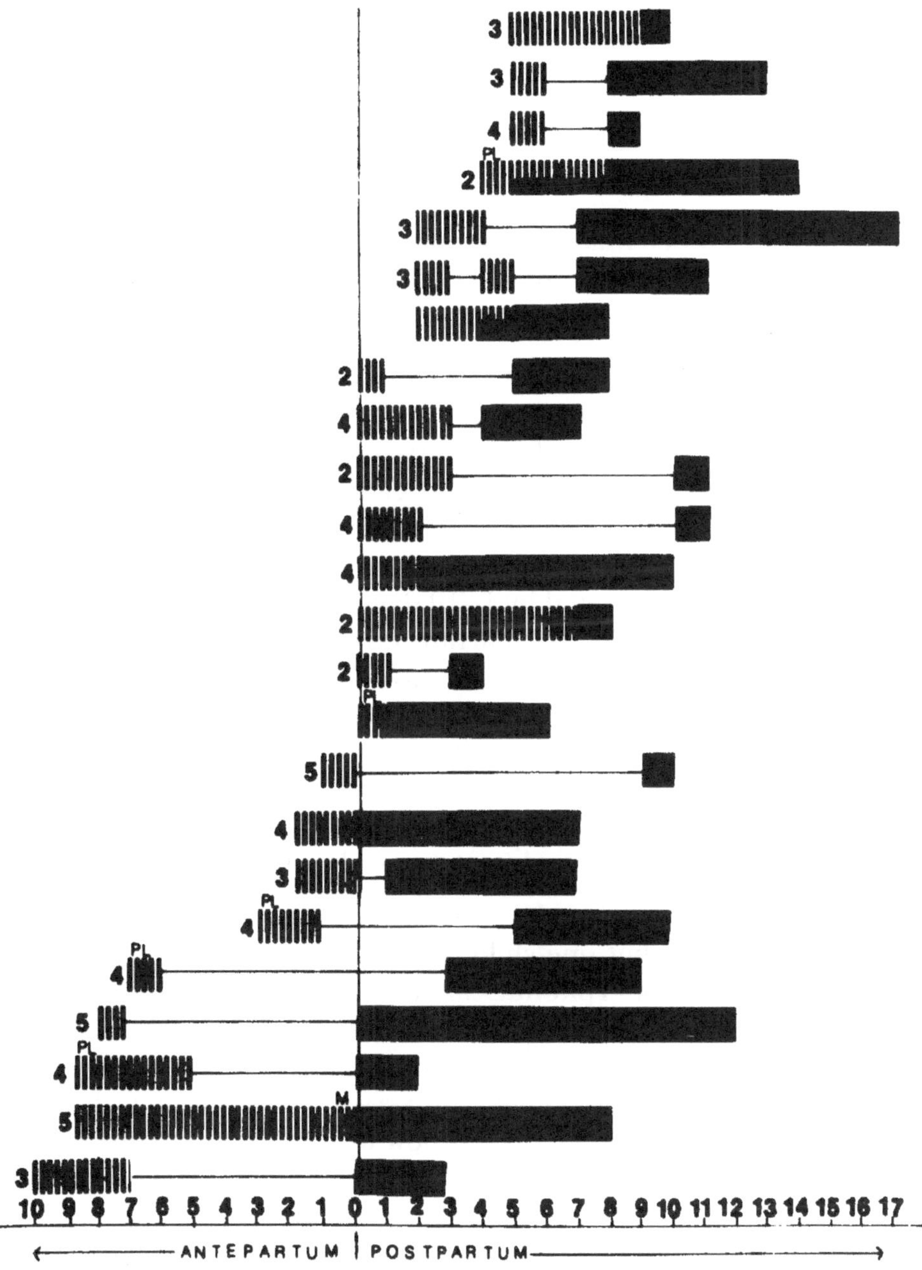

FIGURE 1. The time of maternal illness (vertical parallel bars) and clinical onset of neonatal infection in fatal coxsackievirus B infection. (From Kaplan *et al.*[1])

Heart	Interstitial pancarditis, PMN infiltration, fiber disruption (36/38)
Pulmonary	Interstitial alveolar inflammation, hemorrhage (30/37)
Liver	Central hemorrhagic necrosis (18/33)
Brain	Leptomeningeal inflammation (18/26)
	Pontine lesions-microglial infiltrate (12)
	White matter destruction (7)
Pancreas	Islet cell destruction (9)
Adrenal	Hemorrhage (9)

FIGURE 2. Pathologic findings in fatal cases of coxsackievirus B infection.

In our review of 41 fatal cases reported in the literature from whom CVB was cultured from clinical specimens, three patterns of illness emerged:

1. *Explosive illness:* Twelve infants had explosive onset of fulminant viremia, succumbing to death within 12 hr. These neonates became ill within 2–17 days postpartum.
2. *Diphasic illness:* Ten children developed a diphasic illness characterized by a mild first phase of infection and then a second phase of profound vascular collapse secondary to myocarditis. The two phases were separated by 1–10 days. Illness began either at birth or by the 14th day postpartum (average 4.4 days). These children died between days 8–25 (average 12.8 days).
3. *Progressive illness:* Seventeen neonates presented with progressive disease starting at birth to day 21 (average 2.8 days after birth) and terminating with myocarditis and vascular collapse and/or bleeding diatheses between days 2–28 (average 11.1 days after birth). The pathologic findings are shown in Fig. 2.

Fatal infection revealed the presence of diffuse interstitial pancarditis at autopsy in all but one child. There was severe myocardial fiber disruption, with polymorphonuclear and monocytic infiltrates within the myocardium. Virus was readily cultured from the myocardium in most cases. Viral inclusions were not seen in myocardial tissue. Cardiac tracing in these fatal cases generally revealed profound tachycardia (>200 beats/min). Voltage was usually low with evidence of diffuse myocarditis. Radiologic studies show sudden dilatation of the heart as shown in Fig. 3. Myocarditis was not always clinically apparent and was frequently discovered only after the child was found in vascular collapse. In one reported case, cardiac catheterization during myocarditis revealed multichamber dysfunction with concomitant cardiogenic shock.[1] Recently, we found that an echocardiogram better defines the presence of cardiac dysfunction in this illness than does either an electrocardiogram (ECG) or catheterization. We try to perform echocardiograms on neonates who acquire fulminant viral infections during the summer in order to look for evident myocardial dysfunction.

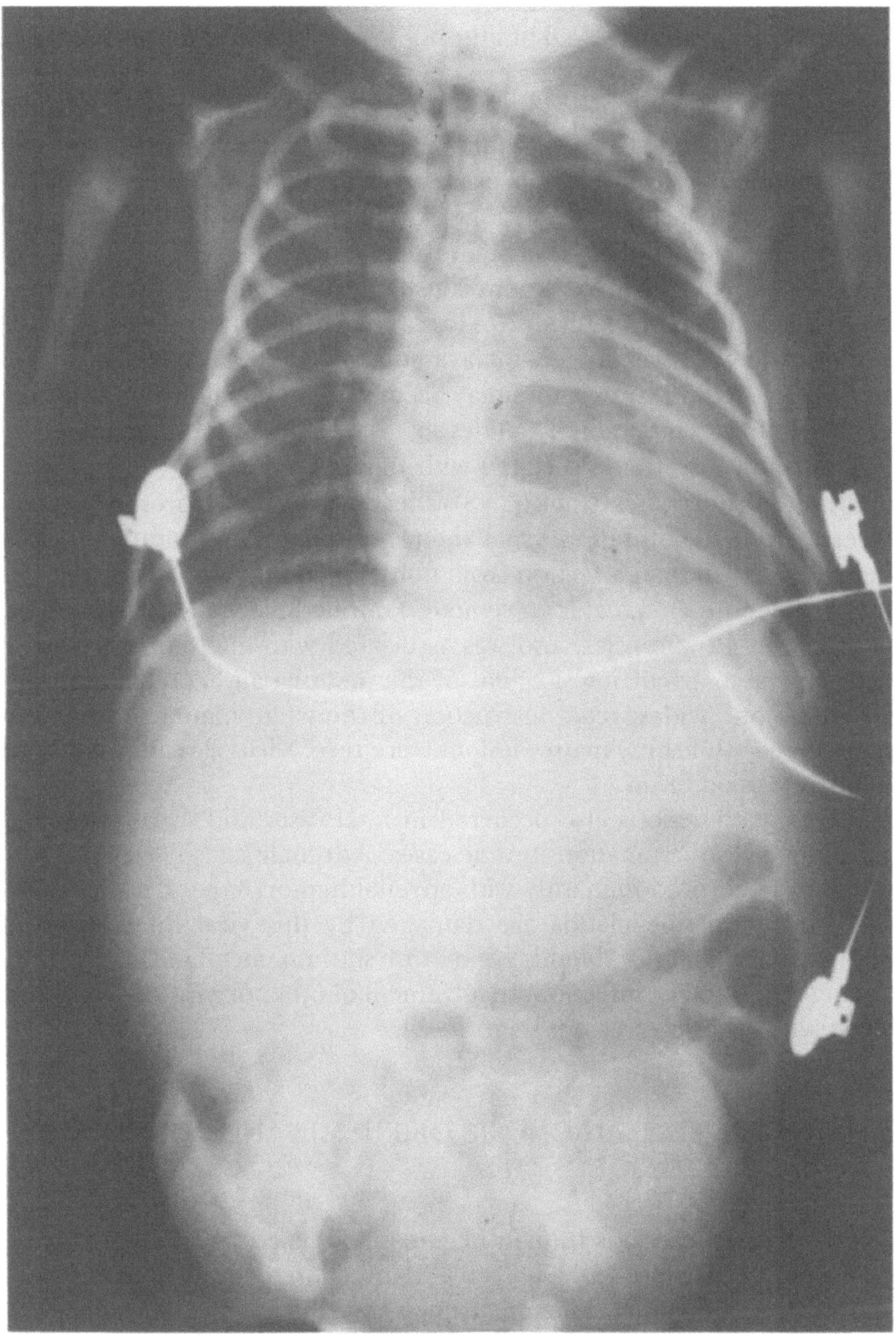

FIGURE 3. Radiograph showing massive cardiomegaly and congestive hepatomegaly in a fatal case of coxsackievirus B myocarditis.

Many children develop progressive pulmonary infiltrates with respiratory failure.[1] In 30 fatal cases, interstitial infiltration with alveolar hemorrhage was discovered at autopsy with alveolar hemorrhage. Neonates who survive myocarditis can develop respiratory failure or serious pulmonary hemorrhage.[1,17] Often the pulmonary disease is overshadowed by congestive heart failure secondary to myocarditis.

Hepatic involvement also occurs in one half of children, characterized pathologically by central hemorrhagic necrosis. Viral culture of the liver has yielded high tissue culture infectious doses ($TCID_{50}$) of 10^5 and 10^6.[7,18,19] Liver involvement is usually attended by severe bleeding diathesis, which is probably due to loss of hepatic-derived clotting factors rather than to disseminated intravascular coagulopathy (DIC), as has been suggested. Liver enzymes may become elevated, along with clotting abnormalities. These abnormalities are harbingers of fulminant infection. Jaundice is often associated with fatal infection.

Central nervous system (CNS) infection is usually present in fatal cases.[1] There is nothing unusual about the spinal fluid parameters that help distinguish fatal infection from nonfatal meningitis. Autopsy shows frequent leptomeningeal involvement. Pontine involvement is common, occurred in 12 autopsies, and was associated with microglial infiltrates with perivascular cuffing. Lesions of the medulla and cerebellum were less common. Widespread destruction of the white matter occurred in seven cases, while gray matter lesions were rare. Virus is readily cultured from autopsied brain.

Pancreatic lesions also occurred in fatal cases and were associated with islet cell inflammation in nine cases. Adrenalitis also occurred and was associated predominantly with adrenal hemorrhage. Because these two vital endocrine glands are damaged by this viral infection, it is important to monitor blood sugar and sodium metabolism carefully during the course of infection. Involvement of other organs was rare, and intestinal pathology was not seen.

4. DIFFERENTIAL DIAGNOSIS OF FULMINANT COXSACKIE B VIRAL INFECTION

This illness can look initially like neonatal sepsis secondary to group B streptococcus or *Escherichia coli* K1 infection. However, the severe abnormalities in spinal fluid chemistry and cellularity seen in these bacterial infections distinguish them from CVB infection. The myocarditis can be initially confused with congenital heart disorders, but echocardiograms and ECG can distinguish these illnesses readily.

This illness can also be confused with neonatal echovirus infection, which in fatal cases generally presents with progressive severe hepatitis with CNS involvement. Fulminant CVB infection can be distinguished from echovirus infection by the presence of severe myocarditis, which is rarely seen with echovirus infection.[20]

5. VIRAL DIAGNOSIS

Children suspected of having CVB infection acquired either during the intrapartum period or within the first 3 months of life should have viral cultures obtained of the throat, rectum, spinal fluid, and blood. Frequently, virus can be isolated within 2–3 days in these children. The isolation of virus is quite useful; we have found that we can often discharge children from the hospital, especially when bacterial cultures are negative. Children under 7 days of age from whom a nonpolioenterovirus is isolated should be watched carefully for overwhelming CVB viral infection. They should have cardiac monitoring with an ECG and careful physical examination. If ECG abnormalies are present, an echocardiogram should be performed to rule out myocarditis. Once myocarditis is found, attempts at cardiac support should be attempted. There should also be liver-function studies and coagulation parameter measurement to determine the extent of hepatic disease. In the event that there appears to be fulminant infection, respiratory support and monitoring of blood sugar should be carried out to watch for pancreatic islet cell infection.

6. TREATMENT OF FULMINANT INFECTION

Currently there is no adequate therapy for fulminant infections other than supportive care. I do not believe that steroids should be used in these children. Some will give these neonates exchange transfusion or immunoglobulin therapy. The value of such therapy is not known. Although we have described mostly fatal cases, it should be remembered that not all infants with transplacentally-acquired infection die. Attention to cardiac support, bleeding disorders, respiratory support, and maintenance of adequate blood sugar levels is essential for survival of these children. There has been documentation that full survival is possible.

7. PREVENTION

Little thought has gone into the prevention of CVB infection in the newborn and in the child under 3 months of age. It is unwarranted for a mother due to deliver in the months of July through September to place herself at risk of CVB infection. Since the best epidemiology suggests that this disease is acquired from fecally contaminated water, I advise pregnant mothers not to go swimming in large public pools during their last 2 weeks of pregnancy. They should also avoid exposure to febrile children during the last week of their pregnancy. During July through September, obstetricians should consider delaying the performance of elective cesarean section in women who have just gotten over a febrile illness. It is clear from case reports that a few more days in the uterus can be life-saving to the neonate born to a mother with recent placental viremia. Finally, hospitals should take special care with proper isolation of febrile children, particularly during the summer, because of the ease of nosocomial transmission of CVB within nurseries. Numerous nosocomial outbreaks of infection have been reported.[21–27] These outbreaks are best controlled by careful handwashing to remove waterborne and fecal borne virus from the hands. Aerosol transmission does not appear to occur.

Vaccine delivered to the mother during the first trimester of pregnancy to prevent fulminant infection in the newborn has not been considered. However, some thought should be given to such an idea, especially for mothers who will come to term during the summer months.

Some thought ought to be given to the use of anti-coxsackie B immunoglobulin for the child suspected of having the onset of fulminant viral infection. Clearly, studies of the effectiveness of such treatment will be useful only if more rapid means of diagnosing these infections evolve. Currently, it is clear that the lack of recognition of this serious disease in the newborn occurs because significant diagnostic technology is not available in most hospitals. Scientists should be encouraged to develop such technology, so that the true extent of this infection in the neonate can be measured.

REFERENCES

1. Kaplan, M. H., Klein, S. W., McPhee, J., and Harper. R. G., 1983, Group B Coxsackievirus infections in infants younger than three months of age: A serious childhood illness, *Rev. Infect. Dis.* **5:**1019–1032.
2. Nahmias, A. J., Alford, C. A., and Korones, S. B., 1970, Infection of the newborn with herpesvirus hominis, *Adv. Pediatr.* **17:**185–226.
3. Morens, D. M., 1978, Enteroviral disease in early infancy, *J. Pediatr.* **92:**374–377.
4. Grist, N. R., Bell, E. J., and Assad, F., 1978, Enteroviruses in human disease, *Prog. Med. Virol.* **24:**114–157.
5. Moore, M., Kaplan, M. H., McPhee, J., Bregman, D. J., and Klein, S. W., 1984, Epidemiologic, clinical, and laboratory features of Coxsackie B1–B5 infections in the United States, 1970–1979, *Public Health Rep.* **99:**515–522.
6. Dagan, R., Jenista, J. A., Prather, S. L., Powell, K. R., and Menegus, M. A., 1985. Viremia in hospitalized children with enterovirus infections, *J. Pediatr.* **106:**397–401.

7. Jahn, C. L., and Cherry, J. D., 1966, Mild neonatal illness associated with heavy enterovirus infection, *N. Engl. J. Med.* **274**:394–395.
8. Jenista, J. A., Menegus, M. A, and Powell, K. R., 1982. Epidemiology of neonatal enterovirus infections, *Pediatr. Res.* **16**(2):430(abst.).
9. McCracken, G. H., Jr., and Mize, S. G., 1976, A controlled study of intrathecal antibiotic therapy in gram-negative enteric meningitis of infancy. Report of the neonatal meningitis cooperative study group, *J. Pediatr.* **89**:66–72.
10. Rantakallio, P., Saukkonen, A.-L., Kruase, U., and Lapinleimu, K., 1970, Follow up study of 17 cases of neonatal Coxsackie B5 meningitis and one with suspected myocarditis, *Scand. J. Infect. Dis.* **2**:25–28.
11. Farmer, K., and Patten, P. T., 1968, An outbreak of coxsackie B5 infection in a special care unit for newborn infants, *NZ Med. J.* **68**:86–89.
12. Sells, C. J., Carpenter, R. L., and Ray, C. G., 1975, Sequelae of central nervous-system enterovirus infections, *N. Engl. J. Med.* **293**:1–4.
13. DeBacker, S., Samuel, K., Carton, D., and Pintelon, J., 1976. Neonatal coxsackie B3 sepsis, *Acta Pediatr. Belg.* **29**:55–57.
14. Wilfert, C. M., Thompson, R. J., Sunder, T. R., O'Quinn, A., Zeller, J., and Blacharsh, J., 1981, Longitudinal assessment of children with enteroviral meningitis during the first three months of life, *Pediatrics* **67**:811–815.
15. Landsdown, A. B. G., 1975, Influence of time of infection during pregnancy with Coxsackievirus B3 on maternal pathology and foetal growth in mice, *Br. J. Exp. Pathol.* **56**:119–123.
16. Modlin, J. F., and Crumpacker, C. S., 1982, Coxsackievirus B infection in pregnant mice and transplacental infection of the fetus, *Infect. Immun.* **37**:222–226.
17. Hurley, R., Norman, A. P., and Pryse-Davies, J., 1969, Massive pulmonary haemorrhage in the newborn associated with coxsackie B virus infection, *Br. Med. J.* **3**:636–637.
18. McLean, D. M., Donohue, W. L., Snelling, C. E., and Wyllie, J. C., 1961, Coxsackie B5 virus as a cause of neonatal encephalitis and myocarditis, *Can. Med. Assoc. J.* **85**:1046–1048.
19. Wright, H. T., Okuyama, K., and McAllister, R. M., 1963, An infant fatality associated with coxsackie B1 virus, *J. Pediatr.* **63**:428–431.
20. Modlin, J. F., 1986, Perinatal ECHOvirus infection: Insights from a literature review of 61 cases of serious infection and 16 outbreaks in nurseries, *Rev. Infect. Dis.* **8**:918–926.
21. Javett, S. N., Heymann, S., Mundel, B., Pepler, W. J., Lurie, H. I., Gear, J., Measroch, V., and Kirsch, Z., 1956, Myocarditis in the newborn infant. A study of an outbreak associated with coxsackie group B virus infections in a maternity home in Johannesburg, *Pediatrics* **48**:1–22.
22. Cherry, J. D., 1976, Enteroviruses, in: *Infectious Diseases of the Fetus and Newborn Infant* (J. S. Remington and J. O. Klein, eds.), pp. 366–413, W. B. Saunders, Philadelphia.
23. Brightman, V. J., Scott, T. F. M., Westphal, M., and Boggs, T. R., 1966, An outbreak of coxsackie B5 virus infection in a newborn nursery, *J. Pediatr.* **69**:179–192.
24. Rantakallio, P., Lapinleimu, K., and Mantyjarvi, R., 1970, Coxsackie B5 outbreak in a newborn nursery with 17 cases of serious meningitis, *Scand. J. Infect. Dis.* **2**:17–23.
25. Swender, P. T., Shott, R. J., and Williams, M. L., 1974, A community and intensive care nursery outbreak of coxsackievirus B5 meningitis, *Am. J. Dis. Child.* **127**:52–55.
26. Marier, R., Rodriguez, W., Chloupek, R. J., Brandt, C. D., Kim, H. W., Baltimore, R. S., Parker, C. L., and Artenstein, M. S., 1975, Coxsackie B5 infection and aseptic meningitis in neonates and children, *Am. J. Dis. Child.* **129**:321–325.
27. Moore, M., 1982, Enteroviral diseases in the United States, 1970–1979, *J. Infect. Dis.* **146**:103–108.

15

Myocarditis

Clinical and Experimental Correlates

MILAGROS P. REYES and A. MARTIN LERNER

1. INTRODUCTION

An etiologic relationship has been established between coxsackievirus (CV) infection and myopericarditis. Etiologic suggestions are based on serologic evidence of prior infection with these viruses in the affected population as compared with age and sex-matched controls, and on isolated case reports during epidemics of CV infection.[1]

Animal models of infection were described after CV discovery in 1948 by Dalldorf and Sickles.[2,3] Pathogenicity and neurotropism decrease with the age of the inoculated mice, and involvement of other organs, such as the heart or pancreas, varies among the several CV types.[4–7] Certain strains of group A coxsackievirus type 9 (CVA9) produce a benign myocarditis in weanling mice. Murine myofiber necrosis is minimal and interstitial inflammation of infected hearts heals completely.[8] Group B coxsackievirus type 3 (CVB3) is the most cardiotropic to weanling mice, whereas CVB4 causes severe necrotizing myocarditis in suckling mice.

This chapter discusses CV-induced cardiopathies in humans and describes murine models of heart disease due to CV that have been developed in our laboratory.

MILAGROS P. REYES and A. MARTIN LERNER • Wayne State University, Division of Infectious Diseases, Hutzel Hospital, Detroit, Michigan 48201.

2. COXSACKIEVIRUS-INDUCED CARDIOPATHIES

Viral infections of the heart often concomitantly involve both the myocardium and pericardium. Clinically, there may be predominantly myocarditis or pericarditis. The acute process may be benign with an interstitial infiltration of inflammatory cells between the myofibers with or without cardiac muscle necrosis. In benign myopericarditis, muscle necrosis is minimal and recovery is usually complete, without anatomical or physiologic sequelae.

Virulent infections show varying degrees of interstitial inflammation and cellular necrosis of the myocardium. Lesions heal, leaving permanent fibrosis (sometimes with calcification) and myofiber disruption. Virus-induced myocardial necrosis may also be focal and transmural, with patent coronary arteries, simulating myocardial infarction secondary to coronary atherosclerosis and thrombosis. Myocardial necrosis can also result in a continuing chronic active cardiomyopathy with subacute mononuclear cellular interstitial infiltrates. Ultimately dilated or hypertrophic cardiomyopathy follows.

2.1. Clinical Manifestations

A summary of human heart diseases associated with CV is shown in Table I. The causative relationship between CV and subendocardial fibroelastosis and congenital malformations is suggested but not established.[9]

Clinical illnesses may be fulminant and rapidly fatal or entirely subclinical.[10,11] The clinical presentation of CV myocarditis varies with the patient's age and virulence of the virus strain. Especially susceptible to severe CVB myocarditis are newborn infants. Cyanosis, respiratory distress, tachycardia, cardiomegaly, and electrocardiographic (ECG) changes are found in most infants. Fever occurs in one half of cases. Lethargy, anorexia, and failure to gain weight may herald the disease. The initial manifestations of myocarditis are usually less fulminant in older infants and children. Involvement of other organs, such as the meninges, brain, liver, pancreas, or adrenals, may be present.

In adults, myocarditis is often subclinical. However, congestive heart failure, arrhythmias, and sudden death are possible. CV myocarditis in adults usually presents with fatigue, dyspnea, palpitations, and occasionally, precordial discomfort. Precordial pain is usually associated with pericarditis, but it may reflect relative coronary insufficiency due to low cardiac output of an injured myocardium. Systemic symptoms of CV infection, such as malaise, myalgia, arthralgia, fever, nausea, vomiting,

TABLE I
Human Heart Diseases Associated with Coxsackieviruses

Myocarditis	Endocardial deformities
Chronic myocardiopathy	Subendocardial fibroelastosis
Pericarditis	Congenital malformations
Constrictive pericarditis	

diarrhea, cough, sore throat, and rash, may be present early in the illness. When myocarditis appears, however, nonspecific symptoms are usually absent.

Common physical findings in patients with myocarditis include sinus tachycardia, fever, pericardial or pleuropericardial rubs, and supraventricular or ventricular arrhythmias. First- and second-degree heart block may be seen. Rarely, syncope or sudden death may occur as a result of complete atrioventricular (AV) block. Systemic or pulmonary embolization may also occur.

Congestive heart failure occurs in more severe cases of myocarditis. CV myocarditis frequently presents as an acute dilated (congestive) cardiomyopathy. Left heart failure occurs more frequently, but biventricular failure may also be present. An enlarged heart, hypotension, narrow pulse pressure, murmurs of mitral and/or tricuspid insufficiency, faint first heart sound, neck vein distention, ventricular gallops, hepatomegaly, and peripheral edema are findings consistent with congestive heart failure. Less commonly, circulatory collapse and clinical shock occur.

Some patients with CV myocarditis may have subacute, recurrent, or chronic courses. Others develop chronic dilated cardiomyopathy.[12,13]

2.2. Laboratory Aids

A moderate leukocytosis during the early phase of the illness is seen in about three fourths of cases. Serum elevations of cardiac enzymes, creatine phosphokinase (CPK), glutamic oxaloacetic acid (SGOT), and lactic dehydrogenase (LDH) vary, depending on the extent of myocardial necrosis. Chest radiography may show an enlarged cardiac silhouette due to either dilatation of the ventricular cavity, pericardial effusion, or both. The electrocardiogram, which is used to consider a diagnosis of viral myocarditis, is usually abnormal. Low-voltage QRS complexes, elevation or depression of ST segments, and elevated, flattened, or inverted T waves are seen. Varied conduction disturbances or arrhythmias are frequent. Myocardial dysfunction and pericardial effusions may be confirmed by echocardiography. Nuclear imaging of the heart with technetium-99m-pyrophosphate, ([^{99m}Tc]-PyP) or with gallium-67 citrate (^{67}Ga) often shows increased myocardial uptake. However, other cardiac conditions, such as myocardial infarction, unstable angina, and ventricular aneurysm, can also be associated with positive scintigrams. Radionuclide angiography may confirm cardiomyopathy. Cardiac catheterization with coronary angiography is necessary in some patients (infants or adults) to exclude congenital anomalies and coronary artery or valvular heart disease.

2.3. Etiologic Diagnosis

Enteroviruses are by far the most frequently implicated human viruses as causative agents of myocarditis. On the basis of studies of type-specific neutralizing antibodies in sera from 385 patients with suspected heart disease conducted over a 6-year period in Scotland, CVB are associated with at least one half of cases of acute myocarditis. One third of cases of acute nonbacterial pericarditis are also associated with these same viruses.[14] Less frequent causes of acute myocarditis are members of the CVA and echovirus groups.

A definitive viral etiology of acute or chronic myocarditis can be established only by isolation of virus from the myocardium, endocardium, or pericardial fluid or by localization of the type-specific virus at these sites by immunofluorescence or peroxidase-labeled antibody. Such evidence, which is possible only at autopsy or myocardial biopsy, establishes a high-order association.[15] Recently, CVB-specific RNA sequences were detected in 9 of 17 myocardial biopsy samples from patients with histologic evidence of active or healing myocarditis or dilated cardiomyopathy with inflammatory changes.[16] Early demonstration of virus in the myocardium would facilitate the diagnosis of myocarditis and, it is hoped, would influence management.

A viral diagnosis can also be made by isolation of virus from serum, buffy coat, pharynx, or feces, along with a demonstration of a concomitant fourfold rise in viral type-specific neutralizing antibodies from paired acute and convalescent sera or a concomitant type-specific IgM titer of >1 of 32. This is referred to as moderate order association but does not establish a definitive diagnosis of virus-induced myocarditis.

Viral isolation from pharynx or feces, a fourfold rise in type-specific antibody, or a single type-specific IgM serum titer of >1 of 32, as single pieces of data represent low-order associations.

3. SPECIFIC MODELS OF HEART MUSCLE DISEASE

Murine models of heart muscle disease due to CV simulate the natural history of disease in humans and are suitable to study the pathogenesis of specific organ damage. Late sequelae of the disease, as well as the ability of antiviral chemotherapeutic agents to modify the disease can also be studied.

Specific susceptibilities to infection of the myocardium vary with age and genetic constitution of the mouse and the type and strain of CV. In

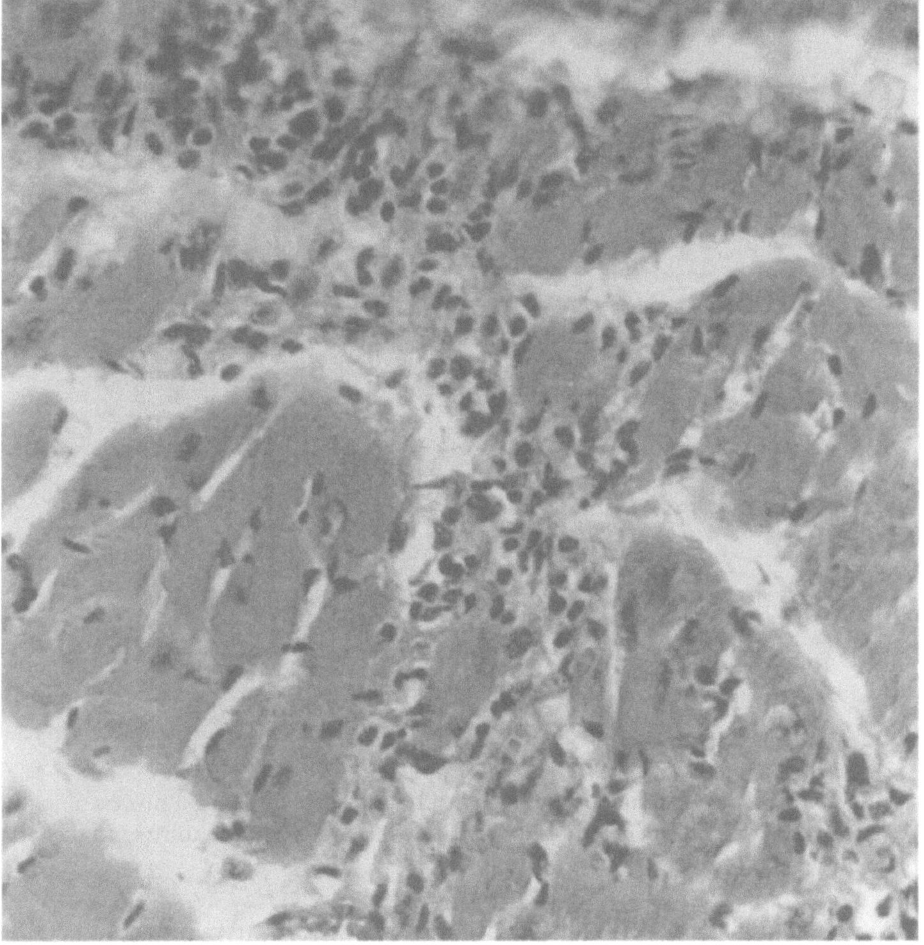

FIGURE 1. Benign CVA9 myocarditis (acute infectious phase of myocardiopathy). With this virus strain, complete healing occurs without sequelae; there is no late noninfectious phase. Myocarditis was induced by CVA9, strain 711, in 8-month-old Swiss mice by intraperitoneal inoculation of 10^5 $TCID_{50}$. An area of moderate focal inflammation was seen after 9 days. (From Lerner and Wilson.[17])

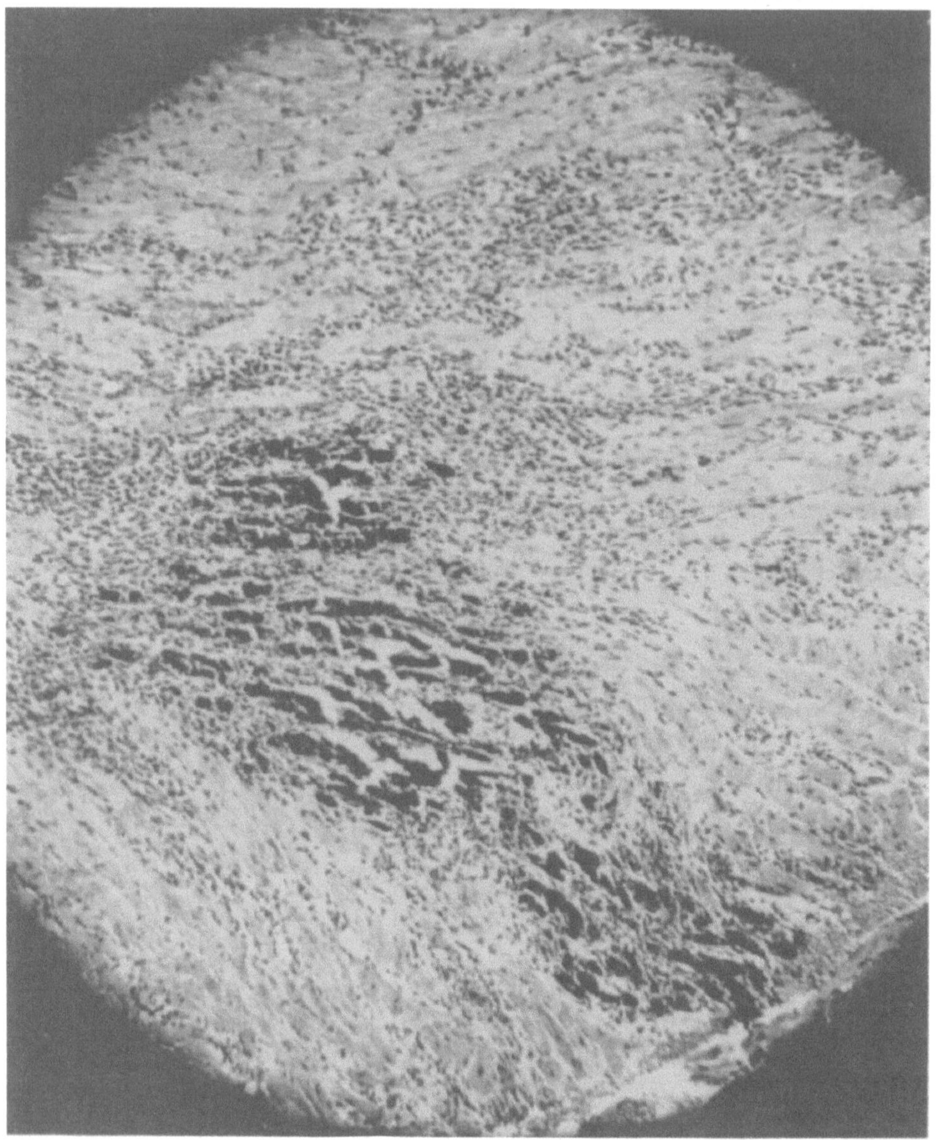

FIGURE 2. Effect of swimming on pathological findings in acute CVB3 myocardiopathy. The extensively involved myocardium of a 27-day-old mouse on the thirteenth day after infection is shown. (From Lerner and Wilson.[17])

adult mice, CVA9 produces a mild focal interstitial inflammatory response in the myocardium (Fig. 1), while CVB3 causes a virulent, diffuse, myonecrotic inflammatory lesion in weanlings[17] (Fig. 2). We describe the acute and chronic effects of benign and virulent heart muscle disease following CV infections in murine models developed in our laboratory. An overview of CV murine myocardiopathy is whown in Fig. 3.[18]

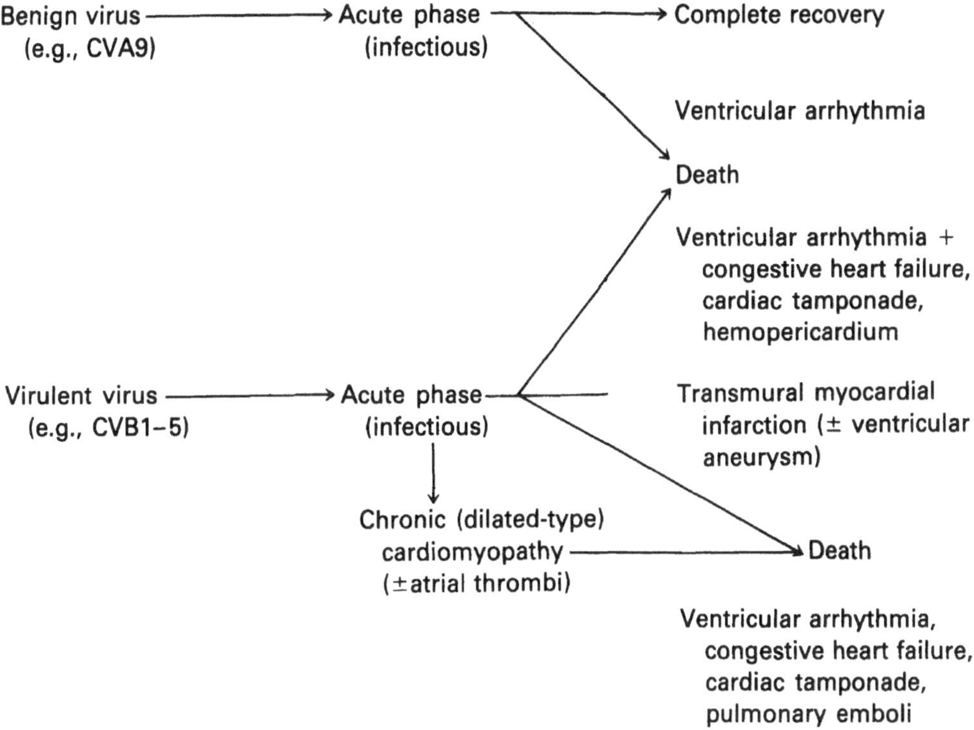

FIGURE 3. Possible courses of CV murine myocardiopathy. (From Lerner.[18])

3.1. Acute Myocarditis

3.1.1. CVB1 and CVB4 Murine Models

Intraperitoneal inoculation of 48-hr-old Swiss-Webster mice with 10^3–10^5 $TCID_{50}$ CVB1 or CVB4 induces transmural myocarditis.[7] Viremias are first detected on the second day postinfection and usually persist for 24 hr. Mean peak titers of virus in blood (CVB1, $10^{2.3}TCID_{50}/g$; CVB4, $10^{3.1}TCID_{50}/g$) occur on day 2 or 3. Virus is first isolated from the myocardium on day 2, usually reaches peak titers within 24 hr (mean peak, CVB2, $10^{2.2}$ $TCID_{50}/g$; CVB4, $10^{4.2}$ $TCID_{50}/g$), and persists longer in this tissue through the fifth to seventh day postinfection (Table II).

Histopathologic changes of myocarditis occur in 65% of CVB1-infected mice and in 78% of CVB4-infected mice less than 48 hr old. Microscopic findings characterized by swollen foci of myocardial fibers and mononuclear infiltrates appear on days 2–3. By days 5–7, necrotic fibers are calcified and fibroblasts appear in involved areas. Mononuclear infiltrates are sparse by day 14. The valves are not affected, but they become thicker with increasing age, both in virus-infected and control mice.

TABLE II
Clinical, Virologic, and Histopathologic Findings in Swiss-Webster Mice Infected with CVB Types 1–5

			Viremia		Virus in myocardium		Histopathology		
Virus type (strain)	Inoculum ($TCID_{50}$)	Age of mice at inoculation (days)	Peak day	Titer ($TCID_{50}$/ml) on peak day	Peak day	Titer ($TCID_{50}$/g) on peak day	No. with myocarditis/ No. examined (%)	Day of maximal pathology	No. died/ No. inoculated (%)
Cytopathic viruses									
1 (6631)	10^3	2	2	$10^{2.3}$	3	$10^{2.2}$	63/96 (66)	3–5	103/224 (46)
		14	N.D.[a]				3/10 (30)	N.D.	0/11
		>30	N.D.				0/16	None	0/16
4 (Dowell)	10^3	2	2	$10^{3.1}$	2	$10^{4.2}$	48/61 (79)	5	53/235 (23)
		14	2	$10^{3.3}$	2	$10^{2.5}$	7/39 (18)	7	0/136
		>30	N.D.	—	N.D.	—	0/36	None	0/52
Immunopathic viruses									
2 (Valery)	10^3	2	2–3	$10^{2.9}$	2–3	$10^{3.5}$	3/30 (10)	N.D.	15/65 (23)
		14	2	$10^{4.3}$	2	$10^{2.3}$	20/32 (63)	5–7	5/41 (12)
		>30					0/24	None	0/27
3 (Nancy)	10^3	2	2–3	$10^{3.2}$	2	$10^{4.0}$	1/9 (11)	N.D.	21/49 (43)
		14	2	$10^{3.5}$	3–5	$10^{2.8}$	34/40 (85)	5–7	3/50 (6)
		>30					11/13 (85)	N.D.	0/12
Nonpathologic virus									
5	10^5–10^8	2	None		None		None		0/105
		14	None		None		None		0/15
		>30	None		None		None		0/10

[a]N.D., not determined.

Mortality rates among mice less than 48 hr old are 46% for CVB1 and 23% for CVB4. Deaths are immediate from fulminant encephalitis. Ventricular aneurysms occur in 14% of surviving mice secondary to transmural necrotizing myocarditis.[19] Myocardial thickness is strikingly reduced at aneurysmal sites. Histology of aneurysms demonstrates severe necrosis of myocardial fibers, mononuclear cell infiltration, and early fibrosis, but calcification is absent (Fig. 4). Pathologic changes in 14-day-old subjects with CVB1 and CVB4 infections are less severe and occur less frequently. Only 30% of 14-day-old mice injected with CVB1 and 18% with CVB4 will develop myocarditis. Hearts from 3–5-month-old animals infected with CVB1 or CVB4 will not develop myocarditis. Regardless of age, myocarditis was never seen in mice inoculated with strains of CVB5.

The mechanisms by which CV induce cardiac injury are incompletely understood. Pathologic changes in the heart are synchronous with maximal cardiac titers of CVB1 and CVB4 viruses. Thus, it is believed that the pathogenesis of myofiber necrosis in these animal models is the direct result of direct viral cytopathogenic effects. When antithymocyte serum is given before CVB4 inoculation, a higher mortality (76%), more extensive necrosis, and mineralization follow.[20] No changes in the occurrence of neutralizing antibodies are seen, however. Thymus-dependent functions are therefore important in the elimination of CV. The thymus may play a critical role in the efficiency of virus phagocytosis by macrophages.

3.1.2. CVB2 and CVB3 Murine Models

CVB2 and CVB3 induce nontransmural necrotizing myocarditis in 14-day-old weanling mice. As in the CVB1 and CVB4 models, viremias first occur on the second day of infection; however, viremias persist for 72 hr in CB3 mice. The mean peaks range from $10^{1.3}$ $TDIC_{50}$/ml (CVB2) to $10^{3.5}$ $TCID_{50}$/ml (CB3). Mean peak virus titers in the hearts range from $10^{2.3}$ $TCID_{50}$/g (CVB2) to $10^{2.8}$ $TCID_{50}$/g (CVB3) and persist for 48 hr in CVB3 infected mice (Table II).

Myocarditis is characteristic in 14-day-old mice infected with CVB2 (62.5%) or CVB3 (85%). Deaths occur in 12% of CVB2-inoculated mice and 6% of CVB3 mice. In baby mice (<48 hr old) infected with CVB2 or CVB3, myocarditis occurs less often, 10% and 11%, respectively. Adult animals exhibit inflammatory changes only after infection with CVB3 (85%). No adult mouse inoculated with CVB2 had histological evidence of myocarditis.

In these models, maximal virus titers that occur on days 2–5 are asynchronous with maximal histopathologic changes appearing on days

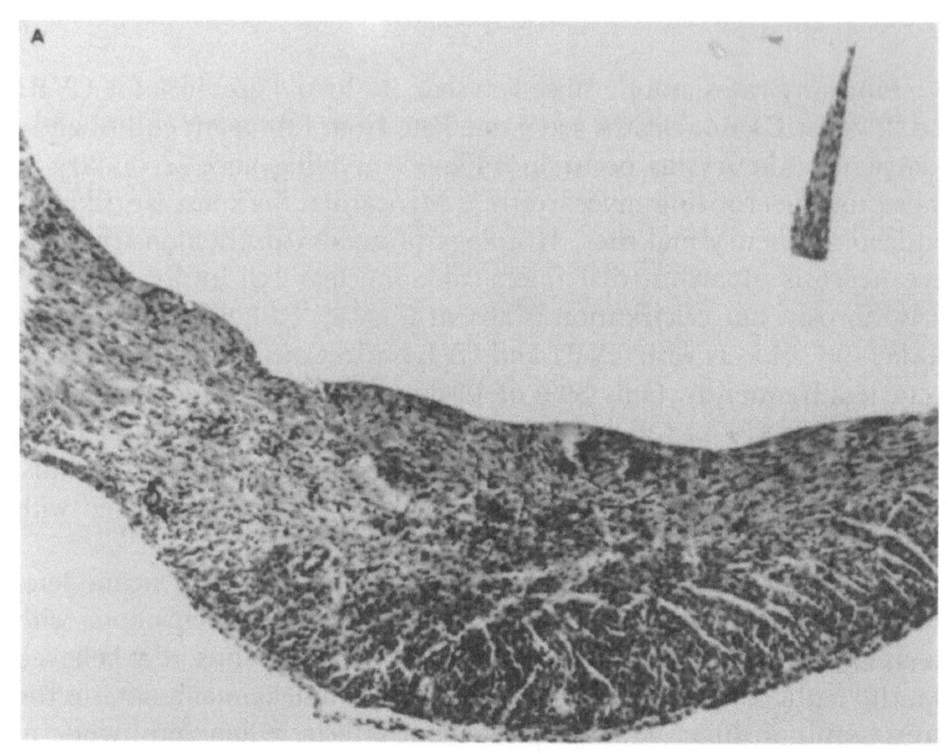
A

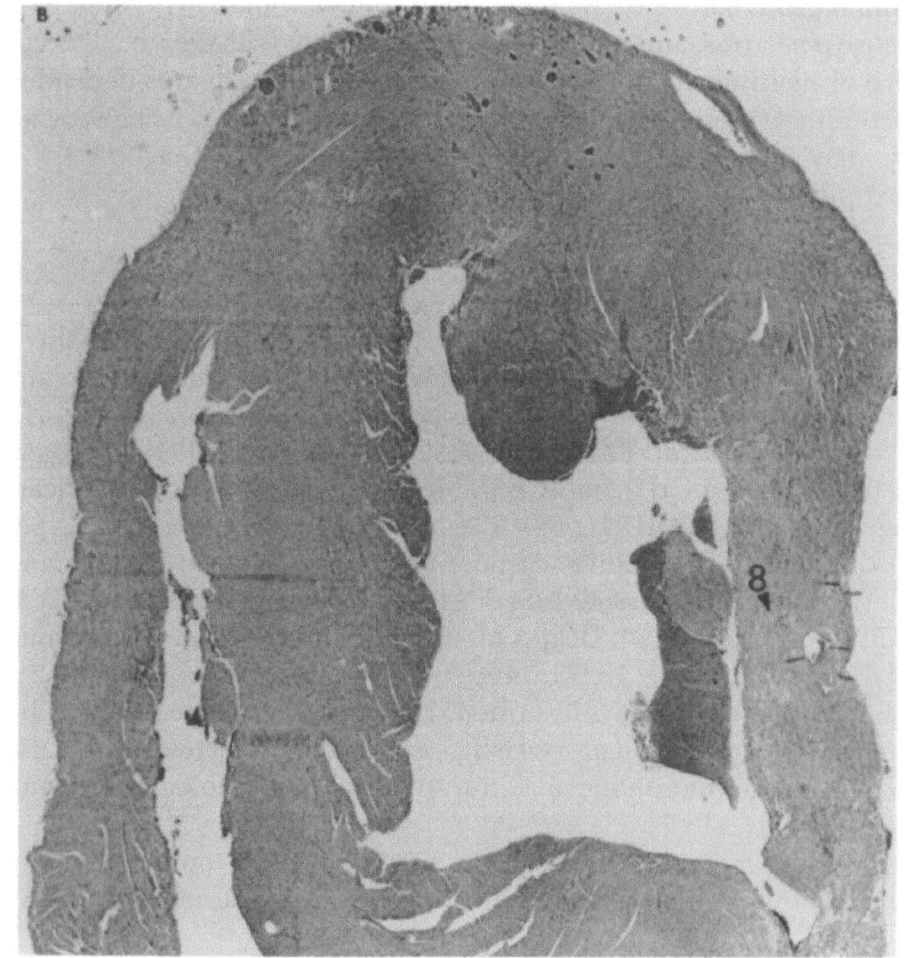
B

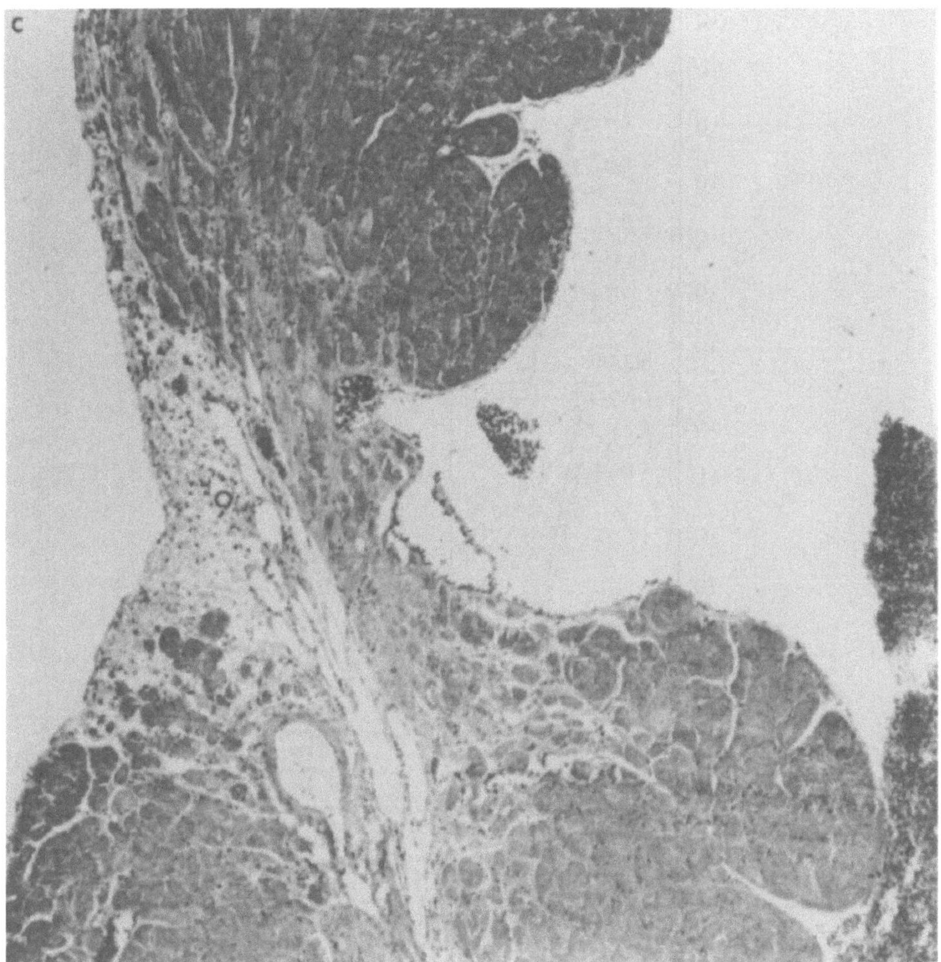

FIGURE 4. (**A**) Seven-day-old mice 5 days after inoculation with 10^4 $TCID_{50}$ CVB4. The right ventricular wall (×80) shows myofiber necrosis, leukocytic infiltration, and thinning (arrow 1). (**B**) Six-month-old animal infected in (**A**). A large centrally located aneurysm of the left ventricle (arrow 2) is shown (×60). Note marked localized thinning of the wall of the left ventricle. (**C**) One-year-old animal infected subject as in (**A, B**). An aneurysm (arrow 3) of the right ventricle (×80) is seen. (From El-Khatib *et al.*[19])

5–7. CVB2 and CVB3 may therefore be immunopathic myocardial viruses, which induce myofiber necrosis via an immunopathologic mechanism.[21] Woodruff *et al.*[22] have now provided direct experimental evidence for a T-cell-mediated immunologic mechanism for CVB3.[22]

3.1.2.a. CVB3 Murine Model and Exercise. Striking augmentation in the virulence of CVB3 myocarditis by swimming during the acute phase of infection is seen.[23] Remarkable increases in quantitative titers of virus (500×) occur in the blood (days 3–6 postinfection) and heart (days 6–9 postinfection) (Fig. 5). Mortality rate increases to 50%, and the myocar-

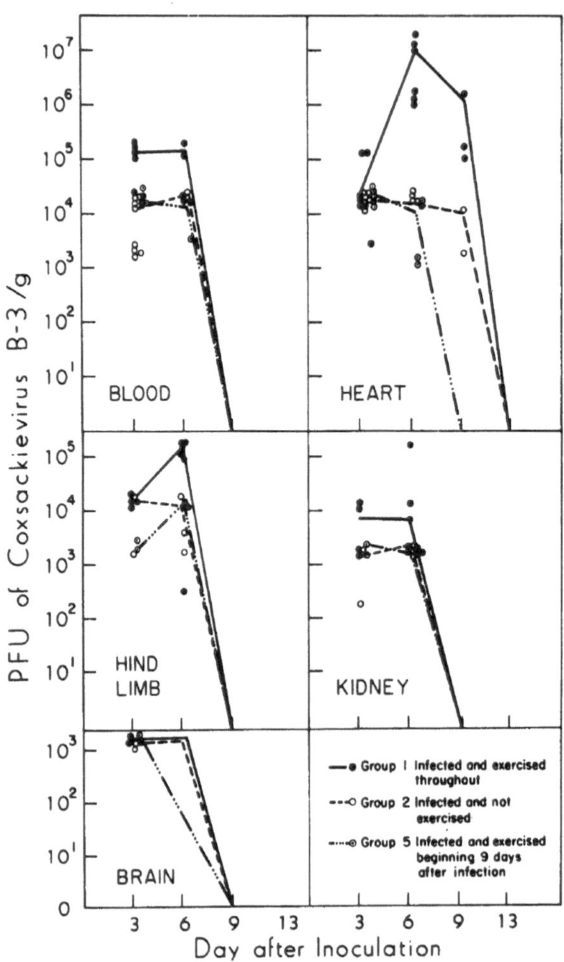

FIGURE 5. Effect of swimming on multiplication of CVB3 in several tissues. PFU, plaque-forming unit. (From Gatmaitan *et al.*[23])

dium is converted into a necrotic calcified mass. Interferon titers increase in the blood and heart; neutralizing antibody to CVB3 appears on day 3 and in unchanged amount.[24] Absolute counts of mononuclear cells (T cells, B cells, macrophage) increase on days 1 and 2 postinfection and become CVB3 positive on days 2–4.[25] Mononuclear cells in the thymus and spleen contain virus as well. Erythrocytes and polymorphonuclear leukocytes remain virus free through day 6 postinfection (Fig. 6).

To explore further the mechanisms of the virulence-enhancing effects of exercise, thymocyte circulation was followed in infected and normal mice. In healthy animals conditioned by swimming, significant decreases in mean thymic index occur on days 6–9. Thymuses are small, and their cortex is thin, but the density and microscopic appearance of retained thymocytes are normal.[26] Thymuses from infected and exer-

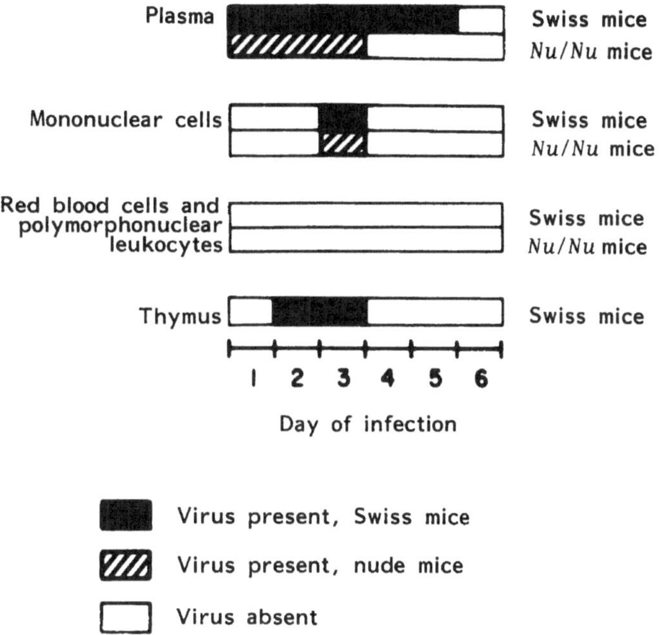

FIGURE 6. Distribution and duration of CVB3 in blood of infected Swiss and nude mice. (From Gomez *et al.*[25])

cised mice are even smaller. We postulate that forced swimming induces enhanced release of T lymphocytes from the thymus.

Levels of both norepinephrine and cyclic guanosine monophosphate (cGMP) were assayed and are increased in isolated thymocytes of infected and exercised subjects on day 9 postinfection.[27] Catecholamines, such as epinephrine and norepinephrine, are released from the adrenal medulla into plasma during exercise. These substances then act as neurohormones to regulate intracellular concentrations of the cyclic nucleotides, which in turn modulate humoral and cellular immune responses. Elevations in intracellular thymocyte cGMP augment the effector functions of these cells.

Migration of antigen-directed thymus-derived lymphocytes was then followed using sensitized and nonsensitized T lymphocytes labeled with ^{51}Cr. T-cell-dependent *in situ* calcification was also assessed by cardiac xeroadiographs. Significant rerouting of sensitized ^{51}Cr-labeled T cells to the heart occurred on day 9 in CVB3-infected but not in exercised subjects. This increased T-cell migration was accompanied by myocardial calcification. A significant augmentation in the rerouting of sensitized ^{51}Cr-labeled T cells to the heart occurred when mice infected with CB3 were forced to swim. Myocardial calcification was much more severe at xeroradiography, and it was now accompanied by cardiac dilatation.[28]

Immunologically mediated myofiber necrosis of CVB3 myocarditis is greatly augmented by exercise. At least three factors may contribute to the exercise-induced augmentation in virulence of CVB3 myocarditis: (1) an increased presence of a CVB3-induced cardiac antigen, (2) increased T-cell traffic, redirected to the heart, and (3) delay in turning off cell-mediated immune response reflected by an increased T-cell cGMP on day 9 postinfection.[10]

3.2. Chronic Heart Muscle Disease in Mice

Group B coxsackievirus is suspected as the cause in some human cases of chronic cardiomyopathy. The first experimental evidence that acute CV myocarditis is followed by permanent heart muscle disease was provided by Wilson *et al.*[29] Subclinical murine CVB3 myocarditis results in a marked permanent and diffuse fibrosis with dystrophic mineralization. In 1981, studies in a murine model from our laboratory provided firm evidence for establishing an etiological relationship between acute CVB3 myocarditis and chronic heart disease.[30] A possible similar relationship for humans is suggested.

In the murine model of CVB3 heart muscle disease, subjects forced to swim during the acute phase of infection and followed over a 15-month period develop changes similar to the findings in humans with congestive cardiomyopathy. Myocardial fiber disintegration with replacement by fibrous scar and heavy deposits of calcium are concen-

TABLE III
Localization of Pathologic Findings (Fibrosis, Mononuclear Infiltrate, Calcium Deposition, and Atrial Thrombi) in Hearts of ICR Swiss Mice with Chronic Cardiomyopathy Caused by CVB3 and in Hearts of Uninfected Mice[a]

			Pathologic findings in hearts[b]			
Group[c]	Infected	Forced to swim	Atria	Left ventricular wall	Septum/atrio-ventricular junction	Mortality at 15 months[d]
1	No	No	0/27 (0)	5/27 (19)	3/27 (11)	2/32 (6)
2	Yes	No	0/42 (0)	20/42 (48)	4/42 (10)	7/53 (13)
3	Yes	Yes	7/54 (13)	50/54 (93)	13/54 (24)	50/111 (45)[e]
4	No	Yes	1/37 (3)	1/37 (3)	5/37 (14)	12/52 (23)

[a]From Reyes *et al.*[30]
[b]Data are the number of hearts with pathologic findings/number of hearts examined (%). The evaluations were made by an observer without knowledge of the experimental design.
[c]All groups were observed for 15 months.
[d]Data are number of deaths/number of mice in group (cumulative mortality).
[e]The cumulative mortality in group 3 was significantly different from that of the other groups (p >0.01). The cumulative mortality among groups 1, 2, and 4 was similar (p >0.05).

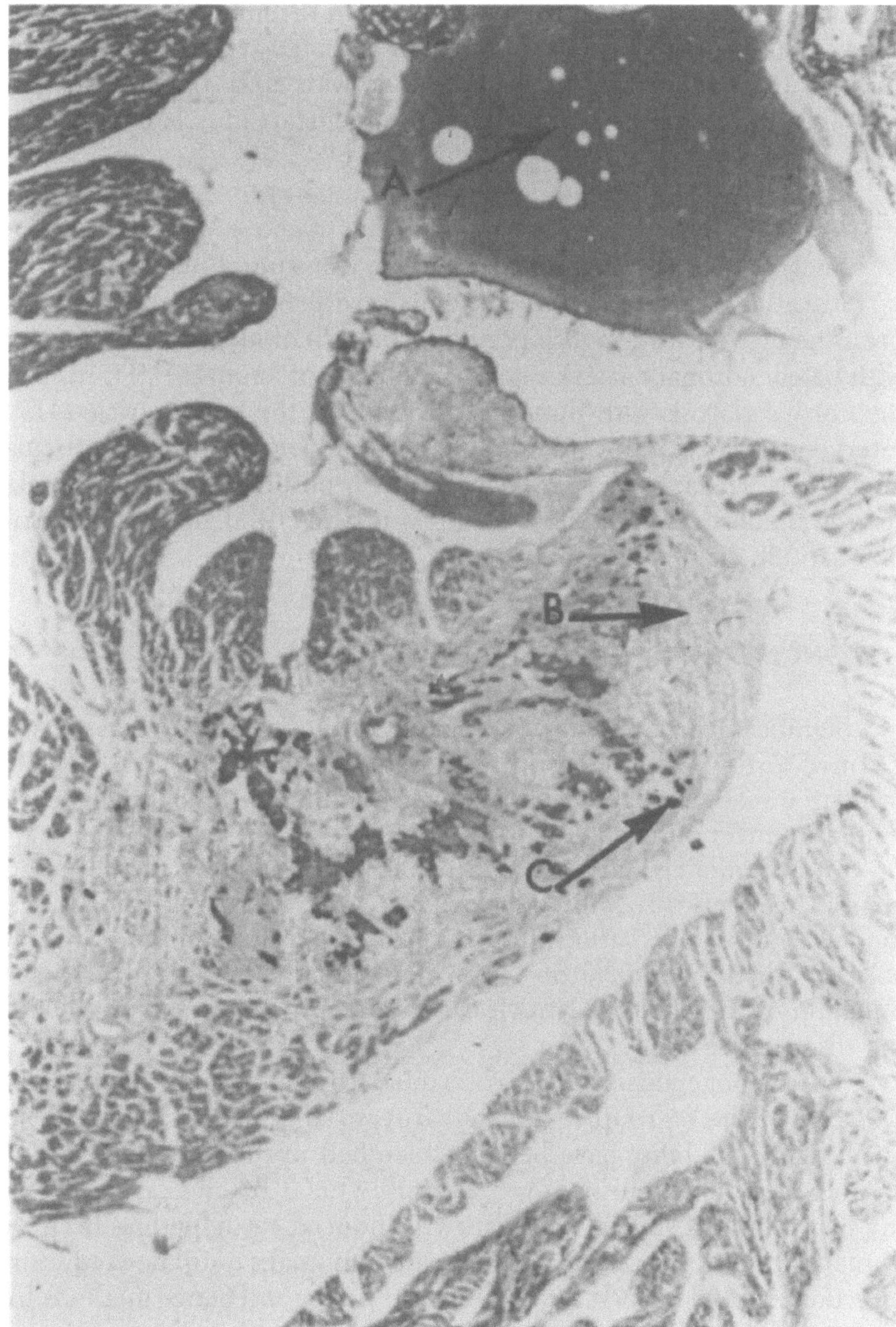

FIGURE 7. Effect of swimming on pathologic findings in chronic CVB3 cardiomyopathy. The extensive scarring, fibrous tissue, and calcium deposition involving the myocardium and atrioventricular septum are shown. (From Reyes and Lerner.[10])

trated in the left ventricle, interventricular septum, and atrioventricular junction (Table III). Atrial hypertrophy with thrombi therein are seen. These findings represent the changes of a dilated-type cardiomyopathy (Fig. 7). Intense scarring of the AV node with atrial thrombi suggests that arrhythmias occur in mice with chronic heart muscle disease.

3.2.1. Other Sequelae

Suckling ICR Swiss mice that survive transmural necrotizing myocarditis after CVB4 infection develop myofiber replacement with fibrous tissue over a 12-month period. A pattern of myocardial infarction with patent coronary arteries develops in 75% of animals.[31] Transmural myocardial fibrosis with fibrous scar occurs in the left ventricle (41%), interventricular septum (57%), and right ventricle (38%). Ventricular aneurysms, which appear as early as day 5 postinfection, are sequelae seen in the left ventricle (62%), right ventricle (19%), and interventricular septum (19%).

4. CONCLUSIONS

Significant progress has been made in the investigative aspects of CV myocarditis over the past three decades. Much of the work done by primary investigators has been performed in murine models and is presented in different chapters of this volume.

Murine models have led to a better understanding of the pathogenesis, acute effects, and sequelae of coxsackievirus myocarditis. These models simulate the natural history of disease in humans and are suitable for the development of newer diagnostic methods and the determination of the ability of antiviral chemotherapeutic agents to modify the disease.

The pathogenesis of CV myocarditis, tropism, role of the different members of the B group of CV and virus variants, and factors influencing host susceptibility have been well studied in the murine model. A link between acute murine myocarditis due to CVB3 and cardiomyopathy has also been shown. Although CV-induced cardiopathies occur in humans, the true incidence of the disease in humans is unknown. Future directions should include development of newer and better methods for diagnosis. The advent of endomyocardial biopsy and, most recently, the development of a hybridization probe to detect the presence of virus nucleic acid sequences in myocardial tissue augur a bright outlook for establishing a definitive diagnosis of viral myocarditis. Search for potential modalities of treatment has become more vital.

REFERENCES

1. Lerner, A. M., 1980, Enteric viruses in: *Principles of Internal Medicine,* Vol. 1 (R. G. Petersdorf, R. D. Adams, E. Braunwald, K. J. Isselbacher, J. B. Martin, and J. D. Wilson, eds.), pp. 1125–1132, McGraw-Hill, New York.
2. Dalldorf, G., and Sickles, G. M., 1948, Unidentified filtrable agent from feces of children with paralysis, *Science* **108:**61–62.
3. Kibrick, S., 1961, Viral infections of the fetus and newborn, *Perspect. Virol. Symp. NY* **2:**140–145.
4. Lerner, A. M., Levin, H. S., and Finland, M., 1962, Age and susceptibility of mice to coxsackie A viruses, *J. Exp. Med.* **115:**745–762.
5. Grodums, E. I., and Dempster, G., 1959, The age factor in experimental coxsackie B3 infection, *Can. J. Microbiol.* **5:**595–604.
6. Grodums, E. I., and Dempster, G., 1962, The pathogenesis of coxsackie group B viruses in experimental infection, *Can. J. Microbiol.* **8:**105–113.
7. Khatib, R., Chason, J. L., Silberberg, B. K., and Lerner, A. M., 1980, Age-dependent pathogenicity of group B coxsackieviruses in Swiss-Webster mice: Infectivity for myocardium and pancreas, *J. Infect. Dis.* **141:**394–403.
8. Lerner, A. M., and Shaka, J. A., 1962, Coxsackie A9 myocarditis in adult mice, *Proc. Soc. Exp. Biol. Med.* **111:**804–808.
9. Reyes, M. P., 1977, Illnesses or syndromes associated with coxsackievirus or echovirus infections, in: *Human Health and Disease* (P. L. Altman and D. D. Katz, eds.), pp. 6–8, Federation of American Societies for Experimental Biology, Bethesda.
10. Reyes, M. P., and Lerner, A. M., 1985, Coxsackievirus myocarditis with special reference to acute and chronic effects, *Prog. Cardiovasc. Dis.* **27:**373–394.
11. Duff, D. F., 1981, Viral and bacterial myocarditis, in: *Textbook of Pediatric Infectious Diseases* (R. D. Feigin and J. D. Cherry, eds.), pp. 255–271, W. B. Saunders, Philadelphia.
12. Woodruff, J. F., 1980, Viral myocarditis: A review, *Am. J. Pathol.* **101:**427–479.
13. Fuster, V., Gersh, B. J., and Giuliani, E. R., 1981, The natural history of idiopathic dilated cardiomyopathy, *Am. J. Cardiol.* **47:**525–531.
14. Grist, N. R., Bell, E. J., 1974, A six-year study of coxsackievirus B infections in heart disease, *J. Hyg.* **73:**165–172.
15. Lerner, A. M., Wilson, F. M., and Reyes, M. P., 1975, Enteroviruses and the heart (with special emphasis on the probable role of coxsackieviruses, group B, types 1–5). II. Observations in humans, *Mod. Concepts Cardiovasc. Dis.* **44:**11–15.
16. Bowles, N. E., Olsen, E. G. J., Richardson, P. J., and Archard, L. C., 1986, Detection of coxsackie-B-virus-specific RNA sequences in myocardial biopsy samples from patients with myocarditis and dilated cardiomyopathy, *Lancet* **1:**1120–1123.
17. Lerner, A. M., and Wilson, F. M., 1973, Virus myocardiopathy, *Prog. Med. Virol.* **15:**63–91.
18. Lerner, A. M., 1985, Myocarditis and pericarditis in: *Principles and Practice of Infectious Diseases* (G. L. Mandell, R. G. Douglas, and J. E. Bennett, eds.), pp. 544–551, John Wiley & Sons, New York.
19. El-Khatib, M. R., Chason, J. L., and Lerner, A. M., 1979, Ventricular aneurysms complicating coxsackievirus B types 1 and 4 murine myocarditis, *Circulation* **59:**412–416.
20. Khatib, R., Khatib, G., Chason, J. L., and Lerner, A. M., 1983, Alterations in coxsackievirus B4 heart muscle disease in ICR Swiss mice by anti-thymocyte serum, *J. Gen. Virol.* **64:**231–236.
21. Lerner, A. M., 1969, Coxsackievirus myocardiopathy, *J. Infect. Dis.* **120:**496–499.

22. Woodruff, J. F., and Woodruff, J. J., 1974, Involvement of T lymphocytes in the pathogenesis of coxsackie virus B3 heart disease, *J. Immunol.* **113:**1726–1734.
23. Gatmaitan, B. G., Chason, J. L., and Lerner, A. M., 1970, Augmentation of the virulence of murine coxsackievirus B3 myocardiopathy by exercise, *J. Exp. Med.* **131:**1121–1136.
24. Reyes, M. P., and Lerner, A. M., 1976, Interferon and neutralizing antibody in sera of exercised mice with coxsackievirus B3 myocarditis, *Proc. Soc. Exp. Biol. Med.* **151:**333–338.
25. Gomez, M. P., Reyes, M. P., Smith, F., Ho, L. K., and Lerner, A. M., 1980, Coxsackievirus B3-positive mononuclear leukocytes in peripheral blood of Swiss and athymic mice during infection, *Proc. Soc. Exp. Biol. Med.* **165:**107–113.
26. Reyes, M. P., Lerner, A. M., and Ho, K. L., 1981, Diminution in the size of the thymus in mice during forced swimming, *J. Infect. Dis.* **143:**292.
27. Reyes, M. P., Thomas, J. A., Ho, K. L., Smith, F. E., and Lerner, A. M., 1982, Elevated thymocyte norepinephrine and cyclic guanosine 3′, 5′-monophosphate in T-lymphocytes from exercised mice with coxsackievirus B3 myocarditis, *Biochem. Biophys. Res. Commun.* **109:**704–708.
28. Reyes, M. P., Smith, F. E., and Lerner, A. M., 1984, An enterovirus-induced murine model of an acute dilated-type cardiomyopathy, *Intervirology* **22:**146–155.
29. Wilson, F. M., Miranda, Q. R., Chason, J., and Lerner, A. M., 1969, Residual pathologic changes following murine coxsackie A and B myocarditis, *Am. J. Pathol.* **55:**253–265.
30. Reyes, M. P., Ho, K. L., Smith, F., and Lerner, A. M., 1981, A mouse model of dilated type cardiomyopathy due to coxsackievirus B3, *J. Infect. Dis.* **144:**232–236.
31. Khatib, R., Chason, J. L., and Lerner, A. M., 1982, A mouse model of transmural myocardial necrosis due to coxsackievirus B4, Observations over twelve months, *Intervirology* **18:**197–202.

16

Relationship of Coxsackievirus to Cardiac Autoimmunity

KIRK W. BEISEL and NOEL R. ROSE

1. VIRUSES AND AUTOIMMUNITY

For many years, a relationship between viral infection and the development of autoimmunity has been suspected in diseases such as diabetes mellitus,[1–3] multiple sclerosis,[4,5] idiopathic thrombocytopenia purpura,[6] chronic active hepatitis,[7] postmeasles encephalitis,[8,9] and myocarditis.[10,12] During the past decade, significant strides have been made in developing an understanding of this intricate and involved relationship. Some of the most studied experimental models are encephalomyocarditic virus-induced and coxsackievirus B4 (CVB4)-induced diabetes mellitus in mice, which have been investigated by Notkins and Yoon and colleagues.[1,13–15] Their studies have shown that both the genetics of the host and the infectious agent determine the host's susceptibility to viral infection, the virus-induced pathology, and the development of autoimmune disease. These investigators have also attempted to clarify the relationship of the virus with organ-specific autoimmunity by hybridizing antibody-forming cells of infected mice. Multiple organ-reactive monoclonal antibodies were produced in the wake of a reovirus infec-

KIRK W. BEISEL • Department of Pathology and Laboratory Medicine, Emory University School of Medicine, Atlanta, Georgia 30322. NOEL R. ROSE • Department of Immunology and Infectious Diseases, The Johns Hopkins University, School of Hygiene and Public Health, Baltimore, Maryland 21205.

tion.[16,17] These studies also demonstrated that there are sometimes shared epitopes between the virus and host tissues.

Several theories have been proposed to explain the development of autoimmunity as a sequel to infection.[18–20] The first mechanism assumes that the virus has an indirect effect by producing an adjuvantlike action. Through virus-induced cellular injury and/or necrosis, local inflammation is established. The cellular infiltrate of macrophages and lymphocytes produces monokines and lymphokines. The presence of these soluble factors is sufficient to initiate antigen-specific autoimmune response. The next theory predicts that the virus functions at the afferent limb of the immune response by affecting the antigen or its presentation. One possible mechanism is the exposure of intracellular antigens, caused by cellular necrosis following the viral infection or by immune-mediated cytolysis of the virus-infected cell; another is the presence of a shared or crossreactive antigenic determinant associated with the virus and the target organ or tissue. Virus-induced alterations of host cellular membranes have also been proposed to induce an autoimmune response. These alterations include production of neoantigens, depression of embryonic antigens, exposure of previously hidden or unrecognized determinants, and the loss of tolerogenic antigenic determinants. Direct effects of virus on the cells of the immune system can cause perturbations of immunoregulatory circuits, leading to development of autoimmunity. Another probability is the development of anti-idiotypic antibodies to virus antibodies that might react with the virus-specific or cell surface.[20] Multiple autoimmune mechanisms may be implicated to participate in concert in the development of autoimmune-mediated pathogenesis. Therefore, each theory explaining the development of postinfectious autoimmunity does not necessarily imply that it is the exclusive cause of disease.

2. CLINICAL EVIDENCE FOR COXSACKIEVIRUS AND AUTOIMMUNITY IN HEART DISEASE

Many different viruses have been implicated in the induction of myocarditis and cardiomyopathies.[21] Coxsackieviruses (CV) are the most common of these causative viral agents of heart disease. Several extensive reviews[21–24] have documented patients with a clinical history of a previous viral illness. Unfortunately, these reports represent isolated case histories. However, the connection between group B coxsackieviruses (CB5) and myocarditis was firmly established by Helin *et al.*[25] These workers reported the clinical findings of 18 patients who developed myopericarditis during a CVB5 epidemic in Finland that occurred during the autumn of 1965. Of the 18 patients, CVB5 was isolated from feces in 12 cases. Epidemiologic analyses suggest that during a CV epidemic about one half the population will be infected, with approximately 49% of cases being subclinical. Since only 5% of the symptomatic individuals exhibit any cardiac problems, we suggest that viral myocarditis is a reflection of the particular genetic constitution of these individuals. Furthermore, only a proportion of these individuals may have autoimmune heart disease.

Evidence of autoimmunity has been reported in a number of disorders in which there is injury to the heart muscle.[26,27] Immunologic factors have been described in patients with postpericardiotomy and postinfarction syndromes. The presence of inflammatory cells within the myocardium, circulating antiheart antibodies, and immune complexes are the immunologic hallmarks of these two syndromes. These autoantibodies to the heart produce both sarcolemmal/subsarcolemmal and cross-striated patterns in indirect immunofluorescence tests.[26,27] Elevated titers of antiactin and antimyosin antibodies correlated with the occurrence of disease.[28] Similar findings have been described by Maisch *et al.*[11,12,27] in their investigations of viral myocarditis and the subsequent primary and secondary cardiomyopathies. These data may suggest that different myocardial antigens are the target of autoimmune responses in patients with cardiac injury. Both cellular and humoral immunopathologic responses have been demonstrated. Heart-reactive antibodies can bind complement and produce lysis of rat myocytes. In patients with viral myocarditis, these antibodies can be absorbed by the respective viral agent such as CV. Other investigators have identified immune complexes in the interstitial spaces of the myocardium.[29] In addition, mitochondrial autoantibodies to the adenine nucleotide translocator protein have been correlated with decreased ejection fraction in dilated cardiomyopathies.[30] Thus, both cell membrane and intracellular antigens have been targeted by the antimyocardial immune response. Autoantibodies to cardiac antigens are indicative of autoimmune myocarditis and may provide valuable evidence of the nature of the autoantigen(s) recognized by the autoimmune response.

3. EXPERIMENTAL STUDIES OF VIRUS-ASSOCIATED AUTOIMMUNE MYOCARDITIS

Initially, Woodruff and subsequently Huber and associates[21,31–34] investigated in detail the immunopathic mechanism(s) in viral myocarditis in BALB/cCUM mice infected with CVB3. Their observations, primarily confined to 7–10 days after infection in young adult mice, have suggested that cell-mediated immunity plays a prominent role in the pathogenesis of CB3-induced myocarditis. Adoptive transfer experiments have demonstrated that Thy 1^+, Lyt 1^-, and Lyt 2^+ lymphocytes (i.e., cytotoxic/suppressor lymphocyte population) were responsible for enhancing the observed myocardial pathology. Besides the expected presence of virus-specific cytotoxic T lymphocytes (CTL), a second CTL population was identified. This population reacted with noninfected myocytes, suggesting that cell-mediated autoimmune mechanisms play a role in cardiac pathology in BALB/c male mice.[32] The lesser pathology in BALB/c females was attributed to the influence of suppressor T cells.[33] Another pathogenic mechanism identified in DBA/2 mice is complement-mediated myocyte injury.[34] Humoral autoimmunity in heart disease has also been implicated by the presence of immune complexes in the focal myocardial lesions. In summary, several immune effector mechanisms may play a role in cardiac pathogenesis.

For the past 5 years, we have studied the genetic role of host susceptibility to virus-induced heart disease. Our premise was that the wide diversity of susceptibility to CV infection and the resulting myocardial disease observed during a CV epidemic is a reflection of genetic diversity in the host population. By identification of the genetic factors regulating the susceptibility to CV infection and virus-induced myocarditis, the mechanisms of cardiac pathogenesis can be analyzed separately.

A number of reasons compelled us to turn to a murine model of CV-induced myocarditis. Several laboratories[21,31–36] had already shown that infection of mice with a myocarditogenic strain of CVB3 parallels in many respects the heart disease observed in patients with a suspected viral myocarditis and/or cardiomyopathy. The advantage of using mice is the availability of a large number of inbred, congenic, and recombinant strains and the extensive genetic knowledge compiled for these mouse strains. In addition, the murine immune system and its genetic control are quite similar to those of humans. A panel of mouse strains was identified and used for subsequent studies. This review reflects our current insights into the development of virus-induced autoimmune myocarditis obtained by using this approach.

In collaboration with L. J. Wolfgram, six mouse strains were selected for a detailed study to determine the response of 2-week-old mice.[51] These animals were infected with 10^5 $TCID_{50}$ of CVB3 and examined 2, 3, 5, 7, 9, 15, 21, and 45 days postinfection. The strains used were A.BY/SnJ (*H-2*^{b}), A.CA/SnJ (*H-2*^{f}), A.SW/SnJ (*H-2*^{s}), B10.S/SgSf (*H-2*^{s}), B10.PL/SgSf (*H-2*^{u}), and C3H.NB/SnJ (*H-2*^{p}). These strains were selected since they differed in their *Mhc* genes, non-*Mhc* genes (i.e., background genome) and in the severity of CB3-induced myocarditis.[37] Analyses of the cardiac pathology revealed that two

TABLE I
Summary of the Two Phases of Cardiac Pathogenesis in Six Inbred Mouse Strains

		Myocarditis[a]				
		Early, focal		Late, diffuse		
Strain	Mhc haplotype	Prevalence[b]	Severity[c]	Interstitial infiltrate	Prevalence	Severity
A.BY/SnJ	*b*	100 (17)	1.9	Present	94 (16)	1.8
A.SW/SnJ	*s*	95 (22)	1.8	Present	85 (20)	1.6
A.CA/SnJ	*f*	36 (11)	0.1	Present	74 (19)	0.8
C3H.NB/SnJ	*p*	59 (17)	1.1	Present	76 (21)	1.2
B10.S/SgSf	*s*	75 (20)	0.6	Absent	41 (22)	0.2
B10.PL/SgSf	*u*	68 (21)	0.5	Absent	52 (21)	0.4

[a]From Wolfgram.[51]
[b]Prevalence is expressed as the percentage of animals with disease. Figures in parentheses are number of animals tested.
[c]Severity was graded on a scale of 0–4.[37]

phases of myocardial disease were separable both by the time of onset and duration and by the histologic appearance of heart (Table I). The first appearance of pathology was noted at day 3 and became apparent 2 days later.[37] The pathologic changes were characteristic in that the myocytes were hypereosinophilic and had contraction band necrosis.[38] Few or no inflammatory cells were present within the myocardium at this time, but by days 5 and 7 after infection inflammatory cells were present. Both the injured myocytes and the inflammatory cells were largely confined to focal sites within the myocardium. Vascular and perivascular inflammation were rarely seen and thus were not prominent abnormalities. There was no qualitative difference in the pathology observed among the six strains examined. However, quantitative differences were found, in that the A.BY/SnJ and A.SW/SnJ strains had the most severe disease with 5–10% of the myocardium being affected. The B10.S/SgSf, B10.PL/SgSf, and C3H.NB/SnJ strains had involvement of 1–5% of the myocardium, whereas the A.CA/SnJ strain displayed minimal disease (<1% of the myocardium) with only a few myocytes being injured.[38] This phase of disease was presumably due to virus-induced myocyte injury and necrosis and correlated with the greatest content of virus in the heart.[37,38] These data further demonstrated that early disease was influenced by multiple genes, which are both *Mhc* and non-*Mhc* associated.[39]

A second, distinct phase of myocarditis became evident at day 9 and peaked in severity 2–3 weeks after infection (Table I). This second phase of myocardial pathology was marked by a diffuse disease, with the presence of an interstitial mononuclear infiltrate.[37,38] The continuing or progressive form of myocarditis was present in the A.BY/SnJ, A.CA/SnJ, A.SW/SnJ, and C3H.NB/SnJ strains (Table I). Unlike the early disease, no quantitative differences were found among these four strains. The late pathology in the strains described above contrasted with that observed in the B10.S/SgSf and B10.PL/SgSf strains, which had healing focal lesions with the deposition of fibrotic tissue. The genetic control of the late disease differed from the early myocarditis in that non-*Mhc* gene(s) determined the predisposition to the development of autoimmune myocarditis.[37,39]

In parallel with these pathologic studies, immunohistochemical studies were performed in collaboration with A. Herskowitz, M. D. Traystman, and W. E. Beschorner (unpublished data) to examine the phenotypes of the cellular infiltrate present during the second or late phase of myocarditis. In the early disease (day 7 postinfection), the cellular infiltrate consisted primarily of polymorphonuclear neutrophils (PMN) with occasional monocytes and both B and T lymphocytes. Minimal deposits of immunoglobulin were present and confined primarily to the area of the focal lesions. The A/J and C57BL/6J strains were similar with respect to the composition of the cellular infiltrate in the heart. In the late phase of myocarditis, the diffuse interstitial cellular infiltrate was present only in the A/J hearts. The vast majority of cells in this infiltrate consisted of monocytes/macrophages, as identified by Mac-1 and Mac-3 monoclonal antibodies. T helper lymphocytes ($L3T4^+$, Lyt 2^-) were prominent at the interface between the focal lesion and normal myocardium, and occasional cells were scattered throughout the diffuse interstitial infiltrate. Low numbers of B lymphocytes and plasma cells were also found throughout. Only an occasional PMN was identified, and cytotoxic/suppressor T lymphocytes ($L3T4^-$, Lyt 2^+) were rarely seen. A striking feature was the presence of heavy deposits of immunoglobulin in both the interstitial spaces and the focal lesions. This immunophenotypic pattern of the chronic disease was present only in animals that developed the autoimmune disease. The hearts of the CVB3-infected C57BL/6J animals did not display these characteristics.

The duration and extent of viremia and CB3 content of the heart, spleen, thymus, and pancreas were studied (Wolfgram[51]) (Table II). The peak of viremia occurred 2–3 days postinfection. Similar peaks of virus were observed in spleen, thymus, and pancreas, with the pancreas having the highest concentration of virus, i.e., 6–9 $\log_{10}$ $TCID_{50}$/g of tissue. Unlike the other organs, peak CVB3 content of the heart occurred 5 days postinfection. For the most part, detectable virus persisted in the blood and spleen less than 1 week postinfection, with the blood being cleared the most rapidly. The rate of clearance of virus from the

TABLE II
Serum and Organ CVB3 Viral Titers[a]

Source	Peak day[b]	Duration[c]	Titer on peak day[d]	Prevalence[e]						
				2	3	5	7	9	15	21
Serum										
A.BY/SnJ	3	3	3.9 ± 0.2	X	X					
A. SW/SNj	3	3	4.1 ± 0.2	X	X					
A.CA/SnJ	2	3	1.6 ± 0.2	X						
B10.S/SgSf	2	3	2.6 ± 0.3	X						
B10.PL/SgSf	3	3	1.7 ± 0.5	X	X					
C3H.NB/SnJ	2	3	2.1 ± 0.2	X	X					
Heart										
A.BY/SnJ	5	9	3.5 ± 0.2	X	X	X	X			
A.SW/SnJ	5	9	3.2 ± 0.2	X	X	X	X	X		
A.CA/SnJ	5	9	3.1 ± 0.2	X	X	X	X	X		
B10.S/SgSf	5	7	3.4 ± 0.3	X	X	X	X			
B10.PL/SgSf	3	9	3.2 ± 0.3	X	X	X	X	X		
C3H.NB/SnJ	5	9	3.3 ± 0.2	X	X	X	X	X		
Pancreas										
A.BY/SnJ	3	15	8.0 ± 0.2	X	X	X	X	X	X	X
A.SW/SnJ	2	15	6.4 ± 0.2	X	X	X	X	X	X	
A.CA/SnJ	2	15	5.2 ± 0.2	X	X	X	X	X	X	X
B10.S/SgSf	2	15	6.0 ± 0.1	X	X	X	X	X	X	X
B10.PL/SgSf	3	15	7.2 ± 0.3	X	X	X	X	X	X	
C3H.NB/SnJ	2	15	6.6 ± 0.2	X	X	X	X	X		
Spleen										
A.BY/SnJ	3	7	5.4 ± 0.4	X	X	X	X			
A.SW/SnJ	2	5	5.9 ± 0.3	X	X					
A.CA/SnJ	2	3	3.5 ± 0.7	X	X					
B10.S/SgSf	2	5	3.5 ± 0.8	X	X	X				
B10.PL/SgSf	3	5	3.6 ± 0.7	X	X					
C3H.NB/SnJ	2	5	2.4 ± 0.2	X	X	X				
Thymus										
A.BY/SnJ	3	9	5.8 ± 0.2	X	X	X	X	X		
A.SW/SnJ	3	9	4.8 ± 0.8	X	X					
A.CA/SnJ	3	7	3.9 ± 0.3	X	X	X				
B10.S/SgSf	3	7	4.3 ± 0.4	X	X	X				
B10.PL/SgSf	3	9	4.0 ± 0.3	X	X	X	X	X		
C3H.NB/SnJ	3	5	2.8 ± 0.3	X	X	X				

[a]From Wolfgram.[51]
[b]Day after infection with maximal viral titer detected.
[c]Last day when infectious virus was detectable at any level.
[d]Mean $\log_{10}$ $TCID_{50}/g$ ±S.E.
[e]Isolation rate ≥50% is marked by X.

heart and thymus was virtually identical and occurred at 1–2 weeks postinfection. The pancreas was the only organ in which infectious virus was found up to 3 weeks postinfection. Susceptible strains could be identified by their higher titers of virus in the blood, spleen, and thymus. The B10.PL/SgSf strain exhibited the greatest concentration of infectious CVB3 in the pancreas. Two points should be noted from these observations. The heart showed a delay in viral replication compared with the other organs examined. This finding suggests that myocyte infection may be different from infection of the other organs. Second, the virus persists much longer in the pancreas than in other organs. By allowing the virus to persist, the pancreas may act as a source for seeding other organs for further infection.

Differences in the time of appearance of neutralizing antibodies were also found among the strains of mice examined.[37] In the strains susceptible to CVB3 infection, antibodies were not detected until day 5 of infection, whereas the resistant strains had neutralizing antibodies in their serum at day 3 (Table III). In general, the A/J and C3H series of congenic strains had higher titers of neutralizing antibodies than did the B10 and BALB/c *H-2* congenic strains. The neutralizing antibodies peaked approximately 3 weeks after infection. Other than the difference in the time of appearance of the anti-CB3 antibodies, the titer of neutralizing antibodies was not useful in distinguishing susceptibility from resistance to viral infection.

Heart-specific antibodies were found only in the serum of animals that developed the late phase of myocarditis.[40] Antibodies localizing on the sarcolemma/subsarcolemma and on the contractile proteins in the myofibers were observed. Absorption studies, using lyophilized muscle homogenates. demonstrated that some of the autoantibodies were heart specific, whereas some crossreacted with skeletal muscle. Autoantibodies to muscle tissue were first observed starting 9 days postinfection and

TABLE III[a]
Neutralizing Antibody Titer[b] **at 3, 21, and 45 Days Postinfection**

Strain	Haplotype	Day 3	Day 21	Day 45
A.BY	*b*	0 (7)	12.6 ± 0.6 (8)[c]	10.0 ± 1.2 (4)
A.SW	*s*	0 (12)	12.1 ± 0.8 (10)	10.2 ± 0.3 (10)
A.CA	*f*	4.6 ± 0.5 (7)	10.2 ± 0.6 (11)	10.7 ± 1.6 (8)
C3H.NB	*p*	4.3 ± 0.6 (10)	7.6 ± 0.6 (12)	9.1 ± 0.8 (11)
B10.S	*s*	2.3 ± 0.5 (10)	8.6 ± 0.8 (12)	7.3 ± 0.3 (10)
B10.PL	*u*	2.2 ± 0.6 (9)	10.1 ± 0.9 (10)	7.4 ± 0.8 (10)

[a]Adapted from Wolfgram *et al.*[37]
[b]Expressed as the $\log_2$ mean neutralizing antibody titer ±SE.
[c]Number of animals tested.

TABLE IV
Strain Differences in the Production of Heart-Specific Autoantibodies and Severity of Cardiac Lesions[a]

			Heart-specific autoantibodies			
		Total animals	Present		Absent	
Strain	Haplotype	(*N*)	*N*	PI[b]	*N*	PI
A.BY	*b*	20	4	1.9 ± 0.7	16	1.7 ± 0.4
A.SW	*s*	30	15	1.9 ± 0.3	15	1.0 ± 0.5
A.CA	*f*	32	10	1.2 ± 0.3	22	0.4 ± 0.3
C3H.NB	*p*	33	5	1.8 ± 0.8	28	0.9 ± 0.3
B10.S	*s*	31	0	0	31	0.3 ± 0.1
B10.PL	*u*	30	0	0	30	0.3 ± 0.1

[a]Adapted from Wolfgram *et al.*[37]
[b]The pathologic index (PI) is a mean score ±SE, using a scale of 0–4+. Pathology scores were given by two independent observers.

remained elevated for many weeks. We then examined the association of the late stage of disease with the pattern of autoantibody.[37] When animals with substantial myocardial lesions (i.e., pathology scores >0.5) were considered, animals with heart-specific autoantibodies had a greater prevalence of disease than did those with the crossreactive antibodies ($p < 0.001$). The third group, which had antibodies to skeletal muscle only, had no significant pathology ($p < 0.001$). These data suggest that the presence of autoantibodies specific for heart tissue correlates with severe myocarditis. Thus, we have used the presence of heart-specific autoantibodies as an indicator of the severity of myocardial lesions and the presence of autoimmune disease.

We then evaluated the correlation between the severity of late-stage myocarditis and the presence of heart-specific autoantibodies. There were strain differences in severity of disease and heart-specific antibody production (Table IV). All four strains that developed heart-specific antibodies—A.BY/SnJ, A.SW/SnJ, A.CA/SnJ, and C3H.NB/SnJ—had significantly greater heart disease than did the two that did not; i.e., the B10.S/SgSf and B10.PL/SgSf strains ($p < 0.001$). A further comparison of the severity of myocardial disease was carried out among those mice that had heart-specific IgG antoantibodies and those that did not. Of the four strains that produced heart-specific IgG autoantibodies, the A.SW/SnJ and A.CA/SnJ animals with these antibodies had significantly greater severity of autoimmune myocarditis than animals not producing these autoantibodies ($p < 0.05$). However, no difference was observed in severity of disease in the animals of the A.BY/SnJ and C3H.NB/SnJ strains with or without detectable heart-specific IgG antibodies. The

C3H.NB/SnJ animals with heart-specific antibodies did show a trend toward increased severity of myocarditis compared to those animals that lacked these autoantibodies. In A.BY/SnJ animals, no difference was observed between the two groups. Since the A.BY/SnJ strain is a poor producer of heart-specific autoantibody,[37] the apparent lack of correlation may be due to an inability to detect the heart-specific autoantibodies in many animals. The heart-specific autoantibody levels may have been below the limits of detection by the indirect immunofluorescence assay and/or the heart itself may have removed much of the circulating heart-specific autoantibody. Direct immunofluorescence and immunohistochemistry studies using hearts from infected animals have shown that immunoglobulin deposits are prominent in the heart tissue of these animals. No differences were found in the severity of myocarditis between animals that produced the skeletal muscle-reactive autoantibodies and animals without muscle autoantibodies. These data again demonstrate that the heart-specific autoantibodies serve as indicators for the severity of disease. The correlation with severity of myocarditis was stronger in animals of the same strain. The presence of heart-specific autoantibodies is therefore a reflection of the genetic predisposition of an individual to develop autoimmune myocarditis. Even if they are not the direct cause of the pathologic changes, the antibodies may play an auxilliary role.

In the initial observations of Wolfgram *et al.*,[37] variations in prevalence of the heart-specific autoantibody response were noted. Experiments were undertaken to extend these observations by determining the kinetics of the influence of sex as well as of age on the autoantibody response in mouse strains which are susceptible to autoimmune myocarditis (F. L. Alvarez and K. W. Beisel, unpublished data). Four susceptible strains of mice were examined for a period of 5 weeks, starting 12 days postinfection. Animals were bled three times a week, and the serum of each individual animal was examined by indirect immunofluorescence for heart-specific autoantibodies. Variations in the prevalence and titer of heart-specific antibodies were observed among the A.BY/SnJ ($H\text{-}2^b$), A/J ($H\text{-}2^a$), A.CA/SnJ ($H\text{-}2^f$), and A.SW/SnJ ($H\text{-}2^s$) strains (Fig. 1). An initial peak of autoantibody occurred 17–21 days postinfection in all four strains. In the A.CA/SnJ strain, two additional peaks were observed on days 29 and 45 postinfection, ~10–14 days apart. A/J and A.SW/SnJ also exhibited additional peaks of autoantibody at 29 and 45 days, respectively. Only the A.BY/SnJ strain showed no evidence of oscillating patterns in myocardial autoantibody production. Individual animals within each of these strains also exhibited these fluctuations in the heart-specific autoantibodies. Two points can be drawn from these observations. First, the differences among these four lines are a reflection of an

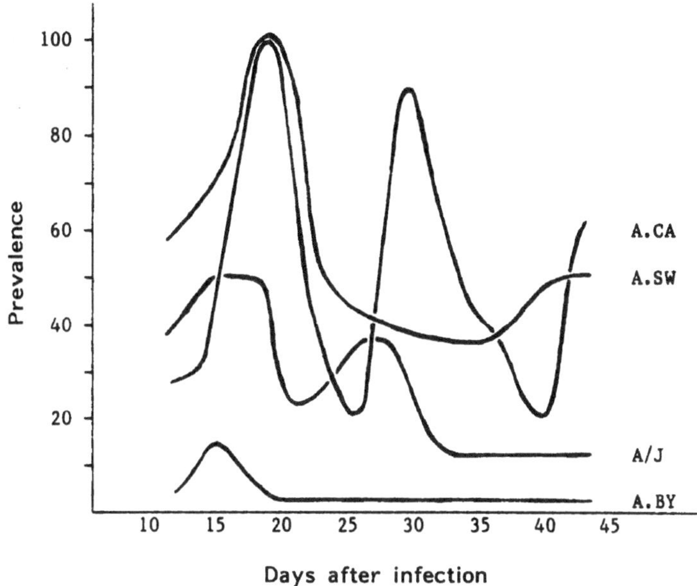

FIGURE 1. Strain differences in myocardial autoantibodies. The autoantibody production is expressed as the prevalence of the heart-specific antibodies in each strain examined over a period of 45 days postinfection. A.CA/SnJ, A.SW/SnJ, A.BY/SnJ, and A/J mouse strains were infected with 10^5 $TCID_{50}$ CVB3 during this time-course study.

Mhc influence on the heart-specific autoantibody response, since these mouse lines are *H-2*-congenic strains and thus differ only at their respective *H-2* complex. The second is that there is some, probably immunoregulatory, process involved in the heart-specific autoimmune response to account for the oscillating pattern of autoantibody levels in the serum.

The incidence of human myocarditis by coxsackievirus infection is well established to be age related. In the murine model of viral myocarditis, the extent of disease induced has also been shown to be age related, where severity of disease decreases with age.[21,35] In experiments performed by Alvarez and Beisel (unpublished data), a relationship with age was also observed in both the humoral response against cardiac antigens and the virus-induced autoimmune pathology. Three groups of A.SW/SnJ mice were studied for the prevalence of autoantibodies using 14-, 21-, and 120-day-old animals. The youngest animals developed the highest incidence of heart-specific autoantibodies with two peaks of antibody production (Fig. 2). The 3-week-old mice had lower responses, as indicated by a diminished prevalence of autoantibody at days 17–21 postinfection and the apparent absence of the second peak. The adult mice did not produce detectable levels of autoantibody. The severity of myocardial disease paralleled the cardiac-specific humoral

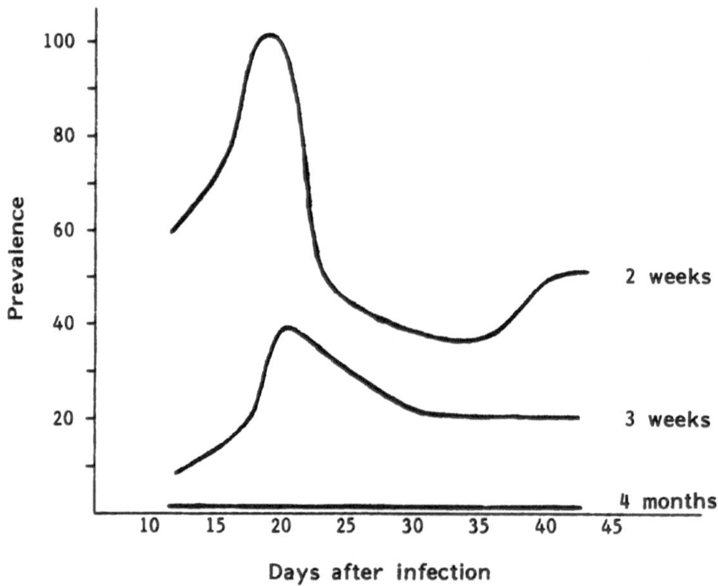

FIGURE 2. Effect of age on myocardial autoantibodies. Three different age groups of A.SW mice were analyzed for the presence and prevalence of heart-specific autoantibodies in their serum. The age groups examined were 2 weeks, 3 weeks, and 4 months old at the time of infection with 10^5 $TCID_{50}$ CVB3. Sera were collected three times per week, for a period of 35 days.

response. Regardless of the age or strain of animals being examined, we have not observed any relationship between the sex of the animal and the autoantibody response or cardiac pathology.

Speculation has been raised that viral myocarditis leads to the development of cardiomyopathy. Evidence from the experiments of Matsumori and Kawai,[41] using EMC-virus-infected DBA/2 mice, and of Reyes *et al.*,[42] using exercised CVB3-infected Swiss ICR mice, have demonstrated that these infected animals develop cardiomyopathy. The relationship between virus-induced autoimmune myocarditis and cardiomyopathy has not been studied. Twenty-one 4-month-old adult mice were infected with CVB3 and then examined 30 days later. The hearts from these A.SW/SnJ, A.BY/SnJ, and A/J animals were arrested in diastole and then measured for left ventricular (LV) wall thickness and the cross-sectional areas of the LV chamber and cavity. Six of the animals from these three strains exhibited myocardial hypertrophy with variable LV area-to-cavity ratios. This variability reflected differences in the reduction of the cavity size. Three of the 21 animals showed extreme thinning of the LV walls and dilation of the LV cavity. This finding was obvious on gross inspection and was clearly demonstrable by a significant reduction in LV area-to-cavity ratio. Histologic abnormalities were also observed in 20 of 21 animals. Inflammatory lesions were mild and confined to scattered interstitial mononuclear infiltrates with rare foci of small organizing fibrous lesions. The striking abnormality seen in light microscopy was the presence of degenerative myocytes showing myocytolytic injury. These myocytotoxic changes were present in 19 of 21 animals and were found in all animals that had hypertrophied or dilated hearts. These injured myocytes showed dissolution of the contractile elements with only the nuclei and sacrolemma remaining intact, giving these cells a ghostlike appearance. These histologic abnormalities have not been detected in the 2- and 3-week-old animals. The pathologic histology in 4-month-old animals is similar to that observed in patients with chronic cardiomyopathy. These observations (A. Herskowitz, F. L. Alvarez, and K. W. Beisel, unpublished data) suggest that this animal model can be useful in studying the relationship between autoimmune heart disease and cardiomyopathy.

The studies described above have demonstrated that certain mouse strains develop a late stage of progressive myocardial disease. Chronic myocarditis is characterized by a diffuse interstitial cellular infiltrate of macrophages, monocytes, and lymphocytes as well as by the presence of heart-specific autoantibodies. Our working hypothesis is that the CVB3 infection results in the concomitant inflammation and release of myocardial antigens. This combination thereby establishes the conditions necessary for development of cardiac autoimmunity. Immunochemical studies were initiated in an effort to determine the target of the heart-specific autoimmune response.[43] A multifaceted approach was taken to identify the pertinent autoantigen(s). The first approach was to use serum pools taken during the early (E) virus-induced disease (i.e., days 5 and 7) and during the later phase (L) of myocarditis (i.e., days 15 and 21). Also, these serum pools were made from individual mouse strains previously identified as predisposed (P) or nonpredisposed (NP) to the autoimmune heart disease. When these four serum pools were tested by indirect immunofluorescence, only the LP pool showed cardiac-specific autoantibodies. The LNP localized on skeletal muscle. These four serum pools were then tested against extracts of whole heart and skeletal muscle that had been run under reducing conditions on 5–17% gradient polyacrylamide sodium dodecyl sulfate (PAGE-SDS) gels and transferred to nitrocellulose.[43] The early serum pools recognized approximately 11 bands in both muscle preparations with the most prominent band being ~200,000 M_r. This complex immunoblot pattern was similar for both the EP and ENP pools. An evident change was observed in the banding pattern with the late pools. Both the LNP and LP pools lost most minor bands, so that five to six bands were apparent. The straining of the 200,000-M_r peptide, the major band, was diminished in the LNP serum. By contrast, the reactivity in the LP pool showed a more intense reaction with the 200,000-M_r band.[43] These data suggested that a major target of the autoimmune response was a 200,000-M_r protein(s). The molecular weight of this protein was the first suggestion that CVB3-infected mouse sera were recognizing the heavy chain of myosin.

In the whole-muscle preparation, the contractile proteins are dominant. Approximately 50% of the protein content of muscle tissue is composed of contractile proteins. Therefore, a sequential fractionation of the muscle tissue was done to enrich for antigens other than for the contractile proteins. Initially, a six-step fraction procedure used a variety of salt concentrations and various detergents.[43] We later found that a three-step sequential extraction, consisting of a low-ionic-strength buffer, a high-salt buffer with pyrophosphate, and an SDS detergent buffer, was sufficient. Comparison of a normal mouse serum pool with the A.BY/SnJ LP pool suggested that the major reaction was directed to proteins in the high-salt and detergent extracts (Fig. 3). The immunoblots of the heart autoantigens, using the LP pool from CVB3-infected A.BY/SnJ mice, showed a strong reaction against a 200,000-M_r protein(s). This band was present in both the high-salt and detergent preparations, and its concentration and relative molecular weight suggested

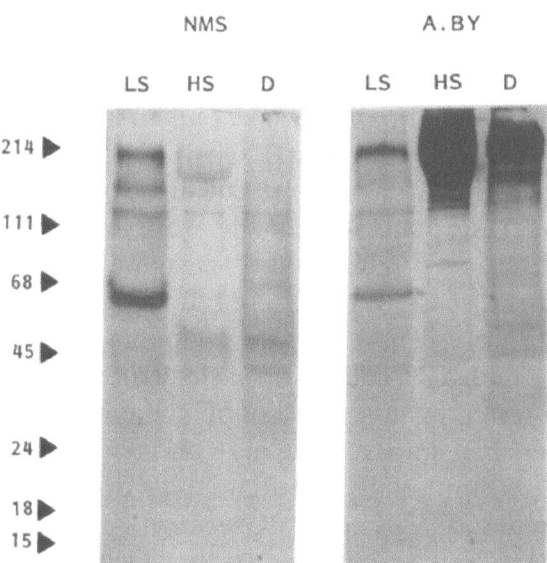

FIGURE 3. Analyses of the reactions of heart-specific autoantibodies with sequential differential extracts from mouse heart tissue. Low-salt (LS), high-salt with pyrophosphate (HS), and SDS detergent (D) extracts were run on 7.5% polyacrylamide gels. Western immunoblotting was performed on these extracts, using pooled samples of normal mouse sera (NMS) and day 21, CVB3-infected A.BY sera. The developing antibody was a horseradish peroxidase-labeled goat antimouse IgG-specific reagent.

TABLE V
Myosin Absorption Analysis of Postinfection Autoimmune Sera[a,b]

	Reactivity						
	Immunofluorescence		Western immunoblot		Myosin ELISA		
Absorbent	Heart	Skeletal	Heart	Skeletal	Heart	Skeletal	Brain
—	+	+	+	+	+	+	+
BSA	+	+	+	+	+	+	+
Cardiac myosin	–	–	–	–	–	–	–
Skeletal myosin	+	–	+	–	+	–	–
Brain myosin	+	–	ND	ND	+	±	–

[a]Adapted from Alvarez *et al.*[43] and Neu *et al.*[44]
[b]BSA, bovine serum albumin; ELISA, enzyme-linked immunosorbent assay.

that it was the heavy chain of myosin. Minor bands were detected and ranged from 120,000 to 195,000 M_r, with the most prominent bands observed at 120,000 and 150,000 M_r. An additional band was observed at 82,000 M_r. These bands were always found in preparations and may represent the breakdown products of myosin. The heavy chain of myosin can be broken down into heavy meromyosin (HMM), head (S1), rod, light meromyosin (LMM), and S2 fragments by limited protolytic digestions. The relative molecular weights of these myosin heavy-chain fragments are 150, 120, 82, 50, and 32, respectively. These findings suggest that the autoantibodies to myosin recognize the HMM, head, and rod portions of this molecule.

From these observations, cardiac myosin was recognized as the most prominent antigen in the CVB3-induced heart-specific autoimmune response. In order to confirm this possibility, a series of experiments were done with N. Neu and S. Craig (Table V). Absorption analysis showed that the antibody reactivity, as determined by immunofluorescence and Western immunoblotting, is directed toward cardiac myosin.[43] Purified cardiac myosin bound to Sepharose 4B beads removed all antibody activity toward heart and skeletal muscle. Even the reactivity to the minor bands, observed in the immunoblots, was removed by absorption. The anticardiac myosin autoantibodies were then eluted from cardiac myosin-bound beads. All the reactivity toward cardiac and skeletal muscle was restored. The previous immunoblotting experiments had indicated that there was enhanced activity toward cardiac myosin compared to the skeletal muscle myosin. An enzyme-linked immunosorbent assay (ELISA) system was developed to identify the presence of cardiac myosin-specific autoantibodies.[44] Pooled serum from infected mice, CB3-

infected B10.PL/SgSf (nonsusceptible to autoimmune myocarditis) and A.CA/SnJ (susceptible), taken 17–21 days postinfection, and from uninfected mice was examined for IgM and IgG antibodies to cardiac, skeletal muscle, and brain isoforms of myosin. Myosin antibodies were present in all three serum pools, with the A.CA/SnJ pool having a higher concentration of myosin antibodies. Comparing isotypic differences in these myosin antibodies, similar antimyosin IgM titers were observed in all three strains. The IgG levels to myosin were significantly greater in the A.CA/SnJ pool than in the B10.PL/SgSf pools and NMS pools. No difference was observed between the B10.PL/SgSf pool and normal mouse serum (NMS). Only the IgG autoantibodies to cardiac myosin distinguished these three sera. Since the immunochemical data of Alvarez *et al.*[43] suggested a preferential reactivity toward cardiac myosin over that of skeletal muscle myosin, competitive inhibition and absorption analyses of the three serum pools were carried out using an ELISA system.[44] The immunoglobulin G (IgG) activity against cardiac myosin in the A.CA/SnJ and A.SW/SnJ strains could be completely inhibited by murine cardiac myosin, whereas the maximal inhibition achieved by skeletal muscle and brain myosin was 74% and 68%, respectively, of the total anticardiac myosin reactivity. Approximately eight times the concentration of skeletal muscle myosin was necessary to obtain a 50% reduction in antibody binding. The concentration of brain myosin needed for a 50% reduction of activity was 80 times greater than that of cardiac myosin. When skeletal muscle myosin was used as the target, all three myosin isoforms were able to inhibit the reaction toward skeletal muscle myosin. The concentrations of heart and skeletal myosins needed to inhibit the skeletal myosin activity were the same; however, approximately 10 times the amount of brain myosin was needed. Similar results were obtained in absorption experiments using various myosin isoforms bound to Sepharose beads. When serum pools from uninfected and infected B10.PL/SgSf were absorbed, antimyosin activity was completely removed both by cardiac myosin and by skeletal muscle myosin. However, absorption with brain myosin was only able to reduce the reactivity with cardiac and skeletal muscle myosin. These data demonstrate that a cardiac-specific IgG present in the serum of A strain mice is a distinguishing characteristic of autoimmune myocarditis. Even though crossreactive myosin IgM antibodies are a common finding in normal and infected individuals, the cardiac myosin-specific IgG autoantibodies are an indicator or marker of disease.

Since cardiac myosin autoantibodies were a prominent feature in CVB3-induced autoimmune myocarditis and appear to be a marker for disease, we predicted that cardiac myosin can induce myocarditis in those mouse strains that are susceptible to the late phase of myocardial

TABLE VI
Induction of Myocarditis by Immunization with Cardiac Myosin[a]

Strain	Immunogen	Myocardial pathology	
		Prevalence	Severity
A.SW/SnJ	Cardiac myosin	100 (11)[b]	2.7[c]
	Skeletal myosin	0 (11)	0
	Pyrophosphate only	0 (5)	0
C57BL/6J	Cardiac myosin	0 (15)	0
	Skeletal myosin	0 (12)	0
	Pyrophosphate only	0 (5)	0

[a]Adapted from Neu *et al.*[45]
[b]Figures in parentheses are numbers of animals tested.
[c]Severity grafted on a scale of 0–4 (see ref. 37).

disease. To test this hypothesis, animals from the A.SW/SnJ and C57BL/6J mouse strains were immunized at days 0 and 7 with either purified cardiac or skeletal muscle myosin in complete Freund's adjuvant.[45,46] Myocarditis was present in only those susceptible animals that had received a total of 200 μg myosin (Table VI). In many respects, the cardiac pathology paralleled the characteristics of the CVB3-induced autoimmune disease.

This series of experiments has determined that a normal intracellular protein, myosin, can act as an autoantigen. The role of CB3 in the induction of the autoimmune disease could be either direct or indirect. A direct relationship between CV and autoantibodies to heart antigens has been suggested from the work of Maisch *et al.*[11,12] The autoantibodies to myolemmal antigen were absorbed by viral capsid proteins. Recently, Beisel *et al.*[47] demonstrated that a monoclonal antibody that neutralizes CVB4 also has a prominent crossreactivity with cardiac myosin and reacts weakly with actin and tropomyosin. A similar phenomenon has been demonstrated to occur in *Streptococcus*-induced rheumatic fever with a common antigen shared between the streptococcal M moiety and cardiac myosin.[48,49] An alternative theory is that the virus-induced anticardiac myosin is a result of the release or exposure of intracellular autoantigens as a result of tissue destruction. To explore the viability of an indirect or direct relationship between an antigenic epitope on cardiac myosin and CVB3, we tested the cardiac myosin autoantibodies for crossreactivity with virus. A serum pool from CVB3-infected A.CA/SnJ mice was absorbed with cardiac myosin. The unabsorbed serum pool, myosin-absorbed serum, and eluate were tested for their reaction with cardiac myosin and CVB3 virions.[50] There was no

TABLE VII
Analysis of Crossreactivity between CVB3 and Cardiac Myosin (Cm) Antibodies[a]

		Treatment of mice				
		CVB3 infection			Cm immunization (unabsorbed serum)	None (unabsorbed serum)
Reactivity	Antigen	Unabsorbed serum	Cm-absorbed serum	Cm Ab		
Immunodot blots	Cm	+	−	+	+	−
	CVB3	+	+	−	−	−
Western immunoblot	Cm	+	−	+	−	−
	CVB3	+	+	−	−	−
CVB3 neutralization	CVB3	+	+	−	−	−

[a]Adapted from Neu *et al.*[50]

apparent crossreactivity between cardiac myosin and CB3 (Table VII). Even the serum of myosin-immunized animals did not react with CB3. Although these data strongly suggest an indirect association with CB3 and the development of autoimmune myocarditis, we cannot exclude the possibilities that other autoantigens in addition to cardiac myosin are involved and that cell-mediated autoimmune reactions recognize a shared determinant.

4. DISCUSSION AND CONCLUSIONS

We have been able to demonstrate that the late phase of CVB3-induced myocarditis is autoimmune and that a prominent autoantigen is cardiac myosin. The ability to induce disease by immunization with cardiac myosin and not with other myosin isoforms lends additional credence to cardiac myosin as an important autoantigen in this disease process. The experimental evidence obtained in our laboratory emphasizes the value of understanding the genetic contribution of the host in determining the extent and severity of CVB3-induced heart disease. This information has provided a mechanism for dissection of the various components in the sequelae of this complex disease. It is apparent that the virus plays a substantial role in the initial cardiac pathology. This viral component of myocarditis is a focal disease and, for the most part, transient, since its duration is approximately 1 week. We have no information to determine the pathogenic role of the resulting generalized inflammation during this phase of myocarditis. However, the work in several laboratories[21,31–34,36] suggests that the anti-CVB3 immune response can and does contribute to exacerbation of disease. In early disease, the severity of lesions is not only determined by the virulence of cardiotrophic viruses but is also strongly influenced by the host's genetic makeup. Our data suggest that susceptibility to viral myocarditis is under multigenic (i.e., both *Mhc* and non-*Mhc* genes) control, distinct from genetic influence on the ongoing autoimmune disease. The predisposition to virus-induced autoimmune myocarditis is determined by non-*Mhc* gene(s), but the *Mhc* influences the prevalence and titer of the heart autoantibodies. It is now clearly understood that the development of progressive myocarditis and quite possibly cardiomyopathy is, for the most part, determined by the immune-response capabilities of infected individuals.

In order to implicate cardiac myosin firmly as the autoantigen in autoimmune myocarditis, this disease must be passively or adoptively transferred into naive animals, using autoimmune serum and/or autoimmune competent cells, respectively. In addition, we would predict that, if cardiac myosin is truly the pertinent autoantigen, it should share that same genetic regulation as that observed for the CB3-induced autoimmune cardiac pathology. These investigations are currently under way.

ACKNOWLEDGMENTS. This work was supported by the American Heart Association, Maryland affiliate, and by grants HL-27932, HL-30144, HL-33878, and HL-38276 from the U.S. Public Health Service.

We wish to thank our colleagues, F. L. Alvarez, W. E. Beschorner, S. W. Craig, A. Herskowitz, J. Modlin, N. Neu, M. D. Traystman, and L. J. Wolfgram, for their active participation in the studies presented in this chapter. We would also like to acknowledge the technical assistance of Janice Larabell, Kathy Tewey, Suzanne Hammen, and Susan Schonlaw.

REFERENCES

1. Yoon, J. W., 1983, Viruses in the pathogenesis of type I diabetes, *Curr. Probl. Clin. Biochem.* **12:**11–44.
2. Brogen, C. H., Baekkeskov, S., Dyrberg, T., Lenmark, A., Marner, B., Nerup, J., and Papadopoulos, G. K., 1983, Role of islet cell antibodies in the pathogenesis of type I diabetes, *Curr. Probl. Clin. Biochem.* **12:**65–95.
3. Gepts, W., 1983, Role of cellular immunity in the pathogenesis of type I diabetes, *Curr. Probl. Clin. Biochem.* **12:**86–107.
4. Waksman, B. H., and Reynolds, W. E., 1984, Multiple sclerosis as a disease of immune regulation, *Proc. Soc. Exp. Biol. Med.* **175:**282–294.
5. Waksman, B. H., 1983, Immunity and nervous system: Basic tenets, *Ann. Neurol.* **13:**517–591.
6. Pini, M., Potli, R., and Dettori, A. G., 1982, Purpura thrombotic thrombocytopenia, *Recent Prog. Med.* **73:**429–442.
7. Sherlock, S., 1984, Chronic hepatitis and cirrhosis, *Hepatology* **4:**25–28.
8. Moench, T. R., and Griffin, D. E., 1984, Immunocytochemical identification and quantification of the mononuclear cells in the cerebrospinal fluid, meninges and brain during acute viral meningoencephalomyelitis, *J. Exp. Med.* **159:**77–88.
9. Johnson, R. T., Griffin, D. E., Hirsh, R. L. Wolinsky, J. S., Roedenbeck, S., Soriano, I. L. de, and Vaisberg, B., 1984, Measles encephalomyelitis—Clinical immunologic studies, *N. Engl. J. Med.* **310:**137–141.
10. Abelman, W. H., 1971, Virus and the heart, *Circulation* **44:**950–956.
11. Maisch, B., Berg, P. A., and Kochsiek, K., 1980, Autoantibodies and serum inhibition factors (SIF) in patients with myocarditis, *Klin. Wochenschr,* **58:**219–225.
12. Maisch, B., Trostel-Soeder, R., Stechemesser, E., Berg, P. D., and Kochsiek, K., 1982, Diagnostic prevalence of humoral and cell-mediated immune reactions in patients with acute viral myocarditis, *Circulation* **70:**149–156.
13. Yoon, J. W., and Notkins, A. L., 1976, Virus-induced diabetes mellitus. VI. Genetically determined host differences in the replication of encephalomyocarditis virus in pancreatic beta cells, *J. Exp. Med.* **143:**1170–1185.
14. Yoon, J. W., Onodera, T., and Notkins, A. L., 1978, Virus-induced diabetes mellitus. XV. Beta cell damage and insulin-dependent hyperglycemia in mice infected with Coxsackie virus B4, *J. Exp. Med.* **148:**1068–1080.
15. Onodera, T., Tonioio, A., Ray, U. R., Jenson, A. B., Knazek, R. A., and Notkins, A. L., 1981, Virus-induced diabetes mellitus. XX. Polyendocrinopathy and autoimmunity, *J. Exp. Med.* **153:**1457–1473.

16. Haspel, M. V., Onodera, T., Prabhakar, B. S., Horita, M. S., and Notkins, A. L., 1983, Virus-induced autoimmunity: Monoclonal antibodies that react with endocrine glands, *Science* **220**:304–306.
17. Notkins, A. L., and Prabhakar, B. S., 1986, Monoclonal autoantibodies that react with multiple organs. Basis for reactivity, *Ann. NY Acad. Sci.* **475**:123–134.
18. Hirsh, M. S., and Proffitt, M. R., 1975, Autoimmunity in viral infections, in: *Viral Immunology and Immunopathology* (A. L. Notkins, ed.), pp. 419–434, Academic, New York.
19. Notkins, A. L., Onodera, T., and Prabhakar, B., 1984, Virus-induced autoimmunity, in: *Concepts in Viral Pathogenesis* (A. L. Notkins and M. B. A. Oldstone, eds.), pp. 210–215, Springer-Verlag, New York.
20. Plotz, P. H., 1983, Autoantibodies are anti-idiotype antibodies to antiviral antibodies, *Lancet* **2**:824–826.
21. Woodruff, J. F., 1970, Viral myocarditis, *Am. J. Pathol.* **101**:427–479.
22. Sainani, G. S., Krompotic, E., and Slodki, S. J., 1968, Adult heart disease due to Coxsackie B infection, *Medicine (Baltimore)* **47**:133–147.
23. Cambridge, G., MacArthur, C. G. C., Waterson, A. P., Goodwin, J. F., and Oakley, C. M., 1979, Antibodies to Coxsackie B viruses in primary congestive cardiomyopathy, *Br. Heart J.* **41**:692–696.
24. MacArthur, C. G. C., Tarin, D., Goodwin, J. F., and Hallidie-Smith, K. A., 1984, The relationship of myocarditis to dilated cardiomyopathy, *Eur. Heart J.* **5**:1023–1035.
25. Helin, M., Savola, J., and Lapinleimu, K., 1968, Cardiac manifestations during a coxsackie B5 epidemic, *Br. Med. J.* **3**:97–99.
26. Kaplan, M. H., and Frengley, J. D., 1969, Autoimmunity to the heart in cardiac disease, current concepts of the relation to autoimmunity to rheumatic fever, postcardiotomy, and postinfarction syndromes and cardiomyopathies, *Am. J. Cardiol.* **24**:459–473.
27. Maisch, B., Deeg, P., Liebau, G., and Kochsiek, K. 1976, Diagnostic relevance of humoral and cytotoxic immune reactions in primary and secondary dilated cardiomyopathy, *Am. J. Cardiol.* **52**:1072–1078.
28. Scheerder, I. de, Vandekerckhove, J., Robbrecht, J., Algoed, L., Buyzere, M. de, Langhe, J. de, Schrijver, G. de, and Clement, D., 1985, Post-cardiac injury syndrome and an increased humoral immune response against the major contractile proteins (actin and myosin), *Am. J. Cardiol.* **56**:631–633.
29. Das, S. K., Callen, J. P., Dodson, V. N., and Cassidy, J. T., 1971, Immunoglobulin binding in cardiomyopathic hearts, *Circulation* **44**:612–616.
30. Schultheiss, H. P., and Bolte, H. D., 1985, Immunological analysis of auto-antibodies against the adenine nucleotide translocator in dilated cardiomyopathy, *J. Mol. Cell. Cardiol.* **17**:603–617.
31. Wong, C. Y., Woodruff, J. J., and Woodruff, J. F., 1977, Generation of cytotoxic T-lymphocytes during Coxsackie B3 infection. II. Characterization of effector cells and demonstration of cytotoxicity against viral-infected myocytes, *J. Immunol.* **118**:1165–1169.
32. Guthrie, M., Lodge, P. A., and Huber, S. A., 1984, Cardiac injury in myocarditis induced by Coxsackievirus group B in Balb/c mice is mediated by Lyt 2^+ cytolytic lymphocytes, *Cell. Immunol.* **88**:558–567.
33. Job, L. P., Lyden, D. C., and Huber, S. A., 1986, Demonstration of suppressor cells in Coxsackievirus group B, type 3 infected female Balb/c mice which prevent myocarditis, *Cell. Immunol.* **9**:104–113.
34. Huber, S. A., and Lodge, P. A., 1986, Coxsackievirus B-3 myocarditis. Identification of different pathogenic mechanisms in DBA/2 and Balb/c mice, *Am. J. Pathol.* **122**:284–291.

35. Khatib, R., Chason, J. L., Silberberg, B. K., and Lerner, A. M., 1980, Age-dependent myocardial pathogenesis of group B Coxsackievirus in Swiss-Webster mice: Comparisons to the pancreas, *J. Infect. Dis.* **141**:394–401.
36. Gauntt, C. J., Trousdale, M. D., LaBadie, D. R. L., Paque, R. E., and Nealson, T., 1979, Properties of Coxsackievirus which are amyocarditic and myocarditic for mice, *J. Med. Virol.* **3**:207–220.
37. Wolfgram, L. J., Beisel, K. W., and Rose, N. R., 1986, Variations in the susceptibility to Coxsackievirus B_3-induced myocarditis among strains of mice, *J. Immunol.* **136**:1846–1852.
38. Herskowitz, A., Wolfgram, L. J., Rose, N. R., and Beisel, K. W., 1987, Coxsackievirus B_3 murine myocarditis—Pathologic spectrum of myocarditis in genetically defined inbred strains, *J. Am. Coll. Cardiol.* **9**:1311–1319.
39. Beisel, K., Wolfgram, L., Herskowitz, A., and Rose, N., 1985, Differences in severity of Coxsackievirus B3-induced myocarditis among H-2 congenic mouse strains, in: *Genetic Control of Resistance to Infection and Malignancy* (E. Skamene, ed.), pp. 195–201, Alan R. Liss, New York.
40. Wolfgram, L. J., Beisel, K. W., and Rose, N. R., 1985, Heart-specific autoantibodies following murine Coxsackievirus B_3 myocarditis, *J. Exp. Med.* **161**:1112–1121.
41. Matsumori, A., and Kawai, C., 1982, Animal model of congestive (dilated) cardiomyopathy: Dilatation and hypertrophy of the heart in the chronic stage in DBA/2 mice with myocarditis caused by encephalomyocarditis virus, *Circulation* **66**:355–360.
42. Reyes, M. R., Khang-Loon, H., Smith, F., and Lerner, A. M., 1981, A mouse model of dilated-type cardiomyopathy due to Coxsackievirus B3, *J. Infect. Dis.* **144**:232–236.
43. Alvarez, F. L., Neu, N., Rose, N. R., Craig, S. W., and Beisel, K. W., 1987, Heart-specific autoantibodies induced by Coxsackievirus B_3: Identification of heart autoantigens, *Clin. Immunol. Immunopathol.* **43**:129–139.
44. Neu, N., Beisel, K. W., Traystman, M. D., Rose, N. R., and Craig, S. W., 1987, Autoantibodies specific for the cardiac myosin isoform are found in mice susceptible to Coxsackievirus B_3-induced myocarditis, *J. Immunol.* **138**:2488–2492.
45. Neu, N., Rose, N. R., Beisel, K. W., Herskowitz, A., Gurri-Glass, G., and Craig, S. W., 1987, Cardiac myosin induces myocarditis in genetically predisposed mice, *J. Immunol.* **139**:3630–3636.
46. Rose, N. R., Beisel, K. W., Herskowitz, A., Neu, N., Wolfgram, L. J., Alvarez, F. L., Traystman, M. D., and Craig, S. W., 1987, *Autoimmunity and Autoimmune Disease: Ciba Foundation Symposium 129* (D. Evered and J. Whelan, eds.), pp. 3–24, John Wiley & Sons, Chichester, England.
47. Beisel, K. W., Tharpe, J. A., Sell, K. W., Notkins, A. L., and Prabhakar, B. S., 1987, A monoclonal antibody against Coxsackievirus B4 which cross-reacts with the V1 isoform of myosin: implication for autoimmunity, J. Am. Coll. Cardiol. **9**:145A.
48. Cunningham, M. W., Hall, N. K., Krisher, K. K., and Spanier, A. M., 1986, A study of anti-group A streptococcal monoclonal antibodies crossreactive with myosin, *J. Immunol.* **136**:293–298.
49. Dale, J. B., and Beachey, E. H., 1985, Epitopes of streptococcal M proteins shared with cardiac myosin, *J. Exp. Med.* **162**:583–591.
50. Neu, N., Craig, S. W., Rose, N. R., Alvarez, F., and Beisel, K. W., 1987, Coxsackievirus-induced myocarditis in mice: Cardiac myosin autoantibodies do not cross-react with the virus, *Clin. Exp. Immunol.* **69**:566–574.
51. Wolfgram, L. J., 1986, Variations in the susceptibility to Coxsackievirus B3 induced myocarditis among different strains of mice, Ph.D. dissertation, Johns Hopkins University, Baltimore, Maryland.

17

The Impact of Recombinant DNA Technology on the Study of Enterovirus Heart Disease

REINHARD KANDOLF

1. INTRODUCTION

Of the numerous viruses associated with acute infectious myocarditis, coxsackieviruses (CV) of group B (CVB) types 1–5 are generally accepted as the most common agents.[1–13] Various coxsackie A viruses (CVA) and echoviruses are also relatively common causes of human viral heart disease.

Although most infections are subclinical, enteroviruses are capable of causing dilated cardiomyopathy of sudden onset, which especially in infants can be fatal, or lead to a variety of cardiac arrhythmias. Some of the acute cases of enterovirus heart disease may also evolve into a chronic form of dilated cardiomyopathy.[7,14]

So far, evidence for a causal linkage of CV with infectious myocarditis has been based mainly on the isolation of these viruses from heart tissue of infected patients in postmortem or biopsy specimens.[15–17] Successful isolation of an enterovirus from the myocardium of a patient with clinical symptoms and signs of myocarditis is a strong indication,

REINHARD KANDOLF • Department of Virology, Max Planck Institute for Biochemistry, D-8033 Martinsried; Department of Internal Medicine I, Klinikum Grosshadern, University of Munich, D-8000 Munich 70, Federal Republic of Germany.

but not direct proof, of cause and effect, since neither the interaction between these viruses and human myocardial cells has been studied nor the definite presence of progeny virus inside human myocardial cells demonstrated.

We recently initiated a project to study the role of viruses in acute and chronic cardiac disease. The aim of the study is to understand the pathogenesis of human viral heart disease at the cellular and molecular level. Also, new therapeutic approaches should be developed. CVB3 was chosen as a model because of the obvious significance of CV in clinical cardiology. Knowledge of the life cycle of this virus in human myocardial cells is essential for a more general understanding of the pathogenesis of human viral heart disease.

Thus far, two approaches to this problem were chosen. The first was to establish a tissue-culture system based on human fetal heart cells that permits simulation of viral heart disease *in vitro*.[18,19] Cultured human fetal myocytes were found to disintegrate upon infection with CVB3, whereas fetal as well as adult human myocardial fibroblasts developed a persistent carrier state infection. In addition, cultured human heart cells offer the unique opportunity of studying the effect of potential antiviral agents, e.g., interferons (IFN).[20]

The second approach was based on molecular cloning of the CVB3 genomic RNA,[21] which is of utmost medical and biological interest. The CVB3 complementary DNA (cDNA) makes it possible to follow the viral genome in myocardial cells, for example, and to study questions concerning the molecular basis of pathogenicity and tissue tropism. In addition, cloned CVB3 cDNA copies provide a powerful means of diagnosing myocardial tissue samples from patients with suspected enteroviral heart disease.

2. DIAGNOSTIC PROBLEMS IN SUSPECTED ENTEROVIRUS HEART DISEASE

Diagnosing viral heart disease by clinical features is an exercise of uncertain validity. In practice, the measurement of serum antibody titers in suspected enterovirus heart disease seldom proves diagnostically useful[22] except for cases that occur as part of an epidemic. Thus, cardiologists lack information on how often enteroviruses are actually the cause of acute dilated cardiomyopathy, what the prognosis is in such cases, and whether specific therapy can affect prognosis. The reason for this fundamental lack of knowledge is the difficulty of establishing an unequivocal diagnosis of viral heart disease.

Confirmation of the clinically suspected diagnosis of viral heart disease demands demonstration of replicating virus inside myocardial cells. Endomyocardial biopsy[22–24] makes it possible to obtain multiple myocardial samples from either ventricle serially. However, the techniques so far used to investigate endomyocardial biopsy samples from patients with suspected enteroviral heart disease seldom yield conclusive results.

The histologic features of biopsy specimens of patients with suspected enteroviral heart disease are nonspecific and indistinguishable from those seen in inflammatory heart disease of nonviral origin,[7] e.g., in bacterial infections or connective tissue diseases. Thus, the histologic findings in myocardial biopsy samples are usually not rewarding in elucidating the etiology. Electron microscopy also fails to yield conclusive results because enteroviruses are indistinguishable in size and density from ribosomes.[25] Only viral crystals can be easily detected by electron microscopy, as demonstrated in cultured human fetal myocytes infected with CVB3.[19] Immunofluorescence techniques using serotype-specific antibodies appear to be impracticable because of the high number of related enteroviral serotypes that might be implicated. Virus-isolation procedures depend on the availability of multiple cell lines as well as on newborn mice[26] and appear to be successful only in the acute phase of enteroviral heart disease with extensive virus replication. With respect to endomyocardial biopsies, virus isolation might also be impaired by the limited amount of myocardial tissue available.

In contrast to these conventional techniques, the technique of recombinant DNA provides a new diagnostic approach for the detection of infectious agents by molecular nucleic acid hybridization,[27,28] using molecularly cloned viral DNA as a diagnostic reagent. Owing to the high sensitivity of the molecular hybridization approach, not only extensive viral replication is detected, but also more elusive virus–cell interactions, e.g., when permissive viral replication is replaced by defective or restricted replication. Thus, nucleic acid hybridization appears especially useful in the study of persistent viral infections, since it can detect even inactive or incomplete viral genomes in a cell.

3. MOLECULAR CLONING AND CHARACTERIZATION OF FULL-LENGTH REVERSE-TRANSCRIBED COXSACKIEVIRUS B3 GENOMIC RNA

One major prerequisite for the introduction of nucleic acid hybridization as a diagnostic tool in suspected enteroviral infections was the molecular cloning and characterization of the complete CVB3 genome.[21] The polyadenylylated viral RNA isolated from a large plaque variant of the Nancy strain[12] of CVB3 served as a template for first-strand cDNA synthesis by using reverse transcriptase under optimized conditions for the generation of large cDNA molecules. Making use of the RNase H and DNA polymerase I mediated second-strand cDNA synthesis,[29,30] an efficient cloning strategy for large RNA molecules was designed and followed (Fig. 1), which represents an improvement on earlier methods used in the molecular cloning of viral genomes. After second-strand cDNA synthesis, double-stranded CVB3 cDNA was directly tailed with dGMP by using terminal deoxynucleotidyl transferase without the necessity of prior trimming of double-stranded cDNA with S1 nuclease. Oligo$(dG)_{10-15}$-tailed CVB3 double-stranded cDNA was annealed to oligo$(dC)_{10-15}$-tailed plasmid vector p2732B, previously linearized with *Cla* I and *Bgl* II in its cloning site (Fig. 1). On transformation of competent *E. coli* BJ5183, at least 4×10^4 bacterial colonies resistant to ampicillin were obtained per microgram of viral RNA used in the initial reverse transcription.

The efficiency of this cloning strategy is given by the fact that full-length reverse-transcribed CVB3 cDNA of ~7500 nucleotides was molecularly cloned. Moreover, full-length reverse-transcribed CVB3 cDNA generated infectious virus antigenically identical to CVB3 upon transfection of recombinant plasmid DNA into mammalian cells,[21] indicating that the cloned CVB3 cDNA is a faithful transcript of the original viral RNA. Interestingly, two initial uridine residues are missing in the infectious CVB3 cDNA that are present in the authentic CVB3 RNA.

The finding of infectious CVB3 cDNA is in agreement with observations by Racaniello and Baltimore[33] as well as others,[34,35] who reported on the infectivity of molecularly cloned poliovirus cDNA copies that had been constructed from subgenomic cDNA clones. In addition, it is of interest to mention that the CVB3 cDNA clone pCB3-M1 also generated infectious CVB3 in newborn mice upon intracerebral transfection of circular recombinant plasmid DNA, providing a unique *in vivo* transfection system.

Since there is no DNA intermediate in the life cycle of picornaviruses, the molecular mechanisms involved in the initiation of an infectious cycle by a cloned CVB3 cDNA will be important to understand. The first event upon transfection of recombinant plasmid DNA into mammalian cells must be the synthesis of a plus-strand RNA transcript, which then serves as a messenger for the synthesis of virus-directed proteins, e.g., viral replicase. However, it is not known how the cloned cDNA gains a promoter that permits the start of transcription by cellular DNA-dependent RNA polymerase. It has been proposed that RNA synthesis may initiate at one or more areas in the plasmid DNA, which may function as a promoter *in vitro*.[33] Alternatively, promoter activity could be provided by integration next to a cellular promoter.

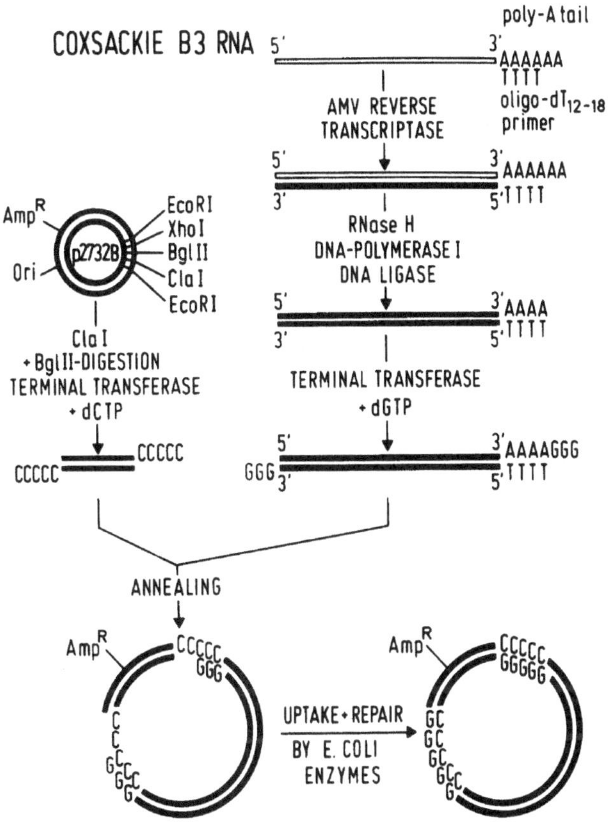

FIGURE 1. Cloning strategy of coxsackievirus B3 (CVB3) cDNA. The polyadenylylated purified CVB3 RNA served as a template for oligo$(dT)_{12-18}$ primed first-strand complementary DNA (cDNA) synthesis, using reverse transcriptase under optimized conditions.[21] The RNA strand of the RNA–cDNA hybrid was replaced by RNase H and DNA polymerase I mediated second-strand synthesis and homopolymer tails of 10–15 residues of dGMP were added directly to double-stranded cDNA by using terminal deoxynucleotidyl transferase. Plasmid vector p2732B, a derivative of pBR322 containing the ampicillin-resistance gene, the plasmid origin of replication and a cloning site array between two *Eco*RI sites, was kindly provided by J. D. Monahan (Roche Institute, Nutley, New Jersey). Plasmid vector p2732B was linearized at its cloning site by digestion with *Cla*I and *Bgl*II, tailed with dCMP, hybridized to the oligo$(dG)_{10-15}$-tailed double-stranded CVB3 cDNA, and used to transform *Escherichia coli* BJ 5183[31] by standard procedures.[32]

The RNA molecules made in this way would probably represent oversized transcripts with extra nonviral sequences at both ends. Precisely how these molecules could replicate remains of focal interest.

In addition, infectious CVB3 cDNA offers the unique opportunity for *in vitro* site-directed mutagenesis and thus for genetic analysis of this cardiotropic virus. Moreover, new constructions involving prokaryotic and eukaryotic gene-expression vectors can be designed for production systems of distinct viral proteins.

4. CLONED COXSACKIEVIRUS B3 cDNA AS A DIAGNOSTIC REAGENT FOR THE DETECTION OF ENTEROVIRUSES

If one addresses nucleic acid hybridization as a diagnostic tool in enteroviral heart disease, a major prerequisite is the demonstration that the radioactively labeled cloned viral cDNA detects specifically viral nucleic acid and does not hybridize to human myocardial total RNA. When radioactively labeled cloned CVB3 cDNA corresponding to the region from 0.06 to 7.2 kilobases (kb) of the viral genome was hybridized to electrophoretically resolved CVB3 RNA and to total RNA isolated from cultured human heart cells, hybridization was found only for the viral RNA and not for human myocardial RNA.[21]

A main advantage of the nucleic acid hybridization approach in suspected enterovirus heart disease is the fact that the complete cloned CVB3 cDNA can be used to detect numerous related enteroviruses in a single hybridization assay. Using a slot blot hybridization assay[36] to examine total RNA from cells infected with different enteroviruses it was shown that the complete CVB3 cDNA detects as a sensitive probe numerous enteroviral serotypes, e.g., CVB types 1 to 6, CVA9, echovirus 11 and 12, and poliovirus type 1. This finding is in agreement with reports by others[37–39] on the detection of various enterovirus types using different cloned CVB cDNA fragments. By contrast, genetically unrelated viruses, e.g., vesicular stomatitis virus, or foot-and-mouth disease virus, a nonhuman picornavirus, do not hybridize with the CVB3 cDNA using stringent conditions.

The molecular basis for enteroviral group-specific hybridization is the high degree of nucleic acid sequence homology between the different enteroviruses. As an example, a comparison of the CVB3 cDNA clone pCB3-M1[21] with the corresponding sequence of poliovirus type 1 (PV1),[40] representing part of the replicase gene, is demonstrated in Fig. 2. For this part of the replicase gene, the homologies were found by us and others[42] to be 69% and 74% at the nucleotide and amino acid sequence levels, respectively. Cross-hybridization experiments appear to predict an even higher degree of nucleotide sequence homology between the different serotypes of CVB and CVA as well as echoviruses. Thus, cloned CVB3 cDNA provides a valuable diagnostic means for the detection of the main etiologic agents in human viral heart disease.

From the clinical point of view, the exact typing of an etiologically implicated enterovirus strain appears to be of secondary importance and can be carried out later by using standard virologic techniques or by hybridization with serotype-specific enterovirus cDNA fragments. Using a CVB3 cDNA fragment corresponding to the region from 0.88 to 2.7 kb of the viral genome encoding for the viral coat proteins VP2 (1B),

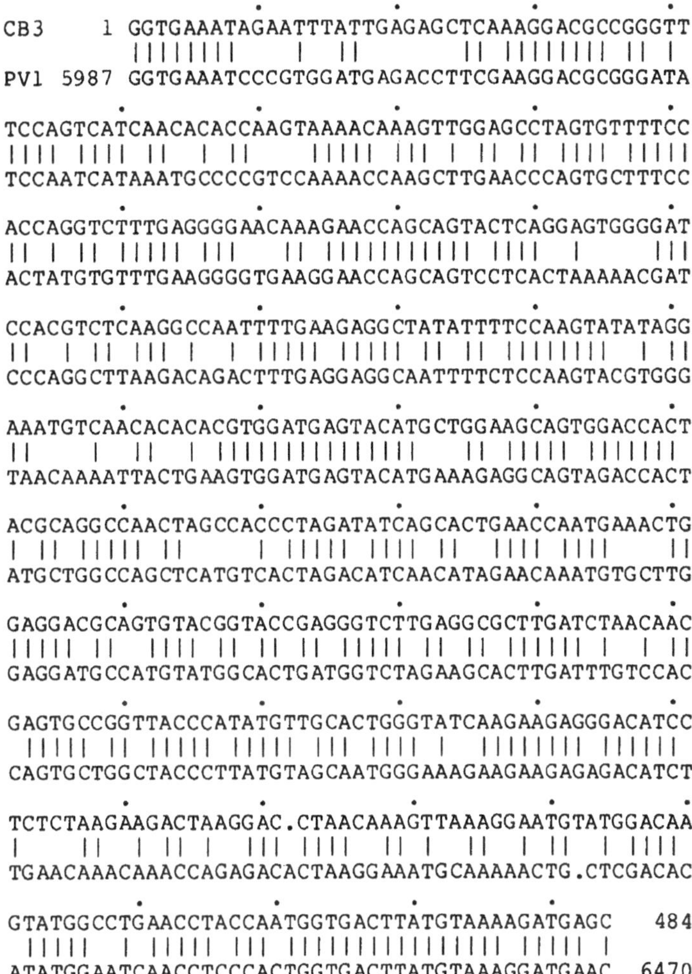

FIGURE 2. Nucleotide sequence of cloned complementary DNA (cDNA) corresponding to nucleotides 1–484 of the coxsackievirus B3 replicase gene. The sequence is compared with the corresponding sequence of poliovirus type 1 (PV1).[40] Matching nucleotides are indicated by vertical lines. Deleted nucleotides are indicated with points. For this part of the replicase gene, the homologies were found to be 69% and 74% at the nucleotide and amino acid sequence level, respectively. Sequence was determined for the insert of pCB3-M1.[41]

VP3 (1C), and part of VP1 (1D),[43,44] type-specific hybridization is possible for CVB3 RNA.

5. DETECTION OF ENTEROVIRUS RNA IN INFECTED CELLS BY *IN SITU* NUCLEIC ACID HYBRIDIZATION

The feasibility of the *in situ* nucleic acid hybridization technique [45–48,64] for the detection of enteroviral RNA was first investigated in cell culture systems and then applied to myocardial tissue of mice infected

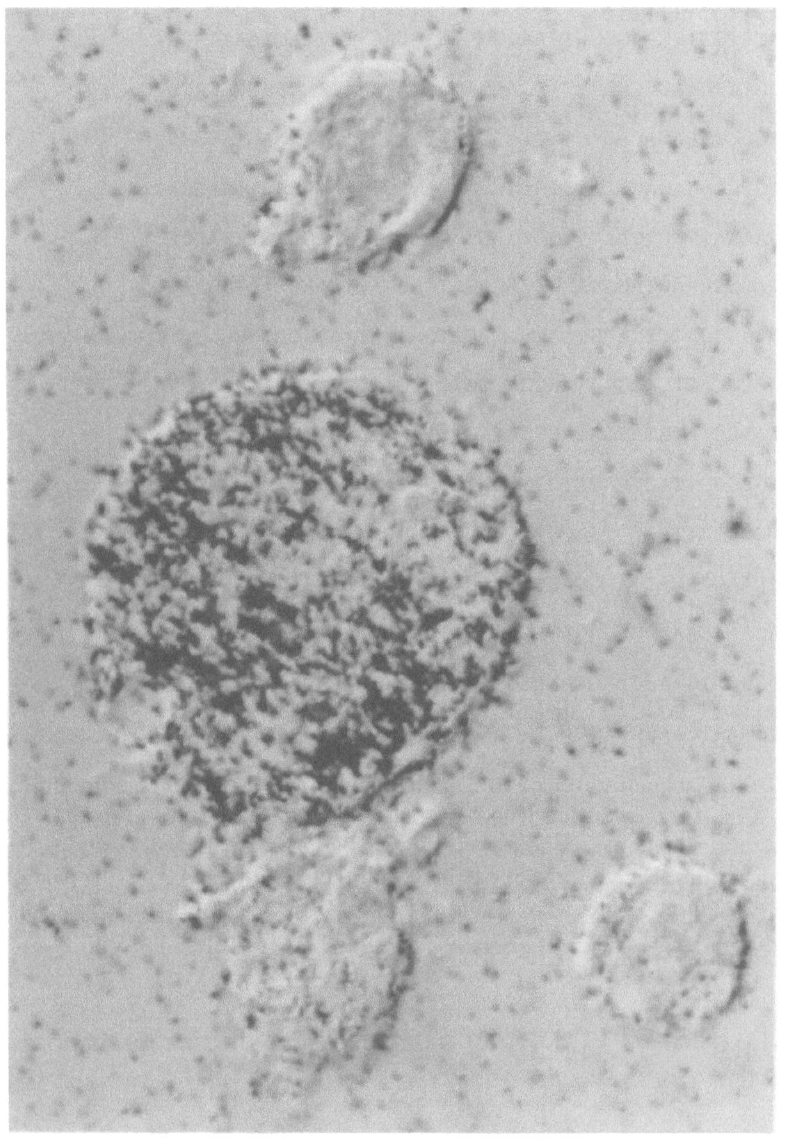

FIGURE 3. *In situ* hybridization of ^{3}H-labeled coxsackievirus B3 complementary DNA (cDNA) probe to persistently infected CVB3 carrier cultures of human myocardial fibroblasts, indicating an asynchronous course of virus replication in this type of infection. Note that since silver grains are positioned at various levels within the photoemulsion, some grains are not observed and appear out of focus. Interference contrast, ×1000.

with CVB3. The specificity of the *in situ* hybridization method was demonstrated by use of Vero cells and cultured human myocardial fibroblasts. Uninfected Vero cells and human myocardial fibroblasts exhibited essentially no grains when hybridized to the ^{3}H-labeled CVB3 cDNA probe corresponding to 0.06–7.2 kb of the infectious CVB3 cDNA clone pCB3-M1. By contrast, highly significant labeling was achieved in Vero cells infected with CVB1, CVB3, CVB5, or echovirus 11. Figure 3 demonstrates, as an example, the detection of CVB3 RNA in persistently infected CVB3 carrier cultures of human myocardial fibroblasts.

Specificity for viral RNA was further demonstrated by hybridization of the ^{3}H-labeled cDNA probe to CVB3 infected cells after digestion with RNase A at 100 μg/ml for several hours. Cells treated with RNase A before hybridization exhibited essentially no label. In addition, when CVB3-infected cells were hybridized with the plasmid p2732B control DNA probe,[21] again no labeling of cells was observed. Thus, detection of grains after hybridization with ^{3}H-labeled pCB3-Ml cDNA correlated conclusively with the presence of CVB3, demonstrating that *in situ* hybridization exhibited specificity for viral sequences.

In a next step, experiments were performed to elucidate the optimal handling and processing conditions for investigation of myocardial tissue.[64] Using cryocuts of the myocardium of persistently CVB3 infected athymic mice (Fig. 4), *in situ* hybridization proved as a powerful tool not only with respect to establishing an unequivocal diagnosis of myocardial infection but also with respect to pathogenicity. The autoradiographic silver grains, which indicate hybridization between viral RNA and radiolabeled cloned CVB3 cDNA are clearly related to distinct infected myocardial cells. Infected myocytes are easily identified by interference contrast microscopy in unstained sections because of their characteristic size and morphology. In this model system of dilated-type cardiomyopathy, enterovirus infection was found to be focal and randomly distributed in the heart muscle.

The sensitivity of the method was determined by analyzing CVB3 infected Vero cells from the same culture by *in situ* hybridization as well as by RNA blot hybridization. Cells were infected at a multiplicity of infection (MOI) of 0.1, 1, 3, and 4 plaque-forming units (PFU) per cell and harvested at the end of viral adsorption 1 hr postinfection and 5 hr postinfection. As estimated by RNA blot hybridization, cells infected at a MOI of one contained an average of 200 copies of viral RNA 1 hr after infection, indicating a particle to PFU ratio of ~200. As expected, copy numbers 1 hr postinfection were proportional to MOI. Using Vero cells harvested 1 hr postinfection with a MOI of 0.1, the sensitivity of *in situ* hybridization was estimated to be about 20 copies of viral RNA detected within 6 weeks of exposure after hybridization with ^{3}H-labeled CVB3

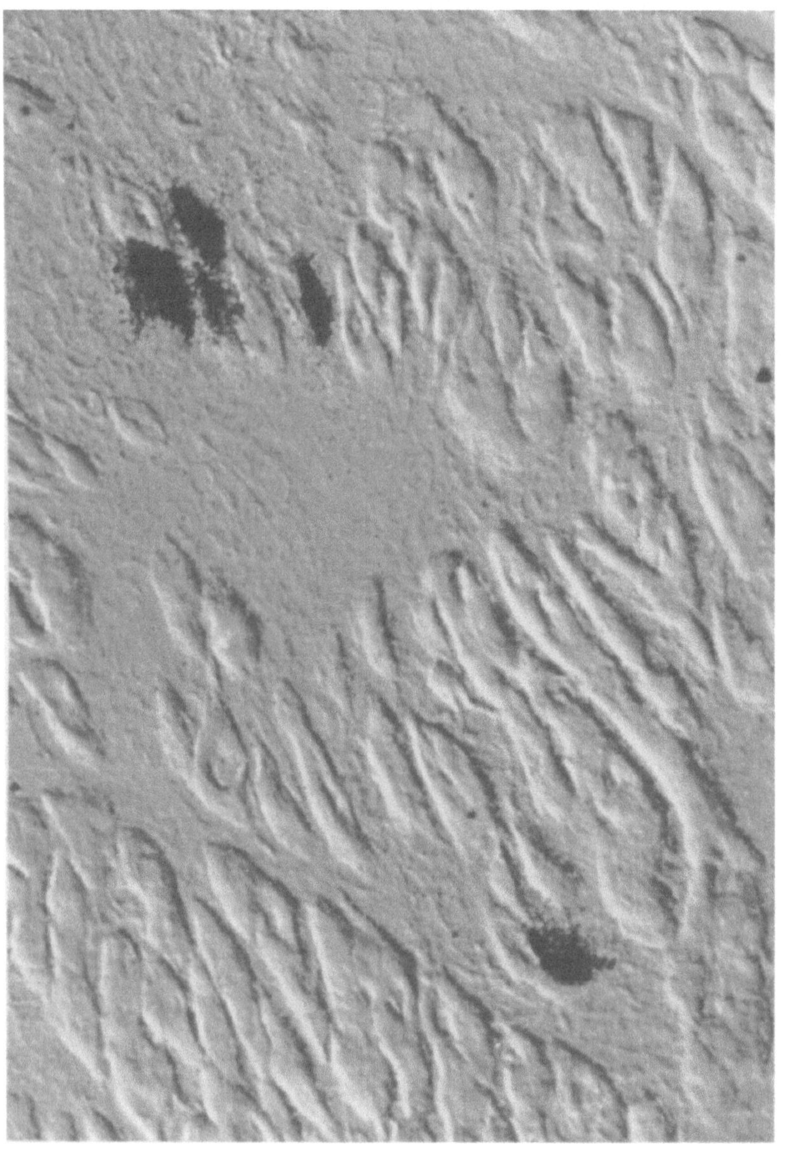

FIGURE 4. *In situ* detection of coxsackievirus B3 RNA in infected mouse myocardial tissue. Virus was administered intraperitoneally to adult athymic mice and the location of viral RNA was determined in cryocut sections of the myocardium 3 weeks postinfection, using the ^{3}H-labeled cDNA as a probe. Note that infected myocytes are easily detected in this longitudinal myocardial tissue section because of their characteristic size and morphology. Interference contrast, ×400.

cDNA at a specific activity of 1×10^8 dpm/μg of DNA. By contrast, 20 copies of viral RNA are detectable within 2 weeks of exposure, using ^{35}S-labeled CVB3 cDNA as a probe. To increase further the sensitivity of the *in situ* hybridization approach, cRNA probes synthesized from the cloned CVB3 cDNA are now being used that can be labeled to high specific activity.[49]

Vero cells harvested 5 hr after infection containing high copy numbers of about 5×10^4–1×10^5 viral genomes per cell, showed clearly positive *in situ* hybridization after exposure times of 4 and 2 days, using ^{3}H- and ^{35}S-labeled CVB3 cDNA, respectively. Clearly positive hybridization signs with infected mouse myocardial tissue was observed already after 2 days of exposure with the ^{35}S-labeled CVB3 cDNA probe, indicating high copy numbers of replicating viral RNA in myocardial cells.

6. DETECTION OF ENTEROVIRUS SEQUENCES IN THE MYOCARDIUM FROM PATIENTS WITH MYOCARDITIS AND DILATED CARDIOMYOPATHY BY *IN SITU* HYBRIDIZATION

Using the *in situ* hybridization approach described, right ventricular endomyocardial biopsy specimens from 29 patients with clinically suspected viral heart disease or dilated cardiomyopathy have been exam-

TABLE I

Detection of Enterovirus RNA Sequences in the Myocardium from Patients with Myocarditis and Dilated Cardiomyopathy by *In Situ* Nucleic Acid Hybridization[a]

	N	Clinical diagnosis	*N*	Detection of viral nucleic acid	*N*
Endomyocardial biopsies (right ventricular)	29	Suspected myocarditis	8	Enteroviruses	2
		Suspected myocarditis in dilated CMP	4	Enteroviruses	1
		Dilated CMP	9	Enteroviruses	1
		Control group[b]	8	Negative	—
Recipient heart	8	Dilated CMP	6	Enteroviruses	2
		Control group[c]	2	Negative	—

[a]CMP, cardiomyopathy; *N*, number of patients.
[b]Hypertrophic CMP ($N = 2$), coronary heart disease ($N = 3$), endocrine CMP ($N = 1$), Fabry disease ($N = 1$), Mönckeberg arteriosclerosis ($N = 1$).
[c]Dilated CMP after threefold mitral valve replacement ($N = 1$), dilated CMP in patent ductus arteriosus ($N = 1$).

FIGURE 5. *In situ* hybridization of the ^{35}S-labeled coxsackievirus B3 cDNA probe to the myocardial explant of a patient who underwent heart transplantation. The autoradiographic silver grains can be clearly related to distinct infected myocytes, thus providing the unique possibility for an unequivocal diagnosis of enteroviral heart disease. No labeling of myocytes was observed when myocardial cryocuts were hybridized with the ^{35}S-labeled plasmid vector p2732B control DNA probe. Interference contrast, ×1000.

ined for the presence of enterovirus RNA. Myocardial tissue samples from recipient hearts of eight patients who underwent heart transplantation have also been examined. Enterovirus-infected myocardial cells have been detected in 3 of 12 patients with clinically suspected viral myocarditis as well as in 3 of 15 patients with dilated cardiomyopathy (Table I). Ten patients with myocardial diseases not consistent with viral heart disease, e.g., coronary heart disease, were used as an internal control group and showed negative *in situ* hybridization. Importantly, the number of infected cells appears to be correlated with the severity of clinical symptoms. Whereas in patients with mild perimyocarditis or healing myocarditis only a few myocardial cells were found to express enterovirus RNA, numerous infected myocardial cells were found in patients with severe dilated cardiomyopathy. Figure 5 shows as an example *in situ* detection of enteroviral sequences in the myocardium of a 49-year-old male patient who underwent heart transplantation.

An important observation is that in this first series positive *in situ* hybridization for enterovirus sequences was obtained by the examination of endomyocardial biopsies usually 2–6 months after onset of clinical symptoms, indicating that virus persists in the myocardium after an initial acute infection. Recently, evidence for persistence of enterovirus sequences in late myocardial biopsies has also been reported by means of a slot blot hybridization assay.[39]

Although persistence of enterovirus RNA is an intriguing concept for the pathogenesis of chronic dilated cardiomyopathy, the possibility has to be considered that the presence of a virus in chronic diseases may be adventitious, or chronic dilated cardiomyopathy may potentiate the growth of the virus as a secondary event. This could actually be the case in the 49-year-old male patient studied in Fig. 5. This patient underwent heart transplantation after an 8-year history of dilated cardiomyopathy. Nonetheless, the finding of positive enterovirus hybridization in endomyocardial biopsy specimens within a period of 2–6 months after onset of the disease suggests persistence of these viruses in human heart muscle, predisposing to the development of ensuing dilated cardiomyopathy. Thus, the molecular basis of enterovirus persistence in human myocardial cells will be important to understand. Preliminary experiments with a mouse model of CVB3 induced dilated-type cardiomyopathy indicate that viral RNA is detectable by *in situ* nucleic acid hybridization even when virus is no longer assayable by virus-isolation procedures.

7. EXPRESSION OF COXSACKIEVIRUS B3 PROTEINS IN *E. COLI* AND GENERATION OF VIRUS-SPECIFIC ANTIBODIES

Besides detection of enteroviral sequences by nucleic acid hybridization, the cloned CVB3 cDNA permits the expression and analysis of viral proteins. As a first step, viral proteins were expressed in *Escherichia coli* and used to elicit antibodies. These antibodies provide an important prerequisite for monitoring distinct viral proteins in different cell systems. For the expression of virus-encoded proteins in *E. coli,* the vector pPLc24 was used[50] that contains the strong left promoter of bacteriophage λ, the ribosomal binding site, start codon, and the following 98 codons of the replicase gene of the MS2-phage. The transcription is controlled by the temperature-sensitive repressor cI857, which is active at 28°C but inactive at 42°C. This vector system is used for the expression of genes lacking expression signals or producing poisonous products, e.g., the 3C protease of foot-and-mouth disease virus, which was shown to be bacteriotoxic.[51,52]

Distinct CVB3 DNA fragments derived from the infectious cDNA clone pCB3-M1[21] were inserted into pPLc24 within the coding sequence of the MS2-replicase, thereby achieving a fusion of both reading frames.

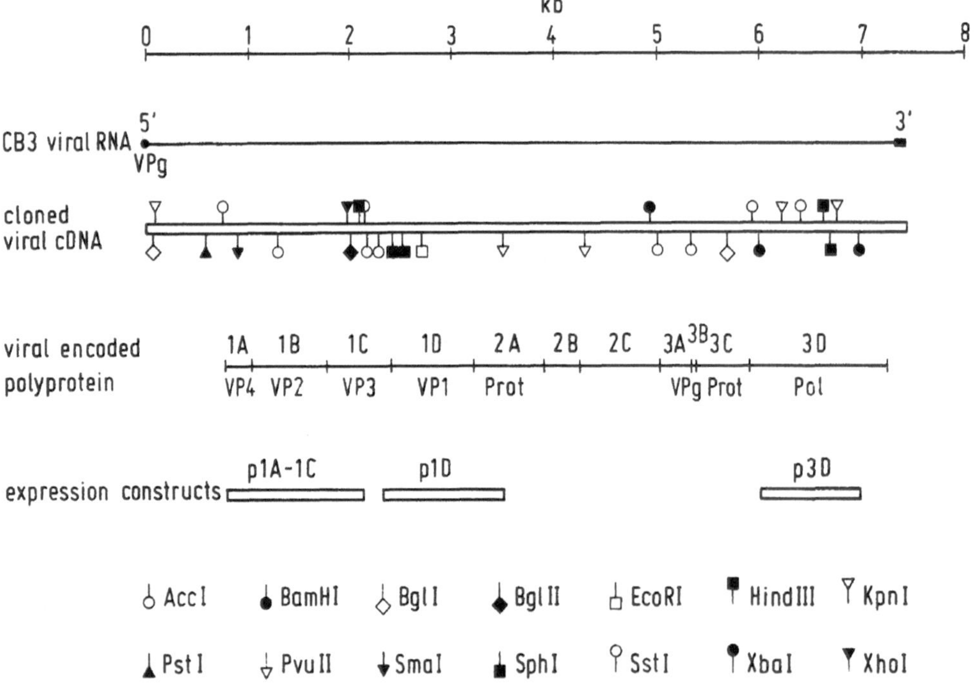

FIGURE 6. Restriction map of the molecularly cloned infectious coxsackievirus B3 complementary DNA (cDNA) in correlation to the genetic map of the virus. Sites for rarely cutting restriction enzymes are indicated with symbols. The corresponding viral RNA is shown with the VPg covalently linked to its 5′ end at the left and with the 3′ prime poly(A) (thicker line) at the right. The viral polyprotein encoded by an open reading frame of about 6.6 kilobases (kb) and the location of distinct viral proteins are correlated to the cloned CVB3 cDNA according to the systematic nomenclature of picornavirus proteins as proposed by Rueckert and Wimmer.[44] cDNA fragments derived from the infectious cDNA clone pCB3-Ml as shown below were inserted into the *Escherichia coli* expression vector pPLc24.

The expressed fusion proteins contain MS2-replicase-specific amino acid sequences at the N-terminal. Using this approach, plasmids were constructed that express either the structural proteins VP4, VP2, and VP3 (p1A-1C) or the structural protein VP1 (p1D) or the RNA-dependent RNA-polymerase (p3D) of CVB3 (Fig. 6).

The fusion proteins expressed from these plasmids were purified from *E. coli* extracts and used for immunizing rabbits. The resulting

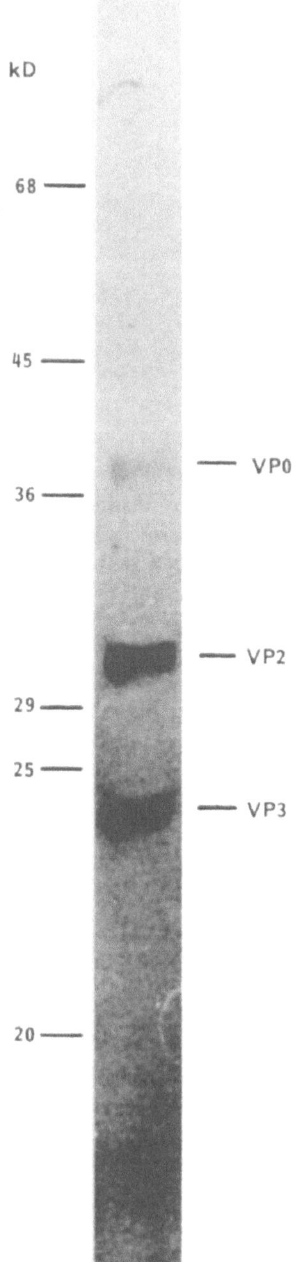

FIGURE 7. Detection of coxsackievirus B3 proteins by immunoblotting using anti-VP4/3-antiserum. Proteins from CsCl gradient purified virus were separated by polyacrylamide gel-electrophoresis and transferred to nitrocellulose filter. The filter was treated with anti-VP4/3-antiserum and antirabbit-IgG conjugated to peroxidase. Virus proteins VP2, VP3, and VP0 (precursor of VP4 and VP2) are detected.

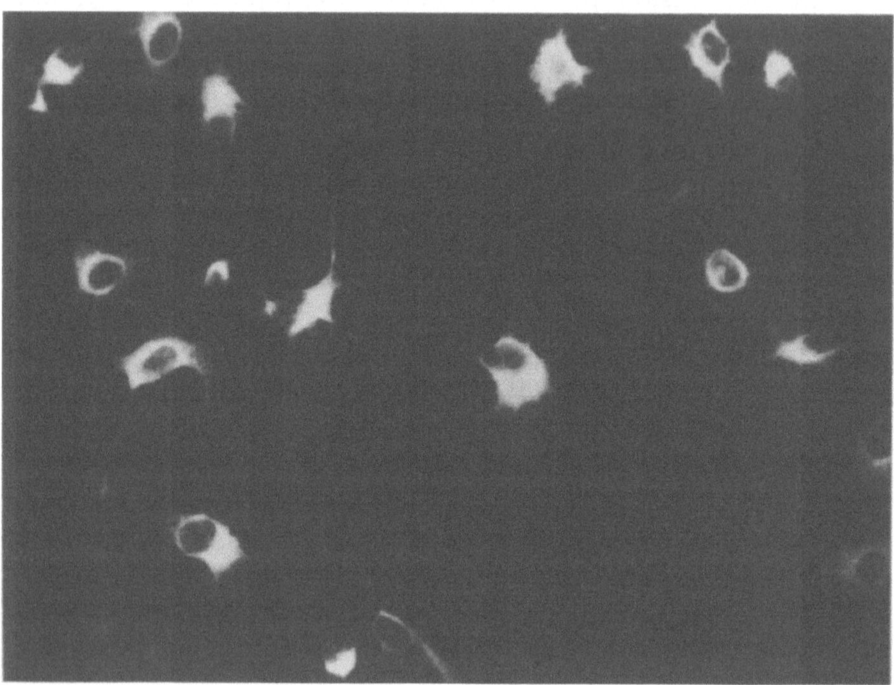

FIGURE 8. Indirect immunofluorescence labeling of coxsackievirus B3-infected Vero cells with antipolymerase antiserum. Vero cells were grown and infected on microscopic slides. At 5 hr postinfection, the cells were fixed with acetone and treated with anti-polymerase-antiserum and antirabbit IgG conjugated to FITC. Infected cells showing a bright fluorescence are easily distinguished from not infected cells, showing background staining. (×250)

antisera were tested for their ability to detect native CVB3 proteins. Using the immunoblot procedure[51] it was shown that the antisera recognize native viral proteins from CsCl gradient-purified CVB3 (Fig. 7). Using an indirect immunofluorescence assay,[53] it was demonstrated that all three antisera detect distinct viral proteins in CVB3-infected cells. Figure 8 shows, as an example, the detection of the CVB3 polymerase in infected cells by the antipolymerase antiserum raised against the viral polymerase expressed in *E. coli.* Preliminary experiments indicate that the above antisera also detect proteins of other CVB serotypes, e.g., CVB1 and CVB5. These antisera are currently used to study the role of CVB3 proteins in cell tropism, viral replication, and persistence.

8. SIMULATION OF COXSACKIEVIRUS CARDIAC DISEASE IN CULTURED HUMAN HEART CELLS

A useful experimental approach to the study of the interaction between CVB3 and human myocardial cells is to simulate coxsackievirus myocarditis in tissue culture. When 8-day-old enriched beating fetal myocytes were infected with a high-input multiplicity of 50 PFU of CVB3 per cell, titers of infectious progeny virus as high as 10^8 PFU/ml culture medium were found within 2 days.[18,19] Virus replication was accompanied by loss of spontaneous contractility of beating networks of myocytes within 9 hr, followed by the classic virus-produced cytopathic effects with cell retraction and severe destruction of formerly beating networks of myocytes within 20 hr of infection. Productive infection of human fetal myocytes could also be demonstrated by the presence of crystalline arrays of virus particles inside myocytes.[19] If the rate of virus assembly is rapid, and many viral particles are formed nearly simultaneously at a circumscribed site, crystallization may occur. The predominant features of ultrastructural pathological changes observed in CVB3-infected human fetal myocytes were disintegration or total dissolution of myofibrils and myofilaments, together with the formation of vesiculated regions which appeared morphologically as the sites of the cytocidal action of the virus.[19] Certain of the observed ultrastructural lesions are in agreement with the findings *in vivo,* both in humans and in animal models,[7,54] e.g., necrosis of myofibers in the absence of inflammatory cells.

9. PERSISTENT COXSACKIEVIRUS B3 CARRIER-STATE INFECTION OF CULTURED HUMAN MYOCARDIAL FIBROBLASTS

During infection of enriched human fetal myocytes, the observation was made that lysed areas of formerly beating networks of myocytes were replaced by overgrowing myocardial fibroblasts, which apparently survived the infection with CVB3.[19] Continued propagation of the latter cells yielded cultures of myocardial fibroblasts that produced CVB3 continuously as observed for periods of up to 9 months, indicating the existence of a persistent carrier state infection. These cultures consisted mainly of myocardial fibroblasts. However, a few heart muscle cells were detected by electron microscopy, which might indicate decreased susceptibility of the differentiated human myocyte, as previously shown for cultured differentiated mouse skeletal muscle cells infected with CVA13.[55] As shown by indirect immunofluorescence studies, viral persistence is based on the production of infectious virus in up to 10% of the total cell population. The virus–cell relationship appears to be similar to the reported carrier state infection of HeLa cells by CV.[56,57] Recently, persistent carrier state infection by group B CV has also been reported for human lymphoid cell lines.[58]

Using the molecular *in situ* hybridization approach, the fate of the viral genome is currently studied in CVB3-infected cultured human heart cells. Results obtained with persistently infected carrier cultures of human myocardial fibroblasts indicate asynchronous replication of viral RNA in this type of infection (see Fig. 3). In these model systems, the distribution of viral genes and their expression in the form of viral proteins can now be studied simultaneously at the single-cell level, using *in situ* nucleic acid hybridization in combination with immunocytochemistry as described for Theiler's murine encephalomyelitis virus.[59] Using antibodies raised against distinct CVB3 proteins after expression in *E. coli*, e.g., viral replicase (3D) or VP1 (1D), this double-labeling approach should be useful to answer questions concerning the molecular basis of restricted or defective CVB3 replication.

10. ANTIVIRAL ACTIVITY OF HUMAN FIBROBLAST INTERFERON IN COXSACKIEVIRUS B3-INFECTED CULTURED HUMAN HEART CELLS

Effective antiviral chemotherapy has not yet been established in the treatment of viral heart disease. The optimal goal of antiviral treatment, i.e., to restore function to the infected cell, usually appears unattainable. A realistic goal would be to inhibit viral replication and thus prevent spread to as yet uninfected cells. This is achieved in part, in many natural viral infections, by the endogenous interferon (IFN) system.[20] With respect to the potential clinical application of exogenous IFN in viral heart disease, one major prerequisite is the demonstration that the virus to be treated is susceptible *in vitro* to the action of a given type of IFN in the specific host cell.

Recently, the major protective role of natural human fibroblast interferon (IFN_β) in CVB3-infected human fetal myocytes has been reported.[18,19] To determine the antiviral potency of human IFN_β, we proceeded as follows. Since IFN causes antiviral resistance not directly but by activating cellular genes for antiviral proteins,[20] permissive cells must be exposed to IFN for several hours before infection. IFN-mediated antiviral resistance was accomplished by exposing 7-day-old cultures of human fetal myocytes to IFN_β 20 hr before CVB3 infection at the multiplicity of 50 PFU/cell. Virus yields could be reduced from 1.2×10^8 to 1.8×10^5 PFU/ml culture medium within 2 days by incubation with IFN_β at 10^3 U/ml, demonstrating the major protective role of IFN_β in CVB3 infection. In unprotected cultures, beating of myocytes ceased within 9 hr of infection, whereas myocytes protected by IFN_β continued to beat rhythmically, as long as IFN_β was given with exchanges of medium every 2 days.[19] In agreement with other virus–cell systems, we found no significant effect when IFN_β was given after viral replication had already begun in most of the highly permissive myocytes, e.g., 9 hr postinfection.

Tissue-culture techniques appear also appropriate to study potential side effects of IFN, which may have influence on clinical usage. For example, a norepinephrine-like stimulation of the beat frequency caused by IFN has been described for cultured mouse myocardial cells.[60] We also observed an increase in the beat frequency from 50/min up to 120/min in infected networks of human myocardial myocytes pretreated with IFN_β at a dose of 10^3 U/ml culture medium. However, no significant increase in the beat frequency was observed at a dose of 300 U/ml. To study the question of whether IFN has a direct cardiotoxic effect, the relatively high dose of 10^3 U IFN_β/ml culture medium was used, although lower levels of 300 U of IFN_β were found sufficient for the induction of the antiviral state. In agreement with other reports,[61] so far no morphologic differences were observed between myocardial cells treated with 10^3 U IFN and untreated cells. At the electron microscopic level, myofibrillar structures had the same morphologic appearance as those in control cultures.[19]

In addition, persistently CVB3 infected carrier cultures of human myocardial fibroblasts offer a particularly useful test system for the study of the antiviral activity of IFN. In contrast to beating myocytes, infected myocardial fibroblasts can be easily grown and maintained in tissue culture by repeated passages over periods of many months. In this model system virus–cell interactions can be followed and the effect of antiviral agents analyzed on well-defined cell and virus populations. Only a small proportion of cells appear to be productively infected, as evidenced by indirect immunofluorescence, giving rise to virus titers of 10^7 PFU/ml culture medium. When persistently infected human myocardial fibroblasts were treated with IFN_β at 300 U/ml every 24 hr, elimination of infectious CVB3 from the supernatant medium was achieved within 11–13 days, demonstrating the potent *in vitro* activity of IFN_β in human myocardial fibroblasts. Another important finding is the observation that the potent antiviral activity of IFN_β is completely blocked in the presence of prednisolone (10 μM).

Using the molecular hybridization method described above, the quantitative impact of therapeutic interventions on viral replication can now be measured at the nucleic acid level. Preliminary results obtained with a slot-blot hybridization assay[36] for the detection of CVB3 RNA indicate that following IFN_β treatment of persistently infected human myocardial fibroblasts viral RNA is still detectable after infectious virus is no longer assayable by biologic tests in the culture medium.

11. DOWNREGULATION OF INTERFERON RECEPTORS

When persistently infected CVB3 carrier cultures of human myocardial fibroblasts were treated with IFN_{β} at 100 U/ml every 24 hr, a reduction of infectious progeny virus was observed from 3×10^7 to 9×10^3 PFU/ml culture medium within 5 days. However, initial virus titers were reached again later on, in spite of continued IFN_{β} treatment at 100 U/ml every 24 hr. This decreased antiviral action of IFN_{β} at a dose of 100 U/ml is presumably mediated by an IFN_{β}-induced downregulation of IFN_{β} receptors within 5 days, since the loss of antiviral activity could be overcome by higher doses of IFN_{β}, e.g., 300 U/ml. The observation of an IFN_{β} mediated induction of a refractory state to IFN_{β} at a dose of 100 U/ml after 5 days, which was not found at a dose of 300 U/ml, is worth further investigation with respect to a possible role of IFN_{β}-receptor downregulation *in vivo,* analogous to certain forms of insulin resistance *in vivo.*[62,63]

12. SUMMARY AND OUTLOOK

The technique of recombinant DNA is now being used successfully to solve a variety of clinical problems. Within the field of infectious diseases, cloned viral DNA provides the unique opportunity for *in situ* detection of viral genomes, whether or not they are associated with viral proteins. Thus, not only extensive viral replication is detected, but also subtle virus–cell interactions, e.g., in persistent infections in which permissive viral replication is replaced by defective or restricted replication.

To study the role of enteroviruses in the pathogenesis of human viral heart disease, the viral genome of a large plaque variant of the CVB3 was molecularly cloned. Full-length reverse-transcribed cloned viral cDNA of ~7500 nucleotides generated infectious virus antigenically identical to CVB3 upon transfection of recombinant plasmid DNA into mammalian cells, demonstrating the molecular cloning of a faithful transcript of the original viral RNA. The cloned CVB3 cDNA provides a valuable diagnostic tool for patients with suspected enterovirus heart disease and is currently being used to study the molecular basis of human enterovirus heart disease.

Owing to the remarkably high degree of sequence homology among different members of the large human enterovirus group, the complete CVB3 cDNA can be used as a diagnostic probe for the numerous related cardiotropic enteroviruses. As demonstrated by cross-hybridization experiments, detection of the most often implicated agents of human viral heart disease including group A and B CV as well as echoviruses is possible in a single hybridization assay. This finding is expected to facilitate diagnosis of suspected enterovirus infections considerably, since from the clinical point of view the exact typing of an etiologically implicated enterovirus strain appears to be of secondary importance and can be carried out later, e.g., by standard virologic techniques or by hybridization with serotype-specific enterovirus cDNA fragments.

The results obtained so far with human endomyocardial biopsies as well as myocardial tissue samples from recipient hearts of patients who underwent heart transplantation indicate a considerable potential of molecular *in situ* hybridization for the diagnosis of suspected enterovirus heart disease. The method developed permits enterovirus RNA to be related in a highly sensitive way to distinct infected myocardial cells, thereby providing a unique possibility for an unequivocal diagnosis. Interestingly, positive *in situ* hybridization for enterovirus sequences was found in endomyocardial biopsies obtained from patients 2–6 months after onset of clinical symptoms, indicating persistence of enterovirus RNA in the human myocardium.

Persistence of viral RNA is an intriguing concept for the pathogenesis of chronic dilated cardiomyopathy, evolving from acute or subacute viral heart disease. However, the presence of enterovirus RNA in chronic myocardial disease does not absolutely establish that the virus is the etiologic agent, since a preexisting dilated cardiomyopathy could also be a predisposing factor for infection of the myocardium. A follow-up study of patients with enterovirus heart disease diagnosed at an early stage of the disease by *in situ* nucleic acid hybridization will permit investigation of the natural course of enterovirus heart disease.

To investigate further the molecular mechanisms of enterovirus replication and persistence, the infectious CVB3 viral cDNA is used as a source for the generation of distinct viral proteins in different expression systems. In a first approach, fusion proteins containing VP4, VP2, and VP3 or VP1 or the RNA-dependent RNA-polymerase were expressed in *E. coli,* purified, and used to elicit antibodies in rabbits. The resulting antisera recognize native viral proteins from purified CVB3 as well as from CVB3 infected cells. In addition, cells infected with CVB1 and CVB5 are detected, indicating a possible broad use of these antisera as diagnostic tools.

The application of these antibodies in combination with *in situ* hybridization is currently pursued. Using a double-labeling assay, the simultaneous *in situ* detection of enteroviral RNA and distinct viral proteins appears to be feasible at the single-cell level. This approach will permit resolution of whether restricted or defective viral replication is implicated in persistent forms of enterovirus infections. Presumably, enterovirus persistence in a myocardial cell is based on continued synthesis of the viral precursor polyprotein and correct processing at least of the virus-encoded polymerase. However, lack of correct protein processing of other enterovirus gene products, e.g., enterovirus coat proteins, could result in restricted or in defective replication.

The nucleotide sequence of the infectious CVB3 cDNA will be completed in the near future. This information makes it possible to specifically mutagenize the cloned CVB3 cDNA, which will allow genetic analysis of this virus, e.g., with respect to virulence or cardiotropism. Moreover, new constructions involving eucaryotic expression vectors are being designed in order to study the interaction between distinct CVB3 proteins and cellular mechanisms.

The most appealing treatment for viral heart disease, which in some cases may conceivably prevent the onset of dilated cardiomyopathy is antiviral chemotherapy. Using the human myocardial tissue-culture systems described, human fibroblast IFN_{β} was found to act as a potent inhibitor of CVB3 replication, with the consequence that myocytes protected by IFN_{β} could be kept alive and beating. Moreover, treatment of persistently infected CVB3 carrier cultures of human myocardial fibroblasts with IFN_{β} at a dose of 300 U/ml culture medium every 24 hr resulted in elimination of infectious virus from the culture medium within 11–13 days. Although the relationship between the described *in vitro* effects of IFN_{β} and its potential *in vivo* activities remains to be determined, these *in vitro* observations are an important prerequisite for correct therapeutic use of exogenous IFN in severe CV infections. When administered at an early stage of the disease, IFN_{β} would probably reduce viral spread in the myocardium by protecting not yet infected cells, and thereby limit the disease.

Acknowledgments. The contributions of Dr. A. Canu, P. Kirschner, Dr. D. Ameis, Dr. W. Klump, and S. Werner are gratefully appreciated. Echovirus 11 and 12 were kindly provided by Professor H. J. Eggers, of the University of Cologne. Endomyocardial biopsy samples were obtained from Professor E. Erdmann and Dr. H. P. Schultheiss of the Department of Internal Medicine, University of Munich. Myocardial tissue samples from recipient hearts were obtained from Professor B. Kemkes, of the Department of Cardiac Surgery, University of Munich. Human fibroblast IFN was prepared by Bioferon (supported by the German Ministry for Research and Technology) and kindly provided by Dr. H. J. Obert. The generous support of Professor P. H. Hofschneider, of the Department of Virology, Max Planck Institute for Biochemistry, Martinsried, and Professor G. Riecker, of the Department of Internal

Medicine, University of Munich, is appreciated. This work was supported in part by grants from the Deutsche Forschungsgemeinschaft.

REFERENCES

1. Abelmann, W. H., 1973. Clinical aspects of viral cardiomyopathy, in: *Myocardial Diseases* (N. O. Fowler, ed.), pp. 253–279, Grune & Stratton, New York.
2. Lerner, A. M., and Wilson, F. M., 1973, Virus myocardiopathy, *Prog. Med. Virol.* **15:**63–91.
3. Grist, N. R., 1977, Coxsackie virus infections of the heart, *Rec. Adv. Clin. Virol.* **1:**141–150.
4. Landsdown, A. B. G., 1978, Viral infections and diseases of the heart, *Prog. Med. Virol.* **24:**70–113.
5. Levine, H. D., 1979, Virus myocarditis: A critique of the literature from clinical, electrocardiographic, and pathologic standpoints, *Am. J. Med. Sci.* **277:**132–143.
6. Woodruff, J. F., 1980, Viral myocarditis, *Am. J. Pathol.* **101:**427–479.
7. Wynne, J., and Braunwald, E., 1980, The cardiomyopathies and myocarditides, in: *Heart Disease. A Textbook of Cardiovascular Medicine* (E. Braunwald, ed.), pp. 1437–1498, W. B. Saunders, Philadelphia.
8. Johnson, R. A., and Palacios, I., 1982, Dilated cardiomyopathies of the adult, *N. Engl. J. Med.* **307:**1119–1126.
9. Melnick, J. L., 1984, Enteroviruses, in: *Viral Infections of Humans,* 2nd ed., (A. S. Evans, ed.), pp. 187–251, Plenum, New York.
10. Bolte, H. D., ed., 1984, *Viral Heart Disease,* Springer-Verlag, New York.
11. Reyes, M. P., and Lerner, A. M., 1985, Coxsackievirus myocarditis—with special reference to acute and chronic effects, *Prog. Cardiovasc. Dis.* **27:**373–394.
12. Melnick, J. L., 1983, Portraits of viruses: The picornaviruses, *Intervirology* **20:**61–100.
13. Putnak, J. R., and Phillips, B. A., 1981, Picornaviral structure and assembly, *Microbiol. Rev.* **45:**287–315.
14. Cambridge, G., MacArthur, C. G. C., Waterson, A. P., Goodwin, J. F., and Oakley, C. M., 1979, Antibodies to coxsackie B viruses in congestive cardiomyopathy, *Br. Heart J.* **41:**692–696.
15. Waterson, A. P., 1980, Coxsackie viruses in acute and chronic cardiac disease, in: *Myocardial Biopsy* (H. D., Bolte, ed.), pp. 116–118, Springer-Verlag, New York.
16. Kuhn, H., Breithardt, G., Knieriem, H. J., Loogen, F., Both, A., Schmidt, W. A. K., Stroobandt, R., and Gleichmann, U., 1975, Die Bedeutung der endomyokardialen Katheterbiopsie für die Diagnostik und die Beurteilung der Prognose der kongestiven Kardiomyopathie, *Dtsch. Med. Wochenschr.* **100:**717–723.
17. Sutton, G. C., Harding, H. B., Trueheart, R. P., and Clark, H. P., 1967, Coxsackie B4 myocarditis in an adult: Successful isolation of virus from ventricular myocardium, *Aerospace Med.* **38:**66–69.
18. Kandolf, R., and Hofschneider, P. H., 1984, Effect of interferon on the replication of coxsackie B3 virus in cultured human fetal heart cells, in: *Viral Heart Disease* (H. D. Bolte, ed.), pp. 57–63, Springer-Verlag, New York.
19. Kandolf, R., Canu, A., and Hofschneider, P. H., 1985, Coxsackie B3 virus can replicate in cultured human foetal heart cells and is inhibited by interferon, *J. Mol. Cell. Cardiol.* **17:**167–181.
20. Stewart II, W. E., 1981, *The Interferon System,* 2nd ed., Springer-Verlag, Wien.

21. Kandolf, R., and Hofschneider, P. H., 1985, Molecular cloning of the genome of a cardiotropic coxsackie B3 virus: Full-length reverse-transcribed recombinant cDNA generates infectious virus in mammalian cells, *Proc. Natl. Acad. Sci. USA* **82:**4818–4822.
22. Dec, G. W., Palacios, I. F., Fallon, J. T., Aretz, H. T., Mills, J., Lee, D. C.-S., and Johnson, R. A., 1985, Active myocarditis in the spectrum of acute dilated cardiomypathies, *N. Engl. J. Med.* **312:**885–890.
23. Mason, J. W., 1978, Techniques for right and left ventricular endomyocardial biopsy, *Am. J. Cardiol.* **41:**887–892.
24. Parrillo, J. E., Aretz, H. T., Palacios, I., Fallon, J. T., and Block, P. C., 1984, The results of transvenous endomyocardial biopsy can frequently be used to diagnose myocardial diseases in patients with idiopathic heart failure, *Circulation* **69:**93–101.
25. Godman, G. C., 1973, Picornaviruses, in: *Ultrastructure of Animal Viruses and Bacteriophages* (A. J. Dalton and F. Haguenau, eds.), pp. 133–153, Academic, New York.
26. Melnick, J. L., Wenner, H. A., and Phillips, C. A., 1979, Enteroviruses, in: *Diagnostic Procedures for Viral, Rickettsial and Chlamydial Infections,* 5th ed. (E. H. Lennette and N. J. Schmidt, eds.), pp. 471–534, American Public Health Assoc., Washington, D.C.
27. Engleberg, N. C., and Eisenstein, B. I., 1984, The impact of new cloning techniques on the diagnosis and treatment of infectious diseases, *N. Engl. J. Med.* **311:**892–901.
28. Bornkamm, G. W., Desgranges, C., and Gissmann, L., 1983, Nucleic acid hybridization for the detection of viral genomes, *Curr. Top. Microbiol. Immunol.* **104:**287–298.
29. Okayama, H., and Berg, P., 1982, High-efficiency cloning of full-length cDNA, *Mol. Cell. Biol.* **2:**161–170.
30. Gubler, U., and Hoffman, B. J., 1983, A simple and very efficient method for generating cDNA libraries, *Gene* **25:**263–269.
31. Hanahan, D., 1983, Studies on transformation of *Escherichia coli* with plasmids, *J. Mol. Biol.* **166:**557–580.
32. Maniatis, T., Fritsch, E. F., and Sambrook, J., 1982, *Molecular Cloning: A Laboratory Manual,* Cold Spring Harbor Laboratory, Cold Spring Harbor, New York.
33. Racaniello, V. R., and Baltimore, D., 1981, Cloned poliovirus complementary DNA is infectious in mammalian cells, *Science* **214:**916–919.
34. Semler, B. L., Dorner, A. J., and Wimmer, E., 1984, Production of infectious poliovirus from cloned cDNA is dramatically increased by SV40 transcription and replication signals, *Nucleic Acids Res.* **12:**5123–5141.
35. Omata, T., Kohara, M., Sakai, Y., Kameda, A., Imura, N., and Nomoto, A., 1984, Cloned infectious complementary DNA of the poliovirus Sabin 1 genome: Biochemical and biological properties of the recovered virus, *Gene* **32:**1–10.
36. Thomas, P. S., 1980, Hybridization of denatured RNA and small DNA fragments transferred to nitrocellulose, *Proc. Natl. Acad. Sci. USA* **77:**5201–5205.
37. Hyypiä, T., Stålhandske, P. Vainionpää, R., and Pettersson, U., 1984, Detection of enteroviruses by spot hybridization, *J. Clin. Microbiol.* **19:**436–438.
38. Tracy, S., 1984, A comparison of genomic homologies among the coxsackievirus B group: Use of fragments of the cloned coxsackievirus B3 genome as probes, *J. Gen. Virol.* **65:**2167–2172.
39. Bowles, N. E., Richardson, P. J., Olsen, E. G. J., and Archard, L. C., 1986, Detection of coxsackie-B-virus-specific RNA sequences in myocardial biopsy samples from patients with myocarditis and dilated cardiomyopathy, *Lancet* **1:**1120–1123.
40. Rancaniello, V. R., and Baltimore, D., 1981, Molecular cloning of poliovirus cDNA

and determination of the complete nucleotide sequence of the viral genome, *Proc. Natl. Acad. Sci. USA* **78**:4887–4891.

41. Maxam, A. M., and Gilbert, W., 1977, A new method for sequencing DNA, *Proc. Natl. Acad. Sci. USA* **74**:560–564.
42. Stålhandske, P. O. K., Lindberg, M., and Pettersson, U., 1984, Replicase gene of coxsackievirus B3, *J. Virol.* **51**:742–746.
43. Crowell, R. L., and Philipson, L., 1971, Specific alterations of coxsackievirus B3 eluted from Hela cells, *J. Virol.* **8**:509–515.
44. Rueckert, R. R., and Wimmer, E., 1984, Systematic nomenclature of picornavirus proteins, *J. Virol.* **50**:957–959.
45. Gall, J. G., and Pardue, M. L., 1969, Formation and detection of RNA–DNA hybrid molecules in cytological preparations, *Proc. Natl. Acad. Sci. USA* **63**:378–391.
46. Wolf, H., zur Hausen, H., and Becker, V., 1973, EB viral genomes in epithelial nasopharyngeal carcinoma cells, *Nature New Biol.* **244**:245–247.
47. Falser, N., Bandtlow, I., Haus, M., and Wolf, H. 1986, Detection of pseudorabies virus DNA in the inner ear of intranasally infected BALB/c mice with nucleic acid hybridization in situ, *J. Virol.* **57**:335–339.
48. Haase, A., Brahic, M., Stowring, L., and Blum, H., 1984. Detection of viral nucleic acids by in situ hybridization, in: *Methods in Virology*, Vol. 7 (K. Maramorosh and H. Koprowski, eds.) pp. 189–226, Academic, New York.
49. Johnson, M. T., and Johnson, B. A., 1984, Efficient synthesis of high specific activity (35S)-labeled human β-globin pre-mRNA, *Biotechniques* **2**:156–162.
50. Remaut, E., Stanssens, P., and Fiers, W., 1981, Plasmid vectors for high-efficiency expression controlled by the pL promoter of coliphage lambda, *Gene* **15**:81–93.
51. Klump, W., Marquardt, O., and Hofschneider, P. H., 1984, Biologically active protease of foot and mouth disease virus is expressed from cloned viral cDNA in *Escherichia coli, Proc. Natl. Acad. Sci. USA* **81**:3351–3355.
52. Klump, W., Soppa, J., Marquardt, O., and Hofschneider, P. H., 1984, Expression of the foot and mouth disease virus protease in *E. coli*, in: *The Impact of Gene Transfer Techniques in Eucaryotic Cell Biology* (J. S. Schell and P. Starlinger, eds.), pp. 202–209, Springer-Verlag, Berlin.
53. Willingham, M. C., Pastan, I., 1985, *An Atlas of Immunofluorescence in Cultured Cells*, Academic, New York.
54. Burch, G. E., 1975, Ultrastructural myocardial changes produced by viruses, *Rec. Adv. Stud. Card. Struct. Metab.* **6**:501–523.
55. Goldberg, R. J., and Crowell, R. L., 1971, Susceptibility of differentiating muscle cells of fetal mouse in culture to coxsackievirus A13, *J. Virol.* **7**:759–769.
56. Crowell, R. L., and Syverton, J. T., 1961, The mammalian cell-virus relationship. VI. Sustained infection of HeLa cells by coxsackie B3 virus and effect on superinfection, *J. Exp. Med.* **113**:419–435.
57. Takemoto, K. K., and Habel, K., 1959, Virus–cell relationship in a carrier culture of HeLa cells and coxsackie A9 virus, *Virology* **7**:28–44.
58. Matteucci, D., Paglianti, M., Giangregorio. A. M., Capobianchi, M. R., Dianzani, F., and Bendinelli, M., 1985, Group B coxsackieviruses readily establish persistent infections in human lymphoid cell lines, *J. Virol.* **56**:651–654.
59. Brahic, M., Haase, A. T., and Cash, E., 1984, Simultaneous *in situ* detection of viral RNA and antigens, *Proc. Natl. Acad. Sci. USA* **81**:5445–5448.
60. Blalock, J. E., and Stanton, J. D., 1980, Common pathways of interferon and hormonal action, *Nature (Lond.)* **283**:406–408.
61. Lampidis, T. J., and Brouty-Boyé, D., 1981, Interferon inhibits cardiac cell function *in vitro, Proc. Soc. Exp. Biol. Med.* **116**:181–185.

62. Aguet, M., Groebke, M., and Dreiding, Ph., 1983, Downregulation of interferon receptors, in: *The Biology of the Interferon System* (E. De Maeyer and M. Schellekens, eds.), pp. 161–170, Elsevier, Amsterdam.
63. Kahn, C. R., 1979, The role of insulin receptors and receptor antibodies in states of altered insulin action, *Proc. Soc. Exp. Biol. Med.* **162:**13–21.
64. Kandolf, R., Ameis, D., Kirschner, P., Canu, A., and Hofschneider, P. H., 1987, *In situ* detection of enteroviral genomes in myocardial cells by nucleic acid hybridization: An approach to the diagnosis of viral heart disease, *Proc. Natl. Acad. Sci. USA* **84:**6272–6276.

18

Neurologic Disorders

ELEANOR J. BELL, FAKHRY ASSAAD, and KARIN ESTEVES

1. INTRODUCTION

The clinical importance of coxsackieviruses (CV) is becoming increasingly more apparent. The role of the six group B viruses (CVB) in aseptic meningitis, respiratory illness, and Bornholm disease is already well established. In addition, these viruses are probably the commonest cause of acute myopericarditis and have recently been implicated as a precipitating factor in congestive cardiomyopathy and acute-onset juvenile diabetes. Good evidence regarding the suspected important role of the 24 group A coxsackieviruses (CVA) in human disease is however scarce. The detection of most of these serotypes still depends mainly on inoculation of newborn mice, a major factor that has discouraged in-depth investigation, except by a few specialized laboratories.

This chapter focuses attention on various aspects of the CV in neurologic disease. These include epidemiologic data at international and regional levels. In addition, recent virus investigations into myalgic encephalomyelitis are described. Improved rapid diagnostic procedures and their potential application in routine diagnostic virology and epidemiologic investigation are also outlined.

ELEANOR J. BELL • Enterovirus Reference (Scotland) Laboratory, Regional Virus Laboratory, Ruchill Hospital, Glasgow G20 9NB, Scotland. FAKHRY ASSAAD • Communicable Diseases Division, World Health Organization, Geneva, Switzerland. (Deceased.) KARIN ESTEVES • Epidemiology and Management Support Services, World Health Organization, Geneva, Switzerland. Dr. Fakhry Assaad, coauthor of this chapter, died on December 28, 1986, after a brief illness. Dr. Assaad possessed unceasing drive but was never too busy to help and care for others. He was a man of sincerity, warmth, and humor. His death has saddened all those who were privileged to know him.

2. EPIDEMIOLOGIC ANALYSIS

Most of the virologic data presented in this section are based on conventional virus isolation from the feces; only occasionally were viruses recovered from the central nervous system (CNS) or cerebrospinal fluid (CSF). Seldom were virus isolation attempts and serologic examinations carried out on the same patient.

Since echovirus type 9 (also known as CVA23) and enterovirus 71 (EV71) share biologic properties with the CVA, these serotypes have been included in the data presented.

2.1. International Data Collected by WHO 1975–1983

In 1963 the World Health Organization (WHO) established a system for the collection and dissemination of information on viral infections. By 1976 laboratories in 49 countries were participating in this. An analysis of 60,000 reports on viral infections in neurologic disease collected during 1967–1976 was published in 1980.[1] This showed a steady increase in the yearly number of reports of viral neurologic disease that closely followed the general increase in the overall reporting of viral diseases.

Computerized WHO data from 42 participating countries are now available for 1975–1983 (Table I). The neurologic disease group comprises three headings: meningitis/encephalitis, paralysis, and other conditions that may not indicate direct involvement of the CNS (e.g., meningism, convulsions). Of the 23,251 reports, 67% were echovirus and 33% were CV infections. Unlike the previous analysis,[1] a downward trend was noted in the number of neurologic infections reported, with the 1983 totals being the lowest recorded since pre-1967 data.[2] This finding is in sharp contrast to the continued general increase in the number of virus infections being reported to WHO irrespective of clinical condition (unpublished WHO data).

To place the CV reports into proper perspective, the main clinical features associated with these viruses and EV71 during this period are shown in Table II. Central nervous system (CNS) involvement predominated especially in EV71 infections. As in an earlier analysis,[3] these data failed to reveal the well-known association of the CVB with cardiac disease. This is not unexpected, since adult viral heart disease is a late manifestation of infection, virus diagnosis being dependent on serologic rather than virus-isolation studies.

Twenty-two of the 24 CVA serotypes were found in association with CNS disease compared with 8 serotypes in the 1967–1974 data[3]; however, in both analyses, CVA9 and CVA23 (echo 9) were the commonest

TABLE I
Reports of Nonpolio Enteroviruses in Neurologic Disease (WHO Data)[a]

Associated viruses	1975	1976	1977	1978	1979	1980	1981	1982	1983	Total	(%)
CVA[b]	459	591	496	254	302	368	192	238	265	3165	(14)
CVB	752	807	413	536	490	368	472	443	204	4485	(19)
Total CV	1211	1398	909	790	792	736	664	681	469	7650	(33)
Echoviruses	2829	1315	861	2164	1781	2710	1093	1630	1116	15,499	(67)
Enterovirus 68–70	0	0	0	0	0	0	4	0	1	5	(0.02)
EV71	0	9	6	16	35	7	6	7	11	97	(0.4)
Total Echo/EV	2829	1324	867	2180	1816	2717	1103	1637	1128	15,601	(67)
Total	4040	2722	1776	2970	2608	3453	1767	2318	1597	23,251	(100)

[a]The difference in total number of CNS infections listed here and in Table II is attributable to the inability of the computer program to deal with some variations in reporting practices.
[b]Includes echovirus 9 = CVA23.

TABLE II
Reports of Coxsackievirus and Enterovirus 71 Infections by Main Clinical Symptoms (WHO Data 1975–1983)

	CVA1[a]		CVB		EV71	
	N	%	*N*	%	*N*	%
Total CNS	3071	35	4364	29	93	54
Cardiac	76	0.9	596	4	0	0
Muscle/joint	46	0.5	302	2	0	0
Skin/mucosa	1425	16	360	2	41	24
Respiratory	1120	13	2880	19	9	5
Gastrointestinal	1389	16	2921	20	10	6
Others	1626	19	3511	23.5	20	12
Total	8753	100	14,934	100	173	100

[a]Includes echovirus 9 = CVA23.

TABLE III
Coxsackievirus A Reports in Neurologic Disease (WHO Data 1975–1983)

	Clinical symptoms			
Virus type	Men/enceph	Paralytic	Other CNS	Total
A1	6	11	1	18
A2	6	2	5	13
A3	2	0	0	2
A4	15	7	9	31
A5	10	1	2	13
A6	5	0	1	6
A7	68	2	11	81
A8	3	4	0	7
A9	1049	15	127	1191
A10	9	2	11	22
A11	2	2	1	5
A13	5	2	1	8
A14	1	0	0	1
A15	3	0	3	6
A16	41	3	7	51
A17	1	1	0	2
A18	5	1	0	6
A20	4	2	1	7
A21	7	0	1	8
A22	1	2	0	3
A23	1304	20	120	1444
A24	9	1	0	10
AUT[a]	65	10	61	136
Total	2621	88	362	3071

[a]Untyped CVA.

TABLE IV
Coxsackievirus B Reports in Neurologic Disease (WHO Data 1975–1983)

Virus type	Clinical symptoms			
	Men/enceph	Paralytic	Other CNS	Total
B1	292	12	113	417
B2	564	21	156	741
B3	648	17	152	817
B4	790	29	225	1044
B5	1070	27	109	1206
B6	81	3	22	106
BUT[a]	26	3	4	33
Total	3471	112	781	4364

[a]Untyped CVB.

serotypes reported (Table III). All six CVB were associated with CNS disease, with CVB5 followed by CVB4 as the most frequent serotypes identified (Table IV). The apparent role of various CVA and CVB viruses in paralytic disease should be interpreted with caution, since most of these reports are based on isolation of the virus from the feces alone. More than 70% of these CV-related neurologic illnesses involved children ≤14 years of age (Table V), a pattern repeated for each of the main serotypes involved.

The seasonal distribution of the CV infections have not been tabulated, since this showed no deviation from the expected; in the Northern

TABLE V
Neurologic Infections and Coxsackievirus Reports by Age (WHO Data 1975–1983)

Age (years)	CVA[a]		CVB	
	N	%	*N*	%
<1	307	10	895	21
1–4	735	24	1274	29
5–14	1143	37	1176	27
All children	2185	71	3345	77
15–24	347	11	333	8
25–59	509	17	618	14
≥60	21	0.7	25	0.6
All adults	877	29	976	23
Total	3062	100	4321	100

[a]Includes echovirus 9 = CVA23.

hemisphere, peak virus activity was seen in July, in the southern hemisphere during December to January.

EV71 was not included in the WHO Reporting System until 1974. Between 1975–1983, a total of 784 EV71 infections were reported to WHO, 85% of which were within the last 3 years of the study period. Analysis of the disease associations of EV71 was possible in only 173 instances (Table II). Most (54%) were CNS disease associated, followed by 24% with the hand-foot-and-mouth syndrome. Eighty-eight percent of infections were in children aged ≤14 years.

The yearly distribution of CV infections diagnosed in CNS conditions (Table I) did not reveal the epidemic periodicity of particular enteroviruses.

Despite its obvious limitations, international data can signal changing trends in viral pathogenicity and because of its vast scope can help confirm illness associations suspected at local level.

2.2. Serial Observations in a Single Virus Laboratory 1975–1985

Despite the sampling bias toward investigation of sick persons, over the years the accumulated results of a single diagnostic laboratory can contribute useful epidemiologic information. The Regional Virus Laboratory (RVL) at Ruchill Hospital in Glasgow provides a diagnostic service to the major population of Scotland and has a particular interest in the

TABLE VI
Enterovirus Isolations in the West of Scotland (RVL Data[a] 1975–1985)

Year	Poliovirus	CVA[b]	CVB	Echovirus	Total
1975	13	7	27	56	103
1976	2	15	13	27	57
1977	4	9	7	29	49
1978	9	0	7	72	88
1979	5	4	14	45	68
1980	27	21	4	191	243
1981	18	14	26	88	146
1982	12	20	15	195	242
1983	7	14	19	37	77
1984	11	11	15	56	93
1985	8	7	76	72	163
Total	116	122	223	868	1329
% Total	9%	9%	17%	65%	100%

[a]RVL data represent virus isolations carried out at the Regional Virus Laboratory, Ruchill Hospital, Glasgow.
[b]Includes echovirus 9 = CVA23.

TABLE VII
Coxsackievirus Infections in the West of Scotland (RVL Data 1975–1985)

Year	A9	A10	A16	A21	A23	AUT[a]	B1	B2	B3	B4	B5	B6
1975	0	0	4	0	0	3	0	7	0	16	1	0
1976	5	0	6	0	1	2	3	0	0	0	8	0
1977	0	0	2	0	7	0	2	0	0	0	4	0
1978	0	0	0	0	0	0	0	6	1	0	0	0
1979	2	1	0	0	1	0	0	3	0	7	0	1
1980	4	0	6	0	4	3	0	0	0	0	4	0
1981	2	0	0	0	7	0	0	17	1	4	2	0
1982	4	0	0	0	12	0	3	2	1	7	0	0
1983	5	0	0	0	2	5	11	0	2	2	0	0
1984	3	0	1	1	5	0	0	1	0	0	14	0
1985	2	0	0	0	4	0	0	0	36	17	13	0
Total	27	1	19	1	43	13	19	36	41	53	46	1

[a]Untyped CVA.

role of enteroviruses in neurologic and cardiac disease. Since the test methods have remained virtually unchanged over the past several years, comparison of results is possible.

In any one year, 15–25 different enteroviruses may circulate in the community. Usually one or two serotypes predominate generally manifesting as summer/autumn outbreaks of aseptic meningitis.

During 1975–1985, 1329 enterovirus isolations were recorded at RVL (Table VI); 65% were echoviruses and 26% CV, figures not dissimilar to those in Table I. The various CV serotypes isolated and their year of detection are listed in Table VII. The predominant viruses encountered were, in descending order, CVB4, CVB5, CVA23, and CVB3. There were no epidemics of CV-associated neurologic illness during this study period. The last major outbreak was in 1965 when CVB5 affected not only Scotland but was also epidemic throughout the United Kingdom and Europe. Tables VIII and IX list the CV infections detected and their main clinical signs. Our findings closely matched those of the WHO data, with CVA9, CVA23, and CVB5 being those most often seen in association with aseptic meningitis. The CVA16 infections together with CVA10 and all the untyped CVA in the rash category were found exclusively in cases of hand-foot-and-mouth disease. Various CVB were isolated from patients with respiratory illness. No CVA were detected in those with cardiac disease; of the eight CVB isolates in this category, two were associated with fatal myocarditis in infants. Investigations of adult heart disease at RVL concentrate chiefly on serologic diagnosis. During 1975–1985, more than 14,000 patients with suspected CVB infections have been tested serologically. Most of these had

TABLE VIII
Coxsackievirus A Infections and Main Clinical Signs (RVL Data 1975–1985)

Clinical signs	A9	A10	A16	A21	A23	AUT[a]	Total
Aseptic meningitis	15	0	0	0	25[b]	1	41
Paralysis	0	0	0	0	0	0	0
Encephalitis	1	0	0	0	0	0	1
Cardiac	0	0	0	0	0	0	0
Muscle/Joint	0	0	0	0	0	0	0
Respiratory	3	0	0	0	3	0	6
Gastrointestinal	3	0	0	0	3	1	7
Rash	1	1[c]	19[c]	0	7	11[c]	39
Other	4	0	0	1	5	0	10
Total	27	1	19	1	43	13	104

[a]Untyped CVA.
[b]Aseptic meningitis often accompanied by maculopapular rash.
[c]All with hand-foot-and-mouth disease.

suspected viral heart disease, of which approximately one third showed evidence of recent infection.[4,5] Until now (see Section 4.3), the serologic approach was not practicable for the investigation of acute neurologic disorders because of their association with a wide variety of enteroviruses, especially the echoviruses.

Of the 100 patients in our study with CV-associated meningitis, the age and sex were known in 96 (Table X). Their age range was 12 days to 65 years. The usual male predominance in the younger age groups was particularly evident here. Fifty-six percent of those affected were ≤14 years of age.

TABLE IX
Coxsackievirus B Infections and Main Clinical Signs (RVL Data 1975–1985)

Clinical signs	B1	B2	B3	B4	B5	B6	Total
Aseptic meningitis	5	2	4	15	33	0	59
Paralysis	0	0	0	1	0	0	1
Encephalitis	0	0	0	0	0	0	0
Cardiac	0	2[a]	1	3	2[a]	0	8
Muscle/Joint	2	0	2	0	2	0	6
Respiratory	5	17	11	12	5	1	51
Gastrointestinal	2	3	8	9	1	0	23
Rash	0	1	4	2	1	0	8
Other[b]	5[c]	11[c]	11	11[c]	2	0	40
Total	19	36	41	53	46	1	196

[a]Two 3-week-old infants with fatal myocarditis (B2, B5).
[b]Mainly headache and fever or febrile convulsions.
[c]Three with sudden infant death syndrome (B1, B2, B4).

TABLE X
Age/Sex Distribution of Patients with Aseptic Meningitis (RVL Data 1975–1985)

	66 Male		30 Female	
Age (yr)	CVA	CVB	CVA	CVB
<1	2	3	0	0
1–4	4	8	3	2
5–14	9	16	3	4
All children	15	27	6	6
15–24	3	4	5	1
25–59	8	9	3	8
≥60	0	0	0	1
All adults	11	13	8	10
Total[a]	26	40	14	16

[a]Age or sex unknown in four patients.

Because there were no major outbreaks of CV infection during 1975–1985, echoviruses being those predominantly isolated, cyclical periodicity of particular viral serotypes was not evident from these RVL data. Nevertheless, the overall picture seen at regional level was similar to that observed at international level.

3. NEUROLOGIC SYNDROMES

The literature on enteroviruses and their disease associations is vast. Most recent journal publications on CV have focused on new laboratory diagnostic tests for their known or suspected etiology in cardiac disease, diabetes, and myalgic encephalomyelitis. A scan of the world literature on their role in aseptic meningitis, paralytic disease, and encephalitis has failed to demonstrate any significant change in pathogenicity since the last comprehensive reviews published in 1978[3] and 1984.[6] Attention is drawn to information reported regularly by the Virus Unit of WHO and to Tables I–V. World data, despite obvious limitations, can be a good indicator of the relationship of viruses to certain diseases and can detect emerging variations in viral pathogenicity.

3.1. Aseptic Meningitis

Nonbacterial aseptic meningitis (AM) continues to be a common sporadic or epidemic manifestation of enterovirus infection. Pleocytosis

TABLE XI
Neurologic Disease Associations of Coxsackievirus and Enterovirus 71

Clinical feature	Outbreaks	Sporadic
Aseptic meningitis	CVA7, 9, 23	CVA1–11, 14, 16–18, 22–24
	CVB1–6	CVB1–6
	EV71	EV71
Paralytic disease	CVA7	CVA4, 6, 7, 9, 11, 14, 21, 23
	EV71	CVB1–6
		EV71
Encephalitis	CVA23	CVA2, 4, 5, 6, 9, 23
	EV71	CVB1–5
		EV71

of the cerebrospinal fluid (CSF) is characteristically lymphocytic but may show transient polymorphonuclear predominance when cell counts are high in the early stage of illness, temporarily mimicking bacterial meningitis. The prognosis is usually good, however, with cases recovering completely within a few weeks, although irritability and fatigue may persist for some weeks, and there is a small risk of serious neurologic damage in the first year of life.[7] Second attacks of AM may occur from infection with different virus types.[8]

Owing partly to the intensive investigations carried out into this type of illness, many enteroviruses have been implicated, with echoviruses predominating. However, outbreaks due to CVA7, CVA9, CVA23, and CVB types 1–5, and occasionally CVB6, have been reported (Table XI). EV71 has emerged as a major sporadic and epidemic cause of AM often accompanied by extensive neurologic complications such as paralysis and encephalitis.

3.2. Paralytic Disease

Polioviruses remain the dominant viral cause of paralysis in most of the world. In countries in which these viruses have been suppressed by efficient immunization programs, sporadic cases may still be seen. Some of the reported associations between paralytic disease and certain enterovirus infections are of doubtful significance, as they are often based on isolation of virus from feces alone.

During the 1950s and 1960s, CVA7 was the only other enterovirus associated with outbreaks of paralytic disease. At that time, it was feared that this virus might supplant the polioviruses in this respect. However, CVA7 has not re-emerged as a serious problem (Table III), although sporadic cases continue to be reported.[9] Among the most recently recognized viruses capable of causing severe CNS disease with persisting flaccid paralysis is EV71, which is biologically similar to the CVA and is difficult to isolate in cell culture. In the original California outbreaks, AM predominated, but cases of encephalitis were also seen. Outbreaks in different regions followed. In some, hand-foot-and-mouth disease predominated, in some AM, and in others the clinical pattern was mixed. In the 1975 Bulgarian outbreak of >705 cases, 149 patients developed paralysis, of whom 44 died; infants and young children were the chief victims. EV71 next appeared as part of a mixed epidemic in Hungary in 1978 when 1550 cases of CNS disease were reported. There were 866 cases of AM and 724 cases of encephalitis with 45 fatalities, 27 of them children. The cases of encephalitis included 13 poliolike cases with flaccid paralysis. At least two agents were responsible for this epidemic; a tick-borne encephalitis accounted for most of the adult cases and EV71 for those in children.[10]

With the exception of EV71 and CVA7, sporadic cases of paralysis due to nonpolio enteroviruses usually have a good prognosis with most patients recovering completely.

3.3. Encephalitis

Brain damage manifested as frank encephalitis or ataxia with or without sequelae has long been recognized as a rare complication of enterovirus infection, especially as part of a generalized neonatal infection with CV.

Again, the dominant enterovirus associated with outbreaks of encephalitis is EV71 (see Section 3.2). Many of the CV may sporadically be associated with encephalitis particularly within the course of an outbreak of AM; CVA types 9 and 23 and CVB types 3, 5, and 6 have all been isolated from CSF and brain tissue.

In cases of encephalitis, it is particularly important to exclude herpes virus infection, since antiviral treatment is available for this condition. The course of enteroviral encephalitis is relatively benign, but instances of neurological sequelae or death have been reported.[10,11]

3.4. Myalgic Encephalomyelitis

Outbreaks and sporadic cases of this bizarre illness have been reported throughout the world during the past 50 years. Various terms used to describe it have included epidemic neuromyasthenia, Iceland Disease, and Royal Free Disease named after a large outbreak within personnel in that London Hospital in 1955. Currently the term myalgic encephalomyelitis (ME) is regarded as that which best encompasses the multiple symptomatology associated with this illness.[12] In most outbreaks, women are more often affected than are men and a curious susceptibility is shown by nursing and medical staff. The afflicted patients have a wide variety of complaints but these always include muscle pain, extreme fatigue following slight physical effort, and psychological upset. The lack of objective physical signs, pronounced emotional lability, and neuroticism and the unknown nature of the disorder have made it easy for some physicians to dismiss the illness as hysterical.[13] ME is not a fatal disease, but recovery may take months or even years; relapses during physical or mental stress are common.

An outbreak of ME occurred in the west of Scotland in 1980–1981. The only positive virologic finding was the detection of significantly high (≥512) CVB-neutralizing antibody titers in 69% of the 16 females affected and in 33% of the six male patients.[14] Epidemiologically, this observation was significant, as a study of healthy adults in the community showed that only 4% had similarly elevated titers.[4] Two further studies of well-documented sporadic cases of ME again revealed a larger number of patients with elevated CVB-neutalizing titers than could be expected by chance[15,16] (Table XII).

Although other viral infections (e.g., influenza, varicella, Epstein–Barr virus) may precipitate ME, our studies suggested that the CVB played an important part in this illness. New techniques for the detection of CVB-specific Igf implying recent or active infection have been applied recently to the study of patients with ME. Between January and August, 1985, 118 sporadic cases of ME were tested; 31% were CVB-IgM positive compared with 9% of 304 healthy adults tested during the

TABLE XII
Comparison of Coxsackievirus B Serologic Studies on Patients with Myalgic Encephalomyelitis (ME) and on Healthy Adults (RVL Data)

Category	No. tested	Year	Neutralization % titer ≥512	% IgM positive	Ref.
ME (epidemic)	22	1980–1981	59	—	14
ME (sporadic)	17	1981–1982	76	—	15
ME (sporadic)	50	1982–1984	70	—	16
Healthy adults	950	1973–1978	4	—	4
Healthy adults	87	1984	5	—	5
ME (sporadic)	118	1985	—	31[a]	5
Healthy adults	304	1985	—	9[a]	5

[a]In 85% of the positive ME patients there were elevated CVB neutralizing antibody titers compared with only 8% of positive healthy adults.[5]

same time period[5] (Table XII). Because of the complex nature of their illness, ME patients are not virologically investigated until some months after onset of illness, which may explain why only one third were CVB IgM positive. However, these results attain greater significance when compared with those of 352 patients with myo/pericarditis tested in 1985; 33% of this group were CVB IgM positive. The role of CVB in this illness is already well established.

Research at other centers to confirm or refute these observations and further clarify the role of viruses in ME is urgently required. Currently, the clinical diagnosis of ME remains difficult without resort to sophisticated tests, such as electromyography (EMG), nuclear magnetic resonance (NMR), and electron microscopic evaluation of muscle biopsy tissue. These tests on selected patients have suggested that ME is associated with disordered regulation of the immune system (triggered by viral infection?) and excessive glycolytic activity during muscle metabolism.[16] The true incidence of ME is unknown but is suspected by some general practitioners to be as frequent as 1 per 1000. No specific treatment is available, only symptomatic relief. Clearly, further work on this distressing and debilitating disease is indicated.

4. NEW VIRUS DIAGNOSTIC TECHNIQUES

The diagnosis of CVA infections has been hampered by the fact that many clinical isolates grow poorly, if at all, in readily available cell cultures. These viruses can be isolated by inoculation of newborn mice, but this facility is not widely available in routine virus diagnostic laboratories. Similarly, their serologic diagnosis has proved impracticable because of the large number of serotypes involved and the fact that many patients exhibit heterotypic responses after infection.

While the CVB grow well in a variety of cell-culture systems, their cytopathic effect on these cells does not permit distinction between serotypes or indeed between others of the enterovirus group. The conventional method of serotype identification using specific neutralizing antibody tests in cell cultures is tedious and time consuming. Other methods of identification such as immunofluorescence and hemagglutination inhibition have not attained widespread usage in virus laboratories. The serologic diagnosis of CVB infections by microneutralization tests has been widely applied, particularly in cases of suspected viral heart disease and other chronic relapsing illnesses in which virus isolation is unlikely. Although epidemiologically useful, interpretation of the significance of static high titers in the individual patient is notoriously difficult.

4.1. Detection of Coxsackieviruses A and B Antigens by ELISA

Yolken and Torsch in 1981[17] described an enzyme-linked immunosorbent assay (ELISA) system for the detection and serotyping of CVA antigens. The test was a double-antibody ELISA that used type-specific monkey and mouse anti-CV antisera. Although some crossreactivity was noted their assay method correctly identified 22 of the 23 CVA complement-fixation (CF) antigens available for testing. The infecting CVA antigen was unequivocally identified in 8 of 11 stool specimens known to contain CVA.

Earlier, these investigators had applied this same double-antibody ELISA technique to the detection and serotyping of CVB antigens in infected cell cultures and rectal swabs from patients.[18] This assay was capable of identifying and distinguishing all six CVB serotypes at concentrations 100–10,000-fold less than could be detected by CF. This assay correctly identified CVB antigen in 19 of 21 cell-culture fluids and in 5 of 9 rectal swab specimens examined. Cell-culture fluids and rectal swabs containing other viruses, such as echovirus, CVA, rhinovirus, rotavirus, and Norwalk virus, were consistently negative in this assay.

While these techniques offered great potential for the rapid detection of CV antigens, there is little evidence so far of their widespread use. The reasons for this are possibly twofold. First, the mouse and monkey antisera or monoclonal antibody substitutes essential for this assay are expensive. Second, each clinical specimen, irrespective of whether virus is present, must be tested against all 30 CV serotypes plus appropriate controls. Currently, this is practically and financially not feasible, except in specialized laboratories.

4.2. Detection of Coxsackievirus Sequences by Hybridization Probes

Coxsackieviruses, despite their considerable morbidity, have not been investigated intensively at molecular level. This failure may be due partly to their having an RNA rather than a DNA genome, making the nucleic acids less amenable to molecular cloning. However, the structure and organization of enterovirus genomic RNA are now well understood, and cDNA cloning techniques make its genetic manipulation feasible. This approach has proved invaluable in analyzing the molecular basis of neurovirulence in poliovirus.

Recently, a cloned DNA complementary to CVB2 RNA was synthesized.[19] This hybridization probe successfully detected CVB sequences in 9 of 17 myocardial biopsy samples taken from patients with histologic evidence of active or healing myocarditis or cardiomyopathy. No virus-specific sequences were found in samples from patients with unrelated disorders.

This probe method is more suitable for the detection of CV in tissues accessible for biopsy. Its application to muscle biopsies from patients with ME could yield invaluable information concerning the suspected viral etiology of this illness. Their use in the detection of virus in postmortem tissues has yet to be evaluated.

Rotbart *et al.* in 1985[20] devised reconstruction experiments in which enteroviruses artificially added to CSF specimens were detected using CVB3 and polio 1 cDNA probes. These investigators admitted that the sensitivity of their probes—10^2–10^5 TCD_{50}—might not be adequate for a practical diagnostic test, since CSF viral titers during enterovirus meningitis may be as little as 10^1–10^3 TCD_{50}/ml. The use of viral probes for the investigation of diseases of suspected CV etiology is an exciting new virologic approach that promises a rich harvest of important data.

4.3. Detection of Coxsackievirus B Specific IgM by ELISA

The recent development of a μ-antibody-capture ELISA technique for the detection of CVB-specific IgM in patients with insulin-dependent juvenile-onset type I diabetes mellitus added a new dimension to the serologic investigation of CVB infections.[21] This technique was adapted and evaluated in this laboratory (RVL) within the context of everyday practice and applied to patients with cardiac and other illnesses of suspected CVB origin.[5] During these studies, strong multiple CVB IgM responses were observed in patients with known systemic CVA and echovirus infections but not in the few poliovirus infections tested. Similar crossreactions had already been noted by other workers.[21,22] It was decided to exploit this apparent disadvantage and in a pilot study at RVL apply this ELISA technique to the rapid detection of systemic enterovirus infection in patients with suspected AM.[23] Conventional virus-isolation studies on CSF and stools were done in parallel. The results obtained on 45 patients are shown in Table XIII. All patients who yielded an enterovirus in cell culture were also CVB IgM positive.

This CVB IgM ELISA test provided a rapid result within 2 days of receipt of the patient's serum at the laboratory. Although CVB IgM responses were detected in acute phase sera, paired sera taken 3–5 days apart were more useful. A positive result was reassuring to the patient and helpful in clinical management. The limitation of the test in identifying the infecting serotype is of more relevance to the virologist and epidemiologist.

The main disadvantage of this test at present is its cost, since all

TABLE XIII
Final Clinical Diagnosis and Enterovirus Results (RVL Data)

Diagnosis	Total number tested	Virus isol. pos.[a]		CVB IgM pos.[b]	
		N	%	N	%
Aseptic meningitis	21	9	43	18	86
Meningism	16	1	6	11	69
Miscellaneous	8	0	0	1	10
Total	45	10	22	30	67

[a]The isolates were CVB5 (1), echo 5 (3), echo 6 (1), echo 7 (1), echo 11 (4).
[b]All patients yielding a virus were also CVB IgM positive.

CVB type 1–5 antigens were essential. The development of a single broadly reactive antigen capable of detecting all nonpolio enteroviruses is desirable. Thus, a rapid, inexpensive diagnostic test, capable of proving systemic infection, would then be practicable for patients with acute neurological and other illnesses of suspected enterovirus etiology.

5. DISCUSSION

Because of the large number of serotypes involved, the rapid diagnosis of enterovirus infections has lagged far behind the advances achieved in other viral infections, such as hepatitis B, rubella, and rotavirus. As a result, most of the data on enterovirus infections in neurologic disorders are based on conventional, time-consuming virus-isolation studies with or without confirmatory serologic tests. This time lag between clinical and laboratory diagnosis, particularly in acute neurologic infections such as aseptic meningitis, has increasingly dissociated the laboratory from the clinical management of the patient.[1] This could explain the fall in viral neurologic disease reports to WHO (Table I). Laboratory investigations have thereby become an epidemiologic rather than a diagnostic tool, and the clinician tends to regard the laboratory investigations as irrelevant to the patient's needs. In the absence of rapid virologic feedback of information to the clinician, the laboratory cannot easily press for more precise clinical details. This breakdown in communication means that the accurate assessment of the role of a given virus in the disease condition of the patient is now not an easy undertaking.

The continued collection of reliable epidemiologic information is further threatened due to financial constraints being experienced by both developed and developing countries alike. Primary health care priorities vary in each country according to need, and virus investigations are expensive in terms of labour and/or materials required. Moreover, the LBM pooled reference antisera, issued for two decades, free to accredited virus laboratories worldwide for enterovirus identification purposes, are now in limited supply.[24] It is recommended that these pools be used mainly to identify the cause of an outbreak; thereafter, identification of other isolates should be achieved using locally prepared antiserum. In individuals, the use of these reference sera should be reserved for the identification of isolates from patients with severe illness, such as paralytic disease or encephalitis. In other situations, it is not essential to identify every enterovirus isolated. Although apparently retrograde, these recommendations are realistic in the current financial climate. Key laboratories throughout the world must continue to receive maximum government and other support for the monitoring of enterovirus infections to permit early detection of emerging new serotypes or changing patterns in antigenicity and pathogenicity of existing known viruses.

While the remit of the larger central laboratories is the provision of reference functions, research facilities and the development, application, and advice on the use of rapid virus diagnostic techniques, local virus laboratories can more usefully concentrate on providing a rapid diagnostic service to referring clinicians. The liaison between these two types of laboratory should be close at all times.

The rapid serologic diagnosis of CVB infections is now feasible using ELISA.[5,21,23,25] Moreover, since detection of virus-specific IgM implies recent or active infection, a more precise viral diagnosis in the individual patient is now possible. Although no specific antiviral therapy is available, a rapid result, positive or negative, can assist the clinician in diagnosis and influence the clinical management of the patient.

In this laboratory, this ELISA has helped confirm the major role of the CVB in acute myo/pericarditis, and their suspected role in ME has been strengthened. More than 4000 patients with suspected CVB infection have been tested at RVL within the past 18 months, which is a measure of the demand for this rapid diagnostic test by clinicians.

Other side benefits of this test have come to light. Because of its known crossreactivity with other nonpolio enteroviruses, it has proved useful for the rapid diagnosis of enterovirus infections in patients with suspected viral mengingitis.[23] Not only was the virus-detection rate increased, a positive report proved the infection was systemic, a definite advantage over isolation of virus from faeces alone.

This test has also proved helpful in the investigations of sudden death in children and adults for whom postmortem tissues were unavailable or unsuitable for virus-isolation studies. A positive result indicated recent infection, providing useful information to the investigating pathologist.

This test was invaluable in the investigation of a child who developed paralysis following the recent administration of oral poliovaccine. It is essential in these rare cases to exclude other intercurrent nonpoliovirus infections. Poliovirus was isolated at RVL from the child who also showed homotypic rising antibody titers; no CV were isolated in newborn mice. The fact that the CVB IgM test was negative (it does not crossreact with the poliovirus group) confirmed that no other enterovirus was involved.

The rapid detection of CV antigens in clinical specimens by ELISA provides another attractive diagnostic approach. However, because of their complexity and expense, they remain for the present within the remit of the specialized laboratory.

The use of virus probes is strictly a research tool particularly suitable for the investigation of such disorders as diabetes, the cardiomyopathies, myalgic encephalomyelitis, and other neurologic diseases, such as motor neuron disease.

All the tests discussed here complement each other. Those that provide a rapid diagnosis require (1) the development of group-reactive reagents to reduce their cost, and (2) enhancement of their sensitivity and specificity before they can provide a substitute for viral isolation.

These exciting new techniques, developed within the 1980s, have already contributed to a greater understanding of the epidemiology and pathogenesis of CV-induced acute and chronic disorders. More will surely follow.

REFERENCES

1. Assaad, F., Gispen, R., Kleemola, M., Syrůček, L., and Esteves, K., 1980, Neurological diseases associated with viral and *Mycoplasma pneumoniae* infections, *Bull. WHO* **58**:297–311.
2. Assaad, F., and Cockburn, W. C., 1972, Four-year study of WHO virus reports on enteroviruses other than poliovirus, *Bull. WHO* **46**:329–336.
3. Grist, N. R., Bell, E. J., and Assaad, F., 1978, Enteroviruses in human disease, *Prog. Med. Virol.* **24**:114–157.
4. Bell, E. J., and McCartney, R. A., 1984, A study of coxsackie B virus infections, 1972–83, *J. Hyg. (Camb.)* **93**:197–203.
5. McCartney, R. A., Banatvala, J. E., and Bell, E. J., 1986, Routine use of μ-antibody-capture ELISA for the serological diagnosis of coxsackie B virus infections, *J. Med. Virol.* **19**:205–212.
6. Moore, M., and Morens, D. M., 1984, Enteroviruses, including polioviruses, in: *Textbook of Human Virology* (R. B. Belshe, ed.), pp. 407–483, P. S. G., Littleton, Massachusetts.

7. Sells, C. J., Carpenter, R. L., and Ray, C. G., 1975, Sequelae of central nervous-system enterovirus infections, *N. Engl. J. Med.* **293**:368–373.
8. Nakao, T., and Miura, R., 1971, Recurrent virus meningitis, *Pediatrics* **47**:773–776.
9. Gear, J. H. S., 1984, Nonpolio causes of polio-like paralytic syndromes, *Rev. Infect. Dis.* **6**:S379–S384.
10. Melnick, J. L., 1984, Enterovirus 71 infections: A varied clinical pattern sometimes mimicking paralytic poliomyelitis, *Rev. Infect. Dis.* **6**:S387–S390.
11. Heathfield, K. W. G., Pilsworth, R., Wall, B. J., and Corsellis, J. A. N., 1967, Coxsackie B5 infections in Essex, 1965 with particular reference to the nervous system, *Q. J. Med.* **36**:579–595.
12. Behan, P. O., 1980, Epidemic myalgic encephalomyelitis, *Practitioner* **224**:805–807.
13. McEvedy, P. C., and Beard, A. W., 1970, Concept of benign myalgic encephalomyelitis, *Br. Med. J.* **1**:11–15.
14. Fegan, K. G., Behan, P. O., and Bell, E. J., 1983, Myalgic encephalomyelitis: Report of an epidemic, *J. R. Coll. Gen. Pract.* **33**:335–337.
15. Keighley, B. D., and Bell, E. J., 1983, Sporadic myalgic encephalomyelitis in a rural practice, *J. R. Coll. Gen Pract.* **33**:339–341.
16. Behan, P. O., Behan, W. M. H., and Bell, E. J., 1985, The post viral fatigue syndrome—An analysis of the findings in 50 cases, *J. Infect.* **10**:211–222.
17. Yolken, R. H., and Torsch, V. M., 1981, Enzyme-linked immunosorbent assay for detection and identification of coxsackieviruses A, *Infect. Immun.* **31**:742–750.
18. Yolken, R. H., and Torsch, V., 1980, Enzyme-linked immunosorbent assay for the detection and identification of coxsackie B antigen in tissue culture and clinical specimens, *J. Med. Virol.* **6**:45–52.
19. Bowles, N. E., Richardson, P. J., Olsen, E. G. J., and Archard, L. C., 1986, Detection of coxsackie-B-virus-specific RNA sequences in myocardial biopsy samples from patients with myocarditis and dilated cardiomyopathy, *Lancet* **1**:1120–1122.
20. Rotbart, H. A., Levin, M. J., Villarreal, L. P., Tracy, S. M., Semler, B. L., and Wimmer, E., 1985, Factors affecting the detection of enteroviruses in cerebrospinal fluid with coxsackie B3 and poliovirus 1 cDNA probes, *J. Clin. Microbiol.* **22**:220–224.
21. King, M. L., Shaikh, A., Bidwell, D., Voller, A., and Banatvala, J. E., 1983, Coxsackie-B-virus-specific IgM responses in children with insulin-dependent juvenile-onset type I diabetes mellitus, *Lancet* **1**:1397–1399.
22. Pugh, S. F., 1984, Heterotypic reactions in a radioimmunoassay for coxsackie B virus specific IgM, *J. Clin. Pathol.* **37**:433–439.
23. Bell, E. J., McCartney, R. A., Basquill, D., and Chaudhuri, A. K. R., 1986, μ-antibody capture ELISA for the rapid diagnosis of enterovirus infections in patients with aseptic meningitis, *J. Med. Virol.* **19**:213–217.
24. Melnick, J. L., and Wimberly, I. L., 1985, Lyophilized combination pools of enterovirus equine antisera: New LBM pools prepared from reserves of antisera stored frozen for two decades, *Bull. WHO* **63**:543–550.
25. Banatvala, J. E., 1983, Coxsackie B virus infections in cardiac disease, in: *Recent Advances in Clinical Virology*, Vol. 3 (A. P. Waterson, ed.), pp. 99–115, Churchill Livingstone, Edinburgh.

19

Mucocutaneous Syndromes

GIOVANNI ROCCHI and ANTONIO VOLPI

1. INTRODUCTION

Coxsackievirus (CV) infections can produce erythematous (maculopapular) and vesicular mucocutaneous lesions. From these lesions as well as from throat secretions or feces, the causative virus can sometimes be isolated. Skin rashes and oropharyngeal lesions may be due to any of several CV. Any of them can cause a variety of mucocutaneous syndromes (Table I) as well as other types of disease. Specific CV diseases are herpangina and hand-foot-and-mouth disease (HFMD), although other enteroviruses have sometimes been isolated in cases of these illnesses.

2. HAND-FOOT-AND-MOUTH DISEASE

Hand-foot-and-mouth disease was first described during an epidemic outbreak in Toronto during the summer of 1957 by Robinson *et al.*[1] It was an acute illness highly communicable from person to person, with short incubation period and duration, characterized by vesiculoulcerative lesions not only on the oropharynx, gingivolabial groove, tongue, and buccal mucosa, but also on the skin. Numerous cases of this disease, sporadic or in epidemic form, have since been reported in almost all parts of the world.[2–4]

Hand-foot-and-mouth disease is predominantly a childhood disease, usually caused by CVA16 or enterovirus type 71, although cases

GIOVANNI ROCCHI and ANTONIO VOLPI • Infectious Diseases Clinic, Department of Public Health, Second University of Rome, 00191 Rome, Italy.

TABLE I
Mucocutaneous Syndromes Associated with CV Infections

Syndromes	CV serotypes	
	Group A	Group B
Herpangina	1–10, 22	1–5
Hand-foot-and-mouth disease	5, 7, 9, 10, 16	2, 5
Lymphonodular pharyngitis	10	—
Erythematous rashes	2, 4, 5, 9, 16	1, 3–5
Epidemic conjunctivitis	24	—

due to CVA5, CVA7, CVA9, CVA10, CVB2, and CVB5 have been occasionally observed.[2–5] Typical cases of HFMD have vesicles and ulcers mainly in the front of the mouth, most frequently on the tongue, and a vesicular exanthem mostly localized on the palms and soles. The vesicular lesions are located subepidermally; extensive acantholysis of overlying epidermidis has been shown by microscopic studies.[6] The intervention of immune response is suggested by a mixed lymphocytic and polymorphonuclear inflammation.[6] Within cells surrounding the dermal vessels, eosinophilic nuclear inclusions and intracytoplasmatic picornavirus particles can be seen.[7–8] Virus can be readily isolated from the vesicle fluid.[9]

During outbreaks, about 60–70% of cases show the complete clinical picture of HFMD, while several cases may be observed of vesicular exanthem without oral lesions (10%) or of vesicular stomatitis without exanthem (30%).[8–10] Virologic studies indicate that clinically evident infections are about 60% of all infections.[11]

Hand-foot-and-mouth disease generally appears after a short incubation period of 3–7 days. However, an incubation of 7–12 days was observed during an outbreak of HFMD due to CVA16 in a study made in Rome in 1973.[10] One investigator in this study developed the typical disease 12 days after her first and only contact with a sick child in a day care center. During the incubation of the disease, 2 days before the onset of symptoms, she occasionally babysat for a 3-year-old child who attended a day care center in another area of town. This child went to school during the following 7 days. On the eighth day, he developed HFMD and was immediately withdrawn from school. Three cases of HFMD occurred among this child's classmates 11–18 days thereafter. The child's father also developed a complete clinical picture of HFMD 7 days later. No other cases occurred among contacts of these patients over a 4-week period. The existence of asymptomatic infections and of long-term virus carriers may, however, have affected the accuracy of calculation of the incubation period in this as well as in other studies.

Hand-foot-and-mouth disease presents with a low-grade fever lasting 2–3 days, as well as malaise and sore throat and mouth. Infants frequently refuse to eat. The enanthem occurs on the oropharynx, tonsillar fauces, buccal mucosa and tongue, and less frequently on the gingiva or palate. The enanthem consists of macules and vesicles 1–6 mm in diameter. Vesicles are surrounded by a thin zone of erythema. They sometimes coalesce, yielding shallow ulcers 1–2 mm diameter (Fig. 1). The number of vesicles or ulcerative lesions generally ranges from 2 to 10. The evolution from vesicles to ulcers occurs in 12–24 hr, so that by the time the patient is first seen by a physician, most mucous lesions are ulcerated.

The cutaneous lesions of HFMD consist of mixed maculopapules and clear vesicles, 3–6 mm in diameter, surrounded by a red areola (Fig. 2). Lesions are more commonly observed on the hands and feet, especially on the dorsum of the fingers and toes. Vesicles are seldom seen on the extensor surface of palms and soles. When present, they appear

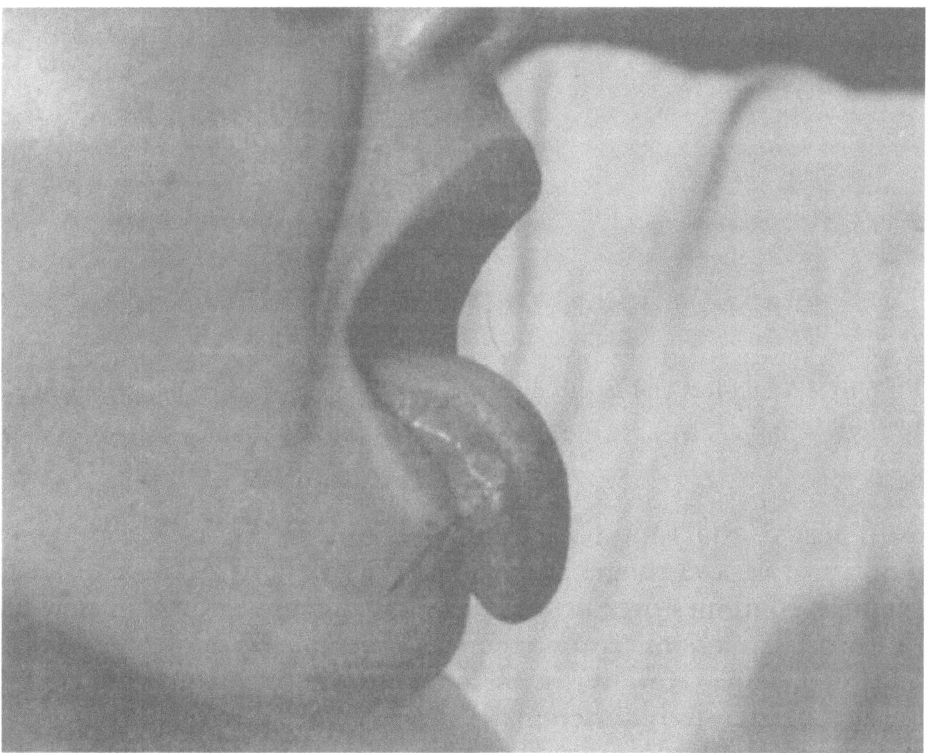

FIGURE 1. Hand-foot-and-mouth disease (CVA16) infection. Note the ulcerative lesion on tongue.

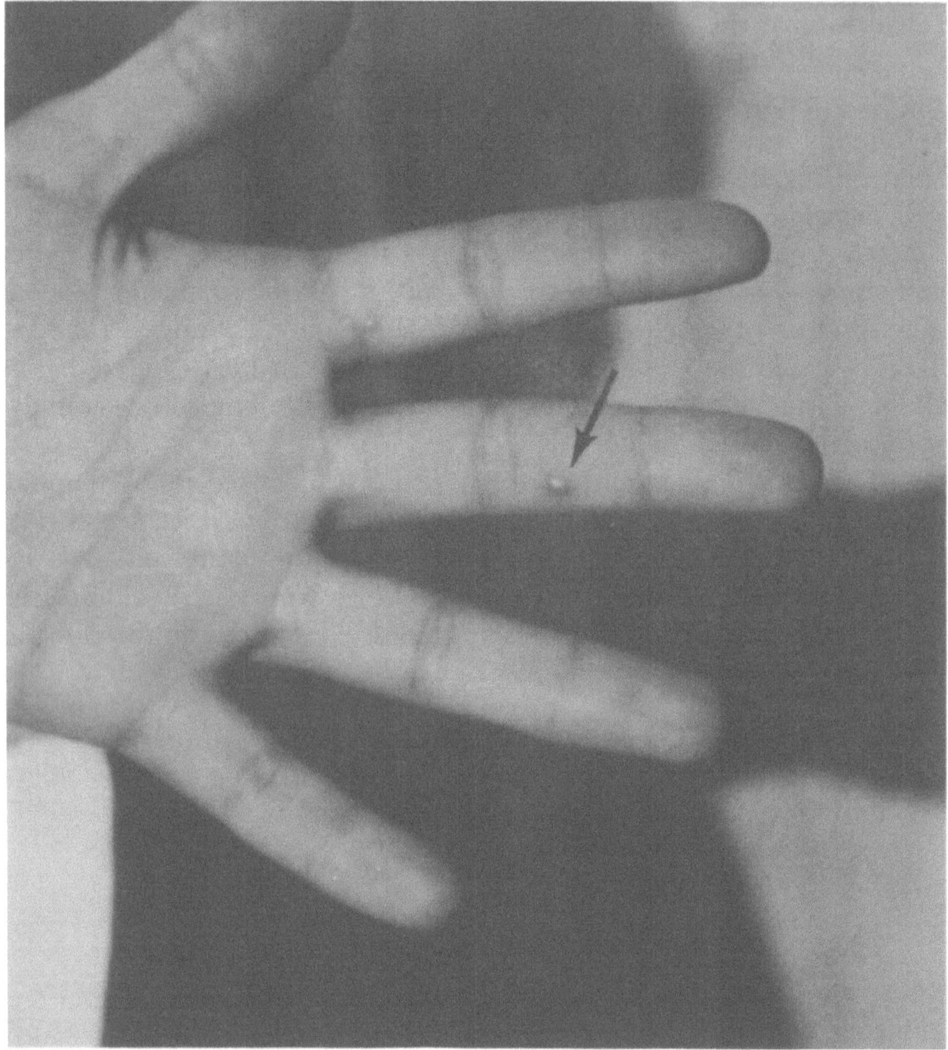

FIGURE 2. Hand-foot-and-mouth disease (CVA16) infection. Note the vesicular lesions on third finger.

deep seated and whitish and may be oval, linear, or crescent. The number of vesicles ranges from 2 to 10 per single area. Sometimes the eruption continues over several days with the appearance of new vesicles. Vesicular lesions are less commonly seen on the buttocks and rarely on the genitalia. However, maculopapules are frequently seen on the buttocks. Extensively disseminated lesions were observed in an infant with pre-existing atopic eczema. The term eczema coxsackium, in analogy to eczema herpeticum, was proposed for this form.[12]

Patients usually do not complain of any pain or pruritus, but the lesions on soles may be moderately painful in the upstanding position.

Mouth and skin lesions develop at the same time and may continue to erupt over a 1–3-day period. Skin vesicles absorb in 2–4 days, leaving a small reddish macula that fades and disappears in a few days. The fever subsides as quickly as the rash, while the enanthem may take more than 1 week to disappear. Other signs and symptoms noted during the illness are abdominal pain, diarrhea, and headache. They subside together with the fever, but adult patients often feel unwell for up to 2 weeks.

A fatal case of CVA7 infection has been reported in a woman with HFMD complicated by pancarditis and pneumonia.[13] Two contacts of this patient had typical HFMD and recovered uneventfully.

Clinical diagnosis of HFMD is usually easy during outbreaks of the disease and in typical cases, once this entity has been seen and considered. Sporadic cases with prevailing cutaneous or buccal lesions may be difficult to differentiate from herpangina, herpes simplex, varicella, or erythema multiforme. The enanthem of herpangina typically involves the faucial pillars and soft palate and only exceptionally extends to the front part of the mouth. Herpetic gingivostomatitis is a severe infection due to herpes simplex virus (HSV) and presents with higher fever and a longer duration than HFMD. In primary HSV infection, cutaneous vesicles are uncommon, and a cervical lymphadenopathy is almost always evident. The vesicular exanthem of varicella is more extended than that of HFMD, usually involving trunk and face with sparing of the palms and soles. The vesicles, unlike those caused by HFMD, evolve to pustules and scabs. Oral lesions are not prominent in the clinical picture of varicella. In erythema multiforme, the cutaneous and mucosal lesions are much larger. The lesion, specifically known as the target, is highly distinctive, since either the erythema or the vesicle is located in a blanched ring, in turn surrounded by an erythematous halo.

Viral isolation may be attempted from throat washings, stool specimens, and vesicle fluid. Throat washings and feces may remain positive for up to 15 and 30 days, respectively. Antibody studies are less reliable diagnostic procedures than viral isolation.

Symptomatic treatment of patients with HFMD is seldom required. Palm and sole lesions are sometimes ruptured to relieve the pain originating from pressure in some of the individual lesions. In this case, it is advisable to paint the corresponding cutaneous area with povidone–iodine solution.

3. HERPANGINA

The term herpangina was introduced by Zahorsky[14] in 1924, 30 years before the recognition of CV, to describe a form of herpetic sore throat. This specific infectious disease is characterized by fever and a vesicular enanthem that typically involves the fauces and soft palate. Herpangina has been causally associated with any one of 12 different CVA (serotypes 1–10, 16, and 22) and less commonly with CVB types 1–5.[4,15–17] Sporadically, echoviruses 3, 6, 9, 16, 17, 25, and 30 have also been isolated from patients with herpangina. The disease occurs during the summer season or late spring and generally involves clusters of patients. Sporadic cases are seldom observed. Herpangina mainly affects young children but is occasionally seen in young adults.

After an incubation period of 2–10 days, the illness begins in a relatively abrupt manner, with fever ranging from 37.4° to 40.5°C in about 90% of patients. Convulsions associated with fever occur in less than 5% of cases. At onset, about one third of patients experience vomiting, and fewer complain of headache or abdominal pain. These symptoms generally subside in 2–4 days. Generalized myalgia has occasionally been reported. Sore throat, dysphagia, and anorexia are the most prominent symptoms. They may precede the appearance of the enanthem by several hours to 1 day. The pharynx is usually hyperemic, and an early inspection in the absence of the characteristic enanthem can lead to a tentative diagnosis of pharyngitis or tonsillitis.

Zahorsky's description of the enanthem on the fauces and adjacent mucuous membranes is classic[14]:

> There are minute vesicles about the size of a millet seed to a small pea situated on the anterior pillars of the fauces or along the free margin of the soft palate. These vesicles are occasionally discovered on the posterior part of the buccal mucous membrane or the roof of the mouth. Much more frequently the blisters are found on the tonsil itself or on pharyngeal mucous membrane. The vesicle seems to begin as a small papule which undergoes vesiculation in 24 hours. This often ruptures and leaves an ulcer having a punched out appearance. The ulcer often becomes covered with a thin exudate and its edges are undermined.

The average number of ulcers is four to six. The duration of the enanthem rarely exceeds 7 days. The illness is self-limited, with complete recovery in a few days. In Zahorsky's description slight scarring, or at least a minute depression in the sites in which ulcers were present during the acute phase of the disease, remained visible in some cases for more than 2–3 weeks after the fever had subsided.[14]

Typical cases of herpangina can be diagnosed on clinical grounds, but viral isolation is needed to identify the causing virus and for epidemiologic purposes. Routine laboratory tests and cytology of smears obtained by scraping the oral lesions do not provide useful data.

Prior to appearance of the oral vesicles and ulcers, herpangina can be confused with bacterial tonsillitis or viral pharyngitis. Among the vesicular and ulcerative diseases, herpangina is typically confined within the posterior oral cavity, in contrast to herpetic gingivostomatitis, aftous stomatitis, and the vesicular enanthem of hand-foot-and-mouth disease.

4. ACUTE LYMPHONODULAR PHARYNGITIS

Steigman *et al.*[18] gave the name acute lymphonodular pharyngitis to an acute febrile disease, in order to distinguish it from herpangina. This disease was characterized by discrete nodular lesions on the mucosa of the posterior oral cavity. Cases were seen during a summer and early fall epidemic caused by CVA10. The patients were predominantly children. Fever, mild headache, malaise, sore throat, and anorexia were usually noted. Enanthem consisted of discrete whitish to yellowish solid nodular lesions, 3–6 mm in diameter, surrounded by a zone of erythema. These lesions did not ulcerate and resolved within 6–10 days. All lesions appeared and regressed at approximately the same time. Lesions were observed more frequently on the uvula, anterior pillars, and pharynx (Fig. 3). Regional lymphadenopathy was extremely mild. Histologic studies on biopsy material showed that the nodules were formed by packed lymphocytes. In several children, conjunctival nodular lesions were observed as well. The disease lasted 4–14 days and subsided uneventfully. On the basis of epidemiologic studies of family clusters of infection, the incubation period was estimated to be about 5 days.

The observation that the lesions of acute lymphonodular pharyngitis have the same distribution as herpangina (even though they do not evolve to vesiculation or ulceration) may lead to consideration of the disease as a variant of herpangina.[5]

5. ERYTHEMATOUS RASHES

Coxsackieviruses have been associated with a variety of maculopapular cutaneous manifestations that can mimic rubella or measles viral infections. Fine maculopapular rubella-like rashes have been reported in cases of CVA9 infection. In these cases, the maculopapular exanthem begins on the face and trunk and spreads to the limbs, where it is located mostly on the extensor surface of the distal parts. Palms and soles are occasionally involved. Pruritus is usually absent. Constitutional symptoms such as fever and malaise are present. Posterior cervical or occipital lymphadenopathy has been reported in about one half of patients.[19] On purely clinical grounds CVA9 rubelliform rash may therefore be easily confused with rubella. The exanthem is also likely to be confused with similar forms caused by echoviruses.[3]

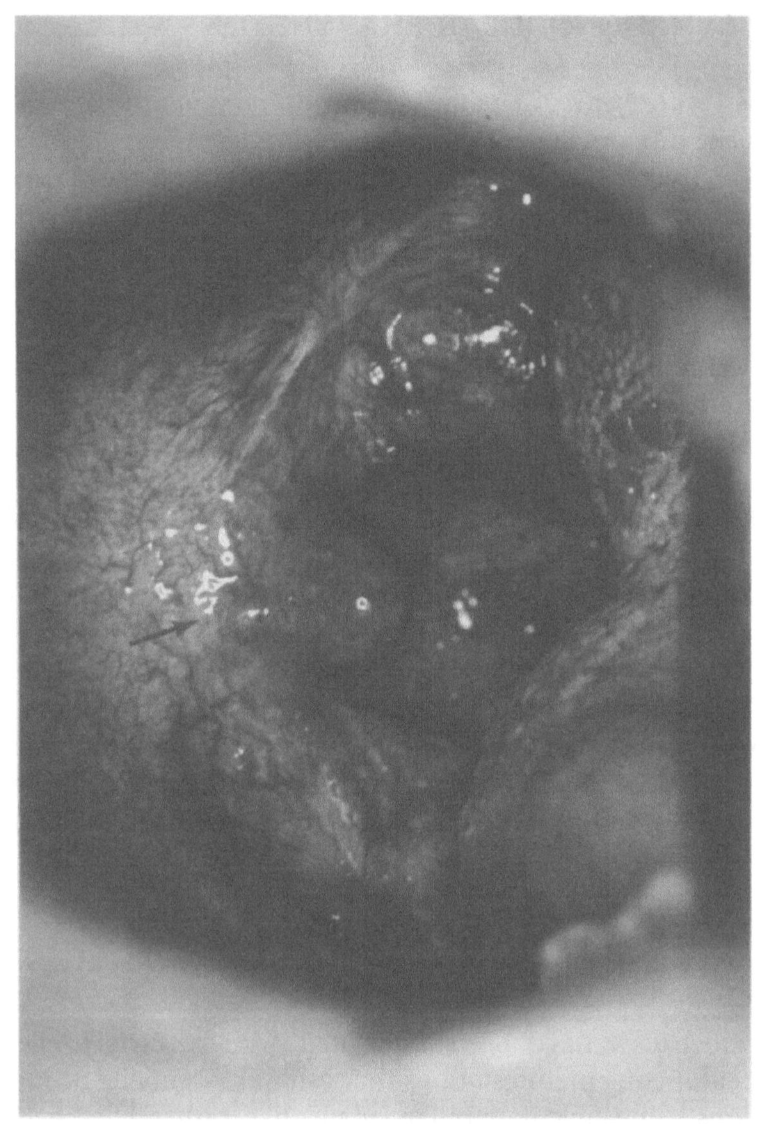

FIGURE 3. Lymphonodular pharyngitis. Note the typical raised lesions on the soft palate (mainly the uvula) and on the anterior pillar.

Morbilliform rash and fever have been reported in association with CVA4 infection. The virus was isolated from the blood and feces during the disease. Moreover, the cutaneous lesions followed typical herpangina and, in some cases, evolved to vesicles that persisted for 1–2 weeks.[20]

The occurrence of petechial and purpuric rashes was occasionally described in association with aseptic meningitis due to CVA9 infection. The syndrome can be easily confused with meningococcal disease.[21] Giannotti–Crosti-like papular acrodermatitis was described in association with CVA16 infection.[22]

Although seldom, hemorrhagic skin manifestations represented the first symptoms of the hemorrhage–hepatitis syndrome, occurring in severe neonatal CVB1–5 infections,[23] as well as in cases of hemolitic–uremic syndrome temporally associated with CVA4 infection.[24]

6. EPIDEMIC CONJUNCTIVITIS

Lim and Yin-Murphy[25] observed an outbreak of acute conjunctivitis in Singapore in 1970 affecting more than 60,000 people. Similar epidemics occurred in Hong Kong 1 year later and in Singapore and Hong Kong in 1975.[26] These epidemics were predominantly due to a variant of CVA24. Thus far, the disease appears to be epidemic mostly in the Far East, probably due to poor sanitation and an overcrowded population.

The disease consists of a mild conjunctivitis that usually subsides completely after 1–2 weeks. Cases of more severe conjunctivitis or associated with subconjunctival hemorrhage are rarely observed. The latter manifestation is frequent in the course of conjunctivitis caused by enterovirus 70.[27] Because it was shown that both CVA24 and enterovirus 70 circulated in the affected population during the cited outbreaks,[26] the term picornavirus epidemic conjunctivitis has been proposed to cover both infections and to better differentiate this condition from the epidemic conjunctivitis due to adenoviruses,[4] which seldom cause subconjunctival hemorrhage.[28]

Epidemic conjunctivitis due to CVA24 is usually self-limiting without local complications. The most prominent ocular manifestations are burning pain, foreign-body sensation, photophobia, and a watery discharge. Fever, malaise, and headache are infrequent, and no neurologic complications have been reported.

The disease is probably spread by direct contact with infected ocular secretions. Fecal–oral transmission is possibly of little importance.

CVA24 has rarely been isolated from feces but is easily isolated from the conjunctiva.

7. CONCLUDING REMARKS

Mucocutaneous syndromes due to CV are markedly seasonal and highly contagious. The ratio of apparent to inapparent infections probably varies depending on virus type and strain. The opportunities for acquiring these infections are greater in early life than in adulthood, when enteroviral illnesses are usually more severe.

Some CV-specific mucocutaneous diseases are recognizable on the basis of typical clinical patterns, seasonal distribution, and knowledge of exposure. Laboratory diagnosis based on viral isolation is, however, always needed because many enteroviruses can be responsible for similar syndromes.

A better knowledge of the immune responses induced by CV would lead to a better understanding of the pathogenesis of the protean mucocutaneous manifestations caused by these viruses. New approaches to laboratory diagnosis that would alleviate the current cumbersome practice of viral isolation in animals would help to make mucocutaneous diseases caused by CV easier to recognize and physicians more aware of their clinical importance.

REFERENCES

1. Robinson, C. R., Doane, F. W., and Rhodes, A. J., 1958, Report of an outbreak of febrile illness with pharyngeal lesions and exanthems. Toronto, Summer 1957—Isolation of group A coxsackievirus, *Can. Med. Assoc. J.* **79:**615–621.
2. Cherry, J. D., and Nelson, D. B., 1966, Enterovirus infections: Their epidemiology and pathogenesis. *Clin. Pediatr.* **5:**659–664.
3. Andreoni, G., and Rocchi, G., 1969, Procedimenti diagnostici nelle malattie esantematiche con particolare riguardo agli esantemi virali, *Giorn. Mal. Inf. Parass.* **7:**548–577.
4. Grist, N. R., Bell, E. J., and Assaad, F., 1978, Enteroviruses in human disease, *Prog. Med. Virol.* **24:**114–157.
5. Modlin, J. F., 1985, Coxsackievirus and echovirus, in: *Principles and Practice of Infectious Diseases* (G. L. Mandell, R. G. Douglas, and J. E. Bennett, eds.), pp. 814–825, Wiley, New York.
6. Miller, G. D., and Tindall, J. P., 1968, Hand-foot-and-mouth disease, *JAMA* **203:**827–830.
7. Kimura, A., Abe, M., and Nakao, T., 1977, Light and electron microscopic study of skin lesions in patients with hand, foot and mouth disease, *Tohoku J. Exp. Med.* **122:**237.
8. Froeschle, J. E., Nahmias, A. J., Feorino, P. M., McCord, G., and Naib, Z., 1967, Hand,

foot and mouth disease (Coxsackievirus A16) in Atlanta, *Am. J. Dis. Child.* **114**:278–283.

9. Tindall, J. P., and Callaway, J. L., 1972, Hand-foot-and-mouth disease. It's more common than you think, *Am. J. Dis. Child.* **124**:372–375.
10. Tosato, G., Rocchi, G., and Archetti, I., 1975, Epidemiological study of a "hand, foot and mouth disease" outbreak observed in Rome in the fall of 1973, *Zentralbl. Bakteriol. Mikrobial. Hyg. [A]* **230**:415–421.
11. Adler, J. L., Mostow, S. R., Mellin, H., Janney, J. H., and Joseph, J. M., 1970, Epidemiologic investigation of hand, foot and mouth disease, *Am. J. Dis. Child.* **120**:309–313.
12. Nahmias, A. J., Froeschle, J. E., Feorino, P. M., and McCord, G., 1968, Generalized eruption in a child with eczema due to coxsackievirus A16, *Arch. Dermatol.* **97**:147–148.
13. Baker, D. A., and Philliphs, C. A., 1979, Fatal hand-foot-mouth disease in an adult caused by coxsackievirus A7, *JAMA* **242**:1065.
14. Zahorsky, J., 1924, Herpangina, a specific infectious disease. *Arch. Pediatr.* **41**:181–184.
15. Huebner, R. J., Cole, R. M., Beeman, E. A., Bell, J. A., and Peers, J. H., 1951, Herpangina: Etiological studies of a specific infectious disease, *JAMA* **145**:628–633.
16. Cherry, J. D., and Jahn, C. L., 1965, Herpangina: The etiologic spectrum, *Pediatrics* **36**:632–634.
17. Rocchi, G., Jemolo, A. M., 1969, Angina vescicolo-ulcerativa da virus coxsackie di gruppo B-Segnalazione di 4 casi, *Giorn. Mal. Inf. Parass.* **10**:811–813.
18. Steigman, A. J., Lipton, M. M., and Braspennickx, H., 1962, Acute lymphonodular pharingitis: A newly described condition due to coxsackie A virus, *J. Pediatr.* **61**:331–336.
19. Lerner, A. M., Klein, J. O., Cherry, J. D., and Finland, M., 1963, New viral exanthems, *N. Engl. J. Med.* **269**:678–685.
20. Forman, M. L., and Cherry, J. D., 1966, Enanthems associated with uncommon viral syndromes, *Pediatrics* **41**:873–882.
21. Cherry, J. D., and Jahn, C. L., 1966, Virologic studies of exanthems, *J. Pediatr.* **68**:204–214.
22. James, W. D., Odom, R. B., and Hatch, M. H., 1982, Giannotti-Crosti-like eruption associated with coxsackievirus A16 infection, *J. Am. Acad. Dermatol.* **6**:862–866.
23. Gear, J. H. S., and Measroch, V., 1973, Coxsackievirus infections of the newborn, *Prog. Med. Virol.* **15**:42–62.
24. Glasgow, L. A., and Balduzzi, P., 1965, Isolation of coxsackie virus group A, type 4, from a patient with hemolytic-uremic syndrome, *N. Engl. J. Med.* **273**:754–756.
25. Lim, K. H., and Yin-Murphy, M., 1971, An epidemic of conjunctivitis in Singapore in 1970, *Singapore Med. J.* **12**:247–249.
26. Yin-Murphy, M., 1984, Acute hemorrhagic conjunctivitis, *Prog. Med. Virol.* **29**:23–44.
27. Kono, R., Sasagawa, A., Miyamura, K., and Tajiri, E., 1975, Serologic characterization and seroepidemiologic studies on acute hemorrhagic conjunctivitis (AHC) virus, *Am. J. Epidemiol.* **101**:444–457.
28. Muzzi, A., Rocchi, G., Lumbroso, B., Tosato, G., and Barbieri, F., 1975, Acute hemorrhagic conjunctivitis during an epidemic outbreak of Adenovirus type 4 infection, *Lancet* **2**:822.

20

Diabetes Mellitus

ANTONIO TONIOLO, GIOVANNI FEDERICO, FULVIO BASOLO, and TAKASHI ONODERA

1. INTRODUCTION

The term diabetes refers to a group of disorders having in common hyperglycemia and an elevated prevalence of serious complications affecting many organ systems. Studies of the natural history and pathogenesis of hyperglycemia have led to a widely accepted classification developed by the National Diabetes Data Group. Among the various diabetic syndromes that are now recognized, insulin-dependent diabetes mellitus (IDDM) and noninsulin-dependent diabetes mellitus (NIDDM) account for the great majority of cases, the former being less prevalent. It is now clear that IDDM derives from a deficiency of insulin production due to the loss of pancreatic β-cells, while NIDDM is mostly due to a reduced effect of this hormone on peripheral tissues. Because coxsackieviruses (CV), and in particular the members of group B (CVB), have been implicated only in the etiopathogenesis of IDDM, this discussion refers exclusively to this form of diabetes.

All patients with IDDM share a common clinical finding: dependence on exogenous insulin for survival. This results from the near-total disappearance of the insulin-producing β-cells of the islets of Langerhans. It is generally believed that genetic, immunologic, and environmental factors all play a role in the irreversible loss of β-cells. There is,

ANTONIO TONIOLO • Institute of Microbiology and Virology, University of Sassari Medical School, 07100 Sassari, Italy. GIOVANNI FEDERICO • Department of Pediatrics, University of Pisa, I-56100 Pisa, Italy. FULVIO BASOLO • Department of Pathology, University of Pisa, I-56100 Pisa, Italy. TAKASHI ONODERA • Laboratory of Viral Immunology, National Institute of Animal Health, Tsukuba, Ibaraki 305, Japan.

however, considerable uncertainty regarding the relative importance of these factors and how they interact to result in what is suspected to be an etiologically heterogeneous disease.[1] Until a few years ago, IDDM was believed to be an acute-onset disease, but recent prospective studies of individuals at high risk of IDDM have indicated that a long euglycemic period, characterized by immune and metabolic abnormalities, precedes the appearance of hyperglycemia. Notably, islet cell antibodies (ICA) and T-lymphocyte abnormalities are present in most patients for months or years before the onset. This, together with the finding of a strong genetic association with the histocompatibility markers HLA-DR 3/4 and with other autoimmune diseases, led to the hypothesis that IDDM is an organ-specific autoimmune disorder[2] and that an autoimmune process is responsible for the chronic and irreversible loss of β-cells. Because genetic factors alone are not sufficient to explain the production of this disease, environmental factors must be implicated.[3] What triggers pancreatic autoimmunity in genetically predisposed individuals is unknown, but many lines of evidence point to the critical role of infectious agents and environmental toxins. More than two decades of study into the mechanisms of virus-induced pancreatic damage have failed to clarify the role of biologic agents in the etiopathogenesis of diabetes.[4] However, many impressive observations indicate that, even if acute virus-induced diabetes is rare, IDDM as other autoimmune diseases may be due to a chronic inapparent viral infection or may represent the consequence of a more classic infectious event. It is the purpose of this chapter to review the large body of evidence linking CVB infections to IDDM.

2. EPIDEMIOLOGY OF INSULIN-DEPENDENT DIABETES MELLITUS

Most studies of the prevalence or incidence of IDDM in young people (i.e., approximately younger than 18) suggest that the prevalence of the disease varies between 5 and 30 in 10,000 and that in Western countries the annual incidence can be estimated at 7–12 in 100,000 in the age group 0–16 years.[5] The highest incidence has been reported from Finland, while Japanese, Indians, Chinese, Eskimos, South African blacks, and Polynesians have a relatively low incidence. Both sexes are approximately equally affected, although some studies indicate a slight preponderance of males in the first 4 years of life. IDDM is extremely rare until 9 months of age, after which its frequency increases abruptly. The peak age of onset is 10–14 years of age, and a secondary peak occurs at 5–8 years. These observations are consistent with the epidemiologic pattern of many infections, which are prevented by maternal immunity and are more frequent at the time of entry into school. The peak frequency at the age of about 12 years (earlier in females) suggests that susceptibility increases at puberty. The higher incidence after puberty of orchitis and paralytic disease due to mumps and polioviruses, respectively, is a well-known example of age-modulated susceptibility to the viral invasion of particular tissues.

Seasonal variation in the onset of IDDM has been reported from several countries,[5–7] with the greatest number of new cases seen in autumn and winter. Seasonality is not very pronounced and apparently does not occur in tropical countries.[8] As before, these observations are consistent with the epidemiologic pattern of several infections; it is unclear, however, as to how seasonality can be maintained supposing that the infectious event precedes diabetes by several months or years. An alternative possibility is that common viruses having a seasonal distribution may merely act as precipitating factors of an ongoing diabetogenic process.

The relative importance of socioeconomic and climatic factors in the incidence of IDDM is unclear, as it is why fluctuations of the incidence rate seem to occur with cycles of several years. All these aspects would merit further investigations in the light of an infectious etiology of IDDM.

3. SEROEPIDEMIOLOGIC STUDIES

Since 1969, when Gamble suggested for the first time the possibility of a link between CVB infections and human diabetes,[9] a large number of serologic investigations were initiated to confirm his findings.[10–54] Two excellent reviews by Gamble *et al.*[5] and Barrett-Connor[11] have attempted a comprehensive interpretation of the available data. We review these findings in an effort to compare those studies that have been done with similar virologic methods and with appropriately selected groups of patients and controls. In addition, some recent data on the immunoglobulin class of antibodies against CVB is reviewed, together with the results of a study we have conducted on our diabetic patients.

3.1. Neutralizing Antibodies

Based on the notion that a growing number of viruses can infect the pancreas of both animals and humans[10] and on sporadic reports attributing juvenile diabetes to mumps or to other infections,[4] Gamble and colleagues[9] initiated a systematic search for antiviral antibodies in the sera of diabetic patients using complement fixation (CF) and comparing the prevalence of these antibodies to that observed in healthy controls. Although the spectrum of infectious agents used for this survey was large (including influenza, parainfluenza, RSV, HSV, mumps, measles, adeno, and CV), significant differences in the prevalence of viral antibodies were found only with regard to influenza C and parainfluenza (reduction), and to CVB (increase). No difference between diabetics and controls was detected in the proportion of sera positive to group A CV (types 2, 5, 10, and 16). Since the CF antibody response to CV infections is often nonspecific, the same sera were examined for neutralizing titers to CVB1–5. Insulin-dependent diabetics within 3 months of onset were found to have higher antibody titers to CVB4 than did either normal subjects or patients with diabetes of longer duration. This finding thus suggested that in the small group of diabetics studied there was in some cases a temporal association of CVB infection (probably type 4) with the onset of diabetes. In this study, the control and patient groups were not appropriately matched and most of the patients were over 15 years of age (i.e., they were not typical IDDM patients).

A second study of 162 patients (with controls matched for age, sex, geographic area, and time of bleeding) was published by Gamble *et al.*[12] in 1973. Again, antibody to CVB4 was found more often in diabetics than in controls, particularly in the age group 10–19 and >20 years. By contrast, children aged 0–9 years had a reduced frequency of antibodies as compared to controls. A reduction of the antibody response to CVB in young diabetic children was later confirmed by two reports from Copenhagen and Seattle[20,21] and is also suggested by a recent study from Pittsburg.[27] This may mean either than individuals with a genetic predisposition to IDDM have a reduced immunologic competence against CVB or that young diabetic children have had less previous experience with CVB infections as compared with nondiabetics and consequently have a more severe disease when they contract CVB infections at later ages (akin to the polio model).

Investigations of this type have been reported from several countries with controversial results.[12,17–23,25,27,40–42,45,47–49,53] It is not surprising, however, that different conclusions have been reached, since it is obviously difficult to establish, by serology alone, cause-and-effect relationships between widespread infectious agents as CVB and a disease with a prevalence of about 1 in 1000 population. In addition, because of the cost and length of these studies, many of the published reports have been rather limited in scope. Table I presents the averaged results obtained by various investigative groups on the prevalence of CVB-neutralizing antibodies in more than 400 newly diagnosed diabetics. Within 4 months of onset, the proportion of patients positive for the CVB4 antibodies is significantly increased as compared with normal

TABLE I

Prevalence of Neutralizing Antibodies to CVB in Insulin-Dependent Diabetics under 20 Years of Age and Matched Controls[a]

Group	Time from onset (months)	% positive and number tested[d]					
		B1	B2	B3	B4	B5	B6
Controls[b]	—	19.9 (347)	39.1 (327)	40.3 (417)	45.1 (483)	34.6 (347)	10.0 (80)
IDDM[b]	<4 (1–4)	23.1 (428)	38.2 (408)	40.8 (441)	59.5 (501)[f]	30.1 (428)	9.4 (203)
IDDM[c]	>6 (6–24)	29.0 (186)[e]	54.8 (186)[f]	46.8 (186)	64.0 (186)[f]	41.9 (186)	3.7 (107)

[a]Cumulated data from the literature.
[b]From refs. 12, 18–20, 25, 39–41, and 52–54.
[c]From refs. 12, 14, and 54.
[d]Percentage of sera with NA titers of ≥1:8 and, in parentheses, the number of individuals tested.
[e]Significantly more prevalent in diabetics than in controls at $p < 0.025$.
[f]$p < 0.001$.

controls, confirming the early data of Gamble's group. When antibodies are measured 6–24 months from onset, the high prevalence of CVB4 infections is still observed, but evidence is also obtained for an elevated incidence of CVB2 and CVB1 infections.

We have carried out a study of this kind on our own patients aged 1–14 years and in a comparable group of healthy children.[54] Because only one method was used for measuring antibody titers and its sensitivity was accurately monitored, we were able to compare not only the proportions of positive individuals in the two groups, but also the mean titers of positives (Table II). Again, within 1 month of onset, evidence for a somewhat increased prevalence of CVB4-neutralizing antibodies was obtained and, more important, significantly altered neutralizing antibody titers were found only to this agent. The finding of a reduced antibody responsiveness to CVB4 was confirmed with sera obtained 6–20 months after diagnosis. Thus, our data suggest that early-onset diabetes is associated with frequent CVB4 infections but that young diabetics may have a reduced ability to produce neutralizing antibodies to this virus. This has been already proposed by others.[12,20,21,27] Interestingly, Tables I and II show that, at onset, the immune response of diabetic children to CVB types other than CVB4 is essentially normal.

The recurring observations of an altered antibody response to CVB4 in these patients has prompted some authors to postulate that antigenic similarities exist between this virus and pancreatic β-cells and that ICA would simulate CVB4 antibodies. This point has been recently addressed by Schernthaner *et al.*,[35] who measured anti-CVB antibodies and ICA in newly diagnosed diabetics. It was concluded that it is unlikely that CVB-specific IgM and CF-ICA antibodies are crossreactive, because by no means every patient who was CVB-IgM positive had CF-ICA and vice versa. It should also be considered that more than 40% of normal children have CVB4 antibodies, whereas less than 1% are ICA positive and that sera of normal individuals with high titers of CVB antibodies do not react with islet cells.[31] These observations, however, do not exclude that some individuals may recognize common epitopes on these antigens.

3.2. Immunoglobulin Class Specificity of CVB Antibodies

Additional serological evidence for a role of CVB in diabetogenesis is provided by recent investigations of the immunoglobulin class of CVB antibodies in young diabetics. Following a report by the group of Banatvala,[29] several investigators studied this point using solid-phase immunoglobulin-capture methods.[27,30–35,37–39,43] These studies provided

TABLE II
Prevalence and Titer of CVB Neutralizing Antibodies by Duration of IDDM in Children under 14 years of Age and Matched Controls[a]

Group	Time from onset (months)	No. tested	% positive and mean titer of positives[b]					
			B1	B2	B3	B4	B5	B6
Controls	—	48	16.7 (43.2)	60.4 (156.3)	35.4 (42.8)	54.2 (328.5)	58.3 (53.1)	4.2 (8.0)
IDDM	0–1	81	25.9 (78.5)	48.1 (134.8)	46.9 (67.0)	65.4 (239.9)[c]	43.2 (72.8)	6.2 (16.7)
IDDM	6–20	107	25.2 (61.4)	45.8 (139.2)	47.7 (44.9)	66.4 (219.1)[c]	46.7 (79.3)	3.7 (19.2)

[a]Patients at onset: mean age 10.2 years (range 1–14); patients several months after onset: mean age 11.3 years (range 2–15).
[b]Percentage of sera with NA titers ≥1:8 and, in parentheses, geometric mean titer of positive sera. Adapted from Toniolo *et al.*[54]
[c]Mean titer of diabetic patients significantly reduced as compared with controls (p <0.025).

TABLE III
CVB Antibodies of Different Immunoglobulin Classes in Children with Newly Diagnosed IDDM[a]

		IDDM	Controls
Virus	Ig class	Positive/no. tested	Positive/no. tested
CVB1–5	IgM[b]	87/247[e]	34/550
CVB4	IgG[c]	97/210	107/210
CVB4	IgM[c]	29/210	26/210
CVB4	IgA[c]	79/210[d]	56/210

[a]Children aged <15 years old; time from diagnosis of less than 3 months. Control subjects matched by age, time, and geographic area. Antibody determinations by Ig-capture ELISA or RIA.
[b]Cumulated data from refs. 29, 30, 32–34, and 39.
[c]Data from Hyoty *et al.*[37]
[d]Significantly more prevalent in diabetics than in controls at $p < 0.025$.
[e]$p < 0.001$.

results that are less controversial than those obtained with tests for neutralizing antibodies.

Table III presents the averaged results obtained by various investigators within a short period of diagnosis in children younger than 15 years. First, it is apparent that most studies have documented a high frequency of CVB-specific IgM antibodies in these patients.[29,30,32–34,39] Second, these antibodies (when homotypic) were shown to react mostly with CVB4, but also with CVB5, CVB2, and rarely CVB3. Third, limited studies suggest that the CVB-specific IgM response of diabetic patients is not unduly prolonged (this tends to exclude the possibility of a persistent infection). Fourth, not all studies support the idea of a central etiologic role of CVB4 in IDDM.[33,37] In fact, a recent study from Finland[37] failed to show an increased frequency of IgM responses to this virus but indicated that diabetic children may have a higher prevalence and mean levels of IgA antibodies to CVB4 as compared with controls. The same has already been observed in young diabetics with regard to mumps virus[55,56] and might reflect the establishment of persistent infections in these patients or, alternatively, the presence of immunoregulatory abnormalities associated with the predisposing genotype.

Thus, collectively, these studies suggest once again that up to one third of new cases of diabetes in children are preceded by an acute CVB infection.

3.3. Genetic Control of the Immune Response to CVB

In 1977, Cudworth *et al.*[16] noted for the first time that diabetic patients who were positive for the HLA antigen B15, and in particular those who were both B8 and B15 positive, had higher neutralizing antibody titers to CVB1–4 than did patients negative for these alleles. This observation stimulated investigations on the influence of HLA loci on the immune response to CVB. It is now clear that the association of IDDM with the HLA class I alleles B8 and B15 is mostly due to a linkage disequilibrium occurring between these markers and the class II alleles DR3 and DR4, which are believed to represent markers of closely linked, but as yet untypable, diabetogenic genes. Thus, it was found that male patients who are negative for DR4 (most of whom are DR3 positive) show a lower frequency of recent CVB infection at diagnosis.[27] This finding may reflect a failure of serologic response to these viruses and seems to confirm indirectly, the results of Cudworth's group. Collectively, these data indicate that the markers B8 and/or DR3 are associated with a reduced antibody response to CVB, whereas patients positive for B15 and DR4 produce increased titers of antibodies to these viruses. Patients with the combined genetic predisposition to IDDM (i.e., DR3 and DR4 positive) seem to have the highest antibody titers. Positivity for either DR3 or DR4, or both, not only appears to influence the levels of neutralizing antibodies to CVB but is also associated with an increased frequency of CVB-specific IgM at onset.[35]

These and other observations stimulated *in vitro* studies of the proliferative response of T lymphocytes to CVB and other diabetogenic viruses. In agreement with what was observed *in vivo*, Bruserud *et al.*[57] found, by limiting-dilution analysis, that DR3-positive individuals had a lower frequency of T lymphocytes proliferating in response to CVB4 and mumps virus as compared with subjects with DR determinants other than DR3. By contrast, DR4-positive individuals had an increased frequency of these lymphocytes. These results were similar in diabetics and healthy individuals, and no differences were found in the frequency of T lymphocytes responding to varicella–zoster virus or tuberculin. Supposing that these differences were mediated by antigen-specific T-suppressor (Ts) cells, peripheral T lymphocytes were deprived of Ts cells and cultured in the presence of antigen-presenting cells. Analysis of the proliferative response to mumps and CVB4 antigens showed that the DR3-mediated hyporesponsiveness is not due to Ts cells.[58] Because the suppression exerted by antigen-presenting cells is believed not to be antigen specific, the reduced proliferative ability of lymphocytes from DR3-positive individuals appears to be most probably caused by an intrinsically low frequency of antigen-reactive helper/inducer T lymphocytes or by a regulatory influence of a subset of these cells.

The significance of this hyporesponsiveness in diabetogenesis, if any, remains unclear, but a recent study from Gorsuch *et al.*[59] tends to exclude the possibility that diabetics are hyporesponsive to picornaviruses in general. Antibody titers to poliovirus types 1–3 were measured in HLA-typed diabetics and their discordant siblings; there was no convincing association of titers with HLA phenotypes. However, statistical analysis suggested that immune-response genes may play some role in viral diabetogenesis.

4. FOLLOW-UP STUDIES

Evidence to associate CVB infections and the clinical onset of IDDM has been also searched for by studying naturally infected populations. Thus, 24 British patients clinically infected with CVB2 or 4 were followed for 3 years, but no one developed diabetes.[48] More interestingly, a negative report came from the Pribilof islands within 5 years of a major CVB4 epidemic.[49] Neither diabetes nor glucose intolerance developed in infected persons aged 5–20 years at the time of the epidemic, nor among any of the approximately 70 children under 5 years at that time. It was therefore concluded that the risk of developing diabetes 5 years after CVB4 infection is less than 3% in Aleuts under 20 years of age. This study also pointed out the extreme difficulty of preventive measures, unless markers are found to identify persons susceptible to diabetes. Four years after an outbreak of CVB3 and CVB4 infection in a children's home, Hierholzer and Farris[51] measured the fasting plasma

TABLE IV
Incidence of Diabetes in Individuals with Recognized CVB Infections: Follow-up Studies

Infection	IDDM cases/ no. infected	Incubation time	Notes	Ref.
CVB2 and CVB4	0/24	—	3-year observation	48
CVB4	0/166	—	5-year observation (OGTT in population)[a]	49
CVB3	0/17	—	4-year observation	51
CVB4	0/32	—	4-year observation	51
CVB3 and CVB4	0/39	—	4-year observation (1 case altered OGTT)	51
CVB4	1/4	>3 year	Family members; patient positive for ICA and CF-ICA	36
CVB4	1/22	10 weeks	U.N. soldiers; patient negative for ICA and CF-ICA	46

[a]OGTT, oral glucose-tolerance test.

glucose and insulin levels of 88 previously infected children. Only one case of suspected chemical diabetes was reported in a girl who had been infected with both CVB3 and CVB4. These data are summarized in Table IV. After clinical infection with CVB4, however, one case of overt diabetes has been described in a family[36] and another in a group of 22 U.N. soldiers serving in Egypt.[46] In the first case, the time between infection and clinical onset was more than 3 years; the child had ICA 3 years before the onset. By contrast, in the second case, the time between infection and hyperglycemia was 10 weeks, and this patient was ICA negative, suggesting an unusual form of diabetes. In none of these cases was it possible to demonstrate that hyperglycemia was indeed related to the infectious event. Investigations of this kind are ill suited to the study of rare diseases such as IDDM and would require the observation of large numbers of infected individuals (possibly more than 1000).

5. CLINICAL OBSERVATIONS

The best way to establish a direct link between viral infections and human diabetes is to isolate viruses from diabetic patients and to show that the recovered agents can cause diabetes in experimental hosts. So far, this goal has been achieved only in two cases. The first is that of a 10-year-old boy who developed diabetic ketoacidosis 3 days after the onset of a flulike illness. Despite intensive therapy, his condition deteriorated, and he died 7 days later. At autopsy, insulitis and necrosis of β-cells were observed. Yoon *et al.*[60] were able to isolate a variant of CVB4 from the pancreas. Studies of the patient's serum showed a rise of neutralizing antibodies to the isolate from a titer of less than 1 : 4 on the second hospital day, to 1 : 32 on the day of death. Viral antigens were detected in various organs, including the child's brain stem, demonstrating a severe CVB infection. Sections of the pancreas were characterized by lymphocytic infiltrates immediately surrounding the islets of Langerhans, often in a perivascular distribution. Complete loss of islet architecture and severe β-cell degeneration were present in scattered islets.

To prove that this agent was truly related to the child's diabetes, several inbred strains of mice were inoculated with the isolate. Mice of strains susceptible to encephalomyocarditis virus-induced diabetes (SJL) developed hyperglycemia. Viral antigens were found by immunofluorescence in pancreatic islets, and insulitis with β-cell necrosis was detected in infected mice. All these data indicated that the patient's diabetes was virus induced. However, subsequent histopathologic studies showed that some pseudoatrophic islets, composed mainly of α- and Δ-cells but lacking β-cells, were present in the child's pancreas. This finding is typical of a long diabetogenic process and might suggest that the CVB4 infection did not cause, but only precipitated, diabetes in this patient.

The second case is that of a 16-month-old French girl who was hospitalized for thrombocytopenic purpura that appeared after 1 week of fever. Laboratory evidence of diabetes was present from day 13 to 25; then a remission ensued, but 2 months later, the child developed overt diabetes. On the eighth day after the onset of fever, CVB5 was isolated from stools.[61,62] Antibody titers to this virus rose from less than 1 : 10 to 1 : 640, showing that in the period preceding the onset of diabetes the baby had been infected with CVB5. When this isolate was inoculated into various strains of mice, glucose intolerance was produced in susceptible strains (DBA and SJL). Since the common strains of CVB5 do not produce diabetes in mice, it was concluded that this child had been infected with a diabetogenic variant of this virus. The girl had the high-risk genetic markers DR3 and BfF1, and at the time of hospitalization high levels of ICA were already present in her serum. Thus, it appears that CVB infection collaborated with genetic and immunologic factors in producing diabetes in this child.

In other cases, proof that the isolated virus could induce diabetes in experimental animals was not obtained. Nevertheless, various CBV have been isolated and their relationship to the disease has been inferred by histopathology. A 5-year-old girl developed myocarditis and acute diabetes 7 days after open-heart surgery; she died of diabetic ketoacidosis a few days later. At necropsy, degeneration and necrosis of islet cells and an associated insulitis were found; more importantly, CVB4 antigens were detected by immunofluorescence in pancreatic islets.[63] As in the previous cases, high titers of neutralizing antibodies were detected, confirming that the girl had had a recent CVB4 infection.

One case of diabetes attributed to CVB1 occurred in a newborn twin who acquired the virus from the mother a few days before birth (transplacental infection has been documented in a few cases). Immediately after birth, the infant's condition deteriorated, and he developed thrombocytopenia and myocardiopathy. On day 12 blood glucose rose to over 500 mg/dl, and he died 4 days later.[64] Pancreatic lesions consisted of insulitis with islet cell degeneration, and CVB1 was recovered from the pancreas. It is of interest that his twin brother, although developing a severe disease due to CVB1, did not become diabetic and could be discharged from the hospital 25 days after birth. This case strongly suggests that CVB1 can indeed produce diabetes in humans, as it is hardly conceivable that the diabetes of this baby could recognize a nonviral etiology. This case also illustrates that clinical infection with a diabetogenic virus does not necessarily produce diabetes; it is therefore unfortunate that these twins and their mother were not HLA typed.

Other cases of IDDM have been associated with CVB infections on less solid grounds. A 4-year-old boy was hospitalized for hyperpyrexia, diarrhea, lymphocytosis, and meningeal involvement. A few days later, he developed mild hyperglycemia and died 20 days after hospitalization.[65] CVB6 was isolated from the spinal fluid and stools. Several serum samples were analyzed for IGM against CVB and other viruses, but only CVB6-specific IgM were found. More interestingly, on the first hospital day the boy was negative for ICA and CF-ICA, but ICA appeared on day 7 and CF-ICA on day 15. In contrast with the French case reported by Champsaur *et al.*,[62] in which ICA were found on the first hospital day (leaving aside the possibility that pancreatic autoantibodies were already present before CVB5 infection), this case strongly suggests that ICA can indeed be induced as a consequence of CVB infection.

An 18-month-old boy developed acute diabetes after 1 week of fever, diarrhea, and polyuria without cardiac involvement.[66] He had lymphocytosis and high titers of neutralizing antibodies to CVB2, but not to other CVB; in addition, low titers of CVB-specific IgM were detected by immunofluorescence in the acute phase of the disease. Both parents and two siblings were also found to have high titers of antibodies to CVB2, confirming a recent familial infection. A similar case was described by Gamble's group[67] in a 10-year-old boy from a diabetic family who developed clinical diabetes 1 month after acute Bornholm disease. Because of the presence of high antibody titers to CVB1 and CVB5, this case was tentatively attributed to these viruses. Interestingly, ICA were not detected in this patient or in family members.

Two additional cases attributed to CVB4 on serologic basis have been reported from Japan in adult patients.[68] One patient underwent a complete remission, while the other developed a mild diabetes, clinically similar to the noninsulin-dependent form.

Although it is impossible to establish whether CVB infections played a direct etiologic role in these patients or acted merely as triggering factors of diabetes in individuals who had an already reduced insulin reserve, these observations are consistent with the view that CVB may damage, at least occasionally, insulin-producing pancreatic β-cells in humans.

6. POSTMORTEM STUDIES

This view is reinforced by numerous histopathologic reports on the effects of CVB on the islets of Langerhans in patients dead of severe infections with or without clinical diabetes. It has long been recognized that CVB can infect and damage the exocrine pancreas;[69,70] more re-

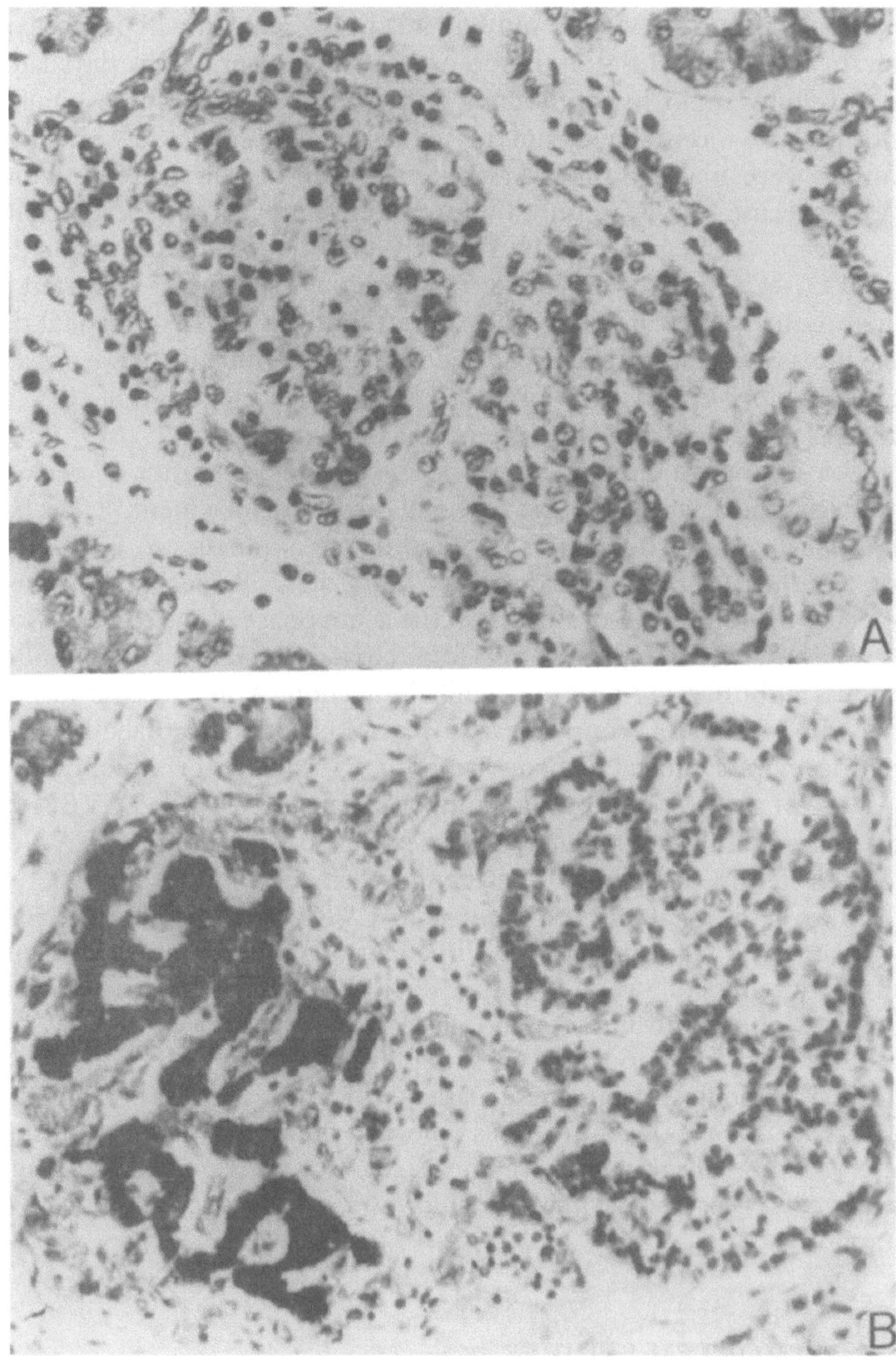

FIGURE 1. Histopathology of pancreatic islets of a neonate who died of disseminated CVB4 infection without evidence of diabetes. (Courtesy of Dr. A. B. Jenson.) (**A**) Degeneration and necrosis without lysis of islet cells is characterized by pyknotic nuclei and condensed eosinophilic cytoplasm. (**B**) Immunologic staining for insulin from the same case (immunoperoxidase method); insulin-containing cells are marked by the dark staining. Note that more than one half the islet is completely degranulated of insulin.

cently it has become apparent that these agents can also damage pancreatic endocrine cells, especially during the neonatal period. Thus Jenson *et al.*,[71] Kaplan *et al.*,[64] and Ujevich and Jaffe[72] examined the pancreas of neonates dying of overwhelming CVB infections and found that 17 of 53 cases had islet cell lesions of probable viral origin. Only one of these babies had hyperglycemia; actually, most of them were hypoglycemic before death. It is known, however, that it may take several days or months for clinical diabetes to develop after CVB infection, while all these babies died within a few days of infection. Another three cases with histopathologic reports were contributed by Yoon *et al.*,[60] Gladisch *et al.*,[63] and Ahmad and Abraham.[73] All these cases were examined during the period of acute disease and were characterized by degeneration and necrosis of islet cells accompanied by the infiltration of lymphocytes and macrophages. Necrotic islet cells were most prominent in islets containing large numbers of neutrophils. In most cases, no more than one third of islets were involved, and the large islets were frequently bisected into segments by a mixed inflammatory cell infiltrate; thus, one part of the islet appeared normal, whereas islet cells of the other segment were degenerated or had condensed eosinophilic cytoplasm and small dark nuclei. This lobular appearance of the damage appears to be characteristic of CVB infections (Fig. 1). When pancreas sections were examined by immunohistochemistry with antibody to insulin, mild to severe loss of insulin (degranulation) was apparent in those cells with eosinophilic cytoplasm and in frankly necrotic cells. Staining for other pancreatic hormones showed, however, that virus-induced damage was not confined to β-cells alone and also that α- and Δ-cells were sometimes lost or degranulated.[72] It is noteworthy that particularly in premature babies, islet cell necrosis was not accompanied by any cellular response. The extent of the exocrine damage was somewhat variable, and in many cases the endocrine damage predominated. Altogether, these cases have been attributed to CVB1 (three cases), CVB2 (two), CVB4 (four), CVB5 (one), and to untyped CVB (seven). Thus, these data suggest that all the CVB types can infect and damage pancreatic islets in newborn children and that, presumably, islet-cell involvement during CVB infection is more widespread than was commonly suspected. However, because only extremely severe infections have been studied, the possible relevance of neonatal CVB infection to the eventual development of diabetes in adulthood is not clear.

7. EXPERIMENTAL MODELS

The diabetogenic properties of CVB have been closely examined in mice because (1) these viruses may be implicated in human diabetes; (2)

unlike mumps and rubella, the host range include both humans and mice; (3) like the D and M variants of encephalomyocarditis virus, they produce a systemic infection in the animal model; and (4) it has long been known that they can infect the pancreas of mice.[74] All these reasons prompted numerous investigators to search for pancreatic endocrine damage in infected animals using histologic or metabolic criteria. Early studies, however, produced conflicting reports, probably because of the pronounced diversity of CVB isolates and of the genetic differences between various strains of test mice.

7.1. CVB-Induced Pancreatitis and Insulitis

In 1951, Pappenheimer and colleagues[74] showed that the Connecticut-5 strain of CVB1 induced pancreatitis in adult mice. This finding could not be confirmed by Gifford and Dalldorf[75] in newborn mice of a different strain, but 1 year later the same workers were able to adapt this virus to the pancreatic tissue of adult mice by repeated passage *in vivo.*[76] Infection with *in vivo* or *in vitro* grown CVB1 caused extensive destruction of pancreatic acinar tissue, but the islets of Langerhans were spared. While the hepatic damage was more pronounced in male than in female CD-1 mice, the degree of pancreatic involvement was independent of sex.[77]

Early electron microscopic studies suggested that CVB could occasionally produce alterations of pancreatic endocrine cells.[78] This finding was later confirmed by various investigators who studied the pancreatic pathology produced by CVB4, CVB1, and CVB3 in neonate and young adult outbred HaM/ICR or CD-1 mice.[79–81,83,84] Collectively, these studies indicated that after infection some islets of Langerhans underwent not only scattered degeneration and necrosis, but subtle atrophic changes as well. Light microscopic examination showed that, especially in areas of extensive exocrine damage, islets were reduced in size, and some islet cells had more eosinophilic cytoplasm with smaller and darker nuclei. Mononuclear cells were seen in the islets, but in no case was this condition pronounced. Electron microscopic changes were noted as early as 1 day postinfection. Minor damage consisted of vacuolation of cytoplasm with formation of membranous vesicles, with the rough endoplasmic reticulum still intact. Severe damage consisted of breakdown of cells, so that their type was difficult to ascertain: the cristae of mitochondria were swollen, the nuclear chromatin was marginated, and a marked loss of electron density was apparent. Loss of specific hormone granules was occasionally seen in β- and α-cells; Δ-cells appeared intact. Viral crystals of CVB1 were seen in a few mice both in acinar cells and in β-cells.[80] The individual particles measured approximately 210 Å in diameter and were separated by a distance of 60 Å. This observation is of great significance in establishing a direct relationship between CVB infection and endocrine damage, as many of the subtle electron microscopic changes are compatible with the local effects of toxins or enzymes released during the acute destruction of acinar tissue. For unknown reasons, crystals of CVB other than type 1 were not detected during the course of these studies.

Not all studies, however, confirmed these findings. Lansdown[82] as well as the Notkins's group[85,86] could not find histopathologic changes in the endocrine pancreas of mice infected with laboratory strains or fresh isolates of the six CVB. Outbred (Swiss and CD-1) and inbred (DBA/2, BALB/c, A, CBA, and C57BL/6) mice were used. Although mild to severe pancreatitis was consistently produced by most virus–host combinations, the islets of Langerhans escaped severe damage. Acutely, capillary dilation and a slight disarray of islet cell architecture were observed. Rarely, β-cells had vacuolated cytoplasms and small pyknotic nuclei. Several months postinfection, the islets remained intact, although the acinar tissue had often been replaced by adipose tissue. Comparable results were obtained in pregnant mice and in mice given cortisone, silica, or other immunosuppressive treatments[82,86] (A. Toniolo, unpublished data). It is unfortunate that these mice were not studied by electron microscopy, as even these negative results suggest that CVB frequently produce minimal pathologic changes in the islets of Langerhans. Fluorescein-labeled polyclonal antibodies to CVB were used to ascertain whether these viruses were replicating in islet cells, but negative results were obtained.[85]

By contrast, islet cell lesions were easily observed when male mice of strains susceptible to the diabetogenic effects of encephalomyocarditis virus (SJL, SWR) were infected with preparations of CVB4 that had repeatedly been passaged in cultures enriched for pancreatic β-cells.[87] Mild inflammatory infiltrates were seen in pancreatic islets within 3–5 days of infection, and the severity of pathologic changes varied considerably among animals and within a single pancreas. Normal-appearing islets were often seen adjacent to islets showing extensive infiltrates. In some cases, only a portion of the islet was involved. Rarely, coagulation necrosis was observed. These findings are reminiscent of those described in the pancreas of neonates dying of severe CVB infections. Within 3 days postinfection, viral antigens were seen in the cytoplasm of islet cells and maximal involvement occurred at 4 days, with some islets containing only a few positive cells; others had viral antigens in most cells. Interestingly, acinar and ductal cells were almost free of infection. Comparable results were obtained in mice inoculated with the CVB4 strain that had been isolated from the acute case of diabetes reported by Yoon *et al.*[60]

That all CVB types can indeed infect and damage the endocrine pancreas of mice was shown by Notkins's group in a subsequent study.[88] Prototype strains of CVB were serially passaged *in vitro* in monkey kidney cells, mouse embryo cells, or mouse β-cells and *in vivo* in the mouse pancreas. When susceptible SJL mice were inoculated with these virus stocks and their pancreata examined by immunofluorescence with type-specific sera, it was apparent that all six types had the potential of infecting some islet cells, had they been passaged in β-cell cultures (Fig. 2). Although the number of infected cells varied considerably among islets, in general 5–15% of islet cells contained viral antigens 4 days postinfection. Double-label technique with antiviral and anti-insulin antibodies showed that most infected cells were indeed insulin-containing β-cells. It is remarkable that only virus stocks grown in β-cell cultures or in the pancreas *in vivo* had the capacity of infecting appreciable numbers of islet cells. Virus preparations grown in monkey cells, in fact, had a remarkable tropism for the acinar tissue, whereas stocks obtained from mouse embryo cultures were somewhat attenuated and failed to induce viral antigens in exocrine and endocrine cells. β cell-passaged viruses produced mild islet damage, ranging from capillary dilation and few pyknotic nuclei to focal necrosis with mononuclear cell infiltration in approximately 10% of islets. Staining with anti-insulin antibody showed that the insulin content of some islets was grossly reduced, even in the absence of major pathologic changes. Islet alterations induced by the different types were indistinguishable.

Recent results obtained with fluorescein-labeled monoclonal antibodies indicate that various laboratory strains of CVB3 have the capacity of infecting minimal numbers of islet cells in mice without causing appreciable histopathologic changes (A. Toniolo and C. Garzelli, unpublished data). Thus, it appears that all CVB may have some tropism for islet cells but that significant endocrine damage is produced only by strongly β-tropic variants.

7.2. Metabolic Studies

In agreement with these pathologic findings, Ross *et al.*[86] reported that infection of different mouse strains with CVB1–6 caused elevations of serum amylase of as much as 10-fold and a reduction of the amylase content of the pancreas by more than 95%. In their studies, however, blood glucose levels were significantly reduced during the acute infection, and at no time was evidence of hyperglycemia found. These findings were in contrast to those of Coleman *et al.*,[83,84] who found CVB4 infection of CD-1 mice to result in a transient elevation of glucose (i.e., 15–20 days postinfection) in the absence of an absolute decrease in the concentration of insulin in blood. For this reason, both Notkins's group

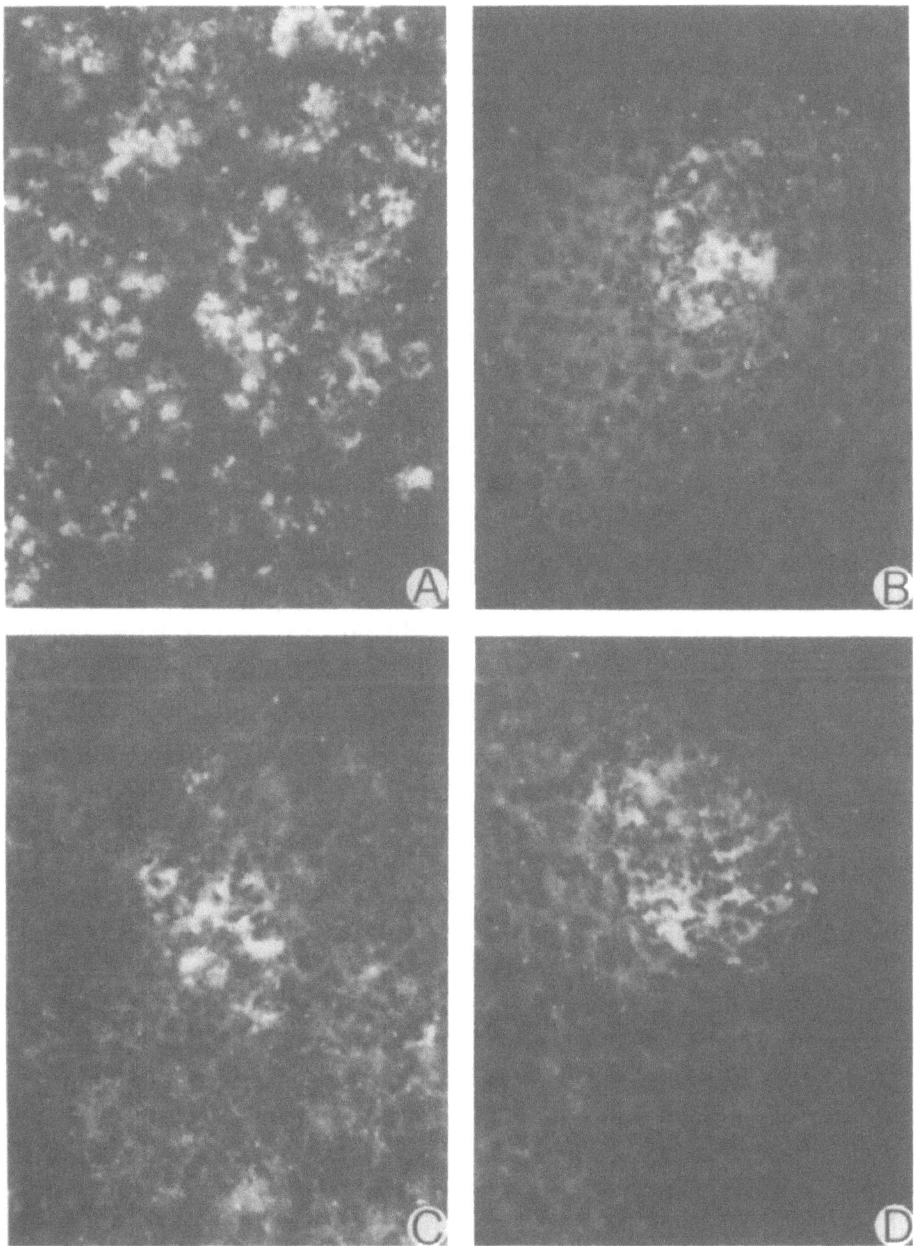

FIGURE 2. Change in the tropism of CVB after passage in mouse β-cell cultures. Four days postinfection, the pancreas was removed from infected mice, and frozen sections were stained with fluorescein-labeled antivirus antibody (×340). Infection with a prototype strain of CVB5 grown in monkey kidney cells. High concentrations of viral antigens in acinar cells (**A**). Antigens of CVB1 (**B**), CVB2 (**C**), and CVB5 (**D**) in the islets of Langerhans. These strains were passaged 15 times in β-cell cultures. Note that the original tropism for acinar cells is virtually lost.

and Craighead and co-workers examined fresh isolates and prototype strains of CVB4 (including the strain employed by Coleman in his studies) for their ability to produce hyperglycemia in mice.[85] Again, nonfasting glucose levels and glucose tolerance tests remained within the normal range. These contradictory results were tentatively attributed to environmental or dietary differences between test mice.

Profiting from the large experience gained in the study of diabetes induced by the encephalomyocarditis virus, Yoon *et al.*[87] used pancreatic β-cell cultures to increase the diabetogenic potential of CVB4 and susceptible male SJL mice as test animals. Four days after infection, the concentration of immunoreactive insulin (IRI) started to decrease both in the pancreas and in blood. Glucose levels correlated inversely with the amount of IRI, and more than 80% of animals were found to be hyperglycemic within 14 days of infection. The degree of hyperglycemia was highly variable, and the percentage of animals with elevated glucose concentrations decreased with time; thus, after 60 days less than 5% of animals were still hyperglycemic. However, many of the euglycemic mice were metabolically abnormal when evaluated by glucose-tolerance tests. It was not investigated whether, in addition to β-cells, CVB4 damaged glucagon- or somatostatin-secreting cells as well. Alterations of the functional capability of these cells, in fact, might also contribute to the production of hyperglycemia.

Precisely how passage in β-cell cultures can increase the diabetogenic capacity of CVB is not known, but alterations in the tropism of viruses after serial passage in animals or tissue culture is a widely recognized phenomenon. Recent studies[89,90] on CVB4 have demonstrated that antigenic changes at the epitope level occur at a frequency of $>10^{-2}$. This points to the possibility that even within the same virus pool there are many genetic variants and that variants with the appropriate tropism for β-cells are selected for during replication in β-cell-enriched cultures. Thus, the great difficulty in obtaining relatively pure β-cell cultures might contribute to explain why experiments with truly diabetogenic strains of CVB were carried out so rarely. It should be borne in mind, however, that about one third of field strains may induce mild glucose abnormalities in susceptible mice.[91]

Only mild alterations of glucose tolerance were produced in mice even by prototype strains of the six CVB that had been serially passaged in pancreatic β-cell cultures.[88] By contrast, remarkable elevations of glucose levels resulted when these strains were injected into mice whose β-cell reserve had been reduced with small doses of streptozotocin, a potent β-cell toxin.[88,92] This observation, together with direct counts of the islet cells expressing viral antigens, indicates that CVB usually do not damage enough β-cells to produce overt diabetes. Hyperglycemia is instead produced in mice with reduced numbers of functional β-cells. This principle was recently confirmed by Yoon *et al.*[93] in patas monkeys. Infection of these animals with CVB4 resulted in transient elevations of glucose-tolerance tests and depressed insulin secretion and glucose in the urine. All these effects were markedly enhanced by pretreatment of monkeys with a subdiabetogenic dose of streptozotocin, but no ICA were detected in these animals. Comparable cumulative effects have been obtained when BALB/c mice, a strain resistant to virus-induced diabetes, were infected with CVB4.[94] Our own experience and that of Jordan *et al.*[91] with mice, together with a few data obtained with monkeys,[93] support the possibility that the diabetogenic potential of CVB not only can be increased by pretreatment with β-cell toxins, but also by sequential infections of the same host with different CVB.

7.3. Genetic Aspects

Numerous studies indicate that the capacity of CVB to induce diabetes is influenced by the genetic background of the host. Yoon *et al.*[87] found that mouse strains susceptible to diabetes caused by encephalomyocarditis virus were also susceptible to CVB4-induced hyperglycemia and that male mice developed more severe diabetes than did females. This finding has been confirmed by others,[88,91,92,95] and it is now believed that strains such as SJL, SWR, and NFS are susceptible and that DBA/2, Swiss, and CD-1 intermediate, C57BL/6, CBA, AKR, BALB/c, and C3H are resistant. Susceptibility does not seem to correlate with MHC type. Recent experiments[93] suggest that even the susceptibility of monkeys to CVB4 is under genetic control and that so far, pancreatic endocrine damage has been induced only in patas monkeys.

Considerable efforts were also spent to assess the role of certain loci conferring predisposition to diabetes in the susceptibility of mice to CVB. Most of these studies were carried out in C57BL mice homozygous or heterozygous for the diabetes-associated genes *db* or *ob*.[95–98] Susceptibility to CVB4-induced pancreatic disease and mortality rate were compared with those of the parental background strains. It was generally seen that genetic traits predisposing to diabetes were associated with an increased mortality after infection and with the development of more severe pancreatic disease. However, because of the low diabetogenic potential of the infecting virus, clear-cut data on the precise relationships between genetic traits and the metabolic response to CVB4 infection were not obtained. No genotypic differences were seen in the virus levels attained in the pancreas, but β-cells of mice carrying the *db* gene underwent severe degranulation during acute infection. This alteration, however, was not followed by overt hyperglycemia, and the most reproducible finding was an acute release of insulin accompanied by hypoglycemia during the early phase of infection.[95,96] Studies of these and other strains of mice demonstrated an inverse relationship between the effects of CVB4 on the exocrine and the endocrine pancreas. In other words, mouse strains susceptible to viral diabetes had less severe acinar pancreatitis than did mice resistant to viral diabetes.[95] The phenotypic basis of this phenomenon is unknown. More recently, it was shown that C57BL/Ks mice homozygous for the *db* gene fail to produce neutralizing antibodies to CVB4 and have other immune abnormalities reminiscent of those described in the prediabetic period in humans.[99]

Progress in these and other genetic aspects of CVB-induced hyperglycemia will depend, however, on the development of reliable methods for obtaining truly diabetogenic variants of CVB. No useful markers for such variants have yet been detected,[100] and empirical methods for the plaque purification of diabetogenic variants based only on tests for the induction of diabetes *in vivo* have been unsuccessful so far (A. Toniolo, unpublished data).

7.4. CVB Infection of Cultured Pancreatic Endocrine Cells

According to one study,[101] mouse pancreatic organ cultures can support CVB replication only if obtained from mice younger than 3 weeks. After that age, no production of infectious virus occurred. Thus, it appears that multiplication in isolated mouse organs ceases at a much earlier age than in the intact animal. In our experience, *in vitro* cultured pancreatic β-cells can support the replication of all CVB, even if taken from mice older than 2 months of age. Within 2–6 days postinfection, CVB produce a rapid lytic effect in these cultures.[87,88] To date, no functional evaluations of CVB-infected mouse β-cell cultures have been reported. By contrast, insulin release has been measured in a continuous line of rat insulinoma cells (RIN cells) persistently infected with CVB4.[101] High titers of infectious virus were continuously released from infected cultures without cytopathology, and the total release of insulin in response to both low and high concentrations of glucose was not reduced by CVB4. These investigators proposed that CVB-induced diabetes might be initiated by a chronic infection, possibly leading to immunopathologic damage of β-cells. However, caution should be exercised in interpreting these data, since persistent CVB infections of human β-cells have not been documented, and it is known that rats are highly resistant to CVB. Thus, this condition seems totally different from that of mice and human subjects, where β-cell permissiveness to CVB is the rule.

An *in vitro* model has been developed to determine whether viruses are capable of infecting human β-cells. By use of double-label immunofluorescence with antibodies to insulin and CVB, it was shown that CVB3 and CVB4 do replicate in insulin-containing β-cells and that these cells are killed by the virus.[60,103] Radioimmunoassay (RIA) showed that intracellular IRI decreased rapidly, beginning at 24 hr after CVB3 infection, and that the decrease in insulin roughly paralleled the increase in viral titer. Only a few human pancreas were used for these studies, precluding any conclusions as to the influence of the HLA phenotype on β-cell susceptibility.

7.5. Mechanisms of Endocrine Damage

What is the pathogenesis of CVB-induced hyperglycemia? Two major pathways may lead from islet cell infection to diabetes: (1) direct lysis of target cells, and (2) virus-induced immunopathology.[104]

The first possibility is suggested by the type of histopathologic damage seen in the pancreas of neonates dying of overwhelming CVB infections and of rare cases of acute diabetes induced by CVB.[60,71] Experimental studies suggest that β-cell lysis is most probably involved as well in the histologic lesions observed in mice and in the acute release of insulin occurring in the early phase of infection.[87,95] That CVB cause only mild metabolic effects is probably due to the fact that these agents usually infect and damage only small numbers of β-cells. It is more difficult to interpret the long-term defects of insulin production seen not only in mice,[87,106] but also *in vitro* in isolated islets of Langerhans[105] and in cultured β-cells[106] obtained from CVB4-infected mice. The studies mentioned above suggest that, after infection, subtle long-term alterations of β-cell functions do occur in the pancreas of mice. Although releasing unchanged or greater than normal amounts of insulin in response to glucose *in vitro,* β-cells isolated from infected mice had a reduced synthesis of total proteins and, in particular, of pro-insulin and insulin. The biochemical basis of these defects is unknown, but it is likely that they contribute to an altered control of glucose homeostasis.

The possibility of an immune-mediated pathogenesis of CVB-induced β-cell damage would strengthen the association of these agents with human diabetes. It is known that most cases of IDDM are preceded by a long prediabetic period characterized by metabolic and immunologic abnormalities and, in particular, by autoimmune reactions to the endocrine pancreas. What triggers pancreatic autoimmunity is, however, unknown.

An hypothesis connecting viruses with the origin of endocrine autoimmunity has been recently proposed.[3] Briefly, it has been postulated that, as a consequence of local damage, an inflammatory response may be induced in endocrine tissues. Interferon or other mediators released

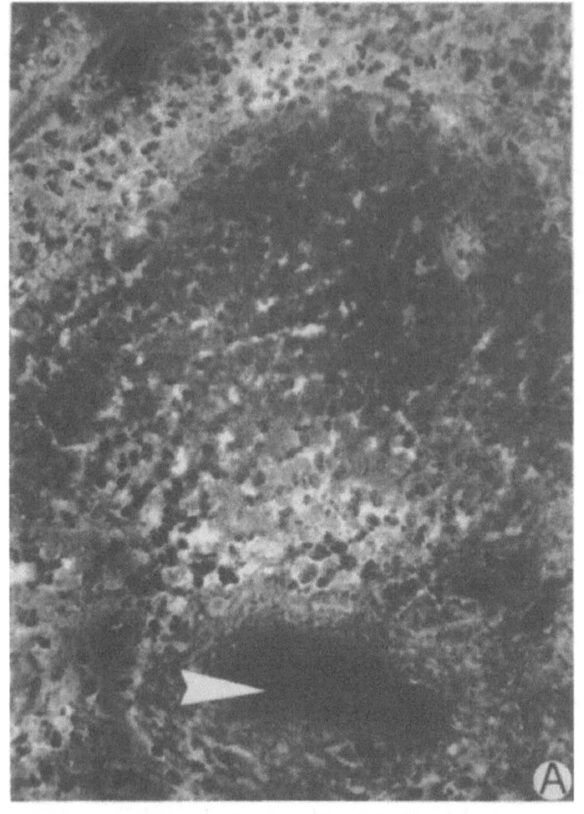

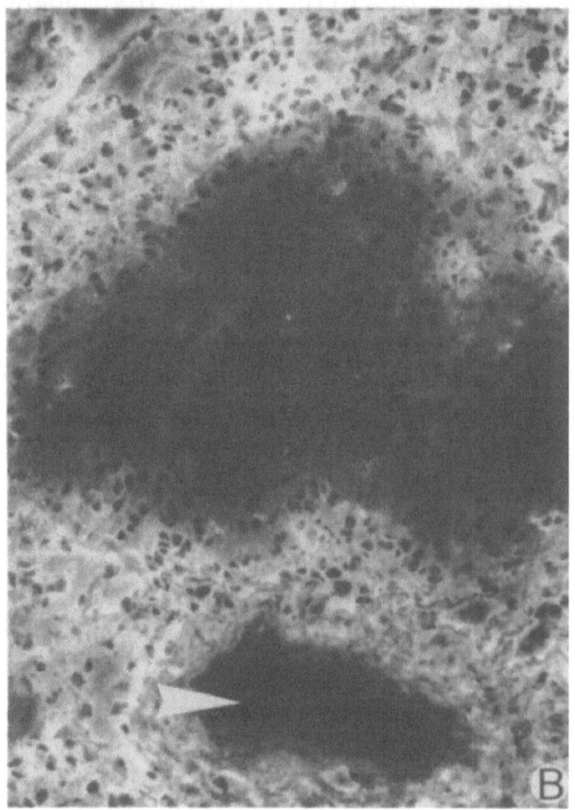

FIGURE 3. Lack of expression of Ia antigens in endocrine pancreatic cells of DBA/2 mice infected with CVB3. Four days postinfection with prototype CVB3, frozen sections of pancreas were stained with fluorescein-labeled monoclonal antibody to CVB3 and rhodamine-labeled monoclonal antibody to murine Ia d (×200). With fluorescein filters, small amounts of viral antigens were seen in islet cells and substantial amounts in the surrounding acinar tissue (**A**). When the same section is seen with rhodamine filters it appears that high concentrations of Ia molecules are expressed in exocrine cells, but not in islet cells (**B**). The arrows point to the lumen of a pancreatic duct.

in situ would induce the aberrant expression of the major histocompatibility complex class II molecules (Ia antigens) on endocrine epithelial cells. Ia-positive epithelial cells, in turn, would become capable of presenting autoantigens to infiltrating lymphocytes, triggering an organ-specific autoimmune response.

It has been shown *in vitro* that this sequence of events may be initiated by coronavirus infection of rat astrocytes[107] and, more importantly, high levels of Ia antigens have been demonstrated *in vivo* in the islets of Langerhans of individuals dying in the early phase of acute diabetes.[108]

To see whether CVB infection of mice is followed by the induction of Ia antigens on islet cells, we infected mice with CVB3 and searched for the expression of viral and murine Ia antigens on frozen sections of pancreas. Direct staining with fluorescein-labeled monoclonal antibody to CVB3 and rhodamine-labeled monoclonals to the Ia antigen of BALB/c and DBA/2 mice showed that moderate CVB3 infection of islet cells does not induce the *de novo* expression of Ia antigens on endocrine cells. By contrast, virus-infected exocrine epithelial cells were strongly Ia positive (Fig. 3). Control studies showed that normal pancreatic acinar cells do not express Ia antigens and that these molecules appear on acinar cells approximately 2 days postinfection. Our observations, however, were limited to the early phase of infection, characterized by the expression of viral antigens in islet cells (2–6 days postinfection). Further studies are in progress. These results are in agreement with those of Bottazzo and colleagues,[109] showing that human endocrine pancreatic cells fail to express Ia molecules under a variety of physiologic stimuli *in vitro*. Under the same conditions, pancreatic exocrine cells were strongly Ia positive.

As it has been reported that ICA may be induced by CVB infections in humans,[65] and it is known that ICA may appear in virus-infected mice,[110] we also searched for ICA in the serum of CVB-infected mice. Pancreatic autoantibodies were not detected. Thus, experimental evidence linking CVB infections to the production of autoimmune diabetes is lacking. However, as already observed in CVB-induced cardiac autoimmunity,[111] it is likely that only a few strains of mice are capable of mounting an autoimmune response against islet cells as a consequence of CVB infection. Newer techniques, and in particular hybridoma technology, will permit assessment of the possible role of molecular mimicry and of anti-idiotypic antibodies in the pathogenesis of CVB-induced endocrine disease.[112]

8. SUMMARY AND CONCLUSIONS

What can we learn from these studies on viral diabetes? First, it has been shown that common human viruses, such as the six CVB, have the potential of infecting and damaging pancreatic endocrine cells causing hyperglycemia. Second, serologic studies demonstrated that up to one third of cases of IDDM in the young are probably preceded by CVB infection. Third, experimental studies have shown that the genetic makeup of the host, as well as phenotypic factors, can modulate the endocrine effects of CVB infections. Fourth, the tropism of CVB to pancreatic islet cells may be increased by *in vitro* passage in β-cell cultures. Fifth, the roles of persistent infections and autoimmunity in CVB-induced endocrine damage have just begun to be investigated. Progress on this point would help explain the reported high frequency of CVB infections in the prediabetic period.

We are confident that newer methods to detect viral antibodies, as well as the use of immunologic and genetic viral probes for the histopathologic analysis of the pancreas from cases of recent-onset diabetes, will permit more precise evaluation of the role of CVB in human diabetes. This knowledge will be of utmost importance should preventive measures for IDDM be worked out. At the very least, these studies on the viral etiology of diabetes should serve as a guide for future investigations on idiopathic diseases of suspect viral origin.

ACKNOWLEDGMENTS. Personal investigations were supported by grants from MPI and by the financial help of Novo Farmaceutici Italia, through the courtesy of Dr. M. Iavicoli. We are grateful to Dr. A. B. Jenson for providing pictures of his histopathologic findings.

REFERENCES

1. Cahill, G. F., and McDevitt, H. O., 1981, Insulin dependent diabetes mellitus. The initial lesion, *N. Engl. J. Med.* **304:**1454–1465.
2. Rossini, A. A., Mordes, J. P., and Like, A. A., 1985, Immunology of insulin-dependent diabetes mellitus, *Annu. Rev. Immunol.* **3:**289–320.
3. Bottazzo, G. F., 1986, Death of a beta cell: Homicide or suicide?, *Diabetic Med.* **3:**119–130.
4. Toniolo, A., and Onodera, T., 1984, Viruses and diabetes, in: *Immunology in Diabetes* (D. Andreani, U. Di Mario, K. F. Federlin, and L. G. Heding, eds.), pp. 71–93, Kimpton, London.
5. Gamble, D. R., 1980, The epidemiology of insulin dependent diabetes, with particular reference to the relationship of virus infection to its etiology, *Epidemiol. Rev.* **2:**49–70.
6. Gamble, D. R., and Taylor, T. W., 1969, Seasonal incidence of diabetes mellitus, *Br. Med. J.* **3:**631–633.
7. Gamble, D. R., and Taylor, K. W., 1973, Seasonal incidence of diabetes mellitus, *Br. Med. J.* **1:**289–290.
8. Gonzalez-Pijem, L., Cintron, C., and Carrion, F., 1985, Insulin-dependent diabetes mellitus in children: Lack of statistically significant seasonal or secular differences in its incidence, in: *Proceedings of the Twelfth Congress of the International Diabetes Federation,* Madrid (abst. 509).

9. Gamble, D. R., Kinsley, M. L., Fitzgerald, M. G., Bolton, R., and Taylor, K. W., 1969, Viral antibodies in diabetes mellitus, *Br. Med. J.* **3:**627–630.
10. Craighead, J. E., 1975, The role of viruses in the pathogenesis of pancreatic disease and diabetes mellitus, *Prog. Med. Virol.* **19:**161–214.
11. Barrett-Connor, E., 1985, Is insulin-dependent diabetes mellitus caused by coxsackievirus B infection? A review of the epidemiologic evidence, *Rev. Infect. Dis.* **7:**207–215.
12. Gamble, D. R., Taylor, K. W., and Cumming, H., 1973, Coxsackie viruses and diabetes mellitus, *Br. Med. J.* **4:**260–262.
13. Baum, J. D., Aynsley-Green, A., and MacCallum, F. O., 1974, Coxsackie viruses and diabetes mellitus, *Br. Med. J.* **3:**172.
14. Nelson, P. G., Pyke, D. A., and Gamble, D. R., 1975, Viruses and the aetiology of diabetes: A study in identical twins, *Br. Med. J.* **4:**249–251.
15. Cudworth, A. G., Woodrow, J. C., and Gamble, D. R., 1975, Coxsackie B4 virus infection and diabetes, *Lancet* **2:**29.
16. Cudworth, A. G., Gamble, D. R., White, G. B. B., Lendrum, R., Woodrow, J. C., and Bloom, A., 1977, Aetiology of juvenile-onset diabetes: A prospective study, *Lancet* **1:**385–388.
17. Di Pietro, C., Del Guercio, M. J., Paolino, G. P., Barbi, M., Ferrante, P., and Chiumello, G., 1979, Type I diabetes and coxsackie virus infection, *Helv. Paediatr. Acta* **34:**557–561.
18. Ray, C. G., Palmer, J. P., Crossley, J. R., and Williams, R. H., 1980, Coxsackie B virus antibody responses in juvenile diabetes mellitus, *Clin. Endocrinol.* **12:**375–378.
19. Hazra, D. K., Singh, R., Wahal, P. K., Lahiri, V., Gupta, M. K., Jain, N. K., and Elhence, B. R., 1980, Coxsackie antibodies in young Asian diabetics, *Lancet* **1:**877.
20. Palmer, J. P., Cooney, M. K., Ward, R. H., Hansen, J. A., Brodsky, J. B., Ray, C. G., Crossley, J. R., Asplin, C. M., and Williams, R. H., 1982, Reduced coxsackie antibody titres in type 1 (insulin-dependent) diabetic patients presenting during an outbreak of coxsackie B3 and B4 infection, *Diabetologia* **22:**426–429.
21. Andersen, O. O., Christy, M., Arnung, K., Buschard, K., Christau, B., Kromann, H., Nerup, J., Platz, P., Ryder, L. P., Svejgaard, A., and Thomsen, M., 1977, Viruses and diabetes, in: *Diabetes* (J. S. Bajaj, ed.), *Proceedings of the Ninth Congress of the International Diabetes Federation*, pp. 294–298, Excerpta Medica, Amsterdam.
22. Mertens, T., Gruneklee, D., and Heggers, H. J., 1983, Neutralizing antibodies against coxsackie B viruses in patients with recent onset of type I diabetes, *Eur. J. Pediatr.* **140:**293–294.
23. Schmidt, W. A. K., Brade, L., Munterfering, H., and Klein, M., 1978, Course of coxsackie B antibodies during juvenile diabetes, *Med. Microbiol. Immunol.* **164:**291–298.
24. Munterfering, H., 1984, Diabetes and virus, in: *The Importance of the Islets of Langerhans for Modern Endocrinology* (K. Federlin and J. Scholtholt, eds.), pp. 111–128, Raven, New York.
25. Sakurami, T., Nabeya, N., Nagaoka, K., Matsumori, A., Kuno, S., and Honda, A., 1982, Antibodies to coxsackie B viruses and HLA in Japanese with juvenile-onset type 1 (insulin-dependent) diabetes mellitus, *Diabetologia* **22:**375–377.
26. Palmer, J. P., Cooney, M. K., Crossley, J. R., Hollander, P. H., and Asplin, C. M., 1981, Antibodies to viruses and to pancreatic islets in nondiabetic and insulin-dependent diabetic patients, *Diabetes Care* **4:**525–528.
27. Orchard, T. J., Atchison, R. W., Becker, D., Rabin, B., Eberhardt, M., Kuller, L. H., La Porte, L. E., and Cavender, D., 1983, Coxsackie infections and diabetes. *Lancet* **2:**631.

28. Eggers, H. J., Mertens, T., and Gruneklee, D., 1983, Coxsackie infection and diabetes, *Lancet* **2:**631.
29. El-Hagrassy, M. M. O., Banatvala, J. E., and Coltart, D. J., 1980, Coxsackie-B-virus-specific IgM responses in patients with cardiac and other diseases, *Lancet* **2:**1160–1162.
30. King, M. L., Bidwell, D., Shaikh, A., Voller, A., and Banatvala, J. E., 1983, Coxsackie B virus specific IgM responses in children with insulin-dependent (juvenile-onset; type 1) diabetes mellitus, *Lancet* **1:**1397–1399.
31. King, M. L., Bidwell, D., Voller, A., Bryant, J., and Banatvala, J. E., 1983, Role of coxsackie B viruses in insulin-dependent diabetes mellitus, *Lancet* **2:**915–916.
32. Banatvala, J. E., Bryant, J., Schernthauer, G., Borkenstein, M., Schober, E., Brown, D., De Silva, L. M., Menser, M. A., and Silink, M., 1985, Coxsackie B, mumps, rubella, and cytomegalovirus specific IgM responses in patients with juvenile-onset insulin-dependent diabetes mellitus in Britain, Austria, and Australia, *Lancet* **1:**1409–1412.
33. Odugbesan, O., and Barnett, A. H., 1985, Coxsackie B virus and juvenile-onset insulin-dependent diabetes, *Lancet* **2:**455.
34. Gamble, D. R., and Cumming, H., 1985, Coxsackie B virus and juvenile-onset insulin-dependent diabetes, *Lancet* **2:**455–456.
35. Schernthaner, G., Scherbaum, W., Borkenstein, M., Banatvala, J. E., Bryant, J., Schober, E., and Mayr, W. R., 1985, Coxsackie-B-virus-specific IgM responses, complement-fixing islet-cell antibodies, HLA DR antigens, and C-peptide secretion in insulin-dependent diabetes mellitus, *Lancet* **2:**630–632.
36. Asplin, C. M., Cooney, M. K., Crossley, J. R., Dornan, T. L., Raghu, P., and Palmer, J. P., 1982, Coxsackie B4 infection and islet cell antibodies three years before overt diabetes, *J. Pediatr.* **101:**398–400.
37. Hyoty, H., Huupponen, T., Kotola, L., and Leinikki, P., 1986, Humoral immunity against viral antigens in Type 1 diabetes: Altered IgA-class immune response against Coxsackie B4 virus, *Acta Pathol., Microbiol., Immunol. Scand. Sect. C: Immunol.* **94:**83–88.
38. Friman, G., Fohlman, J., Frisk, G., Diderholm, H., Ewald, U., Kobbah, M., and Tuvemo, T., 1985, An incidence peak of juvenile diabetes. Relation to Coxsackie B virus immune response, *Acta Paediatr. Scand.* **320** (suppl.):14–19.
39. Frisk, G., Fohlman, J., Kobbah, M., Ewald, U., Tuvemo, T., Diderholm, H., and Friman, G., 1985, High frequency of Coxsackie-B-virus-specific IgM in children developing type I diabetes during a period of high diabetes morbidity, *J. Med. Virol.* **17:**219–227.
40. Alberti, A. M., Amato, C., Candela, A., Costantino, F., Grandolfo, M. E., Lombardi, F., Novello, F., Orsini, M., and Santoro, R., 1985, Serum antibodies against Coxsackie B1-6 viruses in type 1 diabetics, *Acta Diabetol. Lat.* **22:**33–38.
41. Mirkovic, R. R., Varma, K. S., and Yoon, J. W., 1984, Incidence of Coxsackievirus B type 4 (CB4) infections concomitant with onset of insulin-dependent diabetes mellitus, *J. Med. Virol.* **14:**9–16.
42. Buschard, K., and Madsbad, S., 1984, A longitudinal study of virus antibodies in patients with newly diagnosed type 1 (insulin-dependent) diabetes mellitus, *J. Clin. Lab. Immunol.* **13:**65–70.
43. Tuvemo, T., Diderholm, H., Ewald, U., Fohlman, J., Friman, G., Frisk, G., and Kobbah, M., 1985, Coxsackie B virus specific IgM antibodies in newly diagnosed IDDM children, in: *Proceedings of the Twelfth Congress of the International Diabetes Federation,* Madrid (abst. 1495).
44. Dorman, J. S., Cruickshanks, K. J., Laporte, R. E., Atchison, R. W., Songer, T. J., Slemenda, C. W., Eberhardt, M. S., Tajima, N., Orchard, T. J., Becker, D. J., and

Drash, A. L., 1985, Racial differences in the descriptive epidemiology of IDDM: A viral connection?, in: *Proceedings of the Twelfth Congress of the International Diabetes Federation*, Madrid (abst. 359).

45. Bartsocas, C. S., Papadatos, C. J., Lab, M., Spyrou, N., Krikelis, B., and Serié, C., 1982, Coxsackie B viruses and autoimmune diabetes, *J. Pediatr.* **101**:647–648.
46. Niklasson, B. S., Dobersen, M. J., Peters, C. J., Ennis, W. H., and Moller, E., 1985, An outbreak of coxsackievirus B infection followed by one case of diabetes mellitus, *Scand. J. Infect. Dis.* **17**:15–18.
47. Hadden, D. R., Connolly, J. H., Montgomery, D. A. D., and Weaver, J. A., 1972, Coxsackie B virus and diabetes, *Br. Med. J.* **4**:729.
48. Wales, J. K., and Hambling, M. H., 1973, Coxsackie B virus and diabetes, *Br. Med. J.* **2**:120.
49. Dippe, S. E., Bennett, P. H., Miller, M., Maynard, J. E., and Berquist, K. R., 1975, Lack of causal association between Coxsackie B4 virus infection and diabetes, *Lancet* **1**:1314–1317.
50. Hierholzer, J. C., Mostow, S. R., and Dowdle, W. R., 1972, Prospective study of a mixed Coxsackie virus B3 and B4 outbreak of upper respiratory illness in a children's home, *Pediatrics* **49**:744–752.
51. Hierholzer, J. C., and Farris, W. A., 1974, Follow-up of children infected in a Coxsackievirus B3 and B4 outbreak: No evidence of diabetes mellitus, *J. Infect. Dis.* **129**:741–746.
52. Huff, J. C., Hierholzer, J. C., and Farris, W. A., 1974, An outbreak of juvenile diabetes mellitus: Consideration of a viral etiology, *Amer. J. Epidemiol.* **100**:277–287.
53. Samantray, S. K., Christopher, S., Mukundau, P., and Johnson, S. C., 1977, Lack of relationship between viruses and human diabetes mellitus, *Aust. N.Z. J. Med.* **7**:139–142.
54. Toniolo, A., Federico, G., Ginsberg-Fellner, F., and Notkins, A. L., (manuscript in preparation).
55. Hyoty, H., Huupponen, T., and Leinikki, P., 1985, Humoral immunity against viral antigens in insulin-dependent diabetes mellitus: Altered IgA-class immune response against mumps virus, *Clin. Exp. Immunol.* **60**:139–144.
56. Toniolo, A., Conaldi, P. G., Garzelli, C., Benedettini, G., Federico, G., Saggese, G., Vettor, R., and Federspil, G., 1985, Role of antecedent mumps and reovirus infections on the development of type 1 (insulin-dependent) diabetes, *Eur. J. Epidemiol.* **1**:172–179.
57. Bruserud, O., Jervell, J., and Thorsby, E., 1985, HLA-DR3 and -DR4 control T lymphocyte responses to mumps and Coxsackie B4 virus: Studies on patients with Type 1 (insulin-dependent) diabetes and healthy subjects, *Diabetologia* **28**:420–426.
58. Bruserud, O., Sollid, L., Gaudernack, G., and Thorsby, E., 1986, The role of CD8-positive subset of T cells in proliferative responses to soluble antigens. II. CD8-positive cells are not responsible for DR-associated differences in responsiveness to mumps and Coxsackie B4, *Scand. J. Immunol.* **23**:469–473.
59. Gorsuch, A. N., Gamble, D. R., and Ingram, D., 1985, HLA-linked susceptibility to type I diabetes (IDDM) and poliovirus antibody titres after polio vaccination, in: *Proceedings of the Twelfth Congress of the International Diabetes Federation*, Madrid (abst. 510).
60. Yoon, J. W., Austin, M., Onodera, T., and Notkins, A. L., 1979, Virus-induced diabetes mellitus: Isolation of a virus from the pancreas of a child with diabetic ketoacidosis, *N. Engl. J. Med.* **300**:1173–1179.
61. Champsaur, H., Dussaix, E., Samolyk, D., Fabre, E., Bach, C., and Assan, R., 1980, Diabetes and coxsackie virus B5 infection, *Lancet* **1**:251.
62. Champsaur, H. F., Bottazzo, G. F., Bertrams, J., Assan, R., and Bach, C., 1982,

Virologic, immunologic, and genetic factors in insulin-dependent diabetes mellitus, *J. Pediat.* **100:**15–20.

63. Gladisch, R., Hofmann, W., and Waldherr, R., 1976, Myokarditis und insulitis nach Coxsackie-virus-infektion, *Z. Kardiol.* **65:**837–849.
64. Kaplan, M. H., Klein, S. W., McPhee, J., and Harper, R. G., 1983, Coxsackievirus infections in infants younger than three months of age: A serious childhood illness, *Rev. Infect. Dis.* **5:**1019–1032.
65. Nigro, G., Pacella, M. E., Patané, E., and Midulla, M., 1986, Multi-system Coxsackievirus B-6 infection with findings suggestive of diabetes mellitus, *Eur. J. Ped.* **145:**557–558.
66. Wilson, C., Connolly, J. H., and Thomson, D., 1977, Coxsackie B2 virus infection and acute-onset diabetes in a child, *Br. Med. J.* **1:**1008.
67. Nelson, P. G., Arthur, L. J. H., Gurling, K. J., Gamble, D. R., and Taylor, K. W., 1977, Familial juvenile onset diabetes, *Br. Med. J.* **2:**1126–1127.
68. Kanazawa, Y., 1984, Some genetic and immunological aspects of type I (insulin-dependent) diabetes in Japan, in: *Immunology in Diabetes* (D. Andreani, U. Di Mario, K. F. Federlin, and L. G. Heding, eds.), pp. 159–169, Kimpton, London.
69. Kibrick, S., and Bernischke, K., 1958, Severe generalized disease (encephalohepatomyocarditis) occurring in the newborn period and due to infection with Coxsackie virus group B. Evidence of intrauterine infection with this agent, *Pediatrics* **22:**857–875.
70. Sussman, M. L., Strauss, L., and Hodes, H. L., 1959, Fatal Coxsackie group B virus infection in the newborn: Report of a case with necropsy findings and brief review of the literature, *Am. J. Dis. Child.* **97:**483–492.
71. Jenson, A. B., Rosenberg, H. S., and Notkins, A. L., 1980, Pancreatic islet cell damage in children with fatal viral infections, *Lancet* **2:**354–358.
72. Ujevich, M. M., and Jaffe, R., 1980, Pancreatic islet cell damage: Its occurrence in neonatal Coxsackievirus encephalomyocarditis, *Arch. Pathol. Lab. Med.* **104:**438–441.
73. Ahmad, N., and Abraham, A. A., 1982, Pancreatic isleitis with coxsackie virus B5 infection, *Human Pathol.* **13:**661–662.
74. Pappenheimer, A. M., Kunz, L. J., and Richardson, S., 1951, Passage of Coxsackie virus (Connecticut-5 strain) in adult mice with production of pancreatic disease, *J. Exp. Med.* **94:**45–64.
75. Gifford, R., and Dalldorf, G., 1951, The morbid anatomy of experimental Coxsackie virus infection, *Am. J. Path.* **27:**1047–1063.
76. Dalldorf, G., and Gifford, R., 1952, Adaptation of Group B Coxsackie virus to adult mouse pancreas, *J. Exp. Med.* **96:**491–497.
77. Minkowitz, S., and Berkovich, S., 1970, Hepatitis produced by Coxsackievirus B1 in adult mice, *Arch. Path.* **89:**427–433.
78. Robertson, J. S., 1954, The pancreatic lesion in adult mice infected with a strain of pleurodynia virus: I. Electron microscopical observations, *Aust. J. Exp. Biol. Med. Sci.* **32:**393–410.
79. Burch, G. E., Tsui, C. Y., Harb, J. M., and Colcolough, H. L., 1971, Pathologic findings in the pancreas of mice infected with Coxsackievirus B4, *Arch. Intern. Med.* **128:**40–47.
80. Tsui, C. Y., Burch, G. E., and Harb, J. M., 1972, Pancreatitis in mice infected with Coxsackie B1, *Arch. Pathol.* **93:**379–389.
81. Harrison, A. K., Bauer, S. P., and Murphy, F. A., 1972, Viral pancreatitis: Ultrastructural pathological effects of Coxsackie virus B3 infection in newborn mouse pancreas, *Exp. Mol. Pathol.* **17:**206–219.
82. Lansdown, A. G., 1976, Pathological changes in the pancreas of mice following infection with Coxsackie B viruses, *Br. J. Exp. Pathol.* **57:**331–342.

83. Coleman, T. J., Gamble, D. R., and Taylor, K. W., 1973, Diabetes in mice after coxsackie B4 virus infection, *Br. Med. J.* **3**:25–27.
84. Coleman, T. J., Taylor, K. W., and Gamble, D. R., 1974, The development of diabetes following Coxsackie B virus infection in mice, *Diabetologia* **10**:755–759.
85. Ross, M. E., and Notkins, A. L., 1974, Coxsackie B viruses and diabetes mellitus, *Br. Med. J.* **2**:226.
86. Ross, M. E., Hayashi, K., and Notkins, A. L., 1974, Virus-induced pancreatic disease: Alterations in concentration of glucose and amylase in blood, *J. Infect. Dis.* **129**:669–676.
87. Yoon, J. W., Onodera, T., and Notkins, A. L., 1978, Virus-induced diabetes mellitus. XV. Beta cell damage and insulin-dependent hyperglycemia in mice infected with Coxsackievirus B4, *J. Exp. Med.* **148**:1068–1080.
88. Toniolo, A., Onodera, T., Jordan, G., Yoon, J. W., and Notkins, A. L., 1982, Virus-induced diabetes mellitus: Glucose abnormalities produced in mice by the six members of the Coxsackie B virus group, *Diabetes* **31**:496–499.
89. Prabakhar, B. S., Menegus, M. A., and Notkins, A. L., 1985, Detection of conserved and non-conserved epitopes on Coxsackie virus B4: frequency of antigenic change, *Virology* **146**:302–306.
90. Webb, S. R., Kearse, K. P., Foulke, C. L., Hartig, P. C., and Prabhakar, B. S., 1986, Neutralization epitope diversity of Coxsackievirus B4 isolates detected by monoclonal antibodies, *J. Med. Virol.* **20**:9–15.
91. Jordan, G. W., Bolton, V., and Schmidt, N. J., 1985, Diabetogenic potential of Coxsackie B viruses in nature, *Arch. Virol.* **86**:213–221.
92. Toniolo, A., Onodera, T., Yoon, J. W., and Notkins, A. L., 1980, Induction of diabetes by cumulative environmental insults from viruses and chemicals, *Nature (Lond.)* **288**:383–385.
93. Yoon, J. W., London, W. T., Curfman, B. L., Brown, R. L., and Notkins, A. L., 1986, Coxsackie virus B4 produces transient diabetes in nonhuman primates, *Diabetes* **35**:712–716.
94. Wegner, V., Kewitsch, A., and Madauss, M., 1985, Hyperglycemia in BALB/c mice after pretreatment with one subdiabetogenic dose of streptozotocin and subsequent infection with a Coxsackie B4 strain, *Biomed. Biochim. Acta* **44**:21–27.
95. Webb, S. R., and Madge, G. E., 1980, The role of host genetics in the pathogenesis of coxsackievirus infection in the pancreas of mice, *J. Infect. Dis.* **141**:47–54.
96. Webb, S. R., Loria, R. M., Madge, G. E., and Kibrick, S., 1976, Susceptibility of mice to group B coxsackie virus is influenced by the diabetic gene, *J. Exp. Med.* **143**:1239–1248.
97. Cook, S. A., Loria, R. M., and Madge, G. E., 1982, Host factors in coxsackievirus B4 induced pancreopathy, *J. Lab. Invest.* **46**:377–383.
98. Loria, R. M., Montgomery, L. B., Corey, L. A., and Chinchilli, V. M., 1984, Influence of diabetes mellitus on susceptibility to Coxsackievirus B4, *Arch. Virol.* **81**:251–262.
99. Montgomery, L. B., and Loria, R. M., 1986, Humoral immune response in hereditary and overt diabetes mellitus, *J. Med. Virol.* **19**:255–268.
100. Chatterjee, N. K., and Nejman, C., 1986, Protein kinase in nondiabetogenic Coxsackievirus B4, *J. Med. Virol.* **19**:353–365.
101. Van der Groen, G., Vanden Berghe, D. A. R., and Pattyn, S. R., 1976, Replication of coxsackie B3 virus in mouse organ cultures, *Arch. Intern. Physiol. Biochim.* **84**:1114.
102. Frank, J. A., Jr., Schmidt, E. V., Smith, R. E., and Wilfert, C. M., 1986, Persistent infections of rat insulinoma cells with Coxsackie B4 virus, *Arch. Virol.* **87**:143–150.
103. Yoon, J. W., Onodera, T., Jenson, A. B., and Notkins, A. L., 1978, Virus-induced diabetes mellitus. XI. Replication of Coxsackie B3 virus in human pancreatic beta cell cultures, *Diabetes* **27**:778–781.

104. Saggese, G., Federico, G., Garzelli, C., and Toniolo, A., 1985, Insulin-dependent diabetes: A possible viral disease, *Pediatrician* **12:**179–193.
105. Szopa, T. M., Gamble, D. R., and Taylor, K. W., 1986, Coxsackie B4 virus induces short-term changes in the metabolic functions of mouse pancreatic islets in vitro, *Cell Biochem. Funct.* **4:**181–188.
106. Chatterjee, N. K., Haley, T. M., and Nejman, C., 1985, Functional alterations in pancreatic beta cells as a factor in virus-induced hyperglycemia in mice, *J. Biol. Chem.* **260:**12786–12791.
107. Massa, P. T., Dorries, R., and ter Meulen, V., 1986, Viral particles induce Ia antigen expression on astrocytes, *Nature (Lond.)* **320:**543–546.
108. Foulis, A. K., and Farquaharson, M. A., 1986, Aberrant expression of HLA-DR antigens by insulin-containing beta-cells in recent-onset type I diabetes mellitus, *Diabetes* **35:**1215–1224.
109. Pujol-Borrel, R., Todd, I., Doshi, M., Gray, D., Feldmann, M., and Bottazzo, G. F., 1986, Differential expression and regulation of MHC products in the endocrine and exocrine cells of the human pancreas, *Clin. Exp. Immunol.* **65:**128–139.
110. Onodera, T., Toniolo, A., Ray, U. R., Jenson, A. B., Knazek, A., and Notkins, A. L., 1981, Virus-induced diabetes mellitus. XX. Polyendocrinopathy and autoimmunity, *J. Exp. Med.* **153:**1457–1473.
111. Wolfgram, L. J., Beisel, K. W., Herskowitz, A., and Rose, N. R., 1986, Variations in the susceptibility to coxsackievirus B3-induced myocarditis among different strains of mice, *J. Immunol.* **136:**1846–1852.
112. Notkins, A. L., 1984, Diabetes: On the track of viruses, *Nature (Lond.)* **311:**209–210.

21

Epidemiology

G. HAMMOND

1. INTRODUCTION

As a member of the enterovirus group within the Picornaviridae family, the epidemiology of coxsackievirus (CV) is similar to that of other enteroviruses. The update on CV epidemiology is discussed under the mechanisms of virus transmission, followed by factors that have been observed to influence the distribution of CV infections. Information is also presented on CV infections in the family, in institutions, and on geographical trends of CV infections.

2. ROUTES OF VIRUS TRANSMISSION

The most important mechanism of transmission of CV is by person-to-person spread. In general, as CV are shed from the gastrointestinal (GI) tract in bowel contents for prolonged periods of time (for as long as 70 days[1]), this favors the fecal–oral route of transmission, likely by means of contaminated hands. CV have been found to be excreted more commonly in respiratory secretions than have echoviruses.[1] Respiratory transmission may occur by droplet spread, especially with tonsillitis, pharyngitis, and pneumonia. Oral-to-oral contact spread of CV must also be considered in these illnesses. Transmission of infection by the aerosol route has been demonstrated experimentally for coxsackievirus A (CVA) infections. Because CV can be excreted simultaneously from both the respiratory and the GI tracts, it is probable that both routes are

G. HAMMOND • Cadham Provincial Laboratory, Winnipeg, Manitoba R3C 3Y1, Canada.

important in the natural transmission of virus. Transplacental transmission of coxsackievirus B (CVB) directly to a neonate may result in serious illness in the infant.[2]

A second mechanism of transmission is through a contaminated environment in which CV are relatively stable. Among young children in a family or an institution, contact with infected fecal material or respiratory secretions or contaminated objects, such as toys, eating utensils, and furniture, may be an important mode of virus transmission. Fecally contaminated drinking water and contaminated food are vehicles of CV transmission. CV have been found in environmental waters such as lakes and swimming pools, but this likely reflects simply the oral and anal shedding of virus, rather than the source of CV infection. CV have been found in flies and mosquitoes, raising the possible mechanism of vector-borne infection.

3. FACTORS THAT INFLUENCE THE DISTRIBUTION OF COXSACKIEVIRUS INFECTIONS

Host immunity to CV results from the presence of neutralizing antibodies directed against each specific serotype. Thus, maternal immunity can be transferred passively to newborns, if the mother herself is immune. A lack of breast-feeding has been associated with an increased incidence of nonpolio enterovirus (NPEV) infection during the first month of life, regardless of socioeconomic status.[3] After the first several months of life and the disappearance of maternally derived antibody, infants remain susceptible to CV, which they may acquire from children at home, in child care centers, or later in school. Thus, the incidence of CV infections is greatest in young children and diminishes thereafter. Several studies have examined the impact of age on CV infections. The peak age group for CVB infection was under 5 years (1837, or 48%, of the CVB isolates were from children under 5 years of age in a 10-year U.S. surveillance),[4] and children under 1 year of age were especially at risk (approximately 30% were under 1 year of age). While NPEV infection was not demonstrated from specimens 1 day after delivery in a study of 666 neonates, it is common in the first month of life, especially during the summer and fall.[3] Children under 3 months of age appear at increased risk of symptomatic disease from CVB.[2]

Inadequate facilities for hygiene, such as an impure water supply and a lack of adequate sewage facilities, are probably the most important determinants of exposure and transmission of CV infection worldwide. A greater incidence of CV infections occurs among children in poor socioeconomic situations.[3]

Another factor related to CV infection includes the increased incidence in the male sex.[4,5] In a report from the United States, the distribution of reported CVB infection was greatest for males, with a male-to-female ratio of approximately 1.5 : 1 for most age groups. Significant exceptions were for the 20–29-year group and the ≥50-year age group.[4] Moore *et al.*[7] enumerated some possible explanations (originally offered by Gelfand *et al.*[6]) for the excess of NPEV found in male patients, including a longer duration and higher titer of virus excretion, a higher clinical illness rate, and increased physical exertion among males.

Race has also been evaluated as a risk factor for CV infections. Infections with NPEV are more common in nonwhite races, although in one study infection was not independent of other indicators of socioeconomic status.[3]

There are major seasonal and geographic worldwide influences in the epidemiology of CV infections, with a striking seasonal distribution of these infections that peaks during the summer and early fall months. An opposite annual distribution has been observed in the northern and southern hemisphere, in keeping with these seasonal trends. There is no adequate explanation for this seasonal effect. It has been suggested that there is a gradual change in time of onset of CV infections versus latitude, as in the United States enterovirus infections have been observed to begin earlier each year in the southern than in the northern latitudes.[8]

Lastly, the epidemiologic pattern of CV may be influenced by genetic variation of CV strains, which may alter attachment to cell receptors, virulence factors, or the effectiveness of host immunity. The possible occurrence of these events requires analysis by molecular epidemiologic techniques, such as by nucleotide fingerprint mapping or epitope mapping with monoclonal antibodies.[9] Recently, 13 different naturally occurring antigenic variants of CVB4 were described using monoclonal antibodies. In addition, the rate of CVB4 mutational frequency has been estimated at $10^{-4.0}$–$10^{-5.9}$.[9] The clinical and epidemiologic significance of this genetic variation and its detection by monoclonal antibodies require further assessment.

4. A MODEL FOR EPIDEMIOLOGIC PATTERNS OF COXSACKIEVIRUS INFECTIONS

Variations in the incidence of CV infections occur from year to year, and outbreaks occur at cyclical intervals. A recent review of factors that influence oscillations in the incidence of infectious disease includes stochastic effects, age-dependent transmission rates, disease incubation periods, and seasonality in transmission.[10]

As a general principle, viruses such as CV, which are only found in human hosts and are thought not to persist following infection, may cause sharp epidemics when a virus strain is reintroduced to susceptible hosts. Such an outbreak is then followed by cessation of new infections as the number of susceptible persons markedly diminishes, until the virus is reintroduced. Extrapolating from the model of Yorke (for poliovirus infections), in which the transmissibility of viral infections varies by season, we could expect that epidemics of CV in large populations are a result of a large number of susceptible persons at the beginning of the seasonal rise in CV prevalence.[11] According to this model, in high-incidence years the rate of disease should peak earlier. This model also predicts that the interaction between the transmissibility of a viral infection (as influenced by seasonality) and the number of susceptible persons is responsible for biannual cycles of viruses, such as measles and poliovirus. However, the seasonal factors that may cause differences in transmissibility are not understood. Also, the role of some individuals as highly effective virus transmitters requires further understanding as a potentially important factor in the overall patterns of CV infections.

For the future, another potentially important extrapolation from the polio model proposed by Yorke *et al.*[11] is that rather modest reductions in susceptibility rates to CV through some future vaccine program may yield a virtual eradication of wild-type infection. The current epidemiology of CV, where a large percentage of the population is believed susceptible, is probably analogous to poliovirus infection during the prevaccine era. Also extrapolating from the model of Yorke is the postulated greater susceptibility of virus eradication during seasonal periods of low transmissibility. To evaluate further the application of Yorke's model, some examples from recent epidemiologic studies of CV infections are examined later.

5. COXSACKIEVIRUS INFECTIONS WITHIN FAMILIES

In early studies, a great deal has been learned about the epidemiology of CV in the family setting. From ongoing surveillance of individuals in a virus-watch program in Metropolitan New York families, several differences were pointed out in the epidemiology of CV and echoviruses.[1] In a surveillance system that sampled individuals biweekly, excretion of CV was observed to be longer than for echoviruses. Some individuals excreted CV for up to 70 days. Simultaneous excretion of virus by both the respiratory and GI routes was present only in CV infections, although CV and echoviruses were excreted in respiratory or fecal specimens alone. Excretion of CV for longer than the sample taken on 1 day occurred in more than 43% of patients versus only 16% for echoviruses. Another difference between echovirus and CV was in the longer duration of echoviruses in the community, as CV were not seen during the first 5 months of the year. There was a greater intrafamilial spread of CV (76%) than for echoviruses (43%), suggesting a greater infectivity for CV. For CV infections, mothers appeared at a higher risk of acquiring infection (78%) than did fathers (47%). Reinfection (isolation of a virus from an individual known to possess homologous neutralizing antibody) was documented more commonly in CV than for echoviruses. Estimated reinfection rates for CV in families were 50% for those under age 10 and 10% among teenagers and adults. This study also demonstrated a high subclinical rate of infection (48%) for CV and the relatively rare occurrence of severe clinical manifestations. Clinical illness rates were higher if both fecal and respiratory shedding of virus occurred (58%) versus fecal shedding only (46%) or respiratory shedding only (44%).[1]

6. COXSACKIEVIRUS INFECTIONS WITHIN INSTITUTIONS

Coxsackievirus can cause nosocomial infections. Often, however, an infected patient, staff member, or visitor who has a relatively mild illness may go unrecognized as the source of the nosocomial infection. Direct infection from the mother may also be the source of newborn infections in nurseries.[2] The usual route of transmission in institutions is via contaminated hands from contaminated environmental sources. Transplacental transmission can occur from the infected mother. If the CV infections occur in very young children[2] or in immunocompromised patients, the outcome of the nosocomially acquired infection may be more clinically apparent.

A severe nosocomial outbreak of CVA1 infectious gastroenteritis in a bone marrow transplantation unit has been described.[12] During a 3-week period in February and March 1980, 7 of 14 patients in a bone marrow transplantation unit were infected with CVA1. CVA1 antigen was detected by the enzyme-linked immunosorbent assay (ELISA) test from the stools or throats of seven patients with diarrhea but not from seven bone marrow transplant patients without this symptom. Six of the seven infected patients died. The CVA1-infected and -noninfected patients differed only in the striking mortality of those infected. Although a diarrheal illness had been described among hospital staff, none was significantly associated with infected patients. None of the staff was examined for CVA1. However, five children admitted in February 1980 to the pediatric service for symptoms of diarrhea were found to have CVA1 antigen in their stool. Although acute diarrhea in bone marrow transplant recipients may be a sign of acute graft-vs.-host disease, in this study acute graft-vs.-host disease diagnosed by biopsy had a similar distribution among both infected and noninfected patients. At postmortem examination, three of the six infected patients had no evidence of GI graft-vs.-host disease. Thus, CV enteritis is important in such immunocompromised patients. Although the source or mechanisms of transmission of this CVA1 outbreak were not defined, the authors speculated that it was acquired from either infected visitors or hospital personnel.[12]

A subsequent prospective study examining the etiology of infectious gastroenteritis in the same bone marrow transplantation unit showed CV infection in 4 of 78 patients studied from September 1980 through June 1, 1981.[13] These four infections were detected during the months of September and November 1980, at a time of the year when one would expect CV in the community. Although three of the four CV-infected bone marrow transplantation patients also had other enteric pathogens, all four of the CV-infected patients died. Because the clinical and demographic features of those infected with pathogens that cause enteritis did not differ from those patients who were not infected, it was assumed that the increased mortality was related to enteric infections. As for the previous study, the route and mechanism of transmission of these viral infections could not be determined.[13]

Analysis of an outbreak of CVB4 infection in an institution for the neurologically handicapped has suggested possible mechanisms of nosocomial transmission of CV.[14] The outbreak was mainly acute upper respiratory infections, and 19 of 24 ill patients had CVB4 recovered from pharyngeal secretions. As most of the patients were young handicapped children confined to their bed or wheelchair, and the pattern of the outbreak suggested a propagated spread of infection, the hospital staff was implicated in the transmission of this infection. One possible route was oral–oral transmission of viruses from the hands of health care workers during routine mouth-care procedures. The mouth-care technique involved the use of disposable glycerine swabs to clean the mouths of those fed by nasogastric tube or those who had gingival hypertrophy secondary to anticonvulsant medication. It was found that during this cleaning procedure, the nurses did not wear gloves, and handwashing facilities were inadequate.[14]

In an outbreak on the second ward of the same institution, in which all the patients had severe underlying neurologic diseases and were confined to bed in individual rooms, a second propagated epidemic of CVB4 infections occurred. In the epidemic on this ward, it was believed that the CVB4 contamination occurred at the time of bathing. In-dwelling nasogastric tubes (present in 10 of 16 children) were possibly contaminated with CVB4 from the hands of staff or directly from the bathing environment of patients soiled with feces. A phenolic-containing germicidal detergent solution was used to wash hands and clean the bathtubs. However, phenolic solutions have been shown to be ineffective virucides in tests with CVB5.[15]

The mechanism of CVB4 transmission in these nosocomial outbreaks may have involved oral–oral contamination either directly from contaminated hands or by environmental contamination of the nasogastric tube. The risk factors of gavage feeding and mouth care have also been shown to be significantly linked to a nursery outbreak of echo 11 enterovirus infection.[16] Thus, medical and nursing staff should be aware that a similar setting may facilitate CV transmission in a neonatal care nursery.

The use of disinfectants to reduce the risk of CV contamination in an institution requires the removal of organic debris and a sufficient amount of time for the germicidal action to occur. Disinfectants such as a 1 : 75 dilution of Javex, 0.25% (w/v) sodium hydroxide, and 1/200 Wescodyne are effective virucides, causing a 10^5 reduction of CVB5 in 30 sec.[15] However, the most important preventive measure for control of nosocomial CV infections is good handwashing before and after patient contact. If outbreaks of enterovirus infections occur or are suspected, the use of enteric precautions, cohort isolation, and even ward closure is recommended.[17]

7. SEROEPIDEMIOLOGIC SURVEYS OF COXSACKIEVIRUS INFECTIONS

Serologic surveys have provided additional information on the epidemiology of CV. In a study of serum from 252 patients from Amori in northern Japan, Sato *et al.*[18] examined the presence of neutralizing antibodies to the six CVB infections. Their data showed a marked variation in the dissemination of CV within the population, depending on the virus type. On the basis of these data, Sato's group defined six patterns of CV dissemination. CVB1, CVB3, and CVB4 showed a rate of high endemicity in this community. Occasionally, outbreaks caused by these strains were recognized. Infections by these viruses were common in all age groups, with a steep rise in the seropositive rate taking place in the 1–4-year-old group, consistent with frequent viral spread in young children. CVB2 showed a relatively low endemicity with extensive dissemination at longer intervals. A steep rise in geometric mean titers to CVB2 occurred in the 5–9-year-old group, suggesting several years when this virus was not present in the community. CVB5 was found to have relatively low endemicity with extensive dissemination in young children at long intervals. A rise in the age-specific incidence of neutralizing antibody to CVB5 occurred in those aged 1–4 years, and an additional steeper rise in those aged 5–19 years. This was followed by a dramatic reduction in the level of antibody titer. This curve of the incidence of neutralizing antibody to CVB5 suggests a long quiescent period between outbreaks of CVB5 infections.[18] CVB6 infections were believed to be of extremely low endemicity, from data based on the neutralizing antibody titer. For the group of CVB, the geometric mean titers fell gradually with age. This may be due to a lack of booster effect, as older persons may not be exposed to CV as frequently as younger persons. In summary, Sato *et al.*[18] present data that suggest that yearly dissemination of CVB serotypes 1, 3, and 4 occurs frequently in young children. These viral infections are readily transmitted and a relatively high percentage of susceptible individuals are infected. Sato *et al.*[18] speculate that CVB types 1, 2, 3, and 4 may maintain their endemicity through long periods of communicability. CVB5, which has a lower endemicity, appears to have extensive dissemination at long intervals, presumably based on a large number of accumulated susceptible persons. The communicability of viruses such as CVB5 may be lower than that for CVB types 1–4. Viruses such as CVB6 were thought to have a lower efficiency of transmission but are able to persist in communities because of the high number of susceptible persons. Thus, the CVB appear to represent a heterogeneous group of viruses in their patterns of transmission throughout the community.

A serologic study in Winnipeg, Canada, compared maternal and neonatal neutralizing antibody titers to CVB types 2–5 in 98 mother–infant pairs.[19] The percentage of mothers showing antibody levels to the above viruses at a titer of $\geq 1:10$ was 54%, 64%, 87%, and 51%, respectively, and the overall geometric mean titers were 15.3, 18.1, 53.5, and 14.1, respectively. Overall, these data suggest that more than one half the pregnant women in this survey had been previously infected by these viruses. The highest percentage of those infected and the greatest geometric mean titers were to CVB4, consistent with the high endemic rate of transmission of this virus. The geometric mean titer of antibodies in cord bloods of newborns closely paralleled that of their mothers, with the exception of CVB5 antibody titers, which were lower than maternal titers. Thus, 49%, 37%, 17%, and 53% of a sample of newborns lacked antibody to CVB2, CVB3, CVB4, and CVB5, respectively, and were susceptible to infection at birth and during the first few months of life.[19]

In 1020 normal persons in New Zealand, the highest antibody geometric mean titers and highest percentage of those showing immunity to CVB were to CVB2 and CVB4, while the lowest was to CVB6.[5]

8. GEOGRAPHIC PATTERNS OF COXSACKIEVIRUS INFECTIONS BY VIRUS IDENTIFICATION

A review of international data on CV epidemiology and data from ongoing enterovirus surveillance in Glasgow has been presented[20] and is updated in Chapter 13, this volume. The temporal and geographic patterns of NPEV types in the United States from 1970 to 1983 were recently reviewed,[8] from voluntary reporting of virus isolates to the Centers for Disease Control (CDC) in Atlanta. From these data, overall patterns of NPEV in the United States have emerged. By the authors' definition, predicted types were virus types that ranked among the top six in frequency from March through May of each year. Late types occurred after this time. An epidemic type of NPEV was defined as a type that accounted for ≥20% of all NPEV isolates in a single year.

The number and types of isolates reported from year to year varied greatly and correlated with the number of reporting laboratories and the number of isolates reported from each laboratory. The 15 most common NPEV types accounted for 81% of all isolates (Table I). An analysis showed NPEV isolates appearing in epidemic years followed by nonepidemic periods (Table II). Of the 11 years for which an epidemic NPEV type was reported, only CVB5 was found to be an epidemic type. Most of the CVB were in a group with a variable frequency of NPEV isolates that did not meet the definition of epidemic type. In addition, some uncommon NPEV types represented <1% of the total number of yearly isolates and included all CVA (except types 4, 5, 6, 9, and 16) and CVB6.[8]

The original observation by Froeschle *et al.*[21] of a temporal pattern of appearance of NPEV isolates, beginning earlier in the southern United States, was confirmed in the 14-year surveillance by the CDC. Of the five most southerly regions in the United States, a higher percentage of isolates were early isolates (9–17%) than the remaining northerly regions of the United States, in which only 2% of NPEV were early isolates.[8]

Data from the early NPEV isolates (March to May) were useful in predicting the top six late isolates of greatest frequency. In most years, the six most frequent early isolate types were nearly the same as the top six late isolates, with a prediction efficiency of 86.6% on average over the 14 years (Table III).[8] Among the 15 most common NPEV types overall in the 14-year study were found the virus types with the greatest number of isolates in any given year. Thus, most of the NPEV activity in the United States was confined to selected types of enteroviruses, with the surveillance system used.[8]

TABLE I

The 15 Most Comon Types of NPEV Reported Isolated in the United States, 1970–1983[a,b]

NPEV type[a]	Percentage of total isolates per year														
	1970–1983	1970	1971	1972	1973	1974	1975	1976	1977	1978	1979	1980	1981	1982	1983
E11	12.2	2.8	3.5	8.2	6.4	8.7	7.5	1.6	1.0	4.5	36.2	19.9	8.2	20.2	8.3
E9	11.3	11.5	29.0	6.8	7.1	9.3	37.0	2.4	6.6	30.0	2.1	3.9	13.1	6.1	3.8
CVB5	8.7	0.4	0.4	40.0	21.3	0.2	0.1	0.6	5.9	4.9	1.7	3.0	3.6	5.1	20.2
E30	6.8	1.2	1.1	0.5	1.2	1.0	1.1	4.7	2.8	6.0	7.8	8.7	19.9	16.9	6.9
E4	6.3	12.5	31.0	6.9	4.5	12.3	6.4	7.4	1.8	10.2	2.4	1.5	2.4	1.3	0.5
E6	5.5	13.1	3.3	7.5	7.1	9.0	4.4	8.8	23.0	4.2	1.8	3.4	1.3	2.2	4.2
CVB2	4.8	2.7	5.6	4.1	5.3	10.3	3.9	15.8	2.9	2.0	6.7	6.9	1.0	4.0	3.3
CVB4	4.6	3.3	3.3	3.6	6.9	4.1	5.5	13.8	1.7	5.1	4.9	3.7	6.1	5.1	1.4
CVB3	4.5	1.6	1.9	2.7	5.5	6.3	2.2	11.3	7.6	2.1	1.1	18.1	1.6	2.3	3.1
CVA9	4.5	2.5	1.1	4.0	11.6	0.6	5.5	8.1	6.4	6.7	1.4	4.6	7.3	2.0	3.8
E3	3.2	33.4	4.4	0.9	0.6	0.0	0.1	0.2	4.9	0.2	0.2	0.7	7.9	1.1	0.7
E7	3.0	0.3	0.9	1.2	1.9	0.4	0.4	0.9	1.0	3.1	15.5	1.0	0.7	0.9	1.4
E5	2.0	0.4	0.4	0.3	0.7	1.8	1.3	0.6	0.5	1.1	0.6	1.5	7.5	6.4	2.5
E24	1.8	0.0	0.1	0.0	0.1	0.1	0.1	0.1	0.0	0.4	3.0	2.8	0.9	2.1	8.4
CVB1	1.6	0.9	1.4	1.8	0.7	0.5	0.1	2.1	8.7	0.4	0.7	2.6	1.6	1.2	1.6
Other	19.1	13.5	12.5	11.2	19.1	35.5	24.2	21.5	25.2	19.2	13.8	17.8	17.1	23.1	27.5
Total no. of isolates	23,481	1100	1589	2062	1376	829	1355	850	1172	1782	3088	2204	2130	1518	2426

[a]From Strikas *et al.*[8]
[b]NPEV, nonpolio enterovirus; E, echovirus; CVB, coxsackievirus group B; CVA, coxsackievirus group A; other, all other types of NPEV.

TABLE II
The Most Frequently Isolated Types of NPEV by Year from 1970 to 1983, United States[a,b]

Region	Top NPEV type (% of total isolates for that year)															Total no.
	1970	1971	1972	1973	1974	1975	1976	1977	1978	1979	1980	1981	1982	1983	1970–1983	of isolates[c]
N.E.	E3/E11	E9		CVB4			CVB3					E11	CVB4		CVB5	1446
M.A.				CVA9	E4		CVB4					E3	E9		E9	4965
E.N.C.		E9		E9	CVB2		CVB4						E30		E11	4666
W.N.C.		E9			CVB2		CVA9		E4					E20	E11	2817
S.A.					E9		E4			E7	E30		E5	E30	CVB5	2720
E.S.C.	E6				E9			CVB1			E11/CVB3		E9		E9	295
W.S.C.			E11	E11	E6		E6	E9			CVB2	CVA9			E11	3298
MT.	E9	E7/E9	CVB5/E9	E9	E18	E14	E9	CVA9	E30			E9	E30		E9	821
PAC.		E9		E9	E9	E4	E6	CVB5			E6			E30	E11	2285
United States	E3 (33)	E4 (31)[d]	CVB5 (40)	CVB5 (21)	E16 (20)	E9 (37)	CVB2 (16)	E6 (23)	E9 (30)	E11 (36)	E11 (20)	E30 (20)	E11 (20)	CVB5 (20)	E11 (12)	23,481

[a]From Strikas *et al.*[8]
[b]The most frequently occurring type(s) of NPEV for a region are given if different from that for the entire United States.
[c]Total no. of isolates for 1970–1983. For 168 isolates, the region was unknown.
[d]In 1971, E9 was also an epidemic type with 29% of the total isolates.

In an analysis by geographic area, the most common NPEV type throughout the United States was the most common type for an individual region only 56% of the time (range 11–89%). In any given region, the most common type was among the six early types most frequently isolated in the United States from March to May of any given year, an average of 88% of the time.[8]

Caution is required in interpreting data from any voluntary surveillance system. As pointed out by Strikas *et al.*,[8] many factors may influence the reported isolation rates of enteroviruses, such as differences in disease incidence, variability in severity of clinical illness, and variations in technical methods. However, aside from these problems implicit in such a survey, the analysis of data from information submitted to the CDC nonpolio enterovirus surveillance program has proved useful. Table III presents a comparison of CVB type 1–5 infections in the United States based on the CDC surveillance in the United States for 1970–1979, with data from the Nassau County Medical Center virus laboratory surveillance conducted over the same 10-year period.[7]

These data have been used to review some extrapolations from the model proposed by Yorke.[11] Yorke's analysis suggests that virus strains associated with disease early in the season will be associated with the greatest epidemic activity in any given year, as appears to have been borne out by the enterovirus surveillance in the United States. Also, in parallel with Yorke's model is the concept that in epidemic years large

TABLE III
Number of Coxsackievirus B1–B5 Isolates Reported through National Surveillance and by the Nassau County Medical Center, by Agent and Year, 1970–1979[a]

	Isolates reported through national surveillance						Isolates reported by Nassau County Medical Center					
Year	B1	B2	B3	B4	B5	Total	B1	B2	B3	B4	B5	Total
1970	10	30	18	36	4	98	—	3	—	—	—	3
1971	22	89	30	53	6	200	2	10	13	5	—	30
1972	66	150	91	107	1367	1781	2	3	4	12	107	128
1973	9	73	75	96	295	548	—	37	34	22	2	95
1974	4	87	54	34	2	181	—	15	3	3	1	22
1975	2	54	30	76	2	164	—	12	8	11	—	31
1976	18	137	99	120	5	379	2	7	57	51	1	118
1977	103	35	90	21	70	319	17	6	11	1	3	38
1978	7	36	38	92	87	260	1	4	10	4	7	26
1979	22	206	31	152	50	461	15	50	3	59	6	133
Total	263	897	556	787	1888	4391	39	147	143	168	127	624

[a]From Moore *et al.*[7]

numbers of cases of NPEV infections reflect the accumulation of a large number of susceptible individuals at the beginning of the seasonal rise of infection, together with a seasonal change in virus transmissibility. Thus, common widespread peak seasonal outbreaks of NPEV, such as for CVB5, suggests the accumulation of a large number of susceptible individuals during interepidemic periods, as shown by Sato *et al.*[18] A decreased transmissibility of viruses would be predicted during the seasonal trough of CV activity. From the CDC surveillance, 87% of CVB isolates were reported from the peak 5 months of the year. All CVB isolates exhibited this pattern of seasonal distribution, but CVB3 were isolated significantly more frequently during the cooler months of the year, from January to April. Comparing the yearly trends of virus isolations, the CDC surveillance data showed that although CVB5 isolates were the most frequent, most of these isolates were reported during two consecutive years of epidemic activity[8] (Table III). This was followed by 3 years with very few isolates of CVB5, especially in comparison with a more constant rate of isolation of CVB2, CVB3, and CVB4. This rather constant level of occurrence of CVB2, CVB3, and CVB4 infections versus CVB1 and CVB5 may explain in part the apparent lack of epidemic activity of the former and the striking association with epidemics of the latter two virus types. A similar epidemic pattern of CVB5 was also reported in a 20-year longitudinal survey of NPEV isolates in Wisconsin.[22] Outbreaks of CVB5 infection were noted in 1967 and 1972, with no detection of this virus during the interepidemic periods 1968–1971 and 1973–1976 (the last reported study year).[22] There appeared to be a more constant level of CVB4, CVB2, and CVB3 activity, with isolates reported in 18 of 20, 17 of 20, and 12 of 20 years for each virus, respectively, versus detection in 10 of 20 and 5 of 20 years for CVB5 and CVB1.

Extrapolating from Yorke's model, widespread epidemics of CVB1 and CVB5 in the community could be expected if these viruses were present in the early months of enterovirus seasonal activity. This was observed after several years in which CVB1 and CVB5 had been infrequently detected, and a large number of susceptibles were likely to have accumulated in the very young childhood age groups most at risk of these infections. One could also speculate that transmissibility may vary over a wide range among CV types, as many types of CV were not associated with widespread epidemics and in view of the rare detection of some CVB types.

Further verification of Yorke's model of patterns of infectious diseases, as applied to CV in the southern hemisphere, was reported in an outbreak of epidemic CVB3 infection in 1984 in Johannesburg.[23] Schoub *et al.*[23] reported that a few isolates of CVB3 were detected in the late winter (mid-August to early September), with the epidemic increasing in the end of September and peaking at the end of October. These investigators speculated that this early appearance of a herald wave predicted an extensive epidemic of CVB3 infection that was subsequently observed. Although no prospective epidemiologic studies were performed and no specific attack rates were determined, information gathered suggested that several hundred persons were infected during this epidemic. In addition, a nosocomial outbreak occurred in a maternity hospital in which 1 of the 11 infected infants died. Thus, in the southern hemisphere as well, the early seasonal appearance of CVB may signify a significant and early epidemic.[23]

Recently, it has been shown that there appears to be a genetic correlation of CVB epidemiology to virus genotype. By sequencing the 5′ noncoding region from over 50 clinical isolates of six serotypes of CVB collected over 20 years, it has been shown that CVB could be divided into two groups, coxsackie B1, B2, B6 and a second group of coxsackie B3, B4, B5. Coxsackievirus B5 (which appears in epidemic patterns) had a single genotype during epidemic years in comparison to multiple genotypes for the other coxsackie B serotypes which circulated in a single year. Genotype analysis of CV will offer a greater understanding of the epidemiology of CV for the future.[24]

9. SUMMARY

It may be possible to predict common nonpolio enterovirus types in the community (and thus in institutions) through laboratory surveillance of NPEV isolates during the early part of the seasonal cycle, as well from more equatorial continental latitudes, at least in the United States. This will increase our awareness of the likely clinical spectrum of nonpolio enterovirus infections (including CV), facilitate their laboratory identification, and prepare us in our attempts to reduce the risks of nosocomial transmission of these viruses.

REFERENCES

1. Kogon, A., Spigland, I., Frothingham, T. E., Elveback, L., Williams, C., Hall, C. E., and Fox, J. P., 1969, The virus watch program: A continuing surveillance of viral infections in metropolitan New York families. VII. Observations on viral excretion, seroimmunity, intrafamilial spread and illness association in coxsackie and echovirus infections, *Am. J. Epidemiol.* **89:**51–61.
2. Kaplan, M. H., Klein, S. W., McPhee, J., and Harper, R. G., 1983, Group B coxsackievirus infections in infants younger than three months of age: A serious childhood illness, *Rev. Infect. Dis.* **5:**101–1032.

3. Jenista, J. A., Powell, K. R., and Menegus, M. A., 1984, Epidemiology of neonatal enterovirus infection, *J. Pediatr.* **104**:685–690.
4. Moore, M., 1982, Enteroviral disease in the United States. 1970–1979, *J. Infect. Dis.* **146**:103–108.
5. Lau, R. C. H., 1983, Coxsackie B virus infections in New Zealand patients with cardiac and non-cardiac diseases, *J. Med. Virol.* **11**:131–137.
6. Gelfand, H. M., and Holquin, A. H., 1962, Enterovirus infections in healthy children. Study during 1960, *Arch. Environ. Health.* **5**:404–411.
7. Moore, M., Kaplan, M. H., McPhee, J., Bregman, D. J., and Klein, S. W., 1984, Epidemiologic, clinical and laboratory features of coxsackie B1-B5 infections in the United States, 1970–1979, *Public Health Rep.* **99**:515–522.
8. Strikas, R. A., Anderson, L. J., and Parker. R. A., 1986, Temporal and geographic patterns of isolates of nonpolio enterovirus in the United States, 1970–1983, *J. Infect. Dis.* **153**:346–351.
9. Prabhakar, B. S., Hospel, M. V., McClintock, P. R., and Notkins, A. L., 1982, High frequency of antigenic variants among naturally occurring human coxsackie B4 virus isolates identified by monoclonal antibodies, *Nature (Lond.)* **300**:374–376.
10. Anderson, R. M., and May, R. M., 1985, Vaccination and herd immunity to infectious diseases, *Nature (Lond.)* **318**:323–329.
11. Yorke, J. A., Nathanson, N., Pianigianti, G., and Martin, J., 1979, Seasonality and the requirements for perpetuation and eradication of viruses in populations, *Am. J. Epidemiol.* **109**:103–123.
12. Townsend, T. R., Bolyard, E. A., Yolken, R. H., Beschorner, W. E., Bishop, C. A., Burns, W. H., Santos, G. W., and R. Soral, 1982, Outbreak of coxsackie A1 gastroenteritis: A complication of bone-marrow transplantation, *Lancet* **1**:820–823.
13. Yolken, R. H., Bishop, C. A., Townsend, T. R., Bolyard, E. A., Bartlett, J., Santos, G. W., and Saral, R., 1982, Infectious gastroenteritis in bone-marrow-transplant recipients, *N. Engl. J. Med.* **306**:1009–1012.
14. Johnson, I., Hammond, G. W., and Verma, M. R., 1985, Nosocomial coxsackie B4 virus infections in two chronic-care pediatric neurological wards, *J. Infect. Dis.* **151**:1153–1156.
15. Drulak, M., Wallbank, N. M., Lebtag, I., Werboski, L., and Poffenroth, L., 1978, The relative effectiveness of commonly used disinfectants in inactivation of coxsackievirus B5, *J. Hyg. (Camb.)* **81**:389–397.
16. Kinney, J. S., McCray, E., Kaplan, J. E., Low, D. E., Hammond, G. W., Harding, G., Pinsky, P. F., Davi, M. J., Kovnats, S. F., Riben, P., Martone, W. J., Schonberger, L. B., and Anderson, L. J., 1986, Risk factors associated with echovirus 11′ infection in a hospital nursery, *Pediatr. Infect. Dis.* **5**:192–197.
17. Valenti, W. M., Hruska, J. F., Menegus, M. A., and Freeburn, M. J., 1981, Nosocomial viral infection. III. Guidelines for prevention and control of exanthematous viruses, gastroenteritis viruses, picornaviruses and uncommonly seen viruses, *Infect. Control* **2**:38–49.
18. Sato, N., Sato, H., Kawana, R., and Matumoto, M., 1972, Ecological behaviour of 6 coxsackie B and 29 echo serotypes as revealed by serological survey of general population in Aomori, Japan, *Jpn. J. Med. Sci. Biol.* **25**:355–368.
19. Hammond, G. W., Lukes, H., Wells, B., Thompson, L., Low, D. E., and Cheang, M., 1985, Maternal and neonatal neutralizing antibody titers to selected enteroviruses, *Pediatr. Infect. Dis.* **4**:32–35.
20. Grist, N. R., Bell, E. J., and Assaad, F., 1978, Enteroviruses in human disease, *Prog. Med. Virol.* **24**:114–157.
21. Froeschle, J. E., Feorino, P. M., and Gelf, H. M., 1966, A continuing surveillance of enterovirus infection in healthy children in six United States cities. II. Surveillance

enterovirus isolates 1960–1963 and comparison with enterovirus isolates from cases of acute central nervous system disease, *Am. J. Epidemiol.* **83**:455–469.

22. Nelson, D., Hiemstra, H., Minor, T., and D'Alessio, D., 1979, Non-polio enterovirus activity in Wisconsin based on a 20-year experience in a diagnostic virology laboratory, *Am. J. Epidemiol.* **109**:352–361.
23. Schoub, B. D., Johnson, S., McAnerney, J. M., Dos Santos, I. L., and Klaassen, K. I. M., 1985, Epidemic coxsackie B virus infection in Johannesburg, South Africa, *J. Hyg. (Camb.)* **95**:447–455.
24. Pallansch, M. A., and Freeman, C. Y., 1987, Genetic variation and phylogenetic relationships among coxsackie B virus isolates and identification of conserved sequences as targets for pan-enterovirus oligonucleotide probes, Abstract R16.25 from the *VIIth International Virology Congress,* Edmonton, Alberta, Canada, Aug. 10–14, 1987.

22

Epidemiology of Group B Coxsackieviruses

MARK A. PALLANSCH

1. INTRODUCTION

1.1. Prior Reviews

The coxsackie B viruses (CVB) are members of the genus *Enterovirus* and share many properties with other members of this group. There are several excellent reviews covering the basic epidemiology and clinical characteristics of enteroviruses and their infections[1–4]; this material is not repeated in detail, except as it relates to recent studies. Most of the conclusions concerning enteroviruses in general are directly applicable to CVB, including seasonality; geographic, age, and sex distribution; incidence and prevalence; transmission; and most clinical syndromes.

1.2. Purpose of Current Review

This chapter first reviews the literature published since the previous reviews as it relates to the epidemiologic patterns of CVB infections and disease. These data comprise individual reports, outbreaks, serologic studies, and surveillance data. Second, several diseases are discussed for which CVB infection is implicated as an etiologic factor, including some

MARK A. PALLANSCH • Division of Viral Diseases, Center for Infectious Diseases, Centers for Disease Control, U.S. Public Health Service, U.S. Department of Health and Human Services, Atlanta, Georgia 30333.

chronic diseases. These areas are currently the focus of most clinical and epidemiologic study. Third, the design of possible future epidemiologic and laboratory studies of these viruses is considered, including possible limitations and new techniques.

2. SURVEILLANCE

2.1. Viral Isolates

2.1.1. Serotype

Most studies of the occurrence or prevalence of CVB have made use of the reported isolation of virus from clinical specimens, using both active and passive surveillance. The basic sources of surveillance data have been reviewed[2] and are primarily passive. Two recent reports summarize surveillance data for enteroviruses in the United States[5] and Scotland.[6] In addition, occurrences of larger regional outbreaks continue to be reported.[7]

In the United States, enterovirus surveillance data are collected and analyzed by the Centers for Disease Control. The data have been reported irregularly since the beginning of the program in 1961.[8–14] Figure 1 summarizes the data for the 25 years from 1961 to 1985 for the CVB (including unpublished data). CVB6 was isolated so infrequently (40 times in 25 years) that these data are not shown. The left half of Fig. 1 shows the number of times each serotype was isolated per year, while the right half shows what fraction of the total number of nonpolio enterovirus (NPEV) isolates each serotype represents for each year. Note that the maximum fraction for CVB1–4 is 20%, whereas the maximum fraction for CVB5 is 60%. The data analysis shown in the right half of Fig. 1 attempts to control for yearly variation in the surveillance system.

Like enteroviruses in general,[3] CVB fall into two patterns of incidence, endemic and epidemic. The epidemic pattern, as typified by CVB5, is characterized by distinct outbreaks with few isolates observed in intervening years. During the period 1961–1985, there were four

→

FIGURE 1. Isolates of coxsackie B virus reported 1961–1985. (**a–e**) The number of isolates for any serotype for the 25-year period. (Left) Data plotted as total number of isolates reported through the enterovirus surveillance system[8–12] (including unpublished data). The scaling varies for each serotype. (Right) Data plotted as number of isolates as a fraction of all nonpolio enterovirus (NPEV) isolates reported for that year, attempting to compensate for yearly variation in the reporting. The scale for the right is the same for CVB1–CVB4, whereas that for CVB5 is three times greater. The number of NPEV isolates varied from 308 in 1966 to 3088 in 1979. The total number of CVB isolates for this period was 8858.

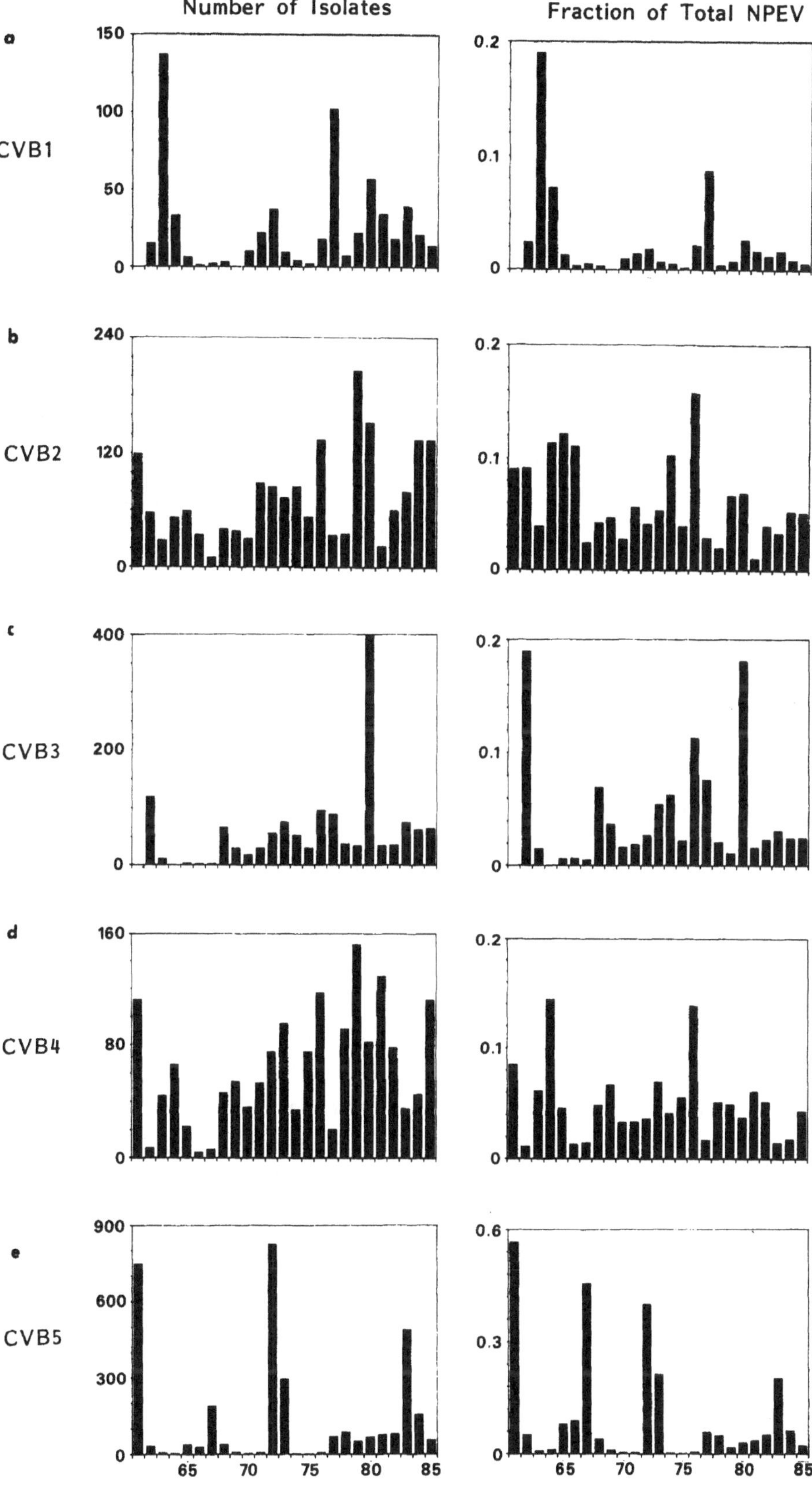

Number of Isolates
Fraction of Total NPEV
a
b
c
d
e
CVB1
CVB2
CVB3
CVB4
CVB5
150
100
50
0
0.2
0.1
0
240
120
400
200
160
80
900
600
300
0.6
0.3
65
70
75
80
85
Year of Isolation

major epidemics of CVB5 in the United States, in 1961, 1967, 1972, and 1983. By contrast, endemic viruses are isolated in nearly every year, with mostly minor yearly differences. This is the pattern seen with CVB2 and CVB4, although larger outbreaks do occur occasionally, as with CVB3.

Enterovirus isolates are routinely characterized by serotype using polyclonal animal sera, but monoclonal antibodies can also be used to examine antigenic variation within a serotype. Studies using monoclonal antibodies prepared against CVB4[15] suggest that there can be considerable variation within a serotype of CVB. Monoclonal antibodies have been used to study prototype CVB4 strains, plaque variants of the prototypes, and clinical isolates.[16,17] In all cases, multiple differences between strains were detected at many epitopes, although no correlation with clinical symptoms was made. Antigenic variation among clinical isolates of CVB4 was also seen by measuring neutralization kinetics with polyclonal sera.[18]

2.1.2. Genotype

The genotype of a virus is defined by the primary sequence of its genome and can be studied by several techniques including sequencing, hybridization, and oligonucleotide fingerprinting. Of the enteroviruses, poliovirus is most commonly studied by fingerprinting.[19–21] From these studies, it is possible to determine whether an isolated poliovirus is related to vaccine virus and to determine similarities or differences between isolates in an epidemic.

Of the CVB, only CVB5 has been extensively fingerprinted.[22] Early studies by this method and others demonstrated the similarity between CVB5 and swine vesicular disease virus, suggesting the existence of a possible animal reservoir, a progenitor for this serotype, or infection of animals with a human virus. Recently, we used this technique to determine the genomic characteristics of isolates from CVB5 epidemics. First, as noted from the surveillance data, CVB5 has an epidemic pattern of occurrence in the United States. Studies on multiple isolates from several of the epidemics have shown that each of the epidemics seems to have been caused by a single genotype. Second, the genotype of CVB5 observed in the 1967 epidemic was nearly identical to the virus observed in the 1983 epidemic. None of the isolates from other epidemics, or interepidemic isolates, had similar genotypes by fingerprinting. These results suggest the 1967 genotype reemerged in 1983 from an unknown reservoir.

An additional approach used to characterize variation of CVB isolates is to measure phenotypes *in vitro*. Studies have included pathogenic potential in mice[23–25] and cell-growth characteristics.[26–28] These studies have so far failed to develop an association between properties of the virus, including antigenic variation, and pathogenesis in humans. A very important set of studies[29,30] demonstrated that even an individual isolate is a collection of variants that have different biologic properties. This has significant implications for studies relating viral isolates to disease.

2.2. Serology

Many studies have been performed to examine the level of antibodies to the CVB in defined populations, including serosurveys in Scotland, Czechoslovakia, India, Israel, and New Zealand.[6,31–37] Although the results of these serosurveys were previously reviewed,[1–4] several important conclusions should be reiterated. First, the number of persons with neutralizing antibody to any of the CVB is high, indicating a high incidence of past infection. A high incidence of recent infection is also suggested by IgM surveys, which typically show 4–6% positivity. Second, infections with one serotype of CVB can boost the antibody titers to other CVB serotypes as measured by either IgM or neutralization. The pattern of the heterotypic response is not consistent with respect to serotype and varies among persons. Other enteroviruses can also cause an anamnestic response toward CVB. Third, the prevalence of antibody to each of the serotypes varies by time and location, thus data from different years and locations are not directly comparable. These three points must be considered when designing and interpreting serologic studies of possible associations between enterovirus infection and a disease.

3. ASSOCIATION BETWEEN INFECTION AND CLINICAL DISEASE

3.1. Review of Previous Findings

Many reports have associated CVB infection with various diseases and syndromes. Most of these associations are based on studies of outbreaks of enterovirus infections. Some are based on one or several clinical case descriptions associated with large numbers of people with similar symptoms who showed evidence of infection with the same serotype of CVB. In outbreaks, it is often possible to determine an etiologic association with confidence. From such studies CVB infection has been ascribed as the cause of aseptic meningitis, encephalitis, pleurodynia, myocarditis, pericarditis, and several others. By far the most common result of CVB infection is asymptomatic infection, an undifferentiated febrile illness, or mild upper respiratory symptoms.[1–4]

Determining the etiology of single clinical cases is difficult. Lerner and Wilson[38] proposed guidelines for evaluating individual cases of myocarditis, which are generally applicable to other enterovirus diseases. A high level of association between the virus and the disease may be obtained when the virus is isolated from the tissue affected, such as the cerebrospinal fluid (CSF) during a case of meningitis, or detected directly, as by *in situ* hybridization or appropriate immunofluorescence. Isolating virus from throat washings or stools supports only a moderate level of association unless supported by evidence of serological conversion. Detecting antibodies supports only a low level of association. Due to the large number of inapparent infections and frequency of prolonged viral excretion, especially in stools, infection does not prove etiology for the disease. If the isolated virus is known to cause the symptoms and no other agent or cause has been identified, then the isolation provides a plausible explanation for the illness.

3.2. Specific Diseases

3.2.1. Central Nervous System Disease

The most frequent serious manifestation of CVB and enterovirus infection is central nervous system (CNS) disease, most commonly aseptic meningitis. The occurrence of this disease with regard to enteroviruses has been reviewed previously.[1–3] Recent literature on CVB in CNS disease has included case reports of meningoencephalitis and liquefactive necrosis in neonates and immune-deficient patients.[39–41] These associations are not always confirmed by viral isolation from CSF but rely on serologic evidence and/or isolation of CVB from other specimens. In a recent report, *in situ* nucleic acid hybridization on brain tissue obtained at autopsy[39] was used to confirm diagnosis. This technique, described in Section 4.1, offers the ability to detect virus when it cannot be isolated and thus extends diagnostic capabilities.

3.2.2. Carditis

Another recognized clinical syndrome associated with CVB infection is acute myocarditis and pericarditis, which have been reviewed.[1–4,42–45] While there continue to be numerous clinical reports of myocarditis and pericarditis associated with CVB infection,[46–48] what fraction of cases of acute idiopathic myocarditis may be due to CVB infection is currently unresolved. One study measured CVB-specific IgM and IgG in a group of patients with acute myocarditis and determined that nearly one half (49%) had IgM antibodies to CVB, while only 8% of controls showed any CVB titers.[49] These and other studies suggest, but do not clearly show, that CVB infection may be associated with a large fraction of cases of acute myocarditis.

Other reports include descriptions of recurrent pericarditis with persistent CVB-specific IgM for periods up to 6 years[50] and detection of CVB sequences by nucleic acid hybridization in myocardial biopsy specimens.[51] In the latter report, the specificity of the probe used is poorly defined because it was not evaluated against other serotypes, and homologies between CVB and other enteroviruses occur.[52]

3.2.3. *Neonatal Infections*

Coxsackie B viruses are associated with severe illness in neonates, including sepsis and generalized disseminated infection. Further reports of severe illness associated with CVB have recently appeared.[53–58] The most systematic study of newborn infections was reported by Kaplan *et al.*[59] covering records over a 10-year period in Nassau County, New York. Considering only cases in which infection was documented by CVB isolation, there were 77 cases and 6 fatalities in more than 153,000 live births. This rate of infection resulting in hospitalization, 1 for every 2000 births during the first 3 months of life, is a conservative estimate because of the recognized insensitivity of viral isolation in detecting viral infection. The contribution of echoviruses to neonatal illness has also been recently reviewed.[60]

The neonate can be infected *in utero,* from its mother during the peripartum period, and through postpartum exposure. In one study, CVB5 was isolated from the cervixes of four women with febrile illness during the third trimester of pregnancy,[61] supporting one of the possible mechanisms by which neonatal infections develop.

3.3. Chronic Disease

In addition to associations with acute disease, several studies have implicated CVB infection with several chronic diseases, such as diabetes mellitus, chronic heart disease, and arthritis.

3.3.1. *Diabetes Mellitus*

The most significant potential association of CVB infection with chronic disease is with juvenile-onset insulin-dependent diabetes mellitus (IDDM). This topic, including most of the published studies, was covered in an excellent review by Barrett-Conner.[62] The studies have neither proved nor disproved the association. Several lines of evidence have maintained interest in this possible association. Continued studies in murine models, as described elsewhere in this volume (see Chapter 10), in which CVB4 infection of certain inbred strains of mice produced diabetes, also contribute to the continued plausibility of CVB-induced IDDM. In humans, serologic case/control studies and several individual case reports[63–68] support an association between CVB infection and diabetes mellitus.

A recent study conducted in Sweden[69,70] found that 67% of 24 patients under 15 years old, newly diagnosed with IDDM, had IgM antibodies to one or more CVB, whereas none of the age-matched controls had IgM antibodies to CVB. Follow-up on the 16 positive patients showed that titers had declined and were below detection in most cases after 6 months, supporting the inference of recent infection. Another study[71] examined serum samples from 22 patients with recently diagnosed IDDM and 46 controls for neutralizing antibodies to CVB1–6 and found that nearly 32% of patients had evidence of CVB4 infection while only 10% of the controls did. CVB2 and CVB5 antibody titers were reduced in the cases compared with controls. A negative correlation between CVB antibodies and infection was observed in a study in Seattle,[72] during which an outbreak of CVB3–5 occurred. In this study, titers to CVB3 and CVB4 were lower in diabetic patients than in the control group. This study points out the critical importance of study timing and knowledge of the viruses circulating in the study area, since the negative correlation of CVB serology with viruses known to be circulating suggests that lack of preexisting antibodies may be an important predisposing factor.

In a study in England,[73] 39% of 28 children (3–14 years old) with IDDM were positive for CVB-specific IgM, whereas only 6% of age-matched controls were positive. In an extension of these studies,[74] which included cases from Austria and Australia, 30% of 122 patients had positive titers while only 6% of 204 age-matched controls were positive. In this study there were no differences between cases and controls in IgM antibodies to mumps, rubella, or cytomegalovirus. In further studies on the patients in Austria,[75] 22 of 23 cases with CVB-specific IgM had human leukocyte antigen (HLA) DR3, DR4, or both, compared with 76% with these HLA types in the viral IgM-negative patients. No other correlations were seen with other HLA types or immunoglobulin allotypes.

A study in Pittsburgh also investigated the association of HLA type, CVB antibody, and diabetes.[76] In this study, HLA type was determined for 172 persons hospitalized for IDDM. The DR3 type was found more often in males and the DR4 type more often in females. In addition, cases with the DR4 type had a greater frequency of CVB antibodies. In summary, several recent studies provide further support for the association between CVB infection and IDDM as well as suggestive evidence for some HLA correlation with this disease. Interpretation of the results of these studies must take into account their choice of controls, timing of serum collection, and the sensitivity of the detection system (see Section 4). Despite these limitations, the evidence clearly warrants further investigation into this potential etiology of an important disease.

3.3.2. Chronic Heart Disease

Five recently published studies examined the specific issue of an association between CVB infection as measured by serology and acute myocardial infarction (MI). In the first study, neutralizing antibodies to CVB1–5 were measured in all patients admitted with chest pain at a single infirmary in Scotland over an 8-month period.[77] Considering a neutralizing titer greater than 1 : 512 in a single serum sample as positive for CVB infection, there was no difference between cases and controls (5% versus 4%). However, there was a difference between a subgroup of nontransmural MI cases and total controls (14% versus 4%), although in this comparison the total numbers are small (3 of 22 versus 4 of 100). In a separate study also done in Scotland,[78] all patients treated at a general medical unit, the majority with chest pain, over a 2-year period were examined for CVB antibodies. Using the same criterion for determining recent CVB infection, 33% of patients were positive compared with 4% of unmatched controls.

In a third study,[79] neutralization titers to CVB1–5 were measured in paired serum samples from 59 patients presenting with acute myocardial infarction and 38 age-matched controls. No virus was isolated from any patient or control. In this study, 15% of cases and 2.6% of the controls had evidence of recent CVB infection, as defined by a fourfold or greater rise in titer.

Reports on a fourth study in New Zealand[80] used as criteria for CVB infection either a fourfold rise in titer, single titers greater than 1 : 512, or viral isolation by culture. In a separate report,[81] IgM antibodies were measured. In both reports, the rates in patients were compared with those in age- and sex-matched normal blood donors. By both measures, cases had significantly higher rates of CVB infection than did normal controls.

The last report[82] followed up a previously studied group in England, where no difference between cases and controls had been seen. When the same serum samples were examined by an ELISA test for

CVB-specific IgM, again no difference between cases and controls was observed. The use of gastroenterology patients for controls, with a possible bias toward enteric infections, might explain the negative results in this study.[81]

The last two series of reports suggest that the choice of techniques used to define recent CVB infection is not as important to the final conclusion as is the design of the study. The evidence for an association between patients presenting with acute MI and recent CVB infection remains unclear, primarily because of variation in case definition and choice of control groups.

3.3.3. Other Chronic Diseases

Several case reports link CVB infection with other chronic conditions, including arthritis[83] and Still's disease.[84] In addition, a case-control study[85] measured complement-fixing (CF) antibodies in a group of 12 children with juvenile dermatomyositis (JDM). While 83% of serum samples from cases were positive for CF antibodies to one or more CVB, only 25% of samples from 24 age-, sex-, and date-matched controls with juvenile rheumatoid arthritis were positive, and 25% of 2192 hospitalized children were positive. No differences were found in antibodies to 13 other viral antigens. This study supports the association between CVB infection and JDM suggested from previous clinical case reports.

A second study[86] found evidence of CNS infection with CVB in ventricular fluid from 4 of 28 newborns with congenital neural tube defects. The infants had neutralizing antibody to only one CVB serotype in the ventricular fluid but to several in serum. In all cases, the mother had antibodies to the same serotype, although all four mothers had antibodies to more than one serotype. No virus could be isolated. The unique distribution of antibodies in the ventricular fluid compared with serum strengthens the observed association; further studies are needed to assess the incidence of these cases.

4. FUTURE TECHNIQUES AND STUDIES

4.1. Techniques

The techniques available for laboratory support of epidemiological investigations of CVB have changed markedly during the past 10 years. Traditional techniques for detecting and characterizing enteroviruses rely on the time-consuming and expensive procedures of viral isolation in tissue culture and neutralization by reference antisera.[87] Isolating

CVB from specimens using appropriate tissue culture cell lines is possible within 2 or 3 days in most cases and remains the most sensitive method for detecting these viruses. Several newer cell lines have been evaluated with regard to their efficiency for enterovirus isolation[88,89] as well as combinations of more traditional cell lines.[90] In almost all cases, CVB appear to grow readily but cannot be reliably differentiated from the other enteroviruses solely on the basis of their cytopathic effect in particular cell lines. Therefore, presumption of an enterovirus infection can be made in a few days by isolation, but serotyping the isolate often requires several weeks.

Several recent technical advances have allowed the development of alternative procedures for detecting and characterizing enteroviruses. Nucleic acid probes have been developed for detecting enterovirus genomes. These probes have been made by synthesizing complementary DNA (cDNA) from viral RNA. Using conventional procedures, cDNA was made to representatives of all six serotypes of CVB.[91] The cDNA can be cloned into bacteria, amplified, and radiochemically labeled to provide a reagent for detecting enterovirus genomes by nucleic acid hybridization.[92] Partial genomic clones were made in this manner to several of the CVB[39,51] and used to detect viral genomes in tissue culture,[93] reconstructed clinical specimens,[94] and tissues.[39,51] This technique is more rapid than isolation and therefore may be clinically helpful; however, it is currently less sensitive than isolation and cannot be used to type enteroviruses.

Enteroviruses are typed by neutralization tests with a collection of reference antisera. The CF test has also been used for typing but is less sensitive and specific than is neutralization.[87] The neutralization test, while more definitive, and the standard for serotype definition, is not rapid and may give equivocal results with antigenic variants of the prototype strains.

Alternatives to the neutralization and CF test have been developed using an ELISA test with either a 96-well microtiter plate[95–99] or nitrocellulose membrane[100] as the solid phase. The isolate can be bound directly to the solid phase or can be captured using bound antiserum in the microtiter plates. These tests require high quality hyperimmune sera with low crossreactivity or monoclonal antibodies. While this approach has been shown to work for typing poliovirus and a few other enteroviruses, it is unclear whether it will work for all enteroviruses.

Another approach to serotyping isolates recently published by Tracy and Latham[101] seems to provide some of the rapidity of the CF test and ELISA, while retaining serotype specificity. This technique involves four basic steps: (1) reacting reference serum with the specimen, (2) concentrating the reacted virus with *Staphylococcus aureus* protein A,

(3) applying the collected virus to nitrocellulose membranes, and (4) detecting the bound viral genome by dot-blot hybridization using a cloned cDNA as described above. The approach should be comparable to an ELISA test in specificity and has potentially greater sensitivity, especially with original specimens. Both the ELISA and the immunoprecipitation hybridization technique offer the potential for more rapid serotyping of enterovirus isolates.

During the past several years, several groups developed an ELISA for CVB-specific IgM.[95,96,98,99] These tests were positive for nearly 90% of culture-confirmed CVB infections and are rapid to perform; however, in most cases the test is not completely serotype specific.[102] Depending on the configuration and sensitivity of the test, 10% to nearly 70% of serum samples show heterotypic response due to other enterovirus infections. In two reports the IgM antibodies were found to persist for only 6–8 weeks, suggesting that the test is a good measure of recent infection.

4.2. Limitations and Study Design Limitations

4.2.1. Considerations Due to the Viruses

In epidemiologic studies of CVB, several important properties of the viruses are critical to study design and data interpretation. First, the grouping of these six enterovirus serotypes is based principally on growth characteristics in newborn mice; they produce a focal myositis and a spastic paralysis.[1,3,87] This growth property in mice has never been uniquely correlated with any human disease. While CVB viruses are more often associated with myocarditis and pleurodynia, other enteroviruses are frequently associated with these illnesses as well.[1–4]

Second, the genes responsible for the pathogenic potential of a virus are not necessarily the same as the genes that code for serotype. Since recombination occurs often with poliovirus,[103] it probably also occurs with the other enteroviruses. Therefore, genes for pathogenic potential found in one serotype may recombine with virus of a different serotype and spread this gene among several serotypes. The number of serotypes is limited only by the range of potential recombination partners.

Third, enterovirus infections are very common. It has been conservatively estimated that there are 10–20 million enterovirus infections yearly in the United States,[2,5] and of these one fourth to one half may be due to CVB. Serosurveys of IgM antibodies in most study control groups typically show around 5% positive for CVB-specific IgM. These numbers support a high incidence of enterovirus infection, and are consistent with the high prevalence of neutralizing antibody to the CVB.

Fourth, the prevalence of any given serotype in any given year can vary dramatically, as shown in Fig. 1. There can also be considerable variation during the same year between different locations within the United States.[5] Studies attempting to correlate virus infection with a given disease must consider what serotypes are prevalent in a particular location in designing the study.

4.2.2. *Considerations Due to Nature of Enterovirus Infections*

In addition to considerations regarding the viruses, several characteristics of enterovirus infections should be reiterated. First, most infections are either subclinical or associated with minor illness. The ratio of infected individuals to reported clinical cases for any serious enterovirus disease is typically quite high,[1–4] for example, around 1000 : 1 for enteroviral aseptic meningitis. This ratio can be significantly lower in certain outbreaks. These large ratios make prospective studies difficult because of the large at-risk population. As many of the factors responsible for the variation in clinical manifestation are unknown, it is not possible to refine the definition of the at-risk population and increase the chance of finding an association. Certain populations, such as neonates, seem to develop serious disease at a higher rate, and other factors such as physical stress have been suggested to be associated with more severe disease.[1–3] For chronic disease in animal models, the genetic makeup of the host is clearly important (see Chapter 20, this volume). The association of certain HLA types with IDDM may be parallel to this genetically defined susceptibility in humans. When host factors are identified that increase the risk of a particular disease, a prospective study could then be designed to the lower numbers in the at-risk population per case.

A second host consideration important to serologic studies, is the variability in the immune response. Heterotypic immune response, as measured by neutralization or IgM ELISA, is widely recognized and seems to be age dependent.[95–98] It is not clear what factors other than age lead to variation in this parameter. If, for example, prior number of enterovirus infections is correlated with heterotypic response, control groups in serologic studies should be matched by family size, socioeconomic status, and other factors correlated with increased rates of enterovirus infection.

A key area for the epidemiologic study of CVB infections is their role in diseases such as diabetes mellitus. Studies of this topic, if they are to be successful, must take into account the characteristics of the viruses and their infections as noted above. For example, if an IgM serologic study had been conducted in Bulgaria in 1975 to determine whether poliovirus was associated with paralysis in a group of nearly 140 patients with recent onset of flaccid paralysis, the results would have shown no significant differences between patients and controls for all three serotypes of poliovirus. The implied conclusion would have been that poliovirus is not a major cause of acute flaccid paralysis. However, Bulgaria in that year had an outbreak of enterovirus 71 that proved to be the etiologic agent of acute flaccid paralysis.[1–3]

Therefore, serologic studies must be sure that controls are matched for factors related to prevalence of enterovirus antibodies such as age, sex, serum collection timing, and socioeconomic status. These studies probably should consider any enterovirus as a possible etiologic agent and should include the circulating enteroviruses in the subset to be examined.

By careful consideration of the unique features of enteroviruses and their infections as well as how they affect study design and data interpretation, as well as use of improved laboratory techniques, it should be possible in the near future to gain a better understanding of the correlations between CVB or enterovirus infection and a number of chronic diseases.

REFERENCES

1. Melnick, J. L., 1982, Enteroviruses, in: *Viral Infections of Humans, Epidemiology and Control* (A. S. Evans, ed.), pp. 187–251, Plenum, New York.
2. Moore, M., and Morens, D. M., 1984, Enteroviruses, including polioviruses, in: *Textbook of Human Virology* (R. B. Belshe, ed.), pp. 407–483, PSG, Littleton, Massachusetts.
3. Melnick, J. L., 1985, Enteroviruses: Polioviruses, coxsackieviruses, echoviruses, and newer enteroviruses, in: *Virology* (B. N. Fields, D. M. Knipe, R. M. Chanock, J. L. Melnick, B. Roizman, and R. E. Shope, eds.), pp. 739–794, Raven, New York.
4. Modlin, J. F., 1985, Coxsackievirus and echovirus, in: *Principles and Practices of Infectious Diseases* (G. L. Mandell, R. G. Douglas, Jr., and J. E. Bennett, eds.), pp. 814–825, Wiley, New York.
5. Strikas, R. A., Anderson, L. J., and Parker, R. A., 1986, Temporal and geographic patterns of isolates of nonpolio enterovirus in the United States, 1970–1983, *J. Infect. Dis.* **153**:346–351.
6. Bell, E. J., and McCartney, R. A., 1984, A study of coxsackie B virus infections, 1972–1983, *J. Hyg. (Lond.)* **93**:197–204.
7. Schoub, B. D., Johnson, S., McAnerney, J. M., Dos Santos, I. L., and Klaassen, K. I., 1985, Epidemic coxsackie B virus infection in Johannesburg, South Africa, *J. Hyg. (Lond.)* **95**:447–455.
8. Centers for Disease Control, 1970, *Neurotropic Viral Diseases Surveillance, Enterovirus Infections, Annual Summary, 1969,* Centers for Disease Control, Atlanta, Georgia.
9. Centers for Disease Control, 1970, *Neurotropic Viral Diseases Surveillance, Enterovirus Infections, January–September, 1970,* Centers for Disease Control, Atlanta, Georgia.
10. Centers for Disease Control, 1971, *Neurotropic Diseases Surveillance, Enterovirus Infections, Annual Summary, 1970,* Centers for Disease Control, Atlanta, Georgia.

11. Centers for Disease Control, 1977, *Enteric and Neurotropic Diseases Surveillance, Enterovirus Surveillance, 1971–1975,* Centers for Disease Control, Atlanta, Georgia.
12. Centers for Disease Control, 1981, *Enterovirus Surveillance, Summary 1970–1979,* Centers for Disease Control, Atlanta, Georgia.
13. Morens, D. M., Zweighaft, R. M., and Bryan, J. M., 1979, Non-polio enterovirus disease in the United States, 1971–1975, *Int. J. Epidemiol.* **8:**49–54.
14. Moore, M., 1982, Enteroviral disease in the United States, 1970–1979, *J. Infect. Dis.* **146:**103–108.
15. Prabhakar, B. S., Haspel, M. V., McClintock, P. R., and Notkins, A. L., 1982, High frequency of antigenic variants among naturally occurring human coxsackie B4 virus isolates identified by monoclonal antibodies, *Nature (Lond.)* **300:**374–376.
16. Prabhakar, B. S., Menegus, M. A., and Notkins, A. L., 1985. Detection of conserved and nonconserved epitopes on coxsackievirus B4: Frequency of antigenic change, *Virology* **146:**302–306.
17. Webb, S. R., Kearse, K. P., Foulke, C. L., Hartig, P. C., and Prabhakar, B. S., 1986, Neutralization epitope diversity of coxsackievirus B4 isolates detected by monoclonal antibodies, *J. Med. Virol.* **20:**9–15.
18. Ash, P., Leong, W. A., Kennett, M. L., and Schnagl, R. D., 1985, Neutralization kinetic analysis of echovirus-30 and coxsackievirus-B4 strains revealed little antigenic variation amongst the echovirus strains, *Aust. J. Exp. Biol. Med. Sci.* **63:**219–222.
19. Nottay, B. K., Kew, O. M., Hatch, M. H., Heyward, J. T., and Obijeski, J. F., 1981, Molecular variation of type 1 vaccine-related and wild polioviruses during replication in humans, *Virology* **108:**405–423.
20. Minor, P. D., 1982, Characterization of strains of type 3 poliovirus by oligonucleotide mapping, *J. Gen. Virol.* **59:**307–317.
21. Kew, O. M., and Nottay, B. K., 1984, Molecular epidemiology of polioviruses, *Rev. Infect. Dis.* **6:**S499–S504.
22. Harris, T. J. R., Doel, T. R., and Brown, F., 1977, Molecular aspects of the antigenic variation of swine vesicular disease and coxsackie B5 viruses, *J. Gen. Virol.* **35:**299–315.
23. Cao, Y., Schnurr, D. P., and Schmidt, N. J., 1984, Monoclonal antibodies for study of antigenic variation in coxsackievirus type B4—Association of antigenic determinants with myocarditic properties of the virus, *J. Gen. Virol.* **65:**925–932.
24. Cao, Y., Schnurr, D. P., and Schmidt, N. J., 1984, Differing cardiotropic and myocarditic properties of group B type 4 coxsackievirus strains, *Arch. Virol.* **80:**119–130.
25. Jordan, G. W., Bolton, V., and Schmidt, N. J., 1985, Diabetogenic potential of coxsackie B viruses in nature, *Arch. Virol.* **86:**213–221.
26. Jimes, S., and Jamison, R. M., 1983, Coxsackievirus B4: *In vitro* genetic markers and virulence, *Arch. Virol.* **77:**1–11.
27. Jimes, S., Jamison, R. M., and Grafton, W. D., 1984, Coxsackievirus B4: *In vitro* genetic markers and cardiovirulence, *Arch. Virol.* **81:**345–352.
28. Jordan, G. W., and Bolton, V., 1985, Interferon-sensitive coxsackievirus variants in nature, *J. Interferon Res.* **5:**289–296.
29. Hartig, P. C., and Webb, S. R., 1983, Heterogeneity of a human isolate of coxsackie B4: Biological differences, *J. Infection.* **6:**43–48.
30. Hartig, P. C., Madge, G. E., and Webb, S. R., 1983, Diversity within a human isolate of coxsackie B4: Relationship to viral-induced diabetes, *J. Med. Virol.* **11:**23–30.
31. Danes, L., and Jaresova, I., 1985, Neutralization microtest with human coxsackievirus and echovirus serotypes, *J. Hyg. Epidemiol. Microbiol. Immunol.* **29:**399–408.
32. Manjunath, N., Balaya, S., and Seth, P., 1982, Serologic survey for neutralizing

antibodies against group B coxsackieviruses in normal population in Delhi area, *Indian J. Med. Res.* **76:**656–661.

33. Mukundan, P., and John, T. J., 1983, Prevalence and titres of neutralizing antibodies to group B coxsackieviruses, *Indian J. Med. Res.* **77:**577–589.
34. Santhanam, S., and Choudhury, D. S., 1984, Antibodies against coxsackie B2 virus in infants and children in Delhi, *J. Commun. Dis.* **16:**304–306.
35. Morag, A., Margalith, M., Shuval, H. I., and Fattal, B., 1984, Acquisition of antibodies to various coxsackie and echo viruses and hepatitis A virus in agricultural communal settlements in Israel, *J. Med. Virol.* **14:**39–48.
36. Margalith, M., Fattal, B., Shuval, H. I., and Morag, A., 1986, Prevalence of antibodies to enteroviruses and varicella–zoster virus among residents and overseas volunteers at agricultural settlements in Israel, *J. Med. Virol.* **20:**189–197.
37. Lau, R. S., 1983, Coxsackie B virus infections in New Zealand patients with cardiac and non-cardiac diseases, *J. Med. Virol.* **11:**131–137.
38. Lerner, A. M., and Wilson, F. M., 1973, Virus myocardiopathy, *Prog. Med. Virol.* **15:**63–91.
39. Hallam, N. F., Eglin, R. P., Holland, P., Bell, E. J., and Squier, M. V., 1986, Fatal coxsackie B meningoencephalitis diagnosed by serology and *in-situ* nucleic acid hybridization, *Lancet* **2:**1213–1214.
40. Cooper, J. B., Pratt, W. R., English, B. K., and Shearer, W. T., 1983, Coxsackievirus B3 producing fatal meningoencephalitis in a patient with X-linked agammaglobulinemia, *Am. J. Dis. Child.* **137:**82–83.
41. Estes, M. L., and Rorke, L. B., 1986, Liquefactive necrosis in coxsackie B encephalitis, *Arch. Pathol. Lab. Med.* **110:**1090–1092.
42. Hirschman, S. Z., and Hammer, G. S., 1974, Coxsackie virus myopericarditis, *Am. J. Cardiol.* **34:**224–232.
43. Woodruff, J. F., 1980, Viral myocarditis. A review, *Am. J. Pathol.* **101:**427–529.
44. Reyes, M. P., and Lerner, A. M., 1985, Coxsackievirus myocarditis—with special reference to acute and chronic effects, *Prog. Cardiovasc. Dis.* **27:**373–394.
45. Banatvala, J. E., 1983, Coxsackie B virus infections in cardiac disease, in: *Recent Advances in Clinical Virology,* Vol. 3 (A. P. Waterson, ed.), pp. 99–115, Churchill Livingstone, Edinburgh.
46. Sundar, S., Shyam, B., Gulathi, A. K., and Somani, P. N., Coxsackie—A viral carditis in eastern India, *J. Assoc. Physicians India* **32:**501–503.
47. Morita, H., Kitaura, Y., Deguchi, H., Kotaka, M., Nakayama, Y., Suwa, M., Kino, M., Hirota, Y., and Kawamura, K., 1983, Pericarditis in a young adult—Clinical course and endomyocardial biopsy findings, *Jpn. Circ. J.* **47:**1077–1083.
48. Read, R. B., Ede, R. J., Morgan-Capner, P., Moscoso, G., Portmann, B., and Williams, R., 1985, Myocarditis and fulminant hepatic failure from coxsackievirus B infection, *Postgrad. Med. J.* **61:**749–752.
49. Frisk, G., Torfason, E. G., and Diderholm, H., 1984, Reverse radioimmunoassays of IgM and IgG antibodies to coxsackie B viruses in patients with acute myopericarditis, *J. Med. Virol.* **14:**191–200.
50. Tilzey, A. J., Signy, M., and Banatvala, J. E., 1986, Persistent coxsackie B virus specific IgM response in patients with recurrent pericarditis. (Letter.) *Lancet* **1:**1491–1492.
51. Bowles, N. E., Richardson, P. J., Olsen, E. G. J., and Archard, L. C., 1986, Detection of coxsackie-B-virus-specific RNA sequences in myocardial biopsy samples from patients with myocarditis and dilated cardiomyopathy, *Lancet* **1:**1120–1122.
52. Rotbart, H. A., Levin, M. J., and Villarreal, L. P., 1984, Use of subgenomic poliovirus

DNA hybridization probes to detect the major subgroups of enteroviruses, *J. Clin. Microbiol.* **20:**1105–1108.
53. Porres, E. R., Werthammer, J., Moss, N., Bernstein, J. M., and Belshe, R. B., 1985, Fatal coxsackievirus B4 infection in a neonate, *South. Med. J.* **78:**1254–1256.
54. Barson, W. J., and Reiner, C. B., 1986, Coxsackievirus B2 infection in a neonate with incontinentia pigmenti, *Pediatrics* **77:**897–900.
55. Rozkovec, A., Cambridge, G., King, M., and Hallidie-Smith, K. A., 1985, Natural history of left ventricular function in neonatal coxsackie myocarditis, *Pediatr. Cardiol.* **6:**151–156.
56. Schurmann, W., Statz, A., Mertens, T., Gladtke, E., and Eggers, H. J., 1983, Two cases of coxsackie B2 infection in neonates: Clinical, virological, and epidemiological aspects, *Eur. J. Pediatr.* **140:**59–63.
57. Iwasaki, T., Monma, N., Satodate, R., Kawana, R., and Kurata, T., 1985, An immunofluorescent study of generalized coxsackie virus B3 infection in a newborn infant, *Acta Pathol. Jpn.* **35:**741–748.
58. Fisher-Hoch, S. P., Gould, J., and Latham, P., 1982, Coxsackie B virus infection associated with hypoparathyroidism in a neonate, *Lancet* **2:**1395–1396.
59. Kaplan, M. H., Klein, S. W., McPhee, J., and Harper, R. G., 1983, Group B coxsackievirus infections in infants younger than three months of age: A serious childhood illness, *Rev. Infect. Dis.* **5:**1019–1032.
60. Modlin, J. F., 1986, Perinatal echovirus infection: Insights from a literature review of 61 cases of serious infection and 16 outbreaks in nurseries, *Rev. Infect. Dis.* **8:**918–926.
61. Reyes, M. P., Zalenski, D., Smith, F., Wilson, F. M., and Lerner, A. M., 1986, Coxsackievirus-positive cervices in women with febrile illnesses during the third trimester in pregnancy, *Am. J. Obstet. Gynecol.* **155:**159–161.
62. Barrett-Conner, E., 1985, Is insulin-dependent diabetes mellitus caused by coxsackievirus B infection? A review of the epidemiologic evidence, *Rev. Infect. Dis.* **7:**207–215.
63. Nihalani, K. D., Pethani, R. R., Menon, P. S., and Mehtalia, S. D., 1982, Coxsackie B4 virus causing insulin dependent diabetes mellitus, myopericarditis and encephalitis—A case report, *J. Assoc. Physicians India* **30:**107–109.
64. Niklasson, B. S., Dobersen, M. J., Peters, C. J., Ennis, W. H., and Möller, E., 1985, An outbreak of coxsackievirus B infection followed by one case of diabetes mellitus, *Scand. J. Infect. Dis.* **17:**15–18.
65. Mirkovic, R. R., Varma, S. K., and Yoon, J. W., 1984, Incidence of coxsackievirus B type 4 (CB4) infections concomitant with onset of insulin-dependent diabetes mellitus, *J. Med. Virol.* **14:**9–16.
66. Asplin, C. M., Cooney, M. K., Crossley, J. R., Dornan, T. L., Raghu, P., and Palmer, J. P., 1982, Coxsackie B4 infection and islet cell antibodies three years before overt diabetes, *J. Pediatr.* **101:**398–400.
67. Champsaur, H., 1980, Diabetes and coxsackie virus B5 infection, *Lancet* **1:**251.
68. Mertens, T., Gruneklee, D., and Eggers, H. J., 1983, Neutralizing antibodies against coxsackie B viruses in patients with recent onset of type I diabetes, *Eur. J. Pediatr.* **140:**293–294.
69. Friman, G., Fohlman, J., Frisk, G., Diderholm, H., Ewald, U., Kobbah, M., and Tuvemo, T., 1985, An incidence peak of juvenile diabetes. Relation to coxsackie B virus immune response, *Acta Paediatr Scand. [Suppl.]* **320:**14–19.
70. Frisk, G., Fohlman, J., Kobbah, M., Ewald, U., Tuvemo, T., Diderholm, H., and Friman, G., 1985, High frequency of coxsackie-B-virus-specific IgM in children de-

veloping type 1 diabetes during a period of high diabetes morbidity, *J. Med. Virol.* **17:**219–227.

71. Alberti, A. M., Amato, C., Candela, A., Costantino. F., Grandolfo, M. E., Lombardi, F., Novello, F., Orsini, M., and Santoro, R., 1985, Serum antibodies against coxsackie B1-6 viruses in type 1 diabetics, *Acta Diabetol. Lat.* **22:**33–38.
72. Palmer, J. P., Cooney, M. K., Ward, R. H., Hansen, J. A., Brodsky, J. B., Ray, C. G., Crossley, J. R., Asplin, C. M., and Williams, R. H., 1982, Reduced coxsackie antibody titers in type 1 (insulin-dependent) diabetic patients presenting during an outbreak of coxsackie B3 and B4 infection, *Diabetologia* **22:**426–429.
73. King, M. L., Shaikh, A., Bidwell, D., Voller, A., and Banatvala, J. E., 1983, Coxsackie-B-virus-specific IgM responses in children with insulin-dependent (juvenile-onset; type I) diabetes mellitus, *Lancet* **1:**1397–1399.
74. Banatvala, J. E., Bryant, J., Schernthaner, G., Borkenstein, M., Schober, E., Brown, D., De Silva, L. M., Menser, M. A., and Silink, M., 1985, Coxsackie B, mumps, rubella, and cytomegalovirus specific IgM responses in patients with juvenile-onset insulin-dependent diabetes mellitus in Britain, Austria, and Australia, *Lancet* **1:**1409–1412.
75. Schernthaner, G., Banatvala, J. E., Scherbaum, W., Bryant, J., Borkenstein, M., Schober, E., and Mayr, W. R., 1985, Coxsackie-B-virus-specific IgM responses, complement-fixing islet-cell antibodies, HLA DR antigens, and C-peptide secretion in insulin-dependent diabetes mellitus, *Lancet* **2:**630–632.
76. Eberhardt, M. S., Wagener, D. K., Orchard, T. J., LaPorte, R. E., Cavender, D. E., Rabin, B. S., Atchison, R. W., Kuller, L. H., Drash, A. L., and Becker, D. J., 1985, HLA heterogeneity of insulin-dependent diabetes mellitus at diagnosis. The Pittsburgh IDDM study, *Diabetes* **34:**1247–1252.
77. O'Neill, D., McArthur, J. D., Kennedy, J. A., and Clements, G., 1983, Coxsackie B virus infection in coronary care unit patients, *J. Clin. Pathol.* **36:**658–661.
78. Bell, E. J., Irvine, K. G., Gardiner, A. J., and Rodger, J. C., 1983, Coxsackie B infection in a general medical unit, *Scott. Med. J.* **28:**157–159.
79. Nikoskelainen, J., Kalliomaki, J. L., Lapinleimu, K., Stenvik, M., and Halonen, P. E., 1983, Coxsackie B virus antibodies in myocardial infarction, *Acta Med. Scand.* **214:**29–32.
80. Lau, R. C., 1982, Coxsackie B virus infection in acute myocardial infarction and adult heart disease, *Med. J. Aust.* **2:**520–522.
81. Lau, R. C., 1986, Coxsackie B virus-specific IgM responses in coronary care unit patients, *J. Med. Virol.* **18:**193–198.
82. Hannington, G., Booth, J. C., Bowes, R. J., and Stern, H., 1986, Coxsackie B virus-specific IgM antibody and myocardial infarction, *J. Med. Microbiol.* **21:**287–291.
83. Hurst, N. P., Martynoga, A. G., Nuki, G., Sewell, J. R., Mitchell, A., and Hughes, G. R., 1983, Coxsackie B infection and arthritis, *Br. Med. J.* **286:**605.
84. Heaton, D. C., and Moller, P. W., 1985, Still's disease associated with coxsackie infection and haemophagocytic syndrome, *Ann. Rheum. Dis.* **44:**341–344.
85. Christensen, M. L., Pachman, L. M., Schneiderman, R., Patel, D. C., and Friedman, J. M., 1986, Prevalence of coxsackie B virus antibodies in patients with juvenile dermatomyositis, *Arthritis Rheum.* **29:**1365–1370.
86. Gauntt, C. J., Gudvangen, R. J., Brans, Y. W., and Marlin, A. E., 1985, Coxsackievirus group B antibodies in the ventricular fluid of infants with severe anatomic defects in the central nervous system, *Pediatrics* **76:**64–68.
87. Melnick, J. L., Wenner, H. A., and Phillips, C. A., 1979, Enteroviruses, in: *Diagnostic Procedures for Viral, Rickettsial and Chlamydial Infections* (E. H. Lennette and N. J. Schmidt, eds.), pp. 471–534, American Public Health Association, Washington, D.C.

88. Patel, J. R., Daniel, J., Mathan, M., and Mathan, V. I., 1984, Isolation and identification of enteroviruses from faecal samples in a differentiated epithelial cell line (HRT-18) derived from human rectal carcinoma, *J. Med. Virol.* **14**:255–261.
89. Patel, J. R., Daniel, J., and Mathan, V. I., 1985, A comparison of the susceptibility of three human gut tumour-derived differentiated epithelial cell lines, primary monkey kidney cells and human rhabdomyosarcoma cell line to 66-prototype strains of human enteroviruses, *J. Virol. Methods* **12**:209–216.
90. Dagan, R., and Menegus, M. A., 1986, A combination of four cell types for rapid detection of enteroviruses in clinical specimens, *J. Med. Virol.* **19**:219–228.
91. Tracy, S., 1985, Comparison of genomic homologies in the coxsackievirus B group by use of cDNA : RNA dot-blot hybridization, *J. Clin. Microbiol.* **21**:371–374.
92. Tracy, S., 1984, A comparison of genomic homologies among the coxsackievirus B group: Use of fragments of the cloned coxsackievirus B3 genome as probes, *J. Gen. Virol.* **65**:2167–2172.
93. Hypiä, T., Stålhandske, P., Vainionpää, R., and Pettersson, U., 1984, Detection of enteroviruses by spot hybridization, *J. Clin. Microbiol.* **19**:436–438.
94. Rotbart, H. A., Levin, M. J., Villarreal, L. P., Tracy, S. M., Semler, B. L., and Wimmer, E., 1985, Factors affecting the detection of enteroviruses in cerebrospinal fluid with coxsackievirus B3 and poliovirus 1 cDNA probes, *J. Clin. Microbiol.* **22**:220–224.
95. Dörries, R., and Ter Meulen, V., 1983, Specificity of IgM antibodies in acute human coxsackievirus B infections, analysed by indirect solid phase enzyme immunoassay and immunoblot technique, *J. Gen. Virol.* **64**:159–167.
96. Morgan-Capner, P., and McSorley, C., 1983, Antibody capture radioimmunoassay (MACRIA) for coxsackievirus B4 and B5-specific IgM, *J. Hyg. (Lond.)* **90**:333–349.
97. Hannington, G., Booth, J. C., Wiblin, C. N., and Stern, H., 1983, Indirect enzyme-linked immunosorbent assay (ELISA) for detection of IgG antibodies against coxsackie B viruses, *J. Med. Microbiol.* **16**:459–465.
98. Chan, D., and Hammond, G. W., 1985, Comparison of serodiagnosis of group B coxsackie virus infections by an immunoglobulin M capture enzvme immunoassay versus microneutralization, *J. Clin. Microbiol.* **21**:830–834.
99. McCartney, R. A., Banatvala, J. E., and Bell, E. J., 1986, Routine use of μ-antibody-capture ELISA for the serological diagnosis of coxsackie B virus infections, *J. Med. Virol.* **19**:205–212.
100. Loh, P. C., Dow, M. A., and Fujioka, R. S., 1985, Use of the nitrocellulose-enzyme immunosorbent assay for rapid, sensitive and quantitative detection of human enteroviruses, *J. Virol. Methods* **12**:225–234.
101. Tracy, S., and Latham, A., 1986, Rapid identification of coxsackie B viruses after immunoprecipitation and nucleic acid hybridization, *Diagn. Microbiol. Infect. Dis.* **4**:327–333.
102. Pattison, J. R., 1983, Tests for coxsackie B virus-specific IgM, *J. Hyg. (Lond.)* **90**:327–332.
103. Kew, O. M., and Nottay, B. K., 1984, Evolution of the oral poliovaccine strains in humans occurs by both mutation and intramolecular recombination, in: *Modern Approaches to Vaccines* (R. M. Chanock and R. A. Lerner, eds.), pp. 357–367, Cold Spring Harbor Laboratory, Cold Spring Harbor, New York.

Index

Actinomycin D, in persistent infection sensitivity, 189
  viral replication and, 191–193
Acute hemorrhagic conjunctivitis, IFN detection in, 75
Acute lymphonodular pharyngitis, 345, 346
Acute myocarditis: *see* Myocarditis
Acute respiratory infections, CV role in, 235
Age, CV infection susceptibility and, 139–142
α-Interferon (IFN_{α}): *see* Interferons (IFN), α
Anti-idiotypic antibodies, 129–131
Anti-inflammatory agents, CV infection and, 150–151
Antibodies, *see also* Antibody response
  anti-idiotypic, 129–131
  in autoimmune myocarditis, 278–279; *see also* Autoantibodies, in autoimmune myocarditis
  and cardiac myosin, crossreactivity, 288
  CV-specific
    detection methods for, 204
    generation of, 305–308
    in IDDM, 356, 358
  heart-reactive (HRA), 109–110
  neutralizing: *see* Neutralizing antibodies
Antibody response, 83–85
  new tools and reagents to detect, 212–216
  in resistance and recovery, 86–87
  in tissue damage, 87–88
Antigen(s)
  cardiac myosin as, 285–289
  CV-specific, EIA to detect, 206–207
Antigens (*cont.*)
  CVA and CVB, detection by ELISA, 332
  Ia, on islet cells, 374–375
  presentation of, pathogenesis mechanisms and, 111
  as target for CTL, 90
Antigenic variants, 161–162
  among CVB4 isolates
  biologic property changes and, 127–128
  characterization of, 121–124
  epitope conservation, 124–125, 126
  naturally occurring, 118–121
  selection of, 121
Antiviral activity, of human fibroblast IFN, 310–311
Arthritis, 236
  CVB infection associated with, 408
Aseptic meningitis, 327–328
Assay, 36; *see also* Immunoassays
  macrophage migration inhibition (MIF), 88, 90
Attachment
  in viral replication, 36–37
  of virion, site identification for, 55
Autoantibodies, in autoimmune myocarditis, 279–281
  age effect on, 282
  production of, 279
  reactions of, 284
  strain differences, 281
Autoimmune myocarditis, 274–275; *see also* Heart muscle disease
  antibodies in, 278–281
  autoantigen theories and, 287–289
  cardiomyopathy development from, 282–283

Autoimmune myocarditis (*cont.*)
  immunochemical studies in, 283–287
  incidence in humans, 281–282
  murine models of, 275–276
  serum and organ CVB3 titers and, 276–278
  susceptibility, 275
Autoimmunity
  in heart disease, 272–273
  molecular mimicry and, 131, 132
  picornavirus-induced, hypothesis for, 112–114
  viruses and, 271–272

β-cells
  in CVB-induced pancreatitis and insulitis, 366–368, 369
  CVB infection of, 372–373
  damage to, mechanisms of, 373–375
  in IDDM experimental models, 368, 370–371
  and viral link to IDDM, 361–362, 365
β-Interferon (IFN_β): *see* Interferons (IFN), β
BGMK: *see* Buffalo green monkey kidney cells (BGMK)
Blood glucose levels, in IDDM experimental models, 368, 370
Body temperature, CV infection susceptibility and, 138–139
Bone marrow transplantation, CVA1 infectious gastroenteritis and, 387–388
Bornholm disease, 236
Botulinum toxin, CV infection and, 151
Breast-feeding, NPEV incidence and, 384
Buffalo green monkey kidney cells (BGMK), CVB3 growth in, 185–186

cAMP, CV infection susceptibility and, 150
Capsid proteins
  and coding regions, of CVB genome: *see* Coding regions
  subunits, of picornaviridae, 10
  in viral replication, 46–47
Cardiac myosin, autoimmune myocarditis and
  absorption analysis, 285
  as antigen, 285–289
Cardiac myosin, autoimmune myocarditis and (*cont.*)
  CVB3 antibody crossreactivity, 288
  induction by, 287
Cardiopathies, CV-induced, 254
  clinical manifestations, 254–255
  etiologic diagnosis, 256
  heart muscle disease: *see* Heart muscle disease
  laboratory aids, 255–256
  possible courses of, 259
Carditis, CVB infection associated with, 404–405; *see also* Myocarditis
Carrier cultures: *see* Persistent infections
Carrier-state infection, CVB3, of cultured human myocardial fibroblasts, 309–310
cDNA
  CVB3
    cloning strategy of, 296–297
    as diagnostic reagent for enterovirus detection, 298–299
    future techniques, 410
  in situ hybridization, 300–303
    myocardial enterovirus sequences detected by, 303–305
  nucleotide sequence of, 299, 314
  restriction map related to viral genetic map, 306
Cell cultures, *see also* Tissue cultures
  CV variants from: *see* Coxsackievirus (CV) variants
  IFN induction and sensitivity in, 74–75
Cell cycles, IFN effect on, 67–68
Cell destruction, 182
Cell-mediated immune response (CMI), 88–89
  in resistance and recovery, 89
  in tissue damage, 89–90
Cellular receptors
  for CV, 55–58
  expression and viral infection susceptibility, 59
  variants, 162
  viral tropism and, 51–52
Central nervous system (CNS)
  disease of, CVB associated with, 404
  infections
    fatal infantile CVB, 248
    persistent, 193–194
cGMP, CV infection susceptibility and, 150

Cloning
  of CVB3 cDNA, 296–297
  of CVB genome, 19–20
    in viral replication, 46
  receptor gene, in *E. coli*, 56
CMI: *see* Cell-mediated immune response (CMI)
CNS: *see* Central nervous system (CNS)
Coding regions, of CVB genome
  P1, 24–26
  P2, 26
  P3, 26–27
Committee on Enteroviruses, CV isolation and, 4–5
Conjunctivitis
  acute hemorrhagic, IFN detection in, 75
  epidemic, 347–348
Cortisone, effect on CV infection susceptibility, 150
Coxsackieviral (CV) infection
  age and aging effect, 139–142
  agents influencing, 150–151
  cardiovascular, 148
  distribution of, factors influencing, 384–385
  dose effect, 138
  epidemiologic pattern model, 385–386
  within families, 386–387
  fetal and neonatal, 142–145
  genetics and, 145–148
  geographic patterns, 391–396
  hormonal effect, 149–150
  in humans, 51
    clinical syndromes, 137
  in infants: *see* Infancy, CV infection in
  within institutions, 387–389
  mucocutaneous infections and, 339, 340
  nutrition and, 148–149
  persistent: *see* Persistent infections
  seroepidemiologic surveys of, 389–390
  temperature effect, 138–139
  transmission mode or route of, 136, 138, 383–384
Coxsackievirus (CV) isolates
  group B
    antigenic variants among: *see* Antigenic variants
    genotype, 402–403
    serotype, 400–402
  monoclonal antibodies to characterize, 211–212
Coxsackievirus (CV) variants
  future goals, 171–172
  group A, 170
  group B
    antigenic, 161–162; *see also* Antigenic variants
    DIP and other aberrant particles, 164–165
    IFN-sensitive, 162–163
    naturally occurring
      diabetes mellitus, 169–170
      myocarditic, 167–168
    receptor, 162
    temperature-sensitive: *see* Temperature-sensitive (*ts*) mutants
    tissue culture derived
      diabetes mellitus, 168–169
      myocarditic, 165–167
    virus particles with new properties, 164
  unaccounted for, 170–171
  virulence and, 161
Coxsackieviruses (CV)
  detection methods, 204
  epidemiology: *see* Epidemiology
  group A
    pathogenicity, 221–222
    prototype strains, 3
  group B
    genomes: *see* Genome
    pathogenicity, 222
    prototype strains, 4
    PV and HRV groups, similarities with, 29
    variants, 161–170; *see also* Antigenic variants; Coxsackievirus (CV) variants
  growth, purification and assay of, 35–36
  history and classification, 1–5
  immunodepression by, 93–95
  isolates: *see* Coxsackievirus (CV) isolates
  isolation and propagation, 5–7
  lymphoid tissue histopathology, 95–96, 97
  mutation frequencies in, 159–160
  nonhuman occurrences, 7–8
  pathogenesis, 52–54
  physicochemical properties, 8–9
  replication and molecular biology of: *see* Viral replication

Coxsackieviruses (CV) (*cont.*)
resistance to, nonspecific mechanisms, 82–83
CPE: *see* Cytopathic effects (CPE)
CTL: *see* Cytolytic T lymphocytes (CTL); Cytotoxic T lymphocytes (CTL)
Cyclic adenosine monophosphate (cAMP), CV infection
susceptibility and, 150
Cyclic guanosine monophosphate (cGMP), CV infection
suspectibility and, 150
Cytolytic T lymphocytes (CTL), pathogenesis mechanisms and, 107–109; *see also* Cytotoxic T lymphocytes (CTL)
Cytopathic effects (CPE)
in cell cultures, 6–7
CVA9 and, 4
CVB4 isolates and, 121–122
neutralization of, 6–7; *see also* Virus neutralization test
Cytotoxic T lymphocytes (CTL)
cardiovascular CV infection, 148
response to CV, 88–89
tissue damage and, 90

Defective-interfering (DI) particles, 41–42
CVB variants and, 164–165
in persistent infection, 184–185, 187
Dermatomyositis, juvenile, CVB infection associated with, 408
DI: *see* Defective-interfering (DI) particles
Diabetes mellitus (DM)
CV infection susceptibility and, 145–148
CVB variants and, 168–170
insulin-dependent (IDDM): *see* Insulin-dependent diabetes mellitus (IDDM)
recognized syndromes, 351
Diarrhea, CV role in, 234
Dilated cardiomyopathy, enterovirus sequence detection by *in situ* hybridization, 303–305
Disease pathogenesis, virus role in, 103–105
Disease syndromes, CV role in, 234–236; *see also individually named disease syndromes*
Disinfectants, and CV infection in institutions, 389
DM: *see* Diabetes mellitus (DM)
DNA
cloned complementary: *see* cDNA
picornavirus pathogenesis and, 53
Dogs, CV in, 7–8
Domesticated animals, CV in, 7–8
Dosage, CV infection susceptibility and, 138
Dot-blot hybridization, 208–209
Downregulation, of IFN receptors, 312

ECG: *see* Echocardiogram (ECG)
Echocardiogram (ECG), in infantile CVB infection, 246
Echoviruses
non-polio (NPEV): *see* Non-polio echoviruses (NPEV)
persistent infection and, 186
RNA hybridization from, 45–46
Eczema coxsackium, 342
EIA: *see* Enzyme immunoassays (EIA)
Electromyography (EMG), and ME diagnosis, 331
ELISA: *see* Enzyme-linked immunosorbent assay (ELISA)
EMCV: *see* Encephalomyocarditis virus (EMCV) infection
EMG: *see* Electromyography (EMG)
Encephalitis, 329
Encephalomyelitis, myalgic: *see* Myalgic encephalomyelitis (ME)
Encephalomyocarditis virus (EMCV) infection, IFN role in, 68–72
and immune system participation, 72–73
Endocrine damage, mechanisms, in IDDM experimental models, 373–375
Enteroviruses (EV)
CV isolation and, 4–5
CVB3 cDNA as diagnostic reagent in detection of, 298–299
and epidemiological study design limitations, 410–412
heart disease, diagnostic problems, 294–295
mutation frequencies in, 159–160
myocardial sequence detection by *in situ* hybridization, 303–305

Enteroviruses (*cont.*)
recombination in, 160–161
71, neurologic associations with, 328
Environmental temperature, CV infection susceptibility and, 138–139
Enzyme immunoassays (EIA), 206–207
Enzyme-linked immunosorbent assay (ELISA)
CVA detection by, 332
CVB
detection by, 332
serologic diagnosis by, 335
specific IgM by, 333–334
future techniques, 409–410
Epidemic conjunctivitis, 347–348
Epidemiology, *see also* Coxsackieviral (CV) infection
CVB, 399–400
and associated clinical disease, 403–408
future techniques, 408–410
Scottish data for, 233–234
study design limitations, 410–412
surveillance studies, 400–403
infection distribution, factors influencing, 384–385
international data, 223; *see also* World Health Organization (WHO)
by age, 225
by clinical condition, 224
in neurologic disease, 320–324
pattern model, 385–386
Scottish data, 223, 226–227
by age and sex, 230
clinical classification, 228
for CVA, 227, 231–233
for CVB, 233–234
for neurologic disease, 324–327
periodic rates, 229
transmission routes, 383–384
United States data, 391, 394–396
nation and Nassau county compared, 394
by region, 393
by viral type, 392
Epitopes
antigenic variant analysis and, 121–124
conservation on CVB4 isolates, 124–125, 126
neutralization, diversity/stability of, 125, 127
Erythematous rashes, 345–347
Escherichia coli
CVB3 proteins expression in, 305–308
dot-blot hybridization and, 208–209
receptor gene cloning in, 56
EV: *see* Enteroviruses (EV)
Exercise, effect on CV-induced cardiopathies, 263–266

Families, CV infections in, 386–387
Fast protein liquid chromatography (FPLC), 206
Fetus, *see also* Pregnancy
CV infection susceptibility and, 141, 142–145
HFD cells, persistent infection of, 181–183
Fever, in infant CVB infection, 243
Fibroblasts
cultured human myocardial, persistent CVB3 carrier-state, infection of, 309–310
IFN, antiviral activity of, 310–311
persistent infection of, 181–183
FPLC: *see* Fast protein liquid chromatography (FPLC)

γ-Interferon (IFN_γ): *see* Interferons (IFN), γ
Gastrointestinal disorders, CV role in, 234
Gender
effect on IFN production, 74
NPEV incidence and, 385
Genetics
CV infection susceptibility and, 145–148
IDDM experimental models and, 371–372
immune response control to CVB in IDDM, 358–360
recombination, in enteroviruses, 160–161
Genome
Coxsackie group B
3′ NTR of, 27–28
characteristics, 20
cloning and sequencing of, 19–20
in viral replication, 46
hybridization probe sequences design and, 29
5′NTR of, 22–24

Genome (*cont.*)
  Coxsackie group B (*cont.*)
    P1 region in, 24–26
    P2 region in, 26
    P3 region in, 26–27
    structure, 20–21
    variations in, 21
    in viral replication, 43–46
Genotype, of CVB isolate, 402–403
Gingivostomatitis, herpetic, 343
Glomerulonephritis, CV and, 236
Glucose levels, in IDDM experimental models, 368, 370
Gold salts, CV infection and, 151

Hand-foot-and-mouth disease, 236, 339–343
HAV: *see* Hepatitis, A virus (HAV)
Heart cells
  cultured human
    CV heart disease simulated in, 308–309
    human fibroblast IFN antiviral activity in, 310–311
  persistent infection of, 181–183
    consequences of, 196–198
Heart disease
  chronic, CVB infection associated with 407–408
  CV and autoimmunity in, evidence for, 272–273
  enteroviral, diagnostic problems, 294–295
  simulation in cultured human heart cells, 308–309
  treatment for, 314
  viral role in, approach to problem, 294
Heart muscle disease, murine models, 257–258
  acute myocarditis
    CVB2, 261, 263
    CVB3, 261, 263
      and exercise, 263–266
    CVB1 and CVB4, 259–261, 262–263
  chronic, 266–268
Heart-reactive antibodies (HRA), pathogenesis mechanisms and, 109–110
HeLa cells
  HRV-14 attachment and, 56–57
HeLa cells (*cont.*)
  persistent infection of, 181–183
  receptor proteins from, 57
Hemagglutination, 9–10
Hemolytic–uremic syndrome, 236
Hemophagocytic syndrome, 236
Hepatitis
  CV role in, 234
  A virus (HAV)
    IFN production and, 75–76
    persistent infections and, 186
Herpangina, 343–345
Herpes simplex virus (HSV), 343
Herpetic gingivostomatitis, 343
HFD: *see* Human fetal diploid (HFD) cells
High-pressure liquid chromatography (HPLC), 206
Hormones
  effect on CV infection susceptibility, 148–149
  effect on IFN production, 73
Hospitals, CV infections in, 387–389
Host cell, in CVB3 persistent infection, 187–193
HPLC: *see* High-pressure liquid chromatography (HPLC)
HRA: *see* Heart-reactive antibodies (HRA)
HRV-14: *see* Human rhinovirus 14 (HRV-14)
HSV: *see* Herpes simplex virus (HSV)
Human fetal diploid (HFD) cells, persistent infection of, 181–183
Human rhinovirus 14 (HRV-14)
  IFN and, 76
  receptors for, 56
  VAS and, 55
Humoral immunity, pathogenesis mechanisms and, 109–110
Hybridization
  dot- or slot-blot, 208–209
  nucleic acid, 207–208
    limited usefulness, 210; *see also* cDNA
  probe sequences, 332–333
    CVB genome and, 29
  in situ, 209–210
Hypercholesteremia, CV infection susceptibility and, 148–149
Hyperglycemia, IDDM experimental models and, 368, 370–371

ICA, CVB-induced, 375
IFN: *see* Interferons (IFN)
Ig: *see* Immunoglobulin (Ig)
IgM antibody-capture immunoassay (MACIA), 213
  enzyme-labeled (MACEIA), 213
Immune pathogenesis, in picornavirus infections, evidence for, 105–107
Immune response, to CVB in IDDM, genetic control of, 358–360
Immune system
  IFN *in vivo* effects, 72–73
  mechanisms and damage significance to, 96, 98
Immunoassays, *see also individually named Immunoassays*
  enzyme (EIA), 206–207
  IgM antibody-capture, 213
  solid-phase, 212
Immunocompetent cells, viral replication in, 91–93
Immunodepression, 93–95
Immunoglobulin (Ig)
  antibody
    pathogenesis mechanisms and, 109
    production and circulation of, 83–84
  IDDM and, CVB antibody class specificity, 356, 358
Immunoglobulin M (IgM)
  CVB specific, detection by ELISA, 333–334
  detection of, 206
  immunoassay, 213
  virus-specific, ISPIA to detect, 212
In situ hybridization, 209–210
Indirect solid-phase immunoassay (ISPIA), 212–213
  enzyme-labeled (ISPEIA), 213
Infancy, CV infection in, 241–242
  diagnosis, 249
  fatal cases pathology, 246
  group B, 405
  household member/nosocomial acquisition, 242–244
  illness pattern, 246
  intrapartum fulminant acquisition, 244–248
    differential diagnosis, 248–249
    prevention, 249–250
    treatment, 249
Infections, *see also* Coxsackieviral (CV) infection; Mucocutaneous syndromes; Persistent infections; Picornavirus infection
  encephalomyocarditis virus (EMCV), 68–73
  fetal: *see* Fetus
  neonatal, 405; *see also* Infancy, CV infection in
  nosocomial, 387–389
Inoculation, viral, mode or route of, 136, 138
Institutions, CV infections in, 387–389
Insulin-dependent diabetes mellitus (IDDM), 351–352
  with CVB infection, 405–407
    clinical observations, 361–363
    experimental models, 365–375
    follow-up studies, 360–361
    postmortem studies, 363–365
    seroepidemiologic studies, 353–360
  epidemiology of, 352–353
Insulitis, CVB-induced experimental model, 366–368
Interferons (IFN)
  α
    antiviral activity of, 66–67
    production of, 66
    respiratory viral infections in children and, 76
    rhinovirus challenge and, 76
  antiviral activity of, 66–67
  β
    antiviral activity of, 66–67
    production of, 65–66
  cell cycles and, 67–68
  CVB variants sensitive to, 162–163
  as defense system, 68
  γ
    antiviral activity of, 66–67
    production of, 66
  human fibroblast, antiviral activity of, 310–311
  pathogenesis mechanisms and, 110
  persistent infection mechanisms and, 184–187
  in picornavirus infections
    host factors affecting production, 73–74
    in humans, 75–76

Interferons (IFN) (*cont.*)
  in picornavirus infections (*cont.*)
    immune system participation and, 72–73
    induction and sensitivity in cell cultures, 74–75
    in vivo experimental models, 68–72
  receptors, downregulation of, 312
  types, 65–66
Islets of Langerhans, *see also* Pancreatic cells
  Ia antigens on, 374–375
  postmortem study of, 364–365
  and viral link to IDDM, 361–363
Isolates: *see* Coxsackievirus (CV) isolates
Isolation, of viruses
  detection methods, 204
  improved techniques, 205
  requirements for, 203
Isoprinosine, IFN action and, 72
ISPIA: *see* Indirect solid-phasse immunoassay (ISPIA)

Jerne's network hypothesis, 129
Juvenile dermatomyositis (JDM), CVB infection associated with, 408

LCMV: *see* Lymphocytic choriomeningitis virus (LCMV)
Liquid chromatography, 206
Lymphocytes
  cytolytic T, 107–109
  cytotoxic T: *see* Cytotoxic T lymphocytes (CTL)
  viral replication in, 91–93
Lymphocytic choriomeningitis virus (LCMV), 196
Lymphocytosis, CV and, 236
Lymphoid cells
  human, CVB infected
    properties of, 93
    response to, 91
  persistent infection of, 181–183
    consequences of, 196
Lymphoid tissues, histopathology during CV infections, 95–96, 97

MACIA: *see* IgM antibody-capture immunoassay (MACIA)
Macrophage migration inhibition assay (MIF), 88, 90
Macrophages
  CV infection control by, 82–83
  pathogenesis mechanisms and, 111–112
  splenic, impairment of, 94
  viral replication in, 93
Major histocompatibility complex (MHC), pathogenesis mechanisms and, 110
  antigen presentation, 111
Malnutrition, CV infection susceptibility and, 148–149
Marasmus, CV infection susceptibility and, 148–149
MBV: *see* Membrane-bound virions (MBV)
ME: *see* Myalgic encephalomyelitis (ME)
Membrane-bound virions (MBV)
  capsid proteins of, 47
  genomic RNA of, 43–45
  in viral replication, 38–42
Meningitis
  aseptic, 327–328
  infant CVB infection and, 243
Metabolism, IDDM experimental models and, 368, 370–371
MIF: *see* Macrophage migration inhibition assay (MIF)
Molecular mimicry, and autoimmunity, 131, 132
Monoclonal antibodies, *see also* Antigenic variants
  anti-CVB4, 125, 127
    anti-idiotypic antibodies to, 129–131
    and cardiac muscle, reactivity with, 132
  to characterize virus isolates, 211–212
  and CV receptor identification, 57–58
  generation and characterization, 118
  neutralization of, CV structure and, 12–13
Mouth care, and CV infection in institutions, 388–389
Mucocutaneous infections, CV associated, 339, 340
Mucocutaneous syndromes
  acute lymphonocular pharyngitis, 345, 346
  and CV infections, 339, 340
  epidemic conjunctivitis, 347–348
  erythematous rashes, 345–347
  hand-foot-and-mouth disease, 339–343
  herpangina, 343–345

Muscle disease
CV role in, 235–236
heart: *see* Heart muscle disease
Mutation, in entero-viruses/coxsackieviruses, 159–160; *see also* Coxsackievirus (CV) variants; Temperature-sensitive (*ts*) mutants
Myalgic encephalomyelitis (ME), 329–331
CVB serologic studies in, and healthy adults compared, 330
Myocardial lesions, and antibody response to CV, 84, 85
Myocarditis, *see also* Heart muscle disease
autoimmune: *see* Autoimmune myocarditis
CV-induced
CTL role in, 108
group B, 29–30, 165–168
hypothetical autoimmunity model for, 112–114
murine models, 259–266
enterovirus sequence detection by *in situ* hybridization, 303–305
infectious, viruses associated with, 293–294
Myocardium, enterovirus sequence detection in, 303–305
Myocyte lysis, in CV infection, possible mechanisms of, 107
Myosin: *see* Cardiac myosin

Nasogastric tubes, and CV infection in institutions, 388–389
Nassau county, and national CV epidemiology compared, 394
Natural killer (NK) cells
in CV resistance, 83
pathogenesis mechanisms and, 111
Neurologic disease
CV-associated, 328
aseptic meningitis, 327–328
encephalitis, 329
epidemiologic analysis, 320
international data, 320–324; by age, 323
Scotland, serial data, 324–327; by age and sex, 327
myalgic encephalomyelitis, 329–331
paralytic disease, 328–329
EV71-associated, 328
Neutralization
of CPE, 6–7; *see also* Virus neutralization test
of CVB4 isolates, 121–122
with monoclonal antibodies, 125, 127
of monoclonal antibodies, CV structure and, 12–13
Neutralizing antibodies, 205–206
in IDDM, 353–356, 357
Newborn, *see also* Infancy, CV infection in
CV infections in, 405
susceptibility, 142–145
fulminant CVB infection of, 244–248
ventricular fluid CVB infection and CNS involvement in, 408
Newcastle disease virus, IFN induction and, 73, 74
NFR mouse cells
CVB3 growth in, 185–186
persistent infections in, 194–195
NK: *see* Natural killer (NK) cells
NMR: *see* Nuclear magnetic resonance (NMR)
N:NIH(S)-II mouse cells, persistent infections in, 194–195
Non-polio echoviruses (NPEV)
breast-feeding and, 384
gender and, 385
United States epidemiology, 391–396
Nondomesticated animals, CV in, 7
Nontranslated regions (NTR), of CVB genome
3′, 27–28
5′, 22–24
Nosocomial infections, CV and, 387–389
NPEV: *see* Non-polio echoviruses (NPEV)
NTR: *see* Nontranslated regions (NTR)
Nuclear magnetic resonance (NMR), and ME diagnosis, 331
Nucleic acid hybridization, 207–208; *see also* cDNA
limited usefulness, 210
Nutrition, CV infection susceptibility and, 148–149

Opsoclonus–myoclonus, CV and, 236
Orchitis, CV and, 236
Overnutrition, CV infection susceptibility and, 148–149

P1-coding region, of CVB genome, 24–26

P2-coding region, of CVB genome, 26
P3-coding region, of CVB genome, 26–27
Pancreatic cells
  cultured endocrine, CVB infection of, 372–373
  persistent infection of, 181–183
    consequences of, 196
  postmortem study of, 364–365
  and viral link to IDDM, 361–363
Pancreatitis
  CV role in, 234–235
  CVB-induced experimental model, 366–368
Paralytic disease, 328–329
Pathogenesis
  disease, virus role in, 103–105
  immune, in picornavirus infections, 105–107
  mechanisms of, 107
    cytolytic T lymphocytes, 107–109
    IFN and NK cells, 110–111
    macrophages, 111–112
    T helper cells and humoral immunity, 109–110
Persistent infections
  cell membrane stability and, 199
  consequences of, 194, 196
    for heart cells, 196–198
    for lymphoid cells, 196
    for pancreatic cells, 196
  CVB3 carrier-state, of cultured human myocardial fibroblasts, 309–310
  in vitro, 181–184
    CVB3, virus and host cell evolution, 187–193
    mechanisms, 184–187
  in vivo, 193–194
Peyer patches, CVB infection and, 242
Pharyngitis, acute lymphonodular, 345, 346
Picornaviridae
  capsid protein subunits of, 10
  enteroviruses and, 5
  physicochemical properties, 8–9
  receptor families for, 56
  serotypes in, 6
Picornavirus infection
  IFN in, 68–76; *see also* Interferons (IFN)
  immune pathogenesis in, evidence for, 105–107
Picornavirus infection (*cont.*)
  pathogenesis
    DNA and, 53
    quantitative aspects, 54
Pigs, CV in, 8; *see also* Swine vesicular disease virus (SVDV)
Plaque morphology, in persistent infection, 188–189
Poliomyelitis, CV isolation and, 1–2
Polioviruses
  pathogenesis in monkeys, 53
  RNA hybridization from, 45–46
Polypeptides, viral, characterization of, 128–129
Pregnancy, CV infection in
  newborn and, 244–248
  susceptibility to, 142–145
Progesterone, effect on CV infection susceptibility, 150
Proteins
  CVB3, expression in *E. coli,* 305–308
  subunits: *see* Capsid proteins, subunits
  synthesis inhibition, in persistent infection, 190
  in viral replication, 37–38
Purification, of CV, 36

Radiography, in infantile CVB infection, 246–247
Rashes, erythematous, 345–347
RD: *see* Rhabdomyosarcoma (RD) cell line
Receptor: *see* Cellular receptors
  variants, 1622
Recombinant DNA technology, *see also* cDNA
  as diagnostic approach, 295
  uses for, 312–313
Recovery
  antibody response role in, 86–87
  CMI role in, 89
Replication: *see* Viral replication
Resistance
  antibody response role in, 86–87
  CMI role in, 89
  nonspecific mechanisms, 82–83
Respiratory infections
  in children, IFN production and, 76
  CV role in, 235
Rhabdomyosarcoma (RD) cell line, CVB serial passage in, 58

RNA
  CVB3 genomic, molecular cloning and characterization of, 295–297
  enterovirus, detection by *in situ* hybridization, 299–303
  hybridization of, 45–46
  nucleic acid hybridization to detect, 207–208
    prerequisite for, 295–296
  persistence of, 313
  synthesis in viral replication, 37–38
    IFN effect on, 75
  of virions and MBV, 43–45

Scotland, enteroviral epidemiologic observations, 223, 226–227
  for CVA, 227, 231–233
  for CVB, 233–234
  in neurologic disease, 324–327
Separation techniques, for neutralizing antibodies, 205–206
Serotype, of CVB isolate, 400–402
Slot-blot hybridization, 208–209
Solid-phase immunoassays, 212
  indirect: *see* Indirect solid-phase immunoassay (ISPIA)
  limited usefulness of, 214–216
Spleen cells
  macrophage impairment, 94
  T suppressor cells in, 95
Staphylococcus aureus, in isolate serotyping, 409–410
Still's disease, CV and, 236
  group B infection, 408
Stress, effect on IFN production, 73–74
Sucrose-gradient centrifugation, 205–206
Surveillance studies, of CVB epidemiology
  serology, 403
  viral isolates, 400–403
SVDV: *See* Swine vesicular disease virus (SVDV)
Swine vesicular disease virus (SVDV)
  CV and, 8
  infective components of, 43
Syndrome(s), *see also* Disease syndromes
  clinical, of CV infection, 137
  in DM, 351
  hemolytic–uremic, 236
  hemophagocytic, 236
  mucocutaneous: *see* Mucocutaneous syndromes

T helper cells, pathogenesis mechanisms and, 109–110
T suppressor cells, activated nonspecific, in spleen cells, 95
Temperature
  CV infection susceptibility and, 138–139
  CVB variants sensitive to: *see* Temperature-sensitive (*ts*) mutants
Temperature-sensitive (*ts*) mutants, 163–164
  in persistent infection, 184–185
Teratogenicity, CV infection susceptibility and, 143–145
Thymostimulin, IFN action and, 73
Thymus
  and antibody response to CV, 84, 85
  in CV-infected hosts, 96, 97
Tissue cultures, variants derived in, *see also* Cell cultures; Coxsackievirus (CV) variants
  diabetes mellitus, 168–169
  myocarditic, 165–167
Tissue damage
  antibody response role in, 87–88
  CMI role in, 89–90
TM: *see* Tunicamycin (TM)
Tonsillitis, CVB infection and, 242
Transmission, viral, routes for, 383–384
Tropism, viral, cellular receptors and, 51–52
ts mutants: *see* Temperature-sensitive (*ts*) mutants
Tunicamycin (TM), IFN action and, 72–73

Uncoating, in viral replication, 36–37
United States, CV epidemiology in, 391, 394–395
  and Nassau county compared, 394
  by region, 393
  by viral type, 392

VAP: *see* Virion, attachment protein (VAP)
Variants: *see* Antigenic variants: Coxsackievirus (CV) variants
VAS: *see* Virion, attachment site (VAS)
Ventricular fluid, CVB infection and CNS involvement in newborns, 408

Vero cells, CVB3-infected
  immunofluorescence labeling of, 308
  in situ hybridization technique and, 301, 303
Viral meningitis, infant CVB infection and, 243
Viral polypeptides, characterization of, 128–129
Viral replication
  attachment and uncoating in, 36–37
  capsid proteins in, 46–47
  CV, schematic representation, 104
  dissimilar density particles in, 42
  genomic RNA in, 43–46
  in immunocompetent cells, 91–93
  protein and RNA synthesis in, 37–38
  virion in
    membrane-bound, 38–42
    structure and stability of, 42–43
Viral transmission, routes for, 383–384
Viral tropism, cellular receptors and, 51–52
Viremia, CVB infection and, 242
Virion
  attachment protein (VAP), 36–37
  attachment site (VAS), structural studies identifying, 55
  genomic RNA of, 43–45
  membrane-bound: *see* Membrane-bound virions (MBV)
  structure and morphology, 10–13
  in viral replication, 38–42
  structure and stability of, 42–43
Virulence, CV variants and, 161
Virus neutralization test, 7
Virus-receptor complex (VRC), 57
VRC: *see* Virus-receptor complex (VRC)

World Health Organization (WHO), enteroviral epidemiologic observations, 223, 224–225, 321
  clinical symptoms, 322
  by neurologic disease group
    by age, 323
    CVA, 322
    CVB, 323